Solid Oxide Fuel Cell Components

Gurbinder Kaur

Solid Oxide Fuel Cell Components

Interfacial Compatibility of SOFC Glass Seals

Springer

Gurbinder Kaur
Thapar University
Patiala
India

ISBN 978-3-319-25596-5 ISBN 978-3-319-25598-9 (eBook)
DOI 10.1007/978-3-319-25598-9

Library of Congress Control Number: 2015953005

Springer Cham Heidelberg New York Dordrecht London

Printed on acid-free paper

Springer International Publishing AG Switzerland is part of Springer Science+Business Media
(www.springer.com)

A dedication to my teachers, who have such big souls ... (Dr. O.P. Pandey, Dr. Kulvir Singh, Dr. B. Chudasama, Dr. Gary Pickrell, Dr. N. Sriranganathan and the Physics Department, GNDU, Amritsar).

Acknowledgments

During my journey of writing this book, there were many people who knowingly and unknowingly helped me in the successful completion of this project. At this overwhelming moment of accomplishment, first of all I am indebted to Drs. O.P. Pandey, Kulvir Singh, B. Chudasama, Gary Pickrell and N. Sriranganathan, whose understanding, encouragement and personal attention have always provided decisive and energetic support. Dr. Pandey stood by me during my struggling days and is one of the strongest pillars of my career. Without his support I would have never been able to accomplish most of the milestones of my career. Dr. Singh always invested extra hours in my work to churn the best out of it and was a wonderful mental booster. He gives reasonable freedom to his students so that they can add wings to their imagination. I feel honoured to have worked as a Ph.D. graduate under such talented supervisors, whose souls are big enough to accommodate every aspect of their students. Their diligence, persistence and vitality are highly admirable.

I have never seen a more dedicated and hardworking supervisor than Dr. Chudasama who always came forward for his students and never left any stone unturned so as to help them reach the zenith. Drs. Gary Pickrell and N. Sriranganathan provided the most conducive and comfortable environment to me during my stay in the United States and were superb mentors; they were an indispensable part of my journey.

A big chunk of thanks goes to Dr. Gopalan, Director, Thapar University for endowing us all with a zeal to move forward. His rational attitude toward problem analysis is really laudable. All the faculties and staff of the Physics Department (Guru Nanak Dev University), SPMS (Thapar University) and MSE (Virginia Tech, USA) are acknowledged, as they never turned me down whenever I approached them for help. I cannot forget the very supportive and cordial attitude of Dr. Manoj Sharma, HOD, SPMS, who is always ready to help his students. This truly demonstrates his excellent administrative qualities. I also cannot forget the support provided by Dr. S.S. Sekhon during the toughest phase of my life. His motivation and zeal for hard work enlightened my path always. I am so grateful to the UGC,

New Delhi for the financial assistance provided to me [F.15-1/2013-14/PDFWM-2013-14-GE-PUN-14803(SA-II)] during the course of my work.

This book emerged amid friendships that provided lasting lessons. It is a pleasure to mention my good friends, especially from the Functional Materials Lab, Ceramic Research Lab, Nanomedicine Lab and NanoMaterials Lab, who made my working atmosphere very conducive.

Finally, my greatest personal source of gratitude is my soul mate and my husband, Dr. Vishal Kumar, who is my enduring strength. He erased the word 'nightmare' from the dictionary of my life. During every downturn of my life, he uplifted my spirits and helped me in innumerable ways. It is only due to him that I could put in a lot of working hours tirelessly. He sheltered me through every situation and took my entire burden off my shoulders. He was full of patience and support when my work schedule became too hectic. This book would not have been possible without him. He was concerned about my important schedules and kept reminding me about them, along with my diet chart!

Dear Mom and Dad, I have let so many years pass without thanking you both for your unconditional love. I am thankful to my parents and to my parents-in-law for their support, encouragement, care, understanding and for creating a pleasant atmosphere for me. My father-in-law, Sh. Surinder Kumar, and my father, Sh. Harbhajan Singh, have both been strong pillars of my career and have always motivated me to fly high and transform my dreams into realities. I am lucky to have such wonderful in-laws to whom I am more like a daughter. I doubt that I will ever be able to convey my appreciation fully, but I owe them my eternal gratitude.

Above all, thanks to the Almighty for bestowing me with his precious blessings! With every passing day, I believe in the Almighty more and more as He has filled my life with wonderful things which I could never have imagined. Thanks for protecting me, sheltering me and for blessing me in the best ways possible!!

Dr. Gurbinder Kaur

Contents

Author Biography

Dr. Gurbinder Kaur received her B.Sc (Hons. Physics) and M.Sc. (Hons. Physics) from Guru Nanak Dev University, Amritsar. Dr. Gurbinder began her teaching career at D.A.V. College, Amritsar, where she taught from 2004 to 2005. Her second academic appointment was at RR Bawa DAV College, Batala, where she served as Head of the Physics Department from 2005 to 2009. Then, she moved to Thapar University, Patiala, to pursue her research work in the field of solid oxide fuel cells (SOFCs) and received her doctorate in 2012. Her Ph.D. dissertation was based on 'Investigations on interfacial interaction of glass sealants with electrolytes and interconnect for solid oxide fuel cells (SOFC)'. She has published more than 35 research papers in the field of materials science. She has published a book on *Modern Physics* by McGraw Hill Pvt. Ltd. She has also carried out research in the field of biomedical engineering and bioglasses. She is a recipient of a research fellowship of the RFSMS scheme of the University Grants Commission (UGC). She also received a fellowship under the Women Scientist Scheme, DST, New Delhi, from 2010 to 2012. After completing her doctorate, she moved to Virginia Tech, USA, to work as a postdoctoral fellow with Dr. Gary Pickrell. She is a recipient of a postdoc scholarship from UGC, New Delhi, for pursuing research work in the field of bioglasses. She works on a variety of different materials and applications including high-temperature energy materials, bioactive materials and optical materials.

Author Biography

Contributor Biography

Dr. Vishal Kumar completed his M.Tech (materials science and engineering) from Thapar University, Patiala, in 2007. He was awarded his Ph.D. in 2010 on the topic 'Study of SiO_2- B_2O_3 based glasses and glass ceramics as sealants'. He has been granted scholarships by funding agencies including CSIR and DST, New Delhi, to pursue his research work. He was awarded a Fulbright Fellowship by USIEF to pursue his research on glasses with Dr. Kathy Lu at Virginia Tech, Blacksburg, USA. His objective was "Study and development of materials for increasing the power generation efficiency of Solid Oxide Fuel Cells (SOFC)."

Chapter 1
Introduction to Fuel Cells

1.1 History of Fuel Cells

With the increasing needs of electricity and power, fuel cells have emerged as a much needed alternative source of energy. Fuel cells possess zero noise pollution, and high efficiency as well as low emissions. Fuel cells are excellent candidates for stationary and mobile applications. Basically, fuel cells are electrochemical devices which convert chemical energy directly into electrical energy. The first demonstration of the fuel cell was given in 1839 by Dr. William Grove, who was a lawyer and a scientist (Fig. 1.1).

Principle

The water molecule H_2O possesses smaller Gibbs energy than the sum total of individual H_2 and $\frac{1}{2}O_2$ molecules. Hence, upon mixing, H_2 and O_2 have a natural tendency to form the water molecule, i.e.,

$$H_2 + \frac{1}{2}O_2 \rightarrow H_2O \tag{1.1}$$

We can also say that combustion of hydrogen takes place, which produces electrical energy instead of heat energy.

G. Kaur, *Solid Oxide Fuel Cell Components*,
DOI 10.1007/978-3-319-25598-9_1

Sir William Grove

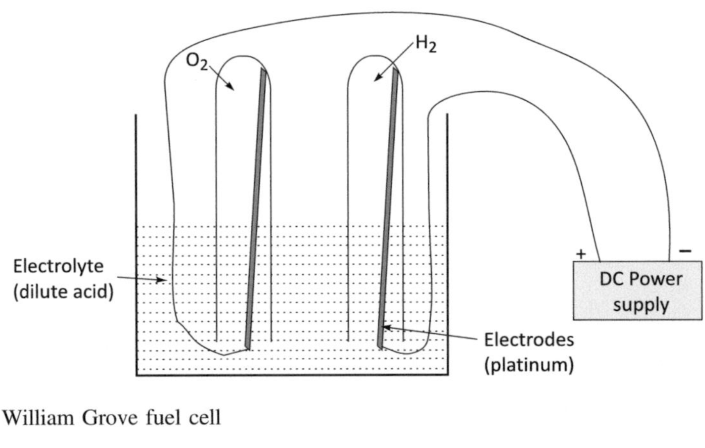

Fig. 1.1 William Grove fuel cell

Working

Current is passed using a DC power supply, which results in the electrolysis of water into hydrogen (H_2) and oxygen (O_2). The electrodes are made of platinum necessarily acting as catalysts. The electrolysis process will run as reverse and current is produced which flows through the external circuit. It should be noted that the magnitude of currents produced is very small. This is attributed to the small contact area between the electrode, electrolyte and the gas. Secondly, the electrolyte provides internal resistance to the cell. Hence, to overcome this problem the electrodes are kept flat and porous, so that the electrolyte and gas can penetrate it. This results in the maximum contact area between electrodes, gas and electrolyte. The phenomena of reaction at electrodes can be explained as follows:

At anode

Oxidation process takes place at anode

$$H_2 \rightarrow 2H^+ + 2e^- \tag{1.2}$$

At the reaction sites hydrogen molecules break into ions and electrons. The electrons migrate through the external circuit and ions diffuse through the electrolyte.

At cathode

A reduction process takes place at cathode

$$O_2 + 2e^- \rightarrow 2O^- \tag{1.3}$$

The oxygen breaks up and reacts with the electrons (from the anode). Furthermore, the protons react with oxygen ions to form water; i.e.,

$$O_2 + 4e^- + 4H^+ \rightarrow 2H_2O \tag{1.4}$$

These reactions proceed continuously due to the smooth flow of protons through the electrolyte and electrons through the external circuit. If the electrons go through the electrolyte instead of the external circuit, then all of them will be lost due to the recombination process. A complete schematic of an electrolyte fuel cell with the electrode reaction is shown in Fig. 1.2.

Hence, a fuel cell's components are almost similar to those of a battery, but there are several aspects which differentiate the fuel cell and battery. Theoretically, the fuel cell works as long as the air and fuel are fed on the cathode and anode, respectively. In contrast to this, the primary battery ceases to produce electrical energy as soon as the chemicals inside it are consumed, whereas the secondary battery can be recharged using an external source. For a fuel cell, any malfunction of components, sulphur poisoning and corrosion at the interfaces may lead to decreased efficiency. There is a significant difference in fuel cell and battery, which seems to be confusing. A cathode is defined as the negative electrodes of an electrolyte fuel cell, and is the positive terminal of a battery. These two definitions seem to be opposite to each other, hence, leading to contradiction. Technically, cations represent positive charges, hence the cathode is the positive terminal for all batteries. On the other hand, electrons always flow from the anode to the cathode for fuel cells.

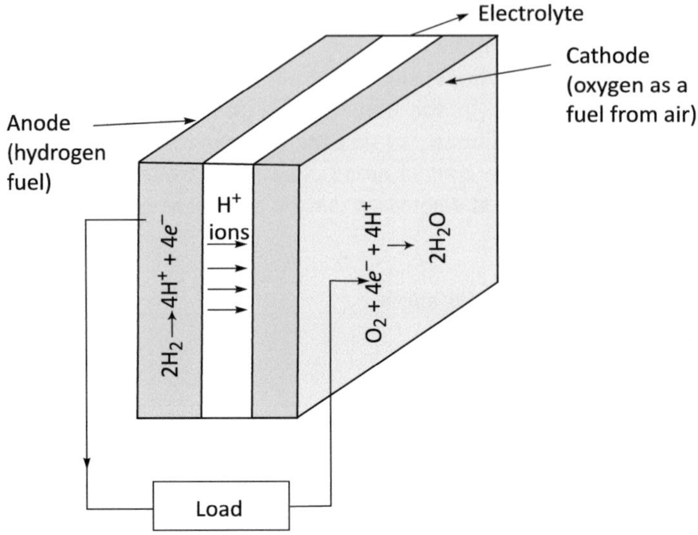

Fig. 1.2 Schematic of electrolyte fuel cell

1.2 Types of Fuel Cells

The fuel cell can be classified on the basis of efficiency, operating temperature, nature of electrolyte used and applications. Some of the commonly used fuel cells are given in Table 1.1. The research is focused on the improvisation of fuel cells and their components in order to obtain long-term performance with high efficiency. The following sections deal with the detailed explanation of some common fuel cells along with their working principle and construction.

1.2.1 Alkaline Fuel Cells (AFC)

J.H. Reid [US Patent 736016017 (1902)] described the alkaline fuel cell system for the first time. These were among the first fuel cells to be developed for the Apollo space vehicles. AFC were used to provide on-board electric power. Furthermore, AFC were used to provide power to offshore navigation equipment, to drive power cars and agricultural vehicles.

1.2.1.1 Basic Principle

In AFC, the hydroxyl ions (OH^-) play a major role. The reactions at the cathode and anode take place as follows (Fig. 1.3):

Table 1.1 Different types of fuels cells

	Name of the fuel cell	Electrolyte	Operating temperature (°C)	Applications
1.	Alkaline fuel cell (AFC)	KOH/NaOH	100–250	Space vehicles
2.	Direct methanol fuel cell (DMFC)	Polymer membrane	90–120	Navigation systems
				Remote monitoring sensing equipment
3.	Polymer electrolyte membrane fuel cell (PEMFC)	Polymer membrane	80–200	Auxiliary power units
				Remote area power supplies
				Transport applications
4.	Phosphoric acid fuel cell (PAFC)	Molten phosphoric acid	150–200	Military bases, leisure centres, factories on-site cogeneration plants
5.	Molten carbonate fuel cell (MCFC)	Molten alkaline carbonate	600–650	Stationary combined heat and power applications
6.	Solid oxide fuel cell (SOFC)	Conducting ceramic oxide	600–1000	Stationary power generation
7.	Microbial fuel cells (MFC)	Polymer membrane or humic acid	<40	Pilot scale reactions treatment of domestic sewage
				Biosensor development
8.	Direct carbon fuel cell (DCFC)	Molten hydroxide	600–850	Power generation
		Molten carbonate solid electrolyte	800–1000	

Based on the type of electrolyte, operating temperature and applications

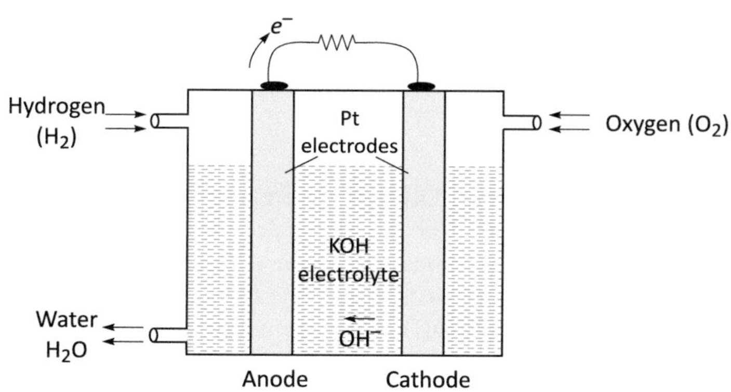

Fig. 1.3 The schematic of alkaline fuel cell

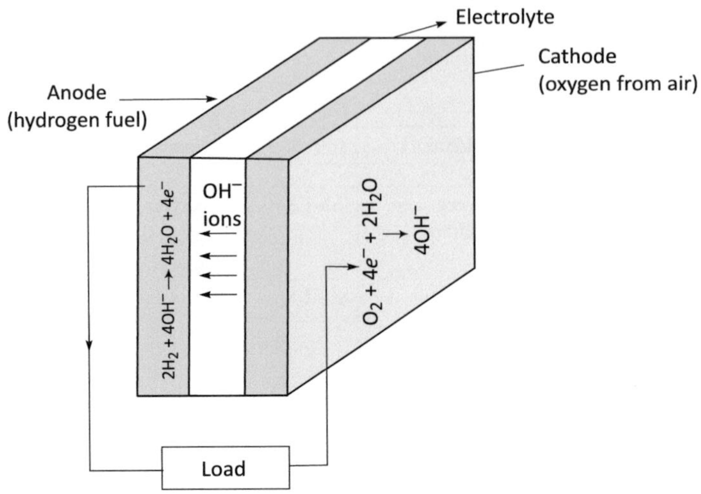

Fig. 1.4 Electrode reactions for AFC

At anode: Hydroxyl ions react with hydrogen producing water and electrons

$$2H_2 + 4OH^- \rightarrow 4H_2O + 4e^- \quad (1.5)$$

Hence, oxidation takes place at the anode.

At cathode: The electrons produced at the anode migrate via an external circuit towards the cathode. The oxygen reacts with those electrons and water leading to the formation of new hydroxyl ions (OH^-)

$$O_2 + 4e^- + 2H_2O \rightarrow 4OH^- \quad (1.6)$$

Figure 1.4 shows the electrode reactions for AFC.

1.2.1.2 Electrolytes for AFC

As the electrode reactions involve OH^- ions, the electrolyte is usually taken to be an alkaline hydroxide solution. For the electrode reactions to proceed without interruption, the hydroxyl ions should be produced and migrated by electrolyte towards the anode. The commonly used electrolyte for AFC is potassium hydroxide (KOH) or sodium hydroxide (NaOH). They are potential candidates due to their low cost, easy availability, least corrosive nature and high solubility. However, it should be noted that if different electrolytes are used, the catalysis, electrodes as well as operating conditions also get changed. Although other fuel cells have reduced the importance of AFC in the recent era, AFC are of technological importance due to their low systems cost and absence of bipolar plates.

The electrodes and electrolyte for AFC are relatively cheap. Secondly, the bipolar plates are expensive, which contributes to the cost of fuel cells.

1.2.1.3 Electrodes for AFC

F.T. Bacon designed AFC in the 1940s using nickel electrodes. The electrodes were made porous by sintering Ni powder at appropriate temperature. The Apollo Mission fuel cells used the two-layer Ni electrode with different particle sizes of nickel. This structure established a good three-phase boundary between the solid electrode, the liquid electrolyte and reactant gas by providing more open pores to the gas side and fine-pore structure on the liquid side. The cathodes were partially lithiated by adding lithium salts to their surface so as to reduce the oxidation of nickel. In contrast to this, the anodes were prepared from the sintering of nickel powder.

Siemens alkaline fuel cells used Raney nickel for the anode and silver for the cathode. Raney metals are prepared by mining the active metal such as nickel with some inactive metal such as aluminium. This mixture is then treated with a strong alkali, so that the aluminium dissolves out of it leaving a porous structure behind with a large surface area. Currently, electrodes use polytetraflouroethylene (PTFE) mixed with carbon-supported catalysts which are then rolled onto a material such as nickel mesh. PTFE is of hydrophobic nature and hence the electrodes do not get overflooded. Sometimes, a thin layer of PTFE is put over the cathode surface to control the porosity of the electrode. Though this helps in preventing the electrolyte passing through the electrode, it makes the cathode surface non-conductive. The carbon dioxide also affects the electrodes and electrolyte which can be demonstrated as follows:

$$2KOH + CO_2 \rightarrow K_2CO_3 + H_2O \tag{1.7}$$

The carbon dioxide in the air and carbon-supported catalysts may react with potassium hydroxide electrolyte leading to the formation of potassium carbonate. Hence, the CO_3^{2-} concentration increases instead of OH^- ion concentration. The lifetime of electrodes is increased if the carbon dioxide is removed continuously using CO_2 scrubber in the cathode air supply. Carbon poisoning can also be avoided using an anode and cathode made of PTFE mixed with Raney nickel and Raney silver, respectively. Hence, the fuel cell performance is improved without the use of any carbon-supported catalysts.

1.2.1.4 Types of Alkaline Fuel Cells

For alkaline fuel cells, the electrolyte is concentrated (85 wt%) KOH at the operating temperature of $\sim$250°C. However, the concentration of KOH is kept between

35 and 50 wt% for lower operating temperatures of <120°C. In addition, spinels, metal oxides, noble metals, nickel and silver are also used as electrocatalysts. The alkaline fuel cells are further classified into three main categories.

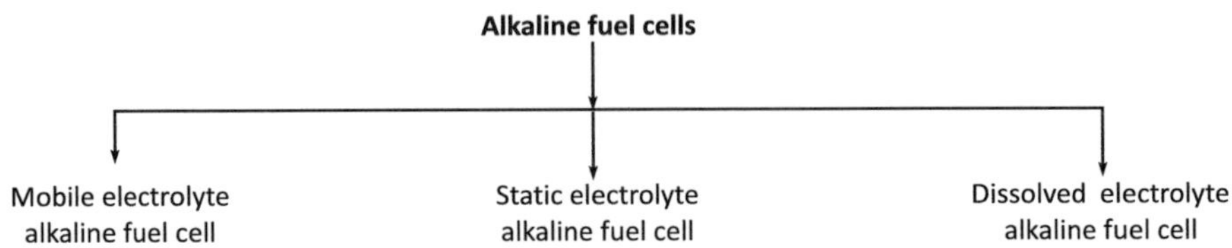

We discuss these three AFC in this section.

(a) *Mobile Electrolyte Alkaline Fuel Cell*

Figure 1.5 shows the schematic of mobile electrolyte AFC. Most alkaline fuel cells are of this type with KOH as the electrolyte.

The air is made free of CO_2 before pumping it to the cathode. As discussed above, the presence of CO_2 might hamper the functioning of fuel cells. In spite of using CO_2 scrubbers, all the CO_2 cannot be removed and it may still be present in the cell as a trace constituent. Hence, the KOH electrolyte deteriorates and needs to be replaced from time to time. Hydrogen is pumped at the anode which is connected to the hydrogen cooler/condenser system. Hydrogen will evaporate the water, subsequently leading to its condensation at the cooling unit. The electrolyte is mobile in nature and serves as a cooling system for the fuel cell. The potassium carbonate (K_2CO_3) formed due to the CO_2 contamination may result in dilution of the electrolyte solution. Hence, the solution should be pumped out. For mobile electrolyte AFC, the pumping out of electrolyte is relatively simple. The pumping process becomes difficult for a stack in which many cells are joined together. Ionic conduction between the cells of a stack may lead to an internal short circuit. To reduce the internal short-circuit, the cells are connected in a series-parallel system. Another point that is important for the working of mobile electrolyte AFC is continuous stirring and circulation of the electrolyte. Circulation of the electrolyte leads to the condensation of water, which may have moved to the electrolyte rather than towards the anode side. Continuous stirring of the electrolyte prevents any kind of solidification at the cathode. Hence, the circulation of water is maintained because water is produced twice as fast at the anode than its consumption at the cathode. For the Apollo Mission fuel cells (1950), the mobile electrolyte AFC was used. Figure 1.6 shows the fuel cell from the Apollo spacecraft and the cooling system basically comprises a water/glycol mixture.

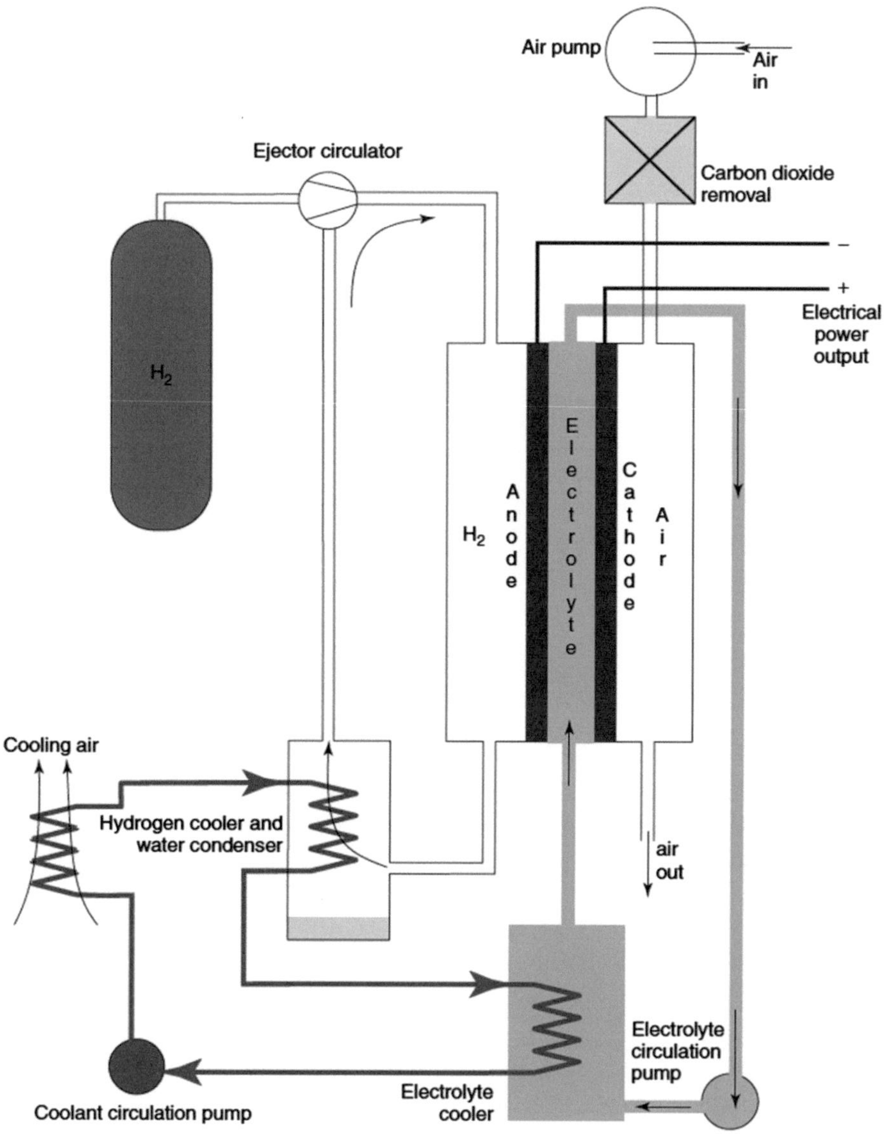

Fig. 1.5 Schematic of alkaline fuel cell with mobile electrolyte (Larminie and Dicks 2003)

Fig. 1.6 Alkaline fuel cell for Apollo command module (Photo courtesy Lindsay B. Johnson center)

(b) ***Static Electrolyte Alkaline Fuel Cells***

Figure 1.7 shows an alkaline fuel cell with stationary electrolyte. In this system, the electrolyte is held in asbestos matrix material.

Hence, unlike mobile electrolyte AFC, the electrolyte does not need to be pumped out. Moreover, there is no problem of an internal short-circuit, which makes it more favourable for space applications. The KOH electrolyte in an asbestos matrix possesses excellent porosity, corrosion resistance and strength. Due to the stationary nature of the electrolyte, it becomes mandatory that pure oxygen should be supplied at the cathode. In case CO_2 is present in the air, carbonate formation takes place and the complete fuel cell needs to be rebuilt. Hydrogen is pumped at the anode and circulated to remove water. The cooling system comprises fluorinated hydrocarbon dielectric liquid. The product water is used for humidification of the cabin, cooking and drinking in space orbiter systems. As the water is used at the cathode, its level must be kept sufficiently high at the anode side.

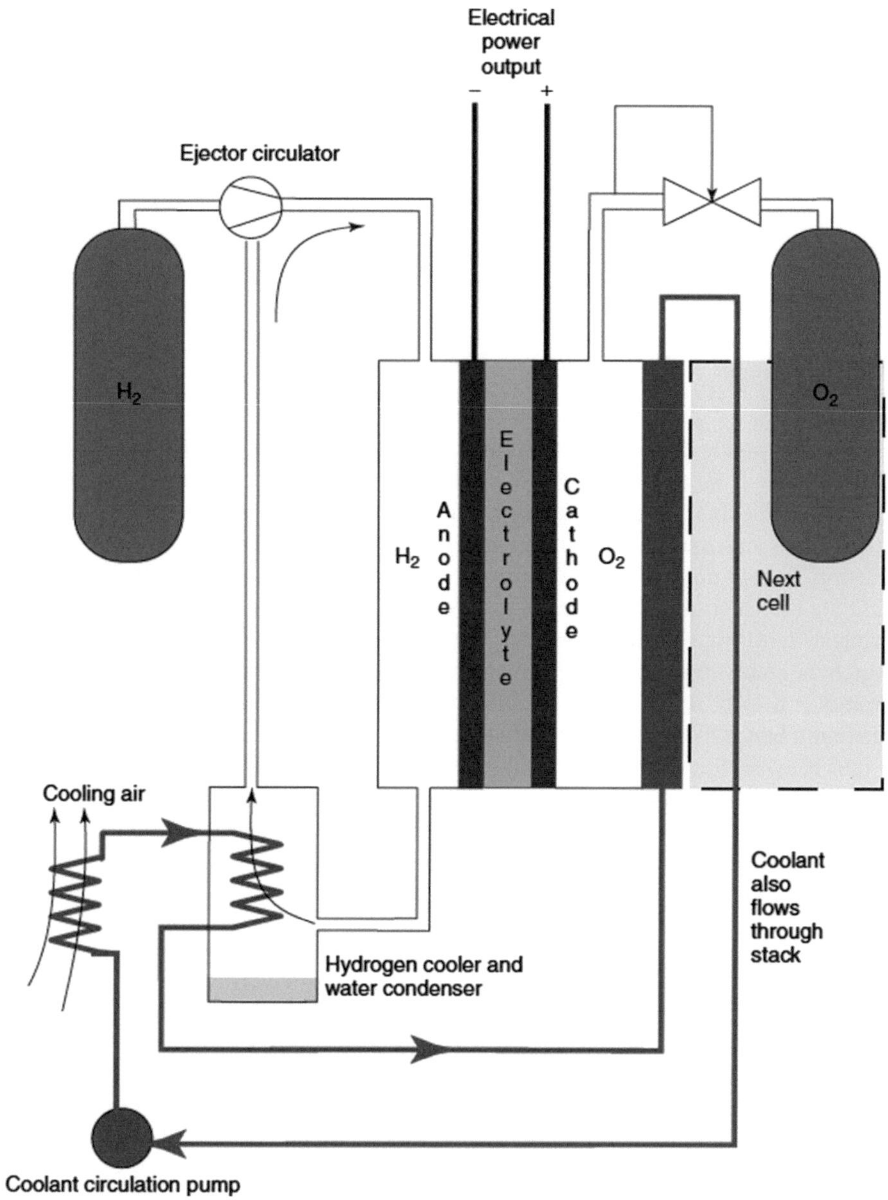

Fig. 1.7 Schematic of alkaline fuel cell with static electrolyte (Larminie and Dicks 2003)

(c) *Dissolved Fuel Alkaline Fuel Cells*

This type of alkaline fuel cell is easy to fabricate and is used for high power generation. The schematic of a dissolved fuel alkaline fuel cell is shown in Fig. 1.8.

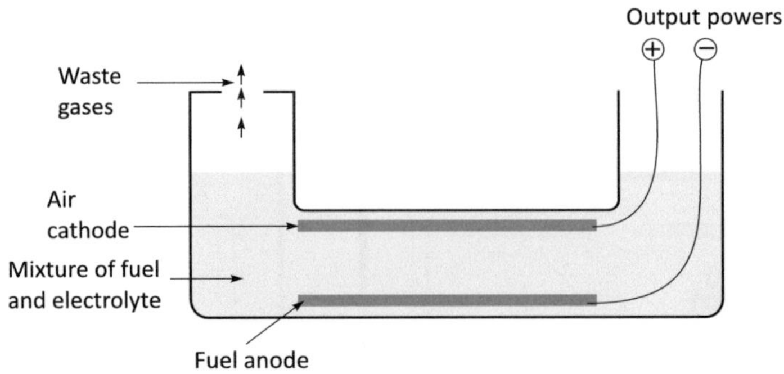

Fig. 1.8 Schematic of dissolved fuel AFC

The fuel anode consists of PTFE/platinum catalyst. The electrolyte is KOH with fuel cells such as ammonia (NH_3), hydrazine (NH_2NH_2) or sodium borohydride ($NaBH_4$). The fuel is also in contact with the cathode. The electrolyte is chosen such that it has maximum ionic conduction. There is always some electron conduction. In addition, some fuel from the anode side will diffuse onto the cathode side via the electrolyte. This leads to direct mixing of fuels (in the presence of the catalyst) with oxygen producing no current from the cell. This small wasted fuel, which migrates through the electrolyte is known as the fuel crossover. The fuel crossover may be a problematic situation in this fuel cell. The cathode catalyst is not platinum, hence the reaction rate at the cathode is low. Usually, hydrazine is used as a fuel because it dissociates into hydrogen and nitrogen. However, the hazardous and toxic nature of NH_2NH_2 imposes restrictions on its usage as a fuel. Hence, sodium borohydride is considered a better fuel thermodynamically and chemically.

The anode and cathode reactions may be written as

At anode:

$$NaBH_4 + 8OH^- \rightarrow NaBO_2 + 6H_2O + 8e^- \tag{1.8}$$

At cathode:

$$O_2 + 4e^- + H_2O \rightarrow 4OH^- \tag{1.9}$$

The overall reaction is written as

$$NaBH_4 + 2O_2 \rightarrow NaBO_2 + 2H_2O \tag{1.10}$$

Hence, $8e^-$ is obtained per the $NaBH_4$ atom, which indicates its remarkable potential as a fuel. The hydrolysis reaction is also supported by the catalysts, which leads to losses in these fuel cells, i.e.,

$$NaBH_4 + 2H_2O \rightarrow NaBO_2 + 4H_2 \quad (1.11)$$

This hydrogen loss makes the system inefficient and hence the performance of the fuel cell is deteriorated. Hydrolysis can be slowed down by reducing the concentration of borohydride in the electrolyte. All the above reactions proceed quickly on the platinum catalysts. Anyhow, platinum is not used in the cathodes. The design of the cell is simple and economical except for the fact that sodium borohydride is expensive.

1.2.2 Polymer Exchange Membrane Fuel Cell (PEMFC)/ Polymer Electrolyte Fuel Cell (PEFC)

These cells are also known as solid polymer fuel cells (SPFC) (Fig. 1.9). These fuel cells were used in NASA Gemini Spacecraft (manned space vehicles) and were developed in 1960 by General Electric (USA). The following key features of PEMFC make them very popular:

- High power density
- Low weight, cost and volume
- Quick start-up
- Absence of corrosive liquids makes them suitable for portable applications
- Can work in any orientation

1.2.2.1 Cell Components

In 1959, William Grubbs used organic cation exchange polymers in fuel cells. As shown in Fig. 1.9, the cell consists of polymer electrolyte sandwiched between platinum impregnated porous electrodes. Teflon® is coated on the electrodes to make them hydrophobic. The operation temperature of PEFMC can be within 80–200°C. The low operation temperature of PEMFC offers quick start-up but platinum catalysts are required to promote the electrochemical reaction. Platinum particles are deposited very finely onto carbon powders to attain the maximum surface area. The contact between gas, electrolyte and platinum catalyst should be good for the uninterrupted function of PEFMC.

(a) ***Electrolyte (Polymer Electrolyte)***

The electrolyte for PEMFC is an ion conduction polymer; H^+ or proton is the mobile ion in the conduction process of PEMFC. Nafion membrane forms the basis of electrolyte for PEMFC. The starting material for Nafion is polyethylene. Polyethylene is modified using fluorine for hydrogen. The process of replacing hydrogen with fluorine is known as perfluorination. The process of perfluorination is shown in Fig. 1.10. All four hydrogen molecules in the ethylene group are replaced by fluorine and the modified polymer is known as polytetrafluoroethylene (PTFE).

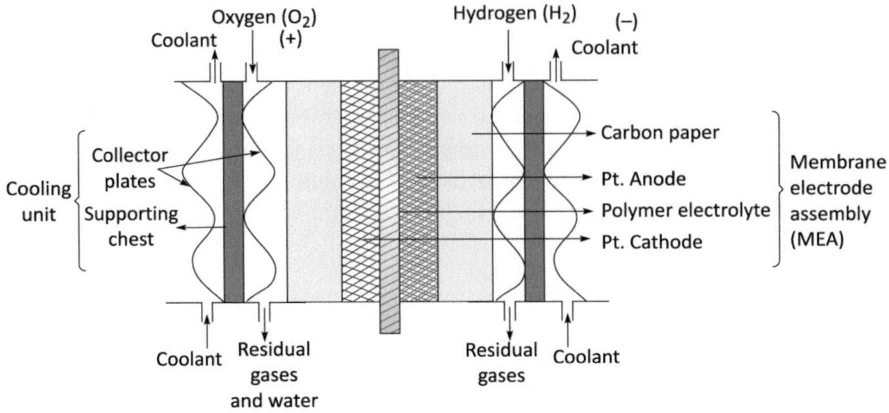

Fig. 1.9 Schematic of PEMFC

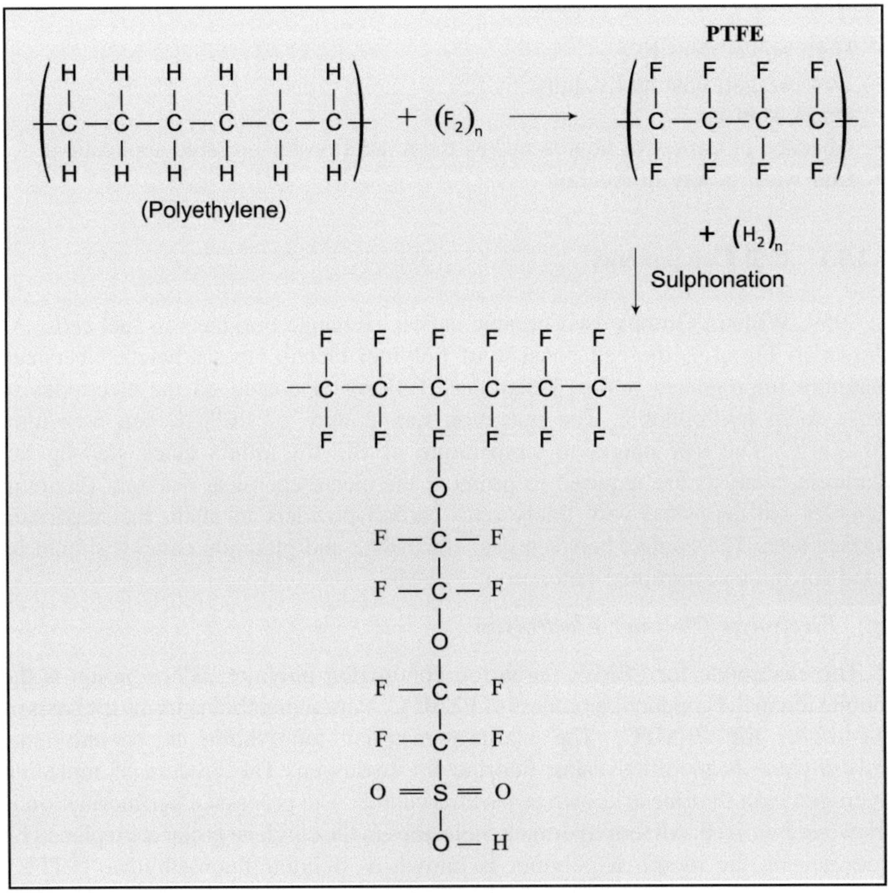

Fig. 1.10 Detailed structure of sulphonated polytetrafluoroethylene

The PTFE is further sulphonated; i.e., a side chain ending with sulphonic acid HSO_3 is added. The HSO_3 group is ionically bonded and hence, the end of the side chain becomes the SO_3^- ion. The sulphonated PTFE is also known as perfluorosulphonic acid PTFE copolymer. PTFE has the following remarkable properties:

(i) PTFE is thermoplastic material possessing high strength.
(ii) Sulphonic acid is highly acidic in nature. This endows them with the tendency to dissociate into anions and protons. The protons thus act as charge carriers.
(iii) PTFE is hydrophobic in nature, whereas sulphonic acid is highly hydrophilic. Hence, the side chain of sulphonic acid tends to cluster within the overall structure of the material. In other words, we can say that the hydrophobic substance contains hydrophilic regions.
(iv) The PTFE/Nafion membranes are highly resilient to chemical attacks.
(v) PTFE membranes quickly absorb water. Hence, the dry weight of the material can be increased to 50%. The H^+ ions are weakly attached to the SO_3^- group. Therefore, H^+ ions move easily leading to good proton conduction. The hydrated water molecule structure inside a hydrophobic polymer gives it micro-phase separated morphology. Hence, an SO_3^- molecule can absorb a $20H_2O$ molecule and the conductivity may increase.

The standard electrolyte material for PEMFC is Nafion, which is produced by E.I. Dupont de Nemours (Membrane No. 117 C20). The Dow Chemical Company has produced an electrolyte membrane (XUS13204.10) that exhibits superior qualities over the Nafion membrane; i.e., it has low electrical resistance and high current densities. But these membranes are relatively more expensive than the Nafion membranes. The Dow membrane is supplied at 2 mils whereas the Nafion is supplied for 7 mils. Ballard Advanced Material Corporation is working on developing trifluorostyrene copolymeric membranes for PEMFC. The membranes used currently are expensive and are available only in specific ranges and conductivity. So, the research is focused on developing low cost membranes with low resistivity.

(b) ***Electrodes***

The platinum catalyst plays an important role in the fabrication of the cathode and the anode. As discussed earlier, platinum particles are made finer on the surface of somewhat large carbon particles (XC72® Cabot). This relatively exposes the high surface area of platinum and is beneficial for electrochemical reactions (Fig. 1.11).

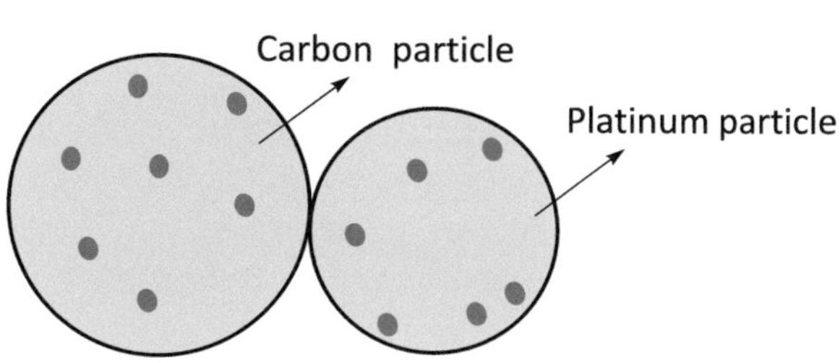

Fig. 1.11 Carbon supported catalysts

There are two routes for fabricating electrodes: electrodes onto the electrolyte or a separate electrode. In the electrode on the electrolyte, the platinum on the carbon catalyst is fixed directly onto the electrolyte using a printing process, spraying or rolling methods. In the separate electrode method, the carbon-supported catalyst is fixed onto porous and conductive carbon cloth or paper. PTFE is also added due to its hydrophobic nature. The carbon cloth/paper makes a gas diffusion layer as it makes the gas diffuse onto the catalyst. Then on the polymer electrolyte membrane sides, electrodes are fixed. The full sulphonation of the membrane is done by cleaning it in 3% H_2O_2 and then putting it in boiling sulphuric acid. The assembly of electrodes and membrane is then hot pressed at 140°C and high pressure for 3 min. This gives a complete membrane electrode assembly (MEA). Usually, the electrodes are coated as thin films and bonded to the membrane.

The Ballard Mark V stack used low platinum electrodes; i.e., for cathode ≤0.60 mg Pt/cm^2 and for anode 0.12 mg Ru/cm^2 and ≤0.25 mg Pt/cm^2 have yielded results as good as with electrodes loaded at 4–8 mg Pt/cm^2. The platinum was carbon supported, which resulted in an increase of the interface between the electrocatalyst and the polymer electrolyte. Originally, the cells consisted of unimpregnated electrode/Nafion electrolyte interface. Later, a proton conductor was impregnated into the electrode. PSI technology has been able to achieve platinum loading as low as 0.05 mg/cm^2. Texas A&M University has checked the performance of a cathode with 0.05 Pt/cm^2 loading. A cathode with 0.12 mg Pt/cm^2 loading has been tested by Los Alamos National Laboratory.

1.2.2.2 Water Retention in the Membranes

For the PEMFC, high water content is required in the electrolyte to maintain high ionic conductivity. The water content should not be overflooding to prevent blocking of pores of the electrodes or gas diffusion layer. The reaction layers at anode and cathode are given as follows:

At anode:

$$H_2 \rightarrow 2H^+ + 2e^- \tag{1.12}$$

At cathode:

$$\frac{1}{2}O_2 + 2H^+ + 2e^- \rightarrow H_2O \tag{1.13}$$

The overall reaction becomes

$$\frac{1}{2}O_2 + H_2 \rightarrow H_2O \tag{1.14}$$

Hence, water is formed at the cathode and it will keep the electrolyte hydrated. Any excess water would be dried by blowing air over the cathode. The water permeates through the electrolyte towards the anode side, hence maintaining the water requirement of the electrolyte. Three processes can cause problems in water management of membranes:

- Water drag through the cell (electroosmotic drag)
- Back diffusion from the cathode
- Diffusion of water in fuel stream through anode

During the operation of PEMFC, the proton's H^+ ions may pull water molecules with them while moving from anode to cathode. This condition is known as electroosmotic drag. This may lead to drying of the anode side while the cathode is well hydrated. Typically, 1–2.5 molecules of water are dragged by each proton. Therefore, the proton turns out to be a hydrated proton or $H(H_2O)_n^+$. This water drag will increase at high current densities. Secondly, at high temperatures, the air will dry the electrodes. Hence, humidification of the air and fuel is required before the fuel enters the fuel cell. If the water management is not kept in check, more water will be consumed than is produced. The electrodes may get overflooded, the polymer may get dehydrated and the reactant gases get diluted by water vapours. The depletion of water may affect the contact between electrode and electrolytes. Usually some parts of the fuel cell may be too wet and others too dry. Conclusively, we can say that the lower humidity, lower pressure, high temperature and reactant flow affect the working of PEMFC. More effects are given for the water management using flow field designs for fuel cells. The water vapour carrying capacity of the gas streams can be managed with proper temperature adjustment.

1.2.2.3 Contamination of PEMFC

There are three main contaminants in PEMFC; i.e., hydrocarbon fuel, carbon dioxide and carbon monoxide. For large PEMFC systems, reformed hydrogen is used as a primary fuel. The reformed hydrogen consists of 1% CO. The reformation reaction can be explained as follows:

$$CH_4 + H_2O \rightarrow CO + 3H_2 \tag{1.15}$$

This carbon monoxide gets adsorbed on the surface of the platinum catalyst, hence blocking the access of hydrogen to the catalyst sites. The remedy for this problem is the conversion of CO into carbon dioxide using the water gas shift reaction:

$$CO + H_2O \rightarrow CO_2 + H_2 \tag{1.16}$$

CO_2 poisoning is less significant than carbon monoxide poisoning, though CO_2 reacts with adsorbed hydrides on platinum at anode sites affecting its performance. Even 10 ppm of CO concentration greatly affects fuel cell performance. Hence, it becomes imperative to reduce the concentration of CO in the fuel gas stream. It has been shown that the platinum/ruthenium catalysts have made the electrodes tolerant up to 200 ppm. Some processes such as selective oxidation, membrane separation during adsorption and methanation are used to cleanup the fuel feed of a fuel cell. In selective oxygen, a small amount of oxygen is added to the fuel stream, so that oxygen removes CO by reacting with it. In case the oxygen reacts with H_2, it leads to the formation of water, hence wasting the fuel. Another approach involves the use of a selective oxidation catalyst which is placed between the anode catalyst and the fuel stream inlet. A gas clean-up unit is also attached to the fuel cell system to reduce CO poisoning. The length of the hydrocarbon chain increases CO poisoning to a significant extent. For example, for CH_4, and C_6H_{14}, we obtain the following products:

$$CH_4 + H_2O \rightarrow 3H_2 + CO \tag{1.17}$$

$$C_6H_{14} + 6H_2O \rightarrow 13H_2 + 6CO \tag{1.18}$$

Hence, the CO constant increases considerably and the functioning of the fuel cell is affected.

1.2.2.4 Bipolar Plates for PEMFC

Individual PEMFC cells produce about 0.7 V electromotive force (EMF). In order to obtain useful voltage, many cells are stacked together using a bipolar plate. It should be noted that the membrane electron assembly (MEA) for PEM is kept very thin, but the bipolar plates constitute almost 80% of the mass of PEMFC. The bipolar plate acts as an interconnect between the anode of one cell and the cathode of the next. Bipolar plates also distribute the fuel gas over the anode and oxygen over the cathode. These bipolar plates also contain cooling fluid and the different flow field patterns of bipolar plates used in PEMFC are shown in Fig. 1.12. Bipolar plates should have the following characteristics:

- Electrical contacts shall be large.
- The bipolar plate must be thin to reduce the stack volume and resistance.
- The production cycle time should be short for plates.
- Light-weight for minimum stack mass.
- It should be cost-effective.
- The flexural strength must be >25 MPa.
- It must be corrosion resistant.
- Electrical conductivity >10 S/cm.
- The gas permeability must be $<10^{-4}$ mbar L/s/cm^2.
- If the cooling fluid is integrated, then heat conductivity must exceed 20 Wm^{-1} K^{-1}.
- If heat is removed from the edges, then the heat conductivity must exceed 100 Wm^{-1} K^{-1}.

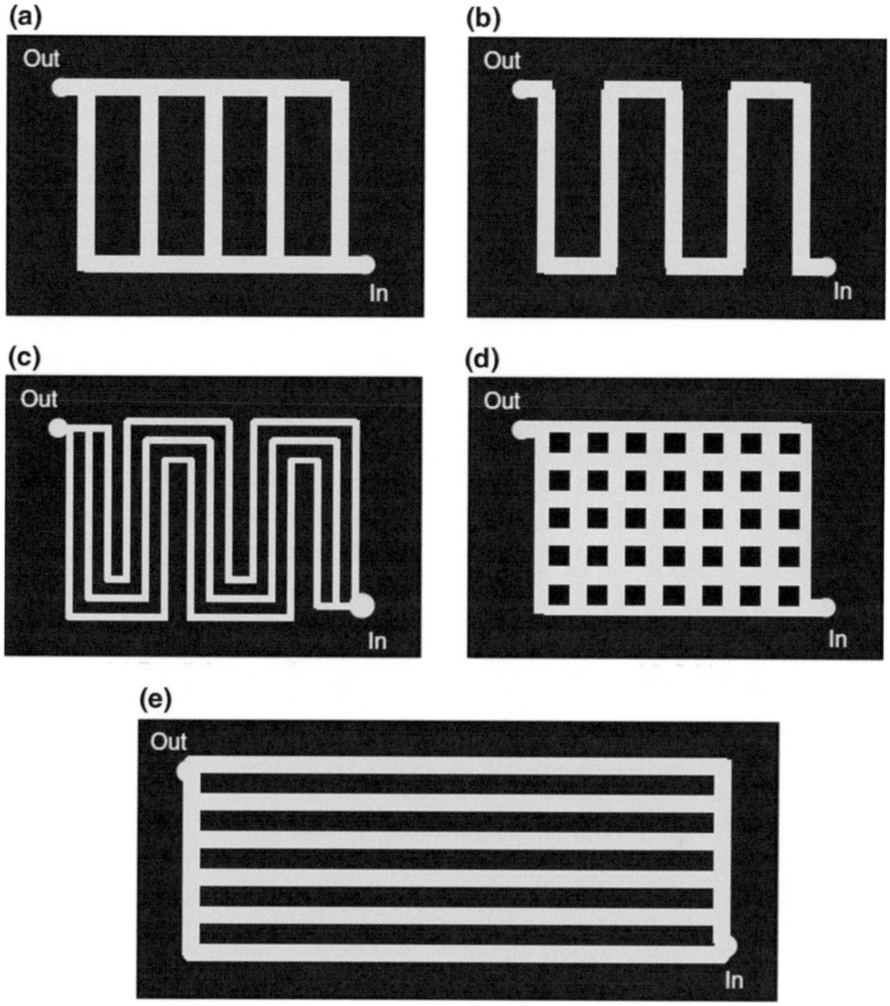

Fig. 1.12 Different flow field patterns used in PEMFC (Larminie and Dicks 2003). **a** Parallel. **b** Serpentine. **c** Parallel serpentine. **d** Grid. **e** Long parallel

1.2.3 Direct Methanol Fuel Cell (DMFC)

The direct methanol fuel cell contains methanol as a fuel. Methanol is economical and readily available. Moreover, its storage is easy compared to hydrogen fuel. Figure 1.13 shows the schematic of a DMFC. Two types of electrolytes can be used for DMFC, i.e., alkaline or polymer membrane. It can be seen that the methanol fuel is supplied at the anode. If alkaline electrolyte is used for DMFC, then the anode reaction becomes:

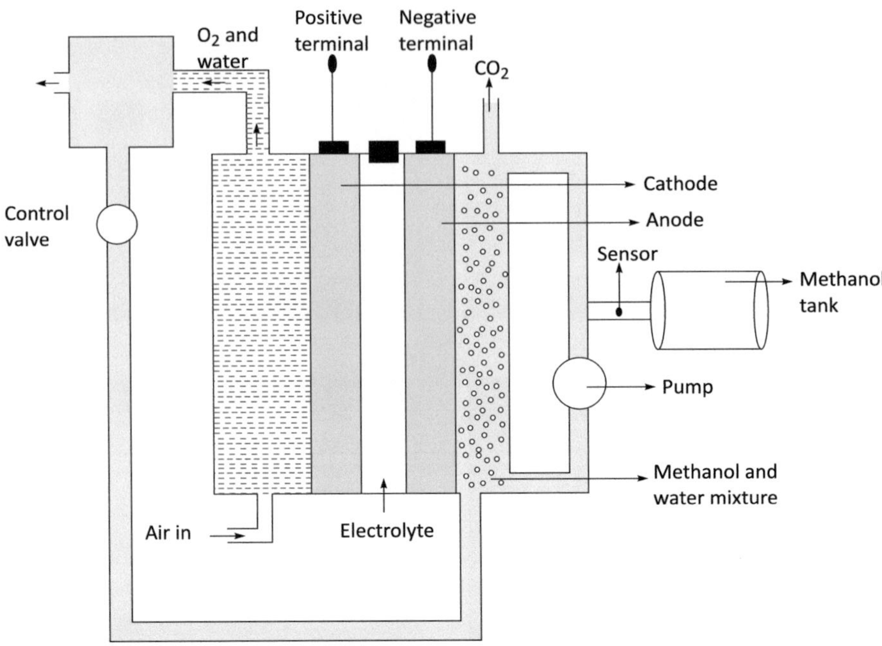

Fig. 1.13 The schematic of DMFC

$$CH_3OH + 6OH^- \rightarrow CO_2 + 5H_2O + 6e^- \tag{1.19}$$

Equation (1.19) clearly implies the efficiency of methanol as a fuel because 6 electrons are produced per methanol molecule. However, as mentioned in Sect. 1.2.1.3, carbon dioxide will react with hydroxide resulting in the formation of the carbonate, which decreases the hydroxide concentration. The major problem for alkaline DMFC is that no carbon dioxide is produced as a product as all of the CO_2 reacts with the hydroxide; i.e.,

$$CO_2 + 2OH^- \rightarrow CO_3^{2-} + H_2O \tag{1.20}$$

Hence, these alkaline DMFC are practically discarded. Therefore, the main focus is on the polymer electrolyte membrane DMFC.

The anode and cathode reactions for PEM DMFC are given as follows:

At anode:

$$CH_3OH + H_2O \rightarrow 6H^+ + 6e^- + CO_2 \tag{1.21}$$

The electrons move via external circuit and the proton migrates through the electrolyte towards the cathode side.

At cathode:

$$\frac{3}{2}O_2 + 6H^+ + 6e^- \rightarrow 3H_2O \tag{1.22}$$

The overall reaction is

$$CH_3OH + \frac{3}{2}O_2 \rightarrow CO_2 + 2H_2O \tag{1.23}$$

The breakdown of methanol into CO_2 takes place in steps and is more complex than observed in Eq. (1.21). Figure 1.14 shows the oxidation of methanol.

We see from Fig. 1.14 that formic acid (HCOOH) is formed as a stable intermediate. If formic acid is used as a fuel then it becomes a direct formic fuel cell. The electron turnover is reduced to 4 or 2 instead of 6 (for DMFC). Hence, we also see that the oxidation of methanol takes time compared to the oxidation of hydrogen fuel. This reduces the performance of the DMFC to a large extent. Therefore, suitable catalysts are required for the anode of DMFC. Bimetal catalysts such as a combination of platinum and ruthenium are used in place of the platinum catalyst only. The catalyst loading is almost 10 times higher than the hydrogen PEMFC because more active catalyst reduces fuel crossover and reduces the activation losses. The carbon-supported catalyst is used for enhancing the surface area. For PEMFC, the gas has to be drawn in and water flooding is required, whereas for DMFC the product CO_2 should be expelled as soon as possible. From Eq. (1.21), it is observed that water is required to proceed with the oxidation of methanol. Therefore, a mixture of methanol and water is fed at the anode instead of pure methanol. Along with the fuel issues, the hydration of PEM is also taken into consideration. At the cathode, water is produced, which evaporates when the air passes over the cathode. The rate of water evaporation should not exceed the rate of

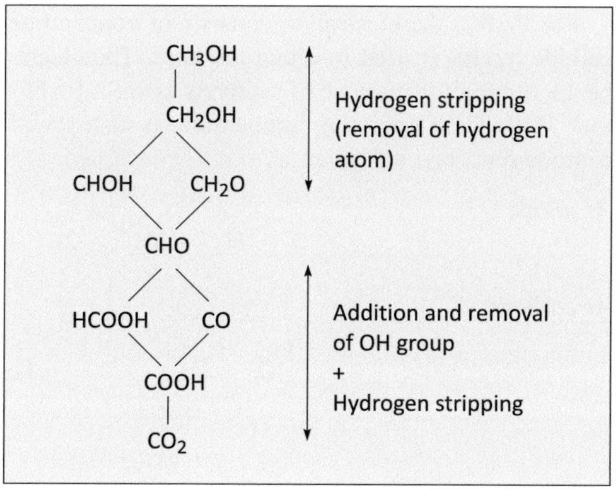

Fig. 1.14 Oxidation of methanol at anode of DMFC

water production, otherwise the condenser unit at the exit will be required to replenish the anode with water. The carbon dioxide produced at the anode is pumped out. It has been shown that the CO_2 is effectively removed faster if air is added to the methanol/water mixture at the cathode.

It is important to monitor the concentration of methane using a sensor as the fuel; crossover may occur. The problem of fuel crossover is severe for the PEM direct methanol fuel cell. Methanol dissolves quickly in water and reaches the cathode. The presence of platinum catalyst on the cathode will oxidise the fuel, though not as effectively as the anode catalysts. Hence, fuel wastage will occur at the cathode reducing the cell voltage. This phenomenon yields mixed potential and the fuel utilisation coefficient η_f is defined as

$$\eta_f = \frac{i_o}{i_o + i_c} \tag{1.24}$$

where i_o is the output current and i_c is the crossover current corresponding to the equivalent current that would have been produced if methanol did not react at the cathode. To reduce the fuel crossover problem, the anode catalysts are kept as active as possible. Secondly, there should not be excess methanol in the fuel, so that the electrolyte does not get flooded with methanol. Moreover, thicker electrolyte should be used to increase the cell resistance and reduce fuel crossover.

1.2.4 Phosphoric Acid Fuel Cell (PAFC)

PAFC is the most commercialised technology all over the world. Mitsubishi Electric Corporation in Japan, Fiji Electric Corporation, Toshiba Corporation and International Fuel Cell Corporation in the United States are actively engaged in improving the PAFC technology. Figure 1.15 shows the schematic of PAFC.

For PAFC, the electrolyte consists of concentrated phosphoric acid and silicon carbide matrix is used to retain the acid. The electrodes are made up of platinum, hence serving the purpose of catalysts as well. PAFC may produce power output up to 1 MW. Their operating temperature is also low, i.e., ~200 °C. The anode and cathode reactions are given as

At anode:

$$H_2 \rightarrow 2H^+ + 2e^- \tag{1.25}$$

At cathode:

$$\frac{1}{2}O_2 + 2H^+ + 2e^- \rightarrow H_2O \tag{1.26}$$

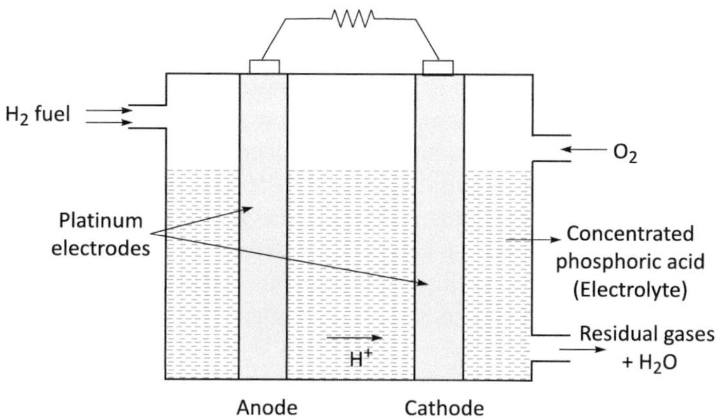

Fig. 1.15 Schematic of PAFC

And the overall reaction is given as

$$H_2 + \frac{1}{2}O_2 \rightarrow H_2O \tag{1.27}$$

Phosphoric acid (H_3PO_3) has thermal, electrochemical stability and low volatility. Hydrogen fuel is also produced from a hydrocarbon through steam reformation; i.e.,

$$CH_4 + H_2O \rightarrow 3H_2 + CO \tag{1.28}$$

Hence, along with hydrogen, carbon monoxide is produced, which is poisonous and a potential problem for the fuel cell (especially for Pt catalysts). Hence, the CO must be processed via water–gas shift reactions, as done for PEMFC cells; i.e.,

$$CO + H_2O \rightarrow CO_2 + H_2O \tag{1.29}$$

Phosphoric acid electrolyte is, however, tolerant to carbon monoxide/carbon dioxide. The acid is contained in the matrix of SiC comprising particles ranging $\approx$1 μm. The H_3PO_3 is contained within the pores of SiC held together with polytetrafluoroethylene (PTFE). Some loss of H_3PO_3 may occur due to its low vapour pressure. Hence, the lost electrolyte must be replenished to avoid any hindrance. To achieve high cell performance, 100 % H_3PO_3 is used at temperatures of about 200°C and pressure of $\sim$8 atm. In the 1960s, the electrodes were polytetrafluroethylene (PTFE) bonded Pt with loading of 9 mg Pt/cm^2. Recently carbon black has been used in place of PTFE as electrocatalyst. The use of carbon black to support platinum has resulted in decreased platinum loading; i.e., the anode is loaded with 0.10 mg Pt/cm^2. The carbon disperses the platinum catalysts and increases electrical conductivity of catalysts. The carbon also endows electrodes with micropores which enhances the electrode/electrolyte surface area as well as the gas diffusion process. The carbon is bonded with PTFE (30–50 wt%) to get an electrode-support structure which is porous in nature. This structure is bonded on a porous carbon paper substrate such that the carbon paper acts as a current collector in addition to providing the structural support. The carbon paper has micropores ~3.4 nm for gas diffusion and macropores ~3–50 μm. Hence, the carbon black/PTFE layer or carbon paper forms a stable, three-phase interface in the fuel cell. The corrosion resistance of carbon can be increased by giving it a heat treatment in a nitrogen atmosphere at very high temperature ~1000–2000 °C. Many other materials such as phthalocynanines, tetraazaanulenes, tetraphenylporphyrins and transition metals (Fe, Co) have been evaluated as O_2 reduction electrocatalysts. The electrocatalysis of O_2 reduction can be increased by alloying platinum with metals such as zirconium, chromium, vanadium, nickel and titanium, etc.; platinum/nickel alloyed catalysts have resulted in 49 wt% increase in the specific activity compared to that of pure platinum. The anode is sensitive to poisoning caused by contamination.

The fuel stream may contain carbon monoxide (CO) and hydrogen sulphide (H_2S). Hence, it becomes important to clean up the fuel stream prior to fuel exposure at the anode. For desulphurisation of fuel, the fuel should be heated at 350°C before entering the desulphuriser. The cell performance may limited due to the electrode flooding and drying. Conventional PTFE particle surfaces may not get completely covered. Hence, the space between agglomerates and PTFE may act as gas channels during the initial stages of operation. However, eventually the cell performance decreases due to filling of channels with the electrolyte resulting in performance degradation. Hence, the wettability and flooding shall be optimised carefully.

1.2.5 Molten Carbonate Fuel Cell (MCFC)

Like PAFC, the molten carbonate fuel cell is also gaining interest in many corporations including MC Power Corporation (United States) and Energy Research Corporation. Other corporations such as Hitachi, Mitsubishi, Toshiba, Deutshe Aerospace and Brandstefel Nederland are engaged in the development and commercialisation of MCFC technology. The schematic of MCFC is given in Fig. 1.16. MCFC have a high operating temperature between 600 and 700°C. The electrolyte is usually a molten mixture of alkali carbonate such as Li^+/Na^+ carbonate or Li^+/K^+ carbonates in $LiAlO_2$ matrix.

As MCFC operates at high temperatures, the carbonate salts melt and conduct carbonate ions from cathode to anode. From Fig. 1.16. it can be seen that both oxygen and carbon dioxide must be supplied at the cathode. The carbon dioxide gets converted into CO_3^{2-} ions which help in the conduction process. At the anode, the

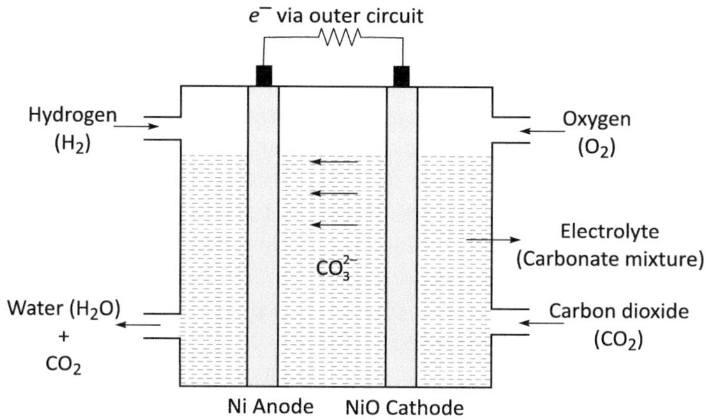

Fig. 1.16 Schematic of MCFC

carbonate ions $CO_3{}^{2-}$ are converted back to CO_2. Hence MCFC require a supply of CO_2 unlike other fuel cells. The electrode reactions for MCFC are as follows:

At anode:

$$2H_2 + 2CO_3^{2-} \rightarrow 2H_2O + 2CO_2 + 4e^- \tag{1.30}$$

At cathode:

$$O_2 + 2CO_2 + 4e^- \rightarrow 2CO_3^{2-} \tag{1.31}$$

The overall reaction becomes

$$\underset{\text{(cathode)}}{2CO_2} + 2H_2 + O_2 \rightarrow 2H_2O + \underset{\text{(anode)}}{2CO_2} \tag{1.32}$$

Hence, the CO_2 generated at the anode needs to be recycled to the cathode. Hydrogen fuel at the anode can be regained by internal reforming. The electrolyte for MCFC consists of 60% carbonate in 40 wt% of $LiAlO_2$ (in fibre form). Almost 70% of the total cell ohmic losses occur in the electrolyte. The matrix resistance can be decreased by increasing the porosity of the electrolyte structure. Secondly Li^+/Na^+ electrolyte replaces Li^+/K^+ electrolyte because Li^+/Na^+ electrolyte offers reduced carbon dissolution and increased ionic conductivity, hence yielding high cell performance. When Li^+/K^+ electrolyte is used, the ratio is kept at 68% Li_2CO_3 + 32% K_2CO_3 (melting point ~550°C). Usually, the loss of electrolyte occurs due to the hydrolysis; i.e.,

$$Me_2CO_3 + H_2O \rightarrow 2MeOH + CO_2 \tag{1.33}$$

Hence, the mixture of Li_2CO_3 + K_2CO_3 exhibits high volatility and dries out quickly. Due to the volatility and segregation of the electrolyte, the electrolyte composition is changed, which influences its melting temperature. Segregation usually occurs due to the difference in mobility of the ions. The development of the MCFC component is a critical issue for attaining long time performance of 40,000–50,000 h. For MCFC, the anode is usually Ni–Cr/Ni–Al alloys. The main function of chromium (Cr) is to prevent anode sintering, but at the same time chromium gets lithiated and then absorbs carbonate from the electrolyte. This leads to the loss of electrolyte, which deteriorates cell performance. Reducing the amount of chromium may reduce electrolyte loss but can increase creep when placed under the torquing load of the stack. Hence, Ni–Al are better anodes than Ni–Cr as they provide minimum electrolyte loss and also provide creep resistance by forming $LiAlO_2$ dispersed in Ni. Ni anodes are expensive and hence replacing Ni by Cu can possess some advantages over the cost. The anode reaction of MCFC is fast and partial flooding of the anode may occur. Partial flooding is advantageous as the carbonate is replenished in the stack and acts as a carbonate reservoir. Some anodes are being tested to tolerate sulphur poisoning. Earlier, a bubble barrier was used between the electrolyte and the anode. The bubble barrier consists of a thin layer of Ni or $LiAlO_2$ consisting of uniform pores. This structure helps in preventing fuel crossover and also prevents flow of electrolyte to the anode.

For cathode materials, structural strength, low dissolution rates in molten electrolyte and good electrical conductivity are required. Nickel oxide is used as the cathode material due to its structural strength and limited solubility in carbonates. Dissolution of nickel ions may occur leading to formation of Ni ions in the electrolyte. Furthermore, the Ni ion may move towards the anode and be precipitated out. The precipitation of Ni may cause an internal short-circuit of the cell and hence loss of power. At high CO_2 pressure, the following reaction may take place:

$$NiO + CO_2 \rightarrow Ni^{2+} + CO_3^{2-} \tag{1.34}$$

Nickel dissolution puts a limiting constraint on the fuel cell performance. The reprecipitation of Ni in the electrode is also known as Ostwald ripening. Ostwald ripening can be reduced if basic carbonates are used instead of acidic carbonates. The acidity of carbonates varies as $K_2CO_3 > Na_2CO_3 > Li_2CO_3$. Hence, working with sodium and lithium carbonates reduces this problem. Secondly, nickel dissolution is prevented by keeping the CO_2 partial pressure low and using a thick electrolyte (which increases the diffusion path for Ni^{2+}). $LiFeO_2$ has been found to be a potential candidate as cathode material because its dissolution is zero. However, slow reaction kinetics of these cathodes at atmospheric pressure is still a concern. It is also shown that the 5 mol% doped-NiO cathode with 0.02 cm thickness yields 43 mV overpotential compared to the standard NiO cathode. When the thickness is increased from 0.5 mm to 1.0 mm, the shorting time increased from 1000–10,000 h. Reducing the partial pressure of CO_2 to 1/3 also decreases the dissolution rate threefold. The use of some additives such as $SrCO_3$, $BaCO_3$, $CaCO_3$ etc., increases the basicity of the electrolyte, which also results in decreased dissolution of nickel. Conclusively, we can say that for acidic melts, NiO dissolution is expressed as

$$NiO \rightarrow Ni^{2+} + O^{2-} \tag{1.35}$$

And for basic melts, nickelate ions are formed.

$$2NiO + O^{2-} + \frac{1}{2}O_2 \rightarrow 2NiO_2^- \tag{1.36}$$

1.2.5.1 Impurities and Their Effects

Impurities always decrease the cell efficiency, hence the clean-up system is required for it. The most common impurities are sulphur compounds, halides, nitrogen compounds etc. The sulphur tolerance of MCFC is influenced by temperature, pressure, cell components, gas composition etc. The anode has relative tolerance for <10 ppm of H_2S, whereas <1 ppm SO_2 is also acceptable. For increasing pressures, these concentration limits decrease, whereas for increasing temperatures these limits tend to increase. Sulphur poisons the catalytic reaction sites for the water gas shift reaction and tends to block the electrochemical sites by chemisorption of Ni surface. The following set of reactions takes place when nickel combines with H_2S:

$$H_2S + CO_2^{2-} \rightarrow H_2O + CO + S^{2-} \tag{1.37}$$

Then

$$Ni + xS^{2-} \rightarrow NiS_x + 2xe^- \tag{1.38}$$

followed by

$$NiS_x + xH_2 \rightarrow Ni + xH_2S \tag{1.39}$$

In reaction (1.39), the anode is sulphided and then gets reduced to nickel. The poisoning due to H_2S interferes with the water gas shift reaction equilibrium. Hence, chromium is used in the anodes as it acts as a sulphur-tolerant catalyst. As CO_2 is required at the cathodes, it is supplied by anode gas recycling. Hence, the sulphur may also contaminate the cathode as it may be present in the exhaust stream of the anode. The sulphur upon entering the cathode may react with the carbonate ions to produce alkali sulphates which are transported towards the anode via the electrolyte. On reaching the anode, the sulphate $SO_4{}^{2-}$ is reduced to S^{2-}.

Halide impurities are also disadvantageous for the functioning of MCFC. When halides react with molten carbonates it leads to the formation of alkali halides, CO_2 and water. Due to the formation of alkali halides (LiCl/KCl/NaCl, etc.), the vapour pressure increases and the rate of electrolyte loss also increases. For tolerance purpose, the HCl level should be kept below 1 ppm in the fuel gas. The nitrogen compounds such as NH_3 and HCN may also react with the electrolyte to form nitrate salts.

1.2.6 Microbial Fuel Cell (MFC)

This is a special category of fuel cells which uses bacteria as catalysts to produce currents by oxidising organic and inorganic matter. Using this technology, the chemical energy in organic compounds is converted into electrical energy by the catalysis action of the microorganisms. Figure 1.17 gives the schematic of microbial fuel cells.

The anodic chamber of MFC contains microbes that oxidise the substrates (glucose, etc.) and electrons/protons are generated along with carbon dioxide as an oxidation product.

The electrons produced by bacteria are transferred to the anode and hence flow to a cathode via a conductive material containing a resistor. The protons enter the cathodic chamber after crossing the polymer membrane. The protons then combine with oxygen in the cathodic chamber to produce water. Chemical oxidisers such as Mn(IV) or ferricyanide can also be used. The bacteria help in catalysing the reoxidation of metal using dissolved oxygen. The typical electrode reactions using acetate substrate are given as

At anode:

$$CH_3COO^- + 2H_2O \xrightarrow{\text{microbes}} 2CO_2 + 7H^+ + 8e^- \quad (1.40)$$

At cathode:

$$O_2 + 4e^- + 4H^+ \rightarrow 2H_2O \quad (1.41)$$

The overall reaction is given by

$$CH_3COO^- + O_2 \xrightarrow{\text{microbes}} 2CO_2 + 3H^+ + 4e^- \quad (1.42)$$

From Eq. (1.42), it can be inferred that the substrate breaks down into carbon dioxide and proton along with electricity as a by-product. The concept of MFC was demonstrated by Potter in 1910, who used platinum electrodes as well as living cultures of *Escherichia coli* and *Saccharomyces*. The anodophile species of the microbes can transfer the electrons directly to the anode. Otherwise, the electron mediators are required in the cell for enhanced power output and increased efficiency. The direct electron transfer to the anode is hindered by a majority of microbial species due to the presence of non-conductive lipopolysaccharides, peptididoglycans and lipid membrane in their outer layers. Hence, mediators are used which capture electrons from the membrane and are reduced. Furthermore, these mediators will again become oxidised once they move across the membrane and release the electrons to the anode. Hence, the electron transfer process keeps the anode replenished which maintains the sustainability of cell. Anthraquinone-2, 6-disulphonate, 2-methylene blue, thionine, 2-hydroxy-1, 4-naphthoquinone, Fe (III) EDTA, Meldola's blue and neutral red are some of the common chemical mediators which enhance electricity production. MFC is known as the

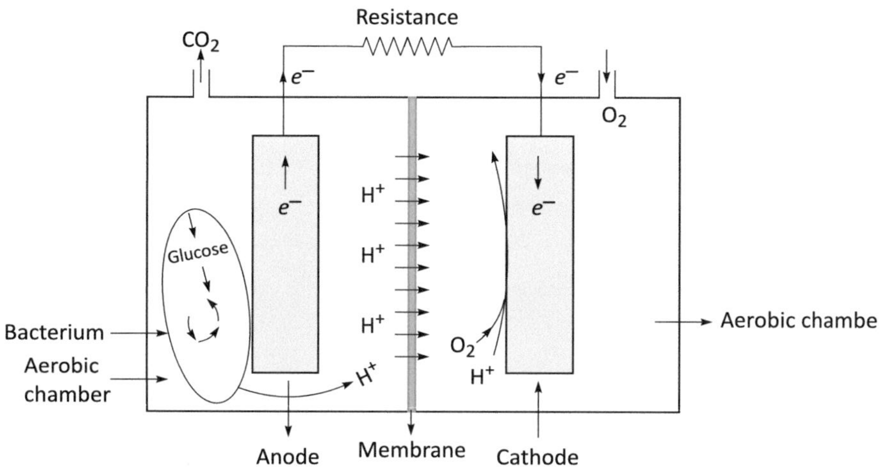

Fig. 1.17 The schematic of microbial fuel cells

"mediator-less" MFC, if no mediators are added to the cell. Some mediators are toxic and unstable for the MFC, hence limiting their applications. The following features are mandatory for the mediators to work in the MFC atmosphere:

- Good stability in anolyte
- Non-toxic to microbes
- Non-biodegradable and low cost
- Permeability across the cell-membrane
- High electrode reaction rate
- Good electron affinity
- Good reducing and oxidising potential

Some microbial metabolites/endogenous mediators are used as mediators by microbes. Some microbes such as *Rhodoferax ferrireducens*, *Geobacter metallireducens*, *Shewanella putrefaciens* and *Geo-bacteriaceae sulferreducens* are bioelectrochemically active. These microbes directly transfer electrons via the membrane. They have the capability to form a biofilm on the anode surface. Hence, in the presence of these microbes the anodes act as an electron acceptor.

1.2.6.1 Microbes and Substrates in MFC

In MFC, the electrons are transferred via microbes constituting an electron transport system. The electron transport system may consist of a series of components in the bacterial extracellular matrix. Figure 1.18 gives the detailed description of the microbes and their respective mediators used for glucose substrates.

Figure 1.19 shows the mediators and microbes for acetate, starch, lactate, pyruvate, sucrose and maltose substrates.

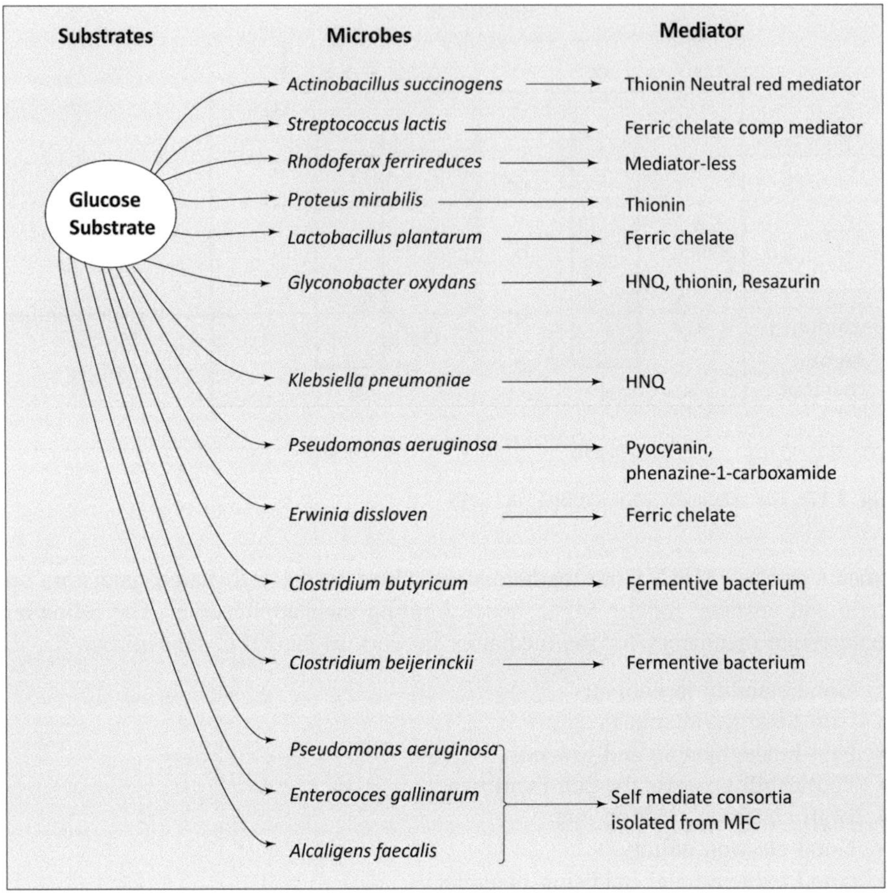

Fig. 1.18 Mediator and microbes for glucose substrate

Mediators play an important role in the process of electron transport, especially for those microbes which cannot transfer the electrons to the anodes. Mediators extract electrons from the microbes and then discharge them onto the anode surface. Hence, mediators shuttle between the bacteria and anode to maintain the electron transfer process.

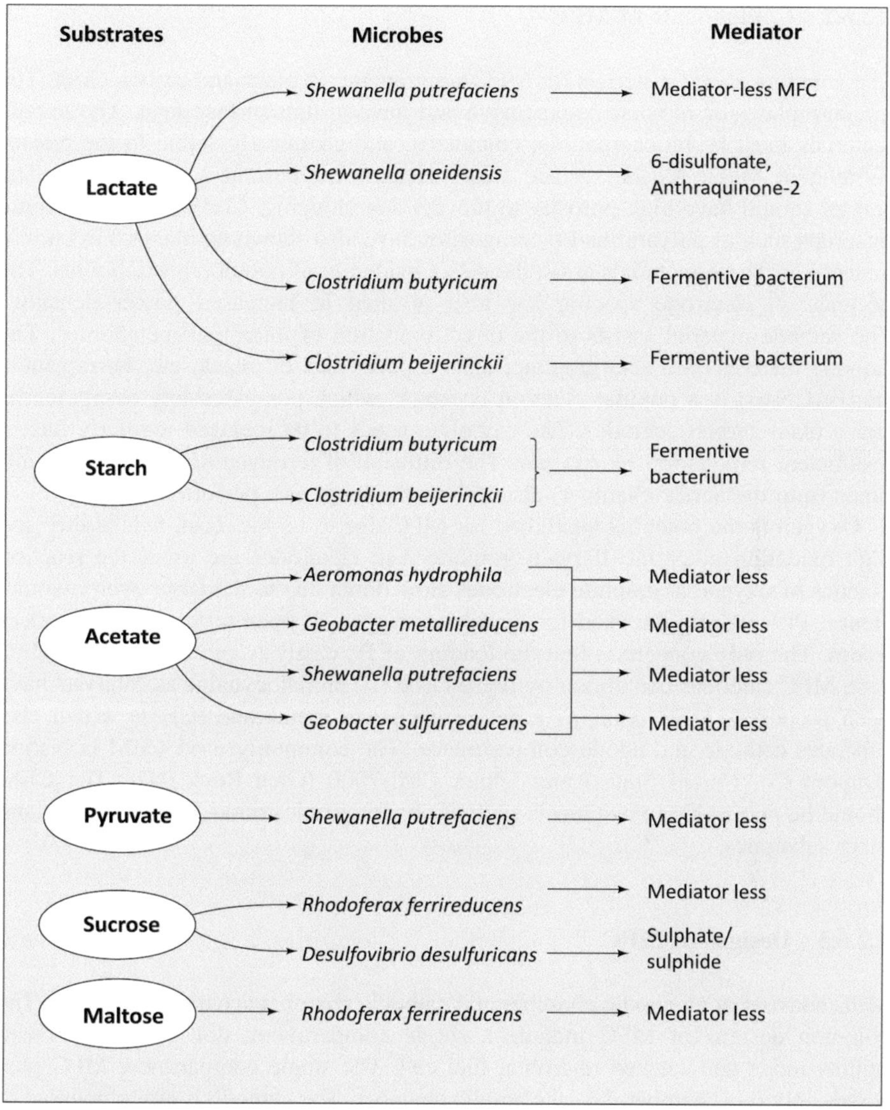

Fig. 1.19 Mediators and microbes for other substrates

1.2.6.2 Components of MFC

The common anode materials for MFC are graphite, Pt black and carbon cloth. The graphite plates or rods are inexpensive and have a high surface area. The anodic materials must be biocompatible, conductive and chemically stable in the reactor solution of MFC. As the surface area increases, the current also increases. The anodes should have high porosity to prevent any clogging. Certain electrocatalytic materials such as polyanilins/Pt. composites have also shown an increase in current generation. These materials assist the direct oxidation of microbial metabolites. The decrease in electrode spacing has also resulted in increased power densities. The cathode material assists in the direct oxidation of microbial metabolites. The cathode materials are also graphite, carbon paper and Pt. black, etc. Ferricyanide ($K_3[Fe(CN)_6]$) is a popular electron acceptor, which possesses low overpotential (for a plain carbon cathode). The catholyte needs to be replaced regularly due to insufficient reoxidation by oxygen. The diffusion of ferricyanide across the membrane (into the anode chamber) also affects the long-term performance of cell.

Oxygen is the potential candidate for MFC due to its low cost, availability and high oxidation potential. If plain graphite disk electrodes are used, the reaction kinetics of oxygen at graphite electrodes slow down due to the large overpotential. Hence, Pt. catalysts are used for dissolved oxygen or open air/gas-diffusion electrodes. The only concern is that the loading of Pt. catalysts can increase the MFC cost. MFC cathodes containing pyrolysed iron (II) phthalocynaine as catalysts have been proposed. MFC contains a cation exchange membrane (CEM) which also separates cathode and anode compartments. The commonly used CEM is Nafion (Dupont Co. United States) and Ultrex CMI-7000 (Glen Rock, NJ). The CEM should be permeable to chemicals such as oxygen, ferricyanide, organic matter and other substrates.

1.2.6.3 Designs of MFC

MFC consists of an anodic chamber and cathodic chamber separated by CEM. The common designs of MFC include a single compartment, double compartment, upflow mode and stacked microbial fuel cell. The single compartment MFC possesses only one chamber, i.e., the anodic chamber. The cathode is either coupled to the anode, fabricated directly on the CEM or in tubular geometry with the anode. Figure 1.20a shows the single compartment MFC.

The two-compartment MFC usually runs on glucose or acetate solution medium to generate electricity (Fig. 1.20b). The two-compartment consists of both cathodic and anodic chambers separated by a CEM or salt bridge to allow the movement of protons and blockage of oxygen. In upflow MFC systems, a plexiglass cylinder is partitioned into two sections using glass bead layers and a glass wall. These two chambers act as cathodic and anodic compartments (Fig. 1.20c). To obtain enhanced current and voltage outputs the microbial cells are connected in series and

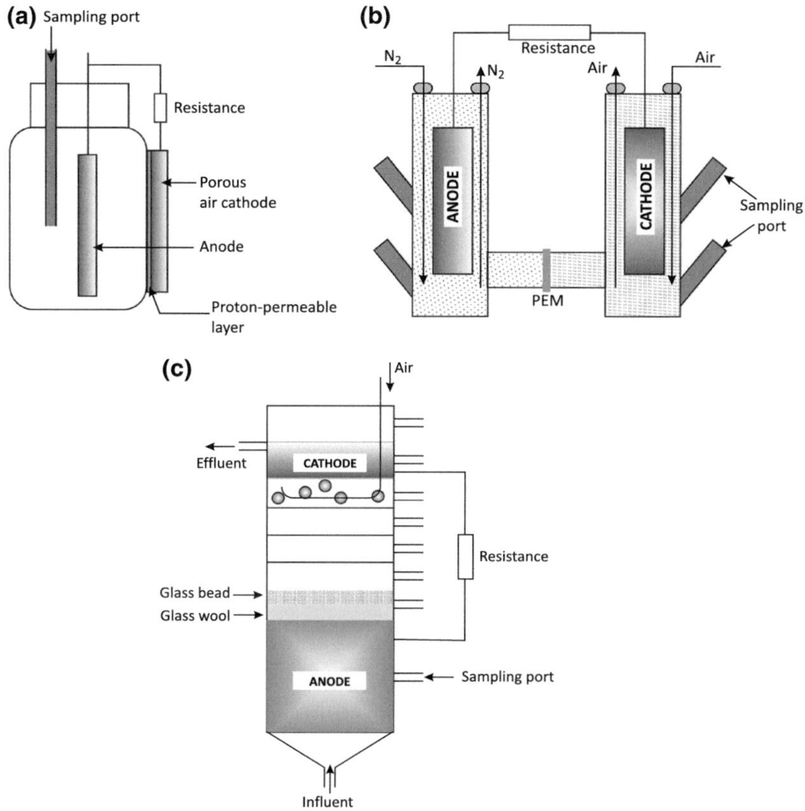

Fig. 1.20 **a** Single chamber MFC with proton permeable membrane (Min et al. 2005a, b). **b** Cylindrical-shaped two-chamber MFC (Park and Zeikus 2003). **c** Upflow MFC with mediator (Jang et al. 2004)

parallel and this arrangement is known as stacking of the MFC (Fig. 1.21). The stack connected in parallel yields a higher short-circuit current compared to the series connected stack. Hence, the parallel stack possesses a higher bioelectrochemical reaction rate. Moreover for chemical oxygen removal, parallel connection is preferred over the series connection in MFC units. The research is ongoing to increase the power density of fuel cells and provide continuous flow through the anode chamber. Systems are also being designed to resemble hydrogen fuel cells, where CEM is sandwiched between the anode and cathode to give the maximum yield and efficiency. The highest power densities are reported for ferricyanide cathodes due to the low internal resistance of the system.

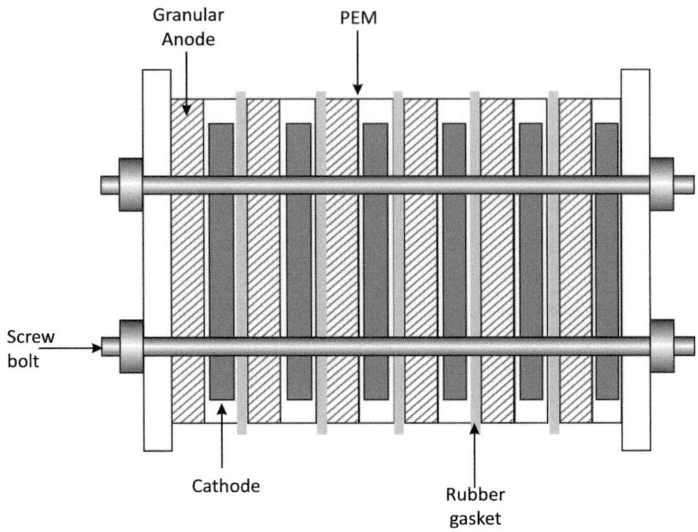

Fig. 1.21 Stacked MFC made of six individual units (Aelterman et al. 2006)

1.2.7 Direct Carbon Fuel Cell (DCFC)

DCFC converts chemical energy into a carbon fuel directly into electricity without the requirement of gasification. DCFC also consists of three basic components, i.e., anode, cathode and electrolyte. Almost pure CO_2 is produced in DCFC when pure carbon is used as a fuel. DCFC offers several attractive features, i.e., lower emissions than coal-firing power plants, higher fuel efficiency than SOFC and MCFC, easy availability of carbon fuels and mechanical simplification due to no reformers and engine requirement. Figure 1.22 shows the schematic of a direct carbon fuel cell.

The solid fuel reacts directly at the electrode (anode) to form a gaseous exhaust product. The cell reactions are given as follows:

At cathode:

$$O_2 + 4e^- \rightarrow 2O^{2-} \tag{1.43}$$

At anode:

$$2O^{2-} + C \rightarrow CO_2 + 4e^- \tag{1.44}$$

The overall reaction is given as

$$C + O_2 \rightarrow CO_2 \tag{1.45}$$

DCFC operates at a high temperature of about 500–900°C and its theoretical efficiency is 100%. The fuel feed and product gases are two distinct phases which

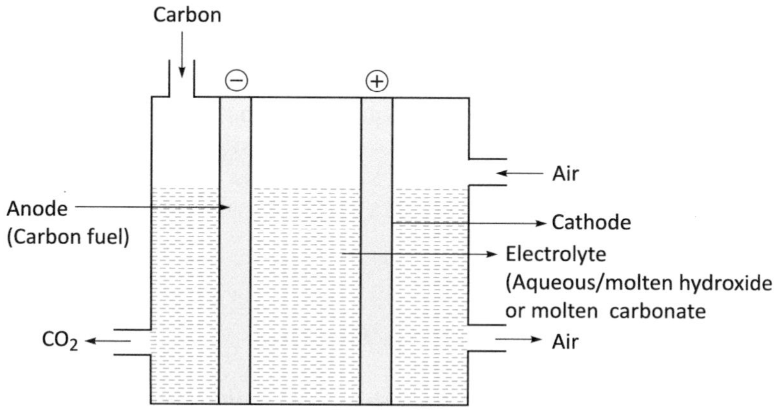

Fig. 1.22 Schematic of DCFC

lead to 100% fuel utilisation. DCFCs is of four types depending on the kind of electrolytes used, i.e., aqueous hydroxide, molten hydroxide, molten carbonate and oxide ion conducting solid electrolyte. It should also be noted that both direct and indirect carbon oxidation reactions occur inside the anode chamber of fuel cell; i.e., the carbon particles coming in direct contact with the electrolyte react with oxygen and are converted to CO_2 as follows:

$$C + O^{2-} \rightarrow CO + 2e^{-} \tag{1.46}$$

$$CO + O^{2-} \rightarrow CO_2 + 2e^{-} \tag{1.47}$$

These carbon particles which do not form direct contact with electrolyte are oxidised by the following reactions:

$$C + CO_2 \rightarrow 2CO \tag{1.48}$$

$$2CO + 2O^{2-} \rightarrow 2CO_2 + 4e^{-} \tag{1.49}$$

CO and CO_2 produced get mixed together in the anodic chamber. Some CO produced may leave the chamber before getting oxidised at the anode, which may lead to a decrease in the efficiency of the DCFC. Equation (1.48) is a chemical reaction and (1.49) is an electrochemical reaction. The DCFC shall be operated continuously to avoid the formation of CO within the anode chamber.

1.2.7.1 DCFC with Aqueous Hydroxide Electrolyte

The aqueous hydroxide electrolyte is used in DCFC with an operating temperature of <250°C and high pressure ~30–30 bar. Raw brown coal was used as fuel in autoclaves operating at 200°C. An iron rod was used as a cathode and the autoclave body worked as an anode. Aqueous alkaline hydroxide was used as the electrolyte in the form of magnesium, sodium, potassium and lithium hydroxides. Recently, cathodes usually been made of silver, platinum, gold and palladium and carbon bed is used as an anode. These designs have not been investigated in detail due to their low performance and efficiency. Hence, these designs could not meet the commercialisation criteria. The operating cell voltage of these cells dropped suddenly and the reasons are not very clear.

1.2.7.2 DCFC with Molten Hydroxide Electrolyte

William Jacques demonstrated DCFC with molten sodium hydroxide as the electrolyte. Figure 1.23 shows the DCFC used by Scientific Applications and Research Associates (SARA) Inc.

Molten KOH or NaOH is used as the electrolyte, which is contained within a metallic container. The metallic container acts as the cathode and a carbon rod dipped into electrolyte acts as both fuel and anode of the cell. Molten hydroxide electrolytes possess advantageous features such as high ionic conductivity, low overpotential and high carbon oxidation rate with a low operation temperature of 600°C for DCFC, which make its components' fabrication economical. The dominant reaction product would be CO_2 instead of CO. The cell reactions are given as follows:

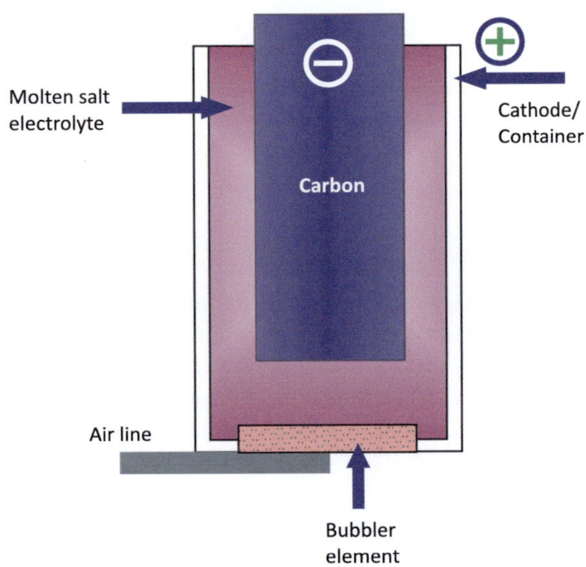

Fig. 1.23 The DCFC used by SARA for most of the investigations

At anode:

$$C + 4OH^- \rightarrow 2H_2O + CO_2 + 4e^- \tag{1.50}$$

At cathode:

$$O_2 + 2H_2O + 4e^- \rightarrow 4OH^- \tag{1.51}$$

Overall reactions:

$$C + O_2 \rightarrow CO_2 \tag{1.52}$$

Carbonate formation still needs to be addressed as it remains the potential problem. Carbonate formation has been explained using chemical and electrochemical process as follows:

Chemical reaction

$$2OH^- + CO_2 \rightarrow CO_3^{2-} + H_2O \tag{1.53}$$

Electrochemical reaction

$$C + 6OH^- \rightarrow CO_3^{2-} + 3H_2O + 4e^- \tag{1.54}$$

Equation (1.54) further consists of two steps as demonstrated below:

$$\begin{aligned} \text{Fast step: } 6OH^- &\rightarrow 3O^{2-} + 3H_2O \\ \text{Slow step: } C + 3O^{2-} &\rightarrow CO_3^{2-} + 4e^- \end{aligned} \tag{1.55}$$

Carbonate formation depends on the water and O^{2-} concentration. If water content is increased, the equilibrium of Eqs. (1.54) and (1.55) shifts towards the left, which would impede the carbonate formation. In addition, the moisture in the electrolyte also increases the ionic conductivity of the melt. There is one major drawback of DCFC: i.e., the air bubbles and air dissolved in the electrolyte come into direct contact with the anode leading to chemical reaction of the fuel rod and hence decreased cell efficiency. Hence SARA has implemented a new design, which has a separator and allows the migration of ions, but electrically separates the cathode and anode chambers. In DCFC, high corrosion rates of metals due to high operating temperatures lead to higher degradation.

1.2.7.3 DCFC with Molten Carbonate Electrolyte

The DCFC with molten carbonate electrolytes are the most commercial type of fuel cells. Molten carbonate as an electrolyte has a number of advantages such as high ionic conductivity long-term stability of CO_2 and catalysis of carbon oxidation. These DCFCs work at high operating temperatures of 600–850°C. Usually, mixed molten

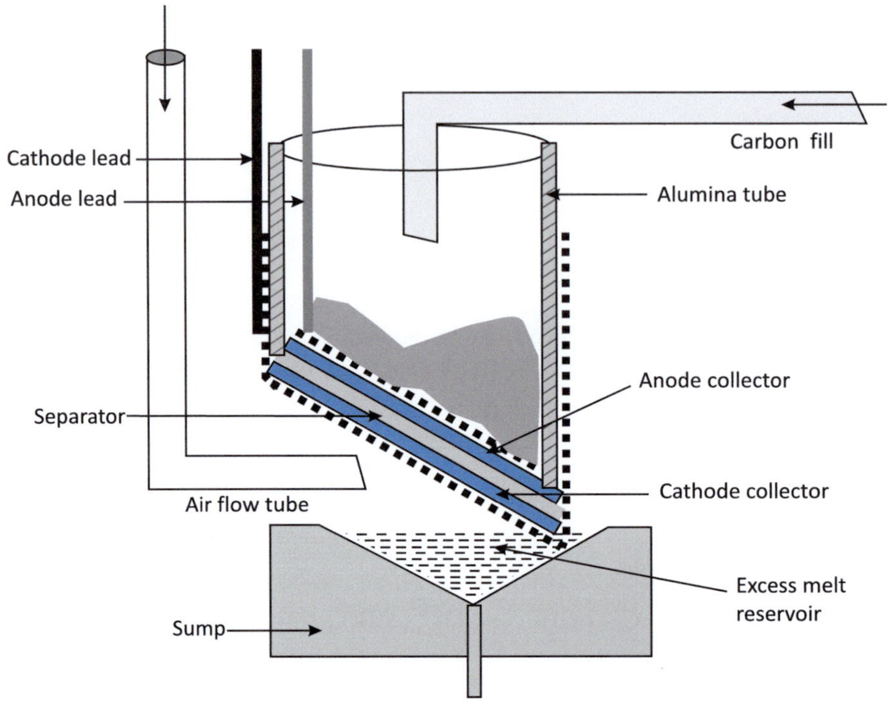

Fig. 1.24 Schematic of the tilted DCFC constructed by LLNL

carbonates of Li_2CO_3 and K_2CO_3 are used as electrolytes; Lawrence Livemore National Laboratory (LLNL) has constructed a DCFC as shown in Fig. 1.24.

In this cell, 68% K_2CO_3 and 32% Li_2CO_3 was used as electrolyte. The cathode is made of sintered frit composed of fine nickel particles, whereas the anode consists of paste of carbon particles (<100 μm). The electrode reactions are given as:

At anode:

$$C + 2CO_3^{2-} \rightarrow 3CO_2 + 4e^- \tag{1.56}$$

At cathode:

$$O_2 + 2CO_2 + 4e^- \rightarrow 2CO_3^{2-} \tag{1.57}$$

Overall Reaction:

$$C + O_2 \rightarrow CO_2 \tag{1.58}$$

Hence 2 carbon dioxide needs to be recycled from the anode to the cathode department. It must be noted that the electrode assembly consists of a separator made of zirconia. This assembly is positioned at an angle of 5°–45° from the horizontal. This configuration prevents flooding of the cathode as excess electrolyte gets drained from the electrode. Molten carbonate DCFC possesses high cathode polarisation corrosion of metal bipolar plates, low power densities and short cell lifetimes. The electrolyte stability and gas flow have been improved by mixing sodium carbonate in it. Many fuels including biological char, thermal black, graphite particles and coal derived pitch have been used as carbon fuel for molten carbon DCFC. In addition, the stack-based issues, fuel delivery systems and fuel processing issues are being investigated. Impurities such as sulphur have been found to degrade the performance of DCFC. Sulphur reacts with the anode leading to the formation of nickel sulphide, which diminishes the current collecting capability.

1.2.7.4 DCFC with Oxide Ion Conducting Solid Electrolyte

This cell uses the same electrolyte as that of the solid oxide fuel cell (SOFC) and operating temperature is between 800 and 1000°C. Scandia-doped zirconia and yttria stabilised zirconia are the most commonly used electrolytes as they have high ionic conductivity. Figure 1.25 shows the oxide ion conducting DCFC (General Electrical Company). ZrO_2–CaO electrolyte was used in a tube of thickness 0.62 mm. The cathode was made of a layer of ZrO_2 flooded with silver and

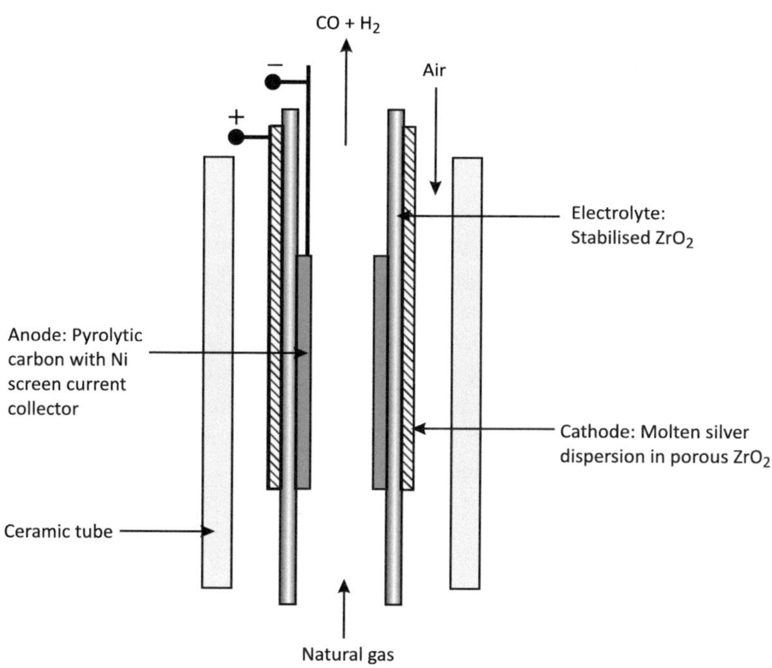

Fig. 1.25 DCFC built by General Electric Company with porous zirconia cathode

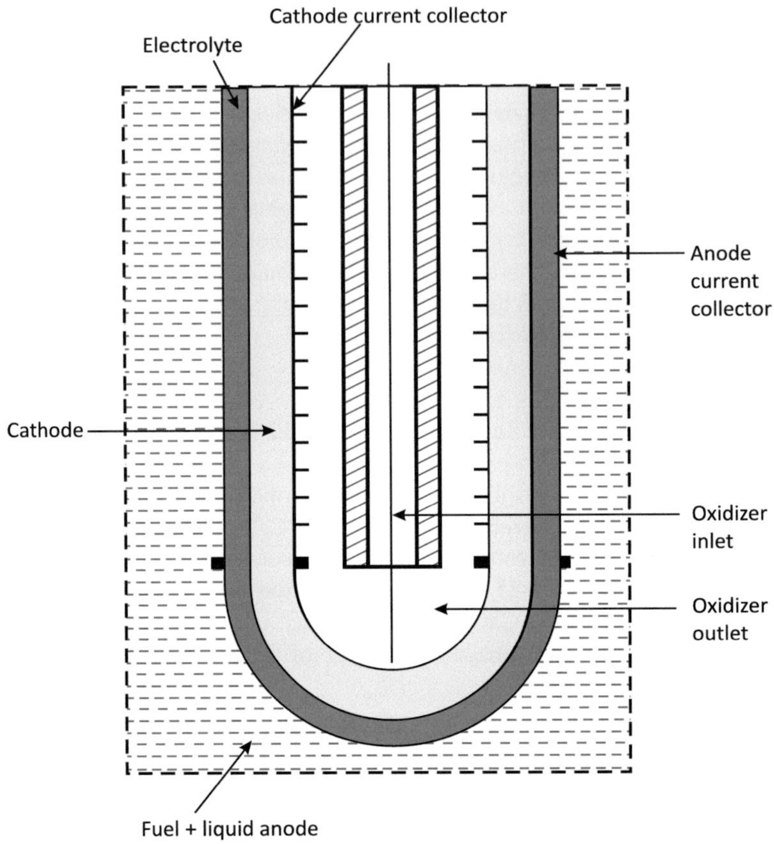

Fig. 1.26 Schematic of DCFC built by SRI

saturated with oxygen. Pyrolytic carbon formed by decomposition of methane was used as the anode. SRI International provided a U-shaped DCFC (Fig. 1.26), where the inner tube was a metal mesh current collector followed by a cathode layer (e.g., lanthanum strontium gallate) an electrolyte layer of solid zirconia and the outer tube was a metal mesh anode current collector. The whole U-tube arrangement was immersed in a mixture of molten carbonates of sodium, lithium and potassium. The power densities achieved by the SRI have been considerably high and comparable to those achieved by MCFC technology.

1.2.8 Direct Borohydride Fuel Cell (DBFC)

These cells used metal borohydrides such as sodium borohydride as a fuel as it contains up to 10.5% of hydrogen by weight. This cell also uses Nafion as the electrolyte membrane. The reaction involves the transport of hydroxyl (OH^-) and cations such as H^+ and Na^+, etc. BH_4^- is the electroactive ion; hydrolysis of BH_4^- causes loss in fuel leading to evolution of hydrogen at the fuel electrode. In addition, the cost of sodium borohydride is a major issue for the commercialisation of this technology.

1.2.9 Solid Oxide Fuel Cell (SOFC)

Solid oxide fuel cells have Ni/YSZ cermet anode and a ceramic composite cathode [usually lathene strontium manganate (LSM)]. The electrolyte is 8% Y_2O_3 doped ZrO_2 which conducts oxygen ions. SOFC operates at high temperatures of about 800–1000°C due to low conductivity of oxide ions at lower temperatures. The two most common designs of SOFC are planar configuration and tubular configuration. A number of organisations and research groups are working on the development of SOFC technology. Large-scale demonstration plants involve Siemens Westinghouse Corporation, Rolls Royce Fuel Cells, Bloom Energy and Mitsubishi Heavy Industries. This book is based on the various fundamentals and development of the SOFC technology. Hence, Chaps. 3, 4 and 5 describe the SOFC technology in detail. In addition, the interfacial interaction among the components of SOFC are discussed in Chaps. 6, 7 and 8. Chapter 9 gives a detailed insight on the fuel cell technology status, i.e., stationary and mobile applications.

Bibliography

Aelterman P, Rabaey K, Pham HT, Boon N, Verstraete W (2006) Environ Sci Technol 40:3388
Jang JK, Pham TH, Chang IS, Kang KH, Moon H, Cho KS, et al (2004) Process Biochem 39:1007
Larminie J, Dicks A (2003) Fuel cells system explained, 2nd edn. Wiley, England
Min B, Cheng S, Logan BE (2005a) Water Res 39:1675
Min B, Kim JR, Oh SE, Regan JM, Logan BE (2005b) Water Res 39:4961
Park DH, Zeikus JG (2003) Biotechnol Bioeng 81:348

Suggested Bibliography

Balachov II, Hornbostel MD, Lipilin AS (2005) Direct coal fuel cells (DCFC): clean electricity from coal and carbon based fuels. The carbon fuel cell seminar, Palm Springs, CA
Cao D, Sun Y, Wang G (2007) J Power Sources 167:250
Cooper JF, Berner K (2005) The carbon/air fuel cell e conversion of coal-derived carbons. In: The carbon fuel cell seminar, Palm Springs, CA, US, 14 Nov 2005
Du Z, Li H, Gu T (2007) Biotechnol Adv 25:464
Giddey S, Badwal SPS, Kulkarni A, Munnings C (2012) Prog Energy Combust Sci 38:360
Gulzow E (2004) Fuel Cells. doi:10.1002/fuce.200400042
Haile SM (2003) Acta Materialia 51:5981

Hirschenhofer JH, Stauffer DB, Engleman RR, Klett MG (1998) Fuel cell handbook, 4th edn
Ieropoulos IA, Greenman J, Melhuish C, Hart J (2005a) Enzyme Microb Tech 37:238
Ieropoulos I, Greenman J, Melhuish C, Hart J (2005b) J Power Sources 145:253
Logan BE et al (2006) Environ Sci Technol 40(17):5181
Min B, Logan BE (2004) Environ Sci Technol 38:5809
Niessen J, Schroder U, Rosenbaum M, Scholz F (2004) Electrochem Commun 6:571
Niessen J, Harnisch F, Rosenbaum M, Schroder U, Scholz F (2006) Electrochem Commun 8:869
Park DH, Zeikus JG (2000) Appl Environ Microb 66:1292
Park DH, Zeikus JG (2002) Appl Microbiol Biotechnol 59:58
Patton E, Zecevic S (2005) Palo Alto: EPRI CA 1011496
Pham CA, Jung SJ, Phung NT, Lee J, Chang IS, Kim BH et al (2003) FEMS Microbiol Lett 223:129
Pham TH, Jang JK, Chang IS, Kim BH (2004) J Microbiol Biotechnol 14:324
Pham TH, Rabaey K, Aelterman P, Clauwaert P, De Schamphelaire L, Boon N, Verstraete W (2006) Eng Life Sci 6:285
Rabaey K, Boon N, Hofte M, Verstraete W (2005a) Environ Sci Technol 39:3401
Rabaey K, Clauwaert P, Aelterman P, Verstraete W (2005b) Environ Sci Technol 39:8077
Rabaey K, Van De Sompel K, Maignien L, Boon N, Aelterman P, Clauwaert P et al (2006) Environ Sci Technol 40:5218
Rhoads A, Beyenal H, Lewandowshi Z (2005) Environ Sci Technol 39:4666
Ringeisen BR, Henderson E, Wu PK, Pietron J, Ray R, Little B et al (2006) Environ Sci Technol 40:2629
Schroder U, Nieben J, Scholz F (2003) Angew Chem Int Ed 42:2880
Singhal SC, Kendall K (2000) High temperature solid-oxide fuel cells: fundamentals. Elsevier, New York
Vega CA, Fernandez I (1987) Bioelectrochem Bioenerg 17:217
Zecevic S, Patton EM, Parhami P (2003) Electrochemistry of direct carbon fuel cell based metal hydroxide electrolyte. Pittsburgh, PA: Direct Carbon Fuel Cell Workshop, NETL. http://www.netl.doe.gov/publications/proceedings/03/dcfcw/Zecevic.pdf

Chapter 2
Cell Voltages, Polarisations and Performances

The fuel cell works under defined conditions such as gas variables, temperature, etc. The operating conditions determine the efficiency of fuel cells and hence require rigorous understanding. Behind the operation of every fuel cell, there are series of thermodynamic and chemical equations that should be well understood. In this chapter, we determine the performance, efficiency and then losses of the fuel cell. The open-circuit voltage (OCV) of a fuel cell is also taken into consideration along with various factors including pressure, temperature, fuel constituents and concentration, etc.

2.1 Electromotive Force of Fuel Cell

Before obtaining the EMF for the cell, we need to have a clear understanding of Gibbs free energy (ΔG). Gibbs free energy is defined as the energy available to do external work. Nowadays, a most common term 'exergy' is used, which is defined as the total external work extracted including that due to pressure and volume changes. We also use the term 'zero-energy point' for the elements in a normal state of standard pressure (0.1 MPa) and temperature (25°C).

Gibbs energy is basically free energy available for conversion into usable work. The change in Gibbs free energy is given by

$$\Delta G = \Delta H - T\Delta S \tag{2.1}$$

ΔH is the enthalpy change, which is the total thermal energy available, ΔS is the entropy change and $T\Delta S$ represents unavailable energy attributed to the entropy or disorder of the system. When the entropy change is positive, then the heat is extracted from the surroundings whereas for the negative entropy change, heat is given out to the surroundings.

G. Kaur, *Solid Oxide Fuel Cell Components*,
DOI 10.1007/978-3-319-25598-9_2

The Gibbs free energy of formation G_f can also be expressed in terms of products and reactants as follows:

$$\Delta G_f = G_f(\text{products}) - G_f(\text{reactants}) \tag{2.2}$$

We can demonstrate the Gibbs energy of formation with the help of the following reaction:

$$H_2 + \frac{1}{2}O_2 \rightarrow H_2O \tag{2.3}$$

$$\Delta G_f = (G_f)_{H_2O} - \left[(G_f)_{H_2} + \frac{1}{2}(G_f)_{O_2}\right] \tag{2.4}$$

However, for a fuel cell the reaction mechanism is not as simple as it appears to be. This can be addressed by considering the electrons moving in an external circuit. For each hydrogen molecule used, 2 electrons pass through the external circuit. Hence, for 1 mol of hydrogen used, $2N$ electrons will migrate through the external circuit [N − Avogadro's number or 6.023×10^{23} atoms]. Hence, the total charge flowing through the external circuit becomes $-2Ne$ or $-2F$ coulombs (F = *Faraday's constant* = Ne = charge on one mole of electrons).

The total electric work done in moving $-2F$ charge around the cell-circuit, for E voltage (ideal potential) of fuel cell is given by

$$\text{Work done} = -2FE \text{ J} \tag{2.5}$$

This work released Gibbs free energy (provided the system is reversible and has no losses).

$$\Rightarrow \Delta G_f = -2FE \tag{2.6}$$

$$\Rightarrow E = -\frac{\Delta G_f}{2F} \tag{2.7}$$

The more general form of Eq. (2.7) is given by

$$\boxed{E = -\frac{\Delta G_f}{nF}} \tag{2.8}$$

where n represents number of electrons participating in the cell reaction. Equation (2.8) is the fundamental electromotive force equation for the fuel cell. 'E' also represents the OCV of cell.

For general cell reaction, the free energy is expressed as:

$$uA + vB \rightarrow xC + yD \tag{2.9}$$

The free energy change is given by

$$\Delta G_{\mathrm{f}} = \Delta G_{\mathrm{f}}^{\bullet} + RT\ln\frac{[C]^x[D]^y}{[A]^u[B]^v} \tag{2.10}$$

R is the gas constant and $\Delta G_{\mathrm{f}}^{\bullet}$ is the free energy change at standard temperature and pressure (STP) conditions. Substituting (2.8) in (2.10), we find

$$E = E^{\circ} + \frac{RT}{nF}\ln\frac{[A]^u[B]^v}{[C]^x[D]^y} \tag{2.11}$$

Equation (2.11) is the general form of the Nernst equation, where $[A]^u$ $[B]^v$ is the reactant activity and $[C]^x$ $[D]^y$ is the product activity. Hence, it is clear from Eq. (2.11), that the increase in concentration of reactants increases the cell potential. The potential can also be obtained by calculating the heat capacities, entropies and enthalpies. The heat capacity (C_p) can be given by

$$C_p = a + bT + cT^2 \tag{2.12}$$

where T is the temperature and a, b and c are empirical constants. Hence, the change in heat capacities of products and reactants can be obtained as

$$\Delta C_p = \Delta a + \Delta(bT) + \Delta\left(cT^2\right) \tag{2.13}$$

or

$$dC_p = da + d(bT) + d\left(cT^2\right) \tag{2.14}$$

The changes in enthalpy and entropy are expressed as:

$$\Delta S = \Delta S^{\circ} + \int_{298}^{T} \frac{\Delta C_p}{T}\,\mathrm{d}T \tag{2.15}$$

and

$$\Delta H = \Delta H^{\circ} + \int_{298}^{T} \Delta C_p\,\mathrm{d}T \tag{2.16}$$

where ΔS° and ΔH° are the changes in entropies and enthalpies at standard temperature (298 K) and pressure. Using Eq. (2.14) in Eqs. (2.15) and (2.16) yields following results:

$$\Delta S = \Delta S^{\circ} + a \ln\left(\frac{T}{298}\right) + b(T - 298) + \frac{1}{2}C(T - 298)^2 \tag{2.17}$$

$$\Delta H = \Delta H^{\circ} + a(T - 298) + \frac{1}{2}(b - 298)^2 + \frac{1}{3}C(T - 298)^3 \tag{2.18}$$

We can obtain OCV for some fuel cells and batteries. For example, for DMFC [Chap. 1, Eq. (2.22)], the cell reaction is given by

$$CH_3OH + \frac{3}{2}O_2 \rightarrow 2H_2O + CO_2 \tag{2.19}$$

$6e^-$ are passing through the external circuit for each methanol molecule. The change in Gibbs free energy for Eq. (2.19) is −698 kJ/mol. By using Eq. (2.8), we obtain

$$E = \frac{698 \times 10^3}{6 \times 96{,}485} = 1.21\ \mathrm{V} \tag{2.20}$$

(F = Faraday constant = Ne = $6.023 \times 10^{23} \times 1.6 \times 10^{-19}$ C = 96,485 C).

For a hydrogen fuel cell, the ΔG_f is 210.3 kJ/mol at temperature of 400°C. The basic reaction is given by:

$$H_2 + \frac{1}{2}O_2 \rightarrow H_2O$$

The OCV, 'E' at 400°C is given by

$$E = \frac{210.3 \times 10^3}{2 \times 96{,}485} = 1.09\ \mathrm{V} \tag{2.21}$$

Similarly for batteries (here alkali battery), the anode and cathode reactions are given by:

At Anode:

$$Zn + 2OH^- \rightarrow ZnO + H_2O + 2e^- \tag{2.22}$$

At Cathode:

$$2MnO_2 + H_2O + 2e^- \rightarrow Mn_2O_3 + 2OH^- \tag{2.23}$$

The overall cell reaction is given by

$$2MnO_2 + Zn \rightarrow ZnO + Mn_2O_3 \tag{2.24}$$

The changes in Gibbs free energy for Eq. (2.24) is −277 kJ/mol, which yields OCV for transfer of $2e^-$ in the outer circuit follows.

$$E = \frac{277 \times 10^3}{2 \times 96{,}485} = 1.44\,\text{V}$$

2.2 Cell Efficiency

There are two processes which may take place inside a fuel cell, i.e., reversible or irreversible. If all energy is converted into electrical energy without any loss, then the process is considered to be reversible. On the other hand, if some of the energy is converted to heat which leads to losses in the system, then the process is considered to be irreversible. The efficiency of reversible system is considered to be 100 %, whereas for an irreversible system the efficiency is <100% attributed to the loss due to some thrown away or wasted energy. The efficiency of an energy conversion system (such as a fuel cell) is defined as electric energy produced per mol of fuel relative to the change in chemical energy of stored fuel; i.e.,

$$\eta = \frac{\Delta G_f}{\Delta H_f} \tag{2.25}$$

ΔH_f is the enthalpy of formation and is also considered to be the calorific value of fuel. The value of ΔH_f is positive when energy is required for the process whereas its negative value represents release of energy. The maximum possible efficiency is defined as:

$$\eta = \frac{\Delta G_f}{\Delta H_f} \times 100 \tag{2.26}$$

The efficiency of a fuel cell is determined by considering the standard free energy of the formation of water

$$H_2 + \frac{1}{2}O_2 \rightarrow H_2O(l) \quad \Delta H_f = -285.84\ \text{kJ/mol}$$

or

$$H_2 + \frac{1}{2}O_2 \rightarrow H_2O\,(g) \quad \Delta H_f = -241.83\ \text{kJ/mol}$$

The difference between the two heats of formation is 44.01 kJ/mol and gives the molar enthalpy of vaporisation of water (latent heat). The efficiency of an ideal fuel cell operation at standard conditions is $\left(\frac{237.30}{285.84}\right) = 0.83$ or 83% for the free energy of 237.3 kJ/mol.

The efficiency depends upon the temperature and pressure, etc., of the system, which is discussed later on in this chapter. The efficiency of a system can also be expressed in terms of the maximum EMF of a cell. Due to the losses in the fuel cells, the efficiency of systems decreases and actual cell voltage (V_{actual}) decreases as compared to the ideal cell voltage (V_{ideal}).

$$\begin{aligned} \eta &= \frac{\text{Useful energy}}{\Delta H_f} = \frac{V_{\text{actual}} \times \text{Current}}{V_{\text{ideal}} \times \frac{\text{Current}}{0.83}} \\ \eta &= \frac{0.83\ V_{\text{actual}}}{V_{\text{ideal}}} \end{aligned} \tag{2.27}$$

For illustration, the efficiency of a hydrogen fuel cell operating at 25°C and after pressure is given by

$$\text{The open circuit voltage}\, E = -\frac{\Delta G_f}{nF}$$

At 25 °C, $\Delta G_f = -237.3$ kJ/mol

$$\Rightarrow E = \frac{237.3 \times 10^3}{2 \times 96{,}485} = 1.22\ V$$

As the OCV is the ideal voltage, hence we can say, $V_{\text{ideal}} = 1.22$ V. Using this value of V_{ideal} in Eq. (2.28), we get

$$\begin{aligned} \eta &= 0.83 \times \frac{V_{\text{actual}}}{1.22} = 0.675\ V_{\text{actual}} \\ \Rightarrow \eta &= 0.675\ V_{\text{actual}} \end{aligned} \tag{2.28}$$

The cell voltage will vary for different temperature and current densities, hence changing the efficiencies. For example, η changes from 8.27 at 100°C to 52% at 1000°C, when we use carbon monoxide, i.e.,

$$CO + \frac{1}{2}O_2 \rightarrow CO_2$$

whereas, η does not change considerably for following reaction

$$CH_4 + 2O_2 \rightarrow CO_2 + 2H_2O$$

This is due to the fact that the ΔG_f changes considerably with temperature for formation of carbon dioxide from the carbon monoxide. In contrast to this, the ΔG_f for formation of CO_2 from methane remains almost constant w.r.t. temperature.

2.3 Polarisation and Cell Losses

In the previous section, we have obtained the open-circuit value (OCV) for a fuel cell as

$$E = -\frac{\Delta G_f}{nF}$$

This value is considered to be maximum for the ideal case. For example, at 200°C, the $\Delta G_f = -220$ kJ, which yields

$$E = \frac{220 \times 10^3}{2 \times 96{,}485} = 1.14\,\text{V}$$

Anyhow, the actual performance of a fuel cell is smaller than this voltage. Figure 2.1 demonstrates the ideal performance and actual performance of the cell.

Fig. 2.1 Graph depicting 'no loss' voltage and 'actual voltage'

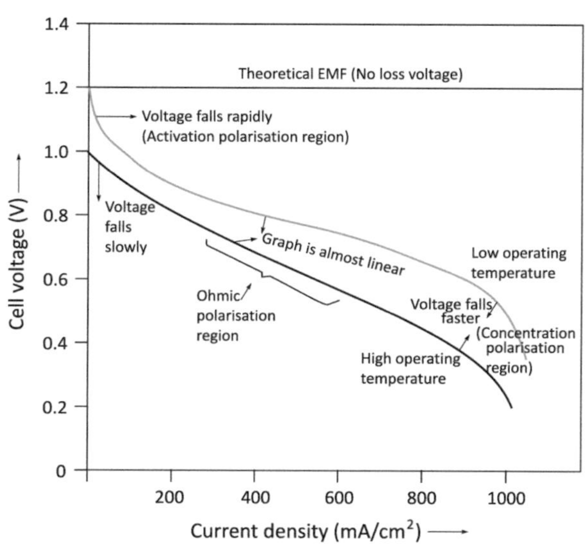

Figure 2.1 shows the variation of cell voltage with current density (current/area) for the cell operating at high and low temperatures. The initial fall in voltage is small for the cells operating at high temperatures, whereas the initial voltage fall is high for the cells operating at low temperatures. In addition to this, the cells operating at higher temperatures exhibit more linear behaviour. The reversible system is considered to be the ideal system because it is a 'no loss' system. The actual potential of a cell decreases from its ideal potential due to irreversible losses in the system. These fuel cell irreversibilities contribute to the voltage drop of the system. There are four major factors which contribute to the decrease in cell performance. The primary four sources of losses are fuel cross-over, activation polarisation, ohmic polarisation and concentration polarisation. We can summarise the total voltage as

$$\text{Total voltage } V = \text{Open circuit voltage } E - \text{Losses} \tag{2.29}$$

The fuel crossover losses are due to the wastage of fuel passing through the electrolyte. Fuel crossover losses are prominent in the fuel cell operation at low temperatures. The activation losses are due to the slow reaction kinetics on the surface of the electrodes. Ohmic losses are due to the resistance offered to the flow of electrons and ions through electrodes and electrolytes, respectively. The concentration losses are due to the change in concentration of the reactants at the surface of electrodes.

2.3.1 Activation Losses

Activation losses are the main cause of voltage drop (mainly at cathode) for the low- and medium-temperature fuel cells. In 1905, Tafel obtained a pattern for the verification of overvoltage against current density as shown in Fig. 2.2.

The voltage drop due to activation polarisation is represented by the Tafel equation as follows:

$$\eta_{\text{act}} = \frac{2.3\,RT}{\alpha nF} \log\left(\frac{i}{i_0}\right) \tag{2.30}$$

where i_0 is the exchange current density.

$$\begin{aligned}
\Rightarrow \quad \eta_{\text{act}} &= \frac{2.3\,RT}{\alpha nF} \log i - \frac{2.3\,RT}{\alpha nF} \log i_0 \\
\Rightarrow \quad \eta_{\text{act}} &= -\frac{2.3\,RT}{\alpha nF} \log i_0 + \frac{2.3\,RT}{\alpha nF} \log i \\
\Rightarrow \quad \eta_{\text{act}} &= a + b \log i
\end{aligned} \tag{2.31}$$

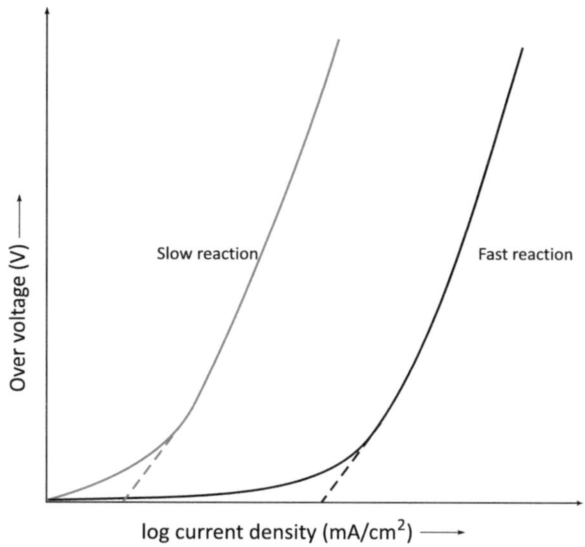

Fig. 2.2 Slow and fast electrochemical reactions represented as Tafel Plots

where

$$a = -\frac{2.3\,RT}{\alpha\,nF}\log i_0 \tag{2.32}$$

and

$$b = \frac{2.3\,RT}{\alpha\,nF} \tag{2.33}$$

α is the electron transfer coefficient and is proportional to the electrical energy, which changes the rate of the electrochemical reaction. The term 'b' represents the Tafel slope. The factor $\frac{RT}{\alpha nF}$ is high for the slow electrochemical reaction, whereas it is small for the fast electrochemical reaction. The exchange current density i_0 is the current density at which overvoltage starts increasing from zero. For a slow electrochemical reaction, i_0 is small and vice versa. If we rearrange Eq. (2.30), then we obtain

$$i = i_0 \exp\left[\frac{\alpha\,nF\,\eta_{\text{act}}}{RT}\right] \tag{2.34}$$

Equation (2.34) is the Butler–Volmer equation and can be used as an alternative for the Tafel equation. It should be noted that the Tafel equation holds good only for $i > i_0$.

Basically, activation polarisation is dominant at low current density because the electronic barriers are to be overcome prior to ion and current flow. Activation polarisation is directly influenced by the rate of electrochemical reaction and is attributed to the sluggish electrode kinetics. When the exchange current density is zero, then it indicates that the forward and reverse reactions are taking place at the some reaction rate; i.e., the flow of electrons takes place at same rate. When the exchange current density is high, then the reaction rate becomes higher in one direction and the electrode surface becomes more active. Current density influences the cell performance crucially; i.e., when i_0 is increased tenfold, the activation polarisation also increases by 100 mV [for slope of 100 mV/log (i_0)]. When the Tafel slope is 50 mV/log current density, then the tenfold increase in the current density yields 50 mV increase of the activation polarisation. If the Tafel slope gets lowered, then the activation polarisation also gets lowered down. Equation (2.30) also gives an impression that the rise in temperature might raise the η_{act}, whereas practically it is observed that the controlling factor for η_{act} is 'i_0'. A smaller value of i_0 yields a greater voltage drop; usually the value of i_0 is smaller at cathode as compared to the anode for hydrogen fed fuel cells. But for the fuel cells, where methane is used as a fuel, the anode overvoltage cannot be neglected. The total voltage drop would be given by:

$$\text{Voltage drop} = b_{\text{anode}} \log\left(\frac{i}{i_{0a}}\right) + b_{\text{cathode}} \log\left(\frac{i}{i_{0c}}\right) \tag{2.35}$$

i_{0a}	anode exchange current density
i_{0c}	cathode exchange current density
$b_{\text{anode}}/b_{\text{cathode}}$	anode and cathode Tafel constants, respectively

Equation (2.35) is the same as Eq. (2.30). Hence, the activation polarisation may be due to one electrode or both electrodes. If the value of i_0 is increased then the activation polarisation is reduced, hence the focus should be on increasing the exchange current. The exchange current density can be increased by using catalysts. In addition to this, increasing the pressure and temperature (Fig. 2.1), would effectively increase the current density. Increasing the reactant concentration as well as electrode roughness would probably lead to an increase in current density due to occupancy of catalyst sites by the reactants.

2.3.2 *Ohmic Losses (Resistive Losses)*

The fuel cell consists of electrodes and electrolytes. Electrodes offer resistance to the flow of electrons whereas the electrolyte provides obstruction to the flow of ions. The ohmic losses occur due to the flow of electrons at cathodes and flow of ions through the electrolyte. The cell interconnects and bipolar plates also contribute to the ohmic losses. Usually, Ohm's law is given by

$$V = IR \tag{2.36}$$

where I is the current and R is the resistance. For a fuel cell, Eq. (2.36) can be written as

$$\eta_{ohm} = ir \tag{2.37}$$

where i is the current density in (mA/cm^2) and r is the area-specific resistance (ASR), i.e., the resistance corresponding to an area of 1 cm^2 of the cell. The ASR includes total cell resistance, i.e., ionic, electronic and contact resistance and the units of r are kΩ cm^2. The cell resistance remains constant; hence as the current increases, the ohmic polarisation also increases. The ohmic polarisation can be reduced by decreasing the electrode separation which decreases the resistance, thereby decreasing the ohmic polarisation. Second, if the electrodes are electrolyte supported, then the electrolyte must be made as this as possible provided the two electrodes should not short circuit. Both the electrodes and electrolyte should exhibit high electronic and ionic conductivity, respectively.

2.3.3 *Concentration Losses (Mass Transport Losses)*

The concentration losses are also known as gas transport losses. During an electrochemical process, the reactant is consumed at the electrode. This leads to the concentration gradient and loss of potential due to the inability of the surrounding material to maintain the initial bulk concentration. The losses due to the concentration gradient and depletion of reactants at the electrode are known as concentration losses. Hydrogen and oxygen are fed at the anode and cathode, respectively. During the working of the fuel cell, the concentration of oxygen and hydrogen will fall on the cathode and anode, respectively. The decrease in concentration depends on the circulation of air/hydrogen around the cathode/anode and the current being taken from the fuel cell. Hence, the partial pressure of oxygen and hydrogen changes the concentration; i.e., the decrease in concentration causes a decrease in partial pressure. This reduction in gas pressure causes a decrease in cell voltage. The main causes of concentration losses are solution of reactants into the electrolyte, dissolution of products out of the electrolyte, slow gas diffusion into electrode pores, diffusion of reactants through the electrolyte to the electrochemical reaction site and diffusion of products through the electrolyte from the electrochemical reaction site. The change in voltage caused by concentration polarisation can be explained with the help of Fick's law.

According to Fick's first law of diffusion, the current density is given by:

$$i = \frac{nFD(C_B - C_S)}{d} \tag{2.38}$$

where n is the number of electrons participating in the reaction, D is the diffusion coefficient of the reactants, C_B and C_S are bulk and surface concentration, respectively, and d is the diffusion layer thickness.

i_l is postulated as the limited current density, i.e., the current density at which the rate of supply of fuel and rate of use of fuel equals. This occurs for the zero surface concentration. i.e., $C_S = 0$. Equation (2.38) becomes

$$i_l = \frac{nFDC_B}{d} \tag{2.39}$$

Equation (2.38) can be rewritten as

$$\begin{aligned} i &= \frac{nFDC_B}{d} - \frac{nFDC_S}{d} \\ \Rightarrow \quad i &= i_l - \frac{nFDC_S}{d} \\ \Rightarrow \quad \frac{i}{i_l} &= 1 - \frac{nFDC_S}{di_l} \\ \Rightarrow \quad \frac{i}{i_l} &= 1 - \frac{C_S}{C_B} \end{aligned} \tag{2.40}$$

$$\Rightarrow \quad \frac{C_S}{C_B} = 1 - \frac{i}{i_l} \tag{2.41}$$

There are two conditions, which may arise in the cell, i.e., zero current, which is the equilibrium condition and when the current is flowing (bulk concentration reduces). The Nernst equations for both conditions can be written as

$$(\text{for} \quad i = 0) \quad E = E^\circ + \frac{RT}{nF} \ln C_B \tag{2.42}$$

$$E = E^\circ + \frac{RT}{nF} \ln C_S \tag{2.43}$$

The potential difference can be written as

$$\eta_{\text{conc.}} = \Delta E = \frac{RT}{nF} \ln \frac{C_B}{C_S} \tag{2.44}$$

$\eta_{\text{conc.}}$ is the concentration polarisation due to the concentration change at the electrode. Using Eq. (2.41) in Eq. (2.44) yields:

$$\eta_{\text{conc.}} = -\frac{RT}{nF}\ln\left(1-\frac{i}{i_1}\right) \quad (2.45)$$

The concentration polarisation is predominant for high temperature fuel cells. When the hydrogen is supplied via reformer, then the rate of supply of hydrogen at the anode decreases as compared to its consumption rate; then the concentration polarisation dominates in the fuel cell. The removal of water can also be a prominent cause for the mass transport or concentration losses.

2.3.4 Fuel Crossover Losses

The fuel cell consists of an ion-conducting electrolyte through which the conduction process takes place. Sometimes, the fuel from the anode side may diffuse to the cathode side. It may react directly with the oxygen at the cathode due to the presence of catalyst. But this reaction yields no current, hence leading to the wastage of fuel, which migrates from cathode to anode through the electrolyte. This loss of fuel is known as the fuel crossover losses. The fuel crossover losses and internal currents are taken to be equivalent. When the cell is in open circuit, the cell current density would not be zero due to the internal current density. For example, if the internal current density is 0.5 mA/cm^2, then the open circuit would yield 1.01 V, which is almost 19% less than the OCV. The gas usage is given by

$$\text{Charge} = 4F \times \text{amount of } H_2$$

By dividing it with time, we obtain
For single cell,

$$H_2 \text{ usage} = \frac{I}{4F} \text{ moles/s} \quad (2.46)$$

For stack of N cells,

$$H_2 \text{ usage} = \frac{IN}{4F} \text{ moles/s} \quad (2.47)$$

For a PEM cell with cell area 10 cm^2, 0.034 cm^2/s of H_2 is consumed. The gas usage becomes 1.4×10^{-7} mol/s corresponding to the volume of 2.43×10^4 cm^3 for one mole of any gas. Hence, from Eq. (2.47), we find

$$H_2 \text{ usage} = 1.4 \times 10^{-7} \times 2 \times 96{,}485 = 27\,\text{mA}$$

For a cell area of 10 cm^2, the current density becomes 2.7 mA/cm^2. The fuel crossover losses can be summarised as

$$\mathrm{H_{fuel}} = \frac{2.3\,RT}{\alpha\,nF}\ln\left(\frac{i_n}{i_0}\right) \tag{2.48}$$

where i_n is the internal current density. For high-temperature fuel cells, the significance of the internal current decreases because exchange current density i_0 is quite high. Anyhow for the fuel cells operating at low temperatures, the fuel crossover and internal currents play a significant role.

2.3.5 *Summation of Losses*

The total effective cell voltage after the losses is given by

$$\begin{aligned}
& V = E - (\text{Total losses}) \\
\Rightarrow\quad & V = E - (\eta_{\text{act}} + \eta_{\text{ohm}} + \eta_{\text{conc.}} + \eta_{\text{fuel}}) \\
\Rightarrow\quad & V = E - \left[\frac{2.3RT}{\alpha nF}\log\left(\frac{i}{i_0}\right) + ir + \frac{RT}{nF}\ln\left(1 - \frac{i}{i_1}\right) + \frac{2.3RT}{\alpha nF}\ln\left(\frac{i_n}{i_0}\right)\right] \\
\Rightarrow\quad & V = E - ir - \frac{2.3RT}{\alpha nF}\log\left(\frac{i}{i_0}\right) - \frac{2.3RT}{\alpha nF}\log\left(\frac{i_n}{i_0}\right) + \frac{RT}{nF}\log\left(1 - \frac{i}{i_1}\right)
\end{aligned} \tag{2.49}$$

It should be noted that the crossover current is very small for high-temperature fuel cells and then Eq. (2.49) is given by:

$$\Rightarrow\quad V = E - ir - \frac{2.3\,RT}{\alpha\,nF}\log\left(\frac{i}{i_0}\right) + \frac{2.3\,RT}{nF}\log\left(1 - \frac{i}{i_1}\right) \tag{2.50}$$

There is an alternative approach to obtain the effective voltage. The total polarisation can also be summed up for the fuel cell. Usually, the total polarisation is predominantly due to concentration losses and activation losses; i.e.,

$$\eta_{\text{anode}} = (\eta_{\text{conc.}} + \eta_{\text{act}})_{\text{anode}} \tag{2.51}$$

and

$$\eta_{\text{cathode}} = (\eta_{\text{conc.}} + \eta_{\text{act}})_{\text{cathode}} \tag{2.52}$$

The general equation can be written can be written as

$$\eta_{\text{electrode}} = (\eta_{\text{conc.}} + \eta_{\text{act}})_{\text{electrode}} \tag{2.53}$$

The polarisation will lead to the change in electrode potential as follows

$$V_{\text{electrode}} = E_{\text{electrode}} \pm |\eta_{\text{electrode}}| \tag{2.54}$$

For the anode, the potential increases whereas the cathode potential shows a decrease. The cell voltage consists of anode and cathode potentials along with ohmic polarisation, i.e.,

$$\begin{aligned} & V = V_{\text{cathode}} - V_{\text{anode}} - ir \\ \Rightarrow \quad & V = (E_{\text{cathode}} - E_{\text{anode}}) - \eta_{\text{cathode}} - \eta_{\text{anode}} - ir \\ \Rightarrow \quad & V = E - (\eta_{\text{conc.}} + \eta_{\text{act}})_{\text{cathode}} - (\eta_{\text{conc.}} + \eta_{\text{act}})_{\text{anode}} - ir \end{aligned} \tag{2.55}$$

For low-temperature fuel cells, the i_n factor is also added; i.e.,

$$V = E - (\eta_{\text{conc.}} + \eta_{\text{act}})_{\text{cathode}} - (\eta_{\text{conc.}} + \eta_{\text{act}})_{\text{anode}} - ir - i_n \tag{2.56}$$

The major loss of fuel cell voltage is due to the ohmic and electrode polarisations. Efforts are made to reduce the polarisations, so that V approaches E and hence maximum efficiency is obtained. This can be done by improving the electrode structure, using highly conducting electrolyte and better electrocatalysts. Hence, the actual and ideal potentials for a fuel cell are different due to various losses.

2.4 Performance of Fuel Cells

For low-temperature fuel cells, electrocatalysts are required to achieve desirable reaction rates at the anode and cathode, whereas the high-temperature fuel cells do not require any electrocatalyst. The most common type of electrode reactions and their corresponding Nernst potential are as follows:

Reaction	Nernst potential
$H_2 + \frac{1}{2}O_2 \longrightarrow H_2O$	$E = E^\circ + \frac{RT}{2F}\left[\ln\left(\frac{p_{H_2}}{p_{H_2O}}\right) + \ln\left[p_{O_2}^{1/2}\right]\right]$ (2.57)
$CO + \frac{1}{2}O_2 \longrightarrow CO_2$	$E = E^\circ + \frac{RT}{2F}\left[\ln\left(\frac{p_{CO}}{p_{CO_2}}\right) + \ln\left[p_{O_2}^{1/2}\right]\right]$ (2.58)
$H_2 + \frac{1}{2}O_2 + CO_{2(c)} \longrightarrow H_2O + CO_{2(a)}$	$E = E^\circ + \frac{RT}{2F}\left[\ln\left(\frac{p_{H_2}}{p_{H_2O}(p_{CO_2})_{(a)}}\right) + \ln\left[p_{O_2}^{1/2}(p_{CO_2})_{(c)}\right]\right]$ (2.59)
$CH_4 + 2O_2 \longrightarrow 2H_2O + CO_2$	$E = E^\circ + \frac{RT}{8F}\left[\ln\left(\frac{p_{CH_4}}{p_{H_2O}^2 p_{CO_2}}\right) + \ln\left[p_{O_2}^2\right]\right]$ (2.60)

There is a simple way to obtain the Nernst equations. For example, consider a reaction

$$aA + bB \rightarrow cC \tag{2.61}$$

which indicates that a moles of A react with b moles of B to form c moles of C. The two reactants A, B and product C are designated different activities, i.e., a_A, a_B and a_C. For an ideal gas, the activity is defined as

$$a = \frac{p}{p^\circ} \tag{2.62}$$

where p° is the standard pressure at 0.1 MPa and p is the partial pressure of gas. For example, the activity of oxygen is given by

$$a = \frac{p_{O_2}}{p^\circ_{O_2}} \tag{2.63}$$

The Gibbs free energy gets modified and can be written as

$$\Delta G_f = G^\circ_f - RT \ln\left[\frac{a^a_A a^b_B}{a^c_C}\right] \tag{2.64}$$

or

$$-\frac{\Delta G_f}{2F} = -\frac{\Delta G^\circ_f}{2F} + \frac{RT}{2F}\ln\left[\frac{a^a_A a^b_B}{a^c_C}\right] \tag{2.65}$$

$$\Rightarrow \quad E = E^\circ + \frac{RT}{2F}\ln\left[\frac{a^a_A a^b_B}{a^c_C}\right] \tag{2.66}$$

Equation (2.66) infers the increase in the potential with increase in the activity of reactants. Equation (2.66) is also referred to as a Nernst equation and can also be written in terms of E.M.F. Let us demonstrate it with the help of the following reaction inside fuel cell (operating at high temperature).

$$H_2 + \frac{1}{2}O_2 \rightarrow H_2O$$

The activities for reactants and products can be given as

$$a_{H_2} = \frac{p_{H_2}}{p^\circ_{H_2}}, \quad a_{O_2} = \frac{p_{O_2}}{p^\circ_{O_2}}, \quad a_{H_2O} = \frac{p_{H_2O}}{p^\circ_{H_2O}}$$

At high temperatures, the steam behaves as an ideal gas and $p^\circ = p^\circ_{O_2} = p^\circ_{H_2O} = 1$ bar.

The E.M.F. is simplified as

$$E = E^{\circ} + \frac{RT}{2F} \ln \left(\frac{p_{H_2} p_{O_2}^{1/2}}{p_{H_2O}} \right) \quad (2.67)$$

Equation (2.67) can be written as

$$E = E^{\circ} + \frac{RT}{2F} \ln \left(\frac{p_{O_2}^{1/2}}{p_{H_2O}} \right) + \frac{RT}{2F} \ln (p_{H_2}) \quad (2.68)$$

If p_{O_2} and p_{H_2O} are unchanged and the partial pressure of hydrogen increases from p_1 to p_2, then the change in voltage will be given by

$$\begin{aligned} \Delta V &= \frac{RT}{2F} \ln p_2 - \frac{RT}{2F} \ln p_1 \\ \Rightarrow \quad \Delta V &= \frac{RT}{2F} \ln \left(\frac{p_2}{p_1} \right) \end{aligned} \quad (2.69)$$

The Nernst equation can also be expressed in another form; let us suppose that the pressure on the anode and cathode are essentially the same and the system pressure can be taken p. Then the p_{H_2}, p_{O_2} and p_{H_2O} are expressed as $n_1 p$, $n_2 p$ and $n_3 p$ where n_1, n_2 and n_3 are constants that depend on concentration and molar mass of hydrogen, oxygen and water, respectively. Equation (2.67) then becomes

$$\begin{aligned} E &= E^{\circ} + \frac{RT}{2F} \left(\frac{n_1 n_2^{1/2} p^{1/2}}{n_3} \right) \\ E &= E^{\circ} + \underbrace{\frac{RT}{2F} \ln \left(\frac{n_1 n_2^{1/2}}{n_3} \right)}_{\text{Ist term}} + \underbrace{\frac{RT}{4F} \ln (p)}_{\text{IInd term}} \end{aligned} \quad (2.70)$$

The first term contributes to the fall in E.M.F. due to the concentration change of reactants and products. During the operation of a fuel cell, the concentration of products will increase at the cost of reactant concentration, which will make the first term smaller. The current density and fuel concentration are lower at the outlet; hence, the E.M.F. is small near the exit. The term ‘RT’ indicates high Nernst voltage at high temperatures. The second term indicates the increase in system pressure which would eventually lead to an increase of cell E.M.F. If pressure gets increased from p_1 to p_2, then the second term increases. The increase in pressure is more significant for the low-temperature fuel cells because high pressure helps in reducing the losses especially for the cathode.

The fuel cell performance is affected by a number of variables including current density, gas composition, fuel utilisation, pressure and temperature. When the operating conditions for a fuel cell are changed; then the system cost may increase.

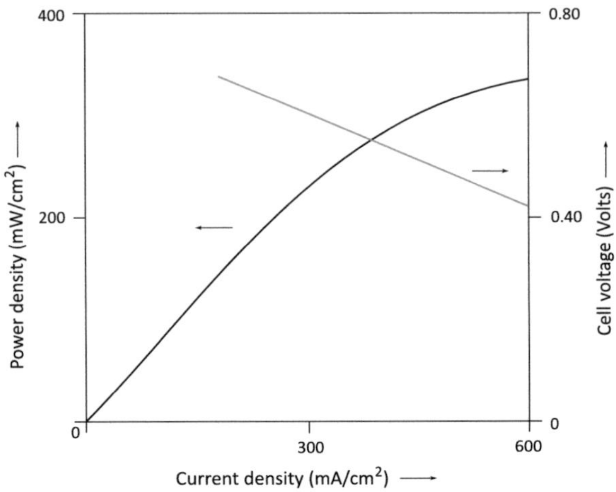

Fig. 2.3 The relation between current density, power density and cell voltage of fuel cell

Ideally, the fuel cell system must possess low cost, high power density, portability and less system weight. When the cell stack size is reduced, the cell cost may also get reduced and the efficiency of cell may decrease. For stationary power applications, higher voltages and hence higher efficiency are required. In contrast to this, high current density or lower cell voltages are apt for mobile applications. Figure 2.3 gives the voltage–power relationship for a fuel cell.

It is clear from Fig. 2.3, that the higher current density and power density usually occur for low cell voltage. Hence, for getting maximum efficiency, the operating conditions should be balanced to get a good compromise between current density, voltage and operating cost. The effect of temperature and pressure on the ideal potential of fuel cell can be described on the basis of Gibbs free energy with pressure and temperature. As we know

$$\begin{aligned} \Delta G_f &= \Delta H - T\Delta S \\ \Rightarrow \quad -nFE &= \Delta H - T\Delta S \\ &\boxed{\left(\frac{\partial E}{\partial p}\right)_T = \frac{\Delta S}{nF}} \end{aligned} \tag{2.71}$$

and

$$\left(\frac{\partial E}{\partial T}\right)_P = \frac{\Delta V}{nF} \tag{2.72}$$

For the reaction $H_2 + \frac{1}{2}O_2 \rightarrow H_2O$, the entropy change is negative due to which the reversible potential decreases with increase in the temperature. In contrast to this, the volume change is negative for this reaction which leads to an increase in reversible cell voltage with an increase in pressure. The pressure increase also increases the gas solubility; mass transfer reacts and reactant partial pressures. Hence, the fuel cell efficiency increases and the electrolyte evaporation rate decreases at elevated pressures in the fuel cell system. Another major factor that determines fuel cell efficiencies is gas composition and utilisation of reactants. The fuel utilisation is given by

$$U_{\text{fuel}} = \frac{\text{Fuel}_{\text{in}} - \text{Fuel}_{\text{out}}}{\text{Fuel}_{\text{in}}} \tag{2.73}$$

It is apparent, if the fuel and oxygen gas have a higher concentration, then the fuel cell will possess higher efficiencies and hence high cell voltages. The difference between fuel provided at the inlet and fuel flow rate at the outlet of the fuel cell represents the fuel consumed. Certain other factors such as fuel leakage and chemical reaction may lead to more consumption of fuel, which increases the apparent utilisation of the fuel. From inlet to outlet, the gas composition keeps on changing, hence leading to reduced cell voltages. The voltage reduction is attributed to the lowest electrode potential (from the Nernst equation) because the electrodes are good ionic conductors and the cell voltage values cannot be greater than the minimum Nernst potential.

The current density and impurities also affect the fuel cell performance. In the initial stages, activation polarisation decreases the cell voltage. When current densities increase, then the concentration losses are predominant and a sharp decrease in cell performance is observed. During normal fuel cell application, ohmic losses are observed due to internal resistances of the fuel cell. This section deals with the effect of different variables on fuel cell performance and efficiency. The different fuel cells and their performances are discussed one by one.

2.4.1 Performance of PAFC

For PAFC, the polarisation contribution of the two half cells is shown in Fig. 2.4. Hydrogen is taken to be the reference point/zero polarisation.

Figure 2.4 indicates the polarisation losses for PAFC with anode and cathode loading of 0.5 mg Pt/cm^2, pressure 1 atm, 100% phosphoric acid and operating temperature of 180°C. The anode losses are very small on pure H_2, whereas the cathode polarisation is greater with air and comparatively small for pure oxygen. As discussed, it is known that an increase in cell pressure enhances the performance of the cell (Eq. 2.72).

$$\Delta V = \frac{RT}{2F} \ln\left(\frac{p_2}{p_1}\right) \tag{2.74}$$

The above equation holds well for the pressure range of 1 atm $\leq p \leq$ 10 atm and temperature range of 180°C $< T <$ 220°C. Figure 2.5 demonstrates that the increase in pressure from 3 to 8 atm increases the cell voltage for H_2 fuel fuel/Air PAFC.

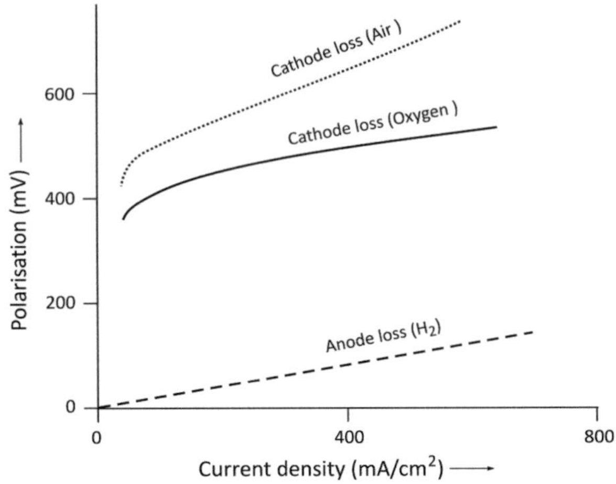

Fig. 2.4 Polarisation of anode and cathode for PAFC

At higher pressure and current density, the diffusion polarisation at the cathode decreases and reversible cell potential increases. At the cathode, increased oxygen and water pressures decrease activation polarisation. As shown in Fig. 2.5, an increase of 44 mV is observed when pressure and temperature are increased by 2.9 atm and 15°C, respectively. An increase in temperature exhibits a beneficial effect on cell performance because the polarisation losses diminish with increased temperature. The relationship between voltage gain and temperature change is given by:

$$\Delta V(\mathrm{mV}) = A\Delta T(^{\circ}\mathrm{C}) \tag{2.75}$$

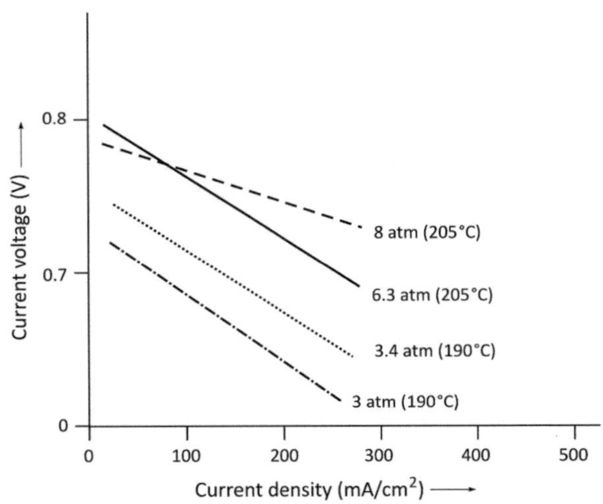

Fig. 2.5 The increase in PAFC cell performance with pressure increase

Fig. 2.6 Effect of temperature and impurities on the cell voltage of PAFC

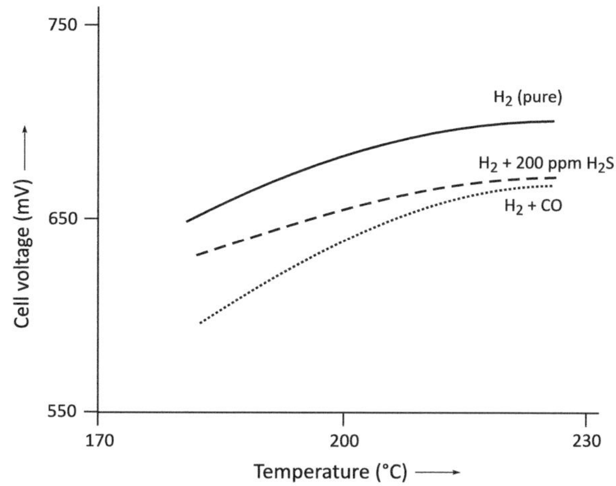

The validity of Eq. (2.75) holds good within a temperature range of 180°C < T < 250°C. The value of A has been given between 0.55 and 0.75. For 250 mA/cm^2 current density, the value of A is given to be 1.15 which means that the voltage increases by 1.15 mV per degree rise in temperature. Increasing the temperature also increases the relative tolerance of the anode for CO poisoning due to reduced adsorption. The impurities such as carbon monoxide and hydrogen sulphide originate while the fuel enters the fuel processor and reformer chamber. The CO in the fuel affects the anode performance because CO poisons the electrocatalytic activity of Pt. electrodes. As discussed in Chap. 1, two carbon monoxide molecules replace 1 H_2 molecule, hence affecting the reaction kinetics of the cell. The effect of temperature and impurity (H_2S and CO) on cell voltage in shown in Fig. 2.6. As evident from Fig. 2.6, cell performance increases with temperature, but problems such as evaporation, electrolyte degradation, corrosion and catalyst sintering also elevate it. The CO poisoning decreases cell performance considerably as compared to the fuel contaminated with sulphide compounds. The fuel processors and coal gasifiers in PAFC power plants usually incorporate hydrogen sulphide and carbonyl sulphide as impurities in the hydrogen fuel. The platinum anodes can tolerate <20 ppm of H_2S and <50 ppm of (H_2S + carbonyl sulphide) concentration within a temperature range of 190–210°C and 9.2 atm. A CO level of up to 1% is acceptable at working temperature above 190°C. If temperature is increased, then the CO absorbed on the anode gets desorbed. Similarly, the sulphur-poisoned anodes can also be reactivated by increasing the temperature or by increasing the cathode potentials.

The cathode performance is also affected by the utilisation and composition oxidant. The use of 21% O_2 instead of pure O_2 decreases the current density. As the oxygen is utilised, the cathode polarisation also increases. The voltage loss due to change in oxidant utilisation at cathode is given by

$$(\Delta V)_{\text{cathode}} = A_c \ln\left(\frac{p_2}{p_1}\right)_{O_2} \quad \begin{bmatrix} A_c = 148 \text{ for } 0.04 \le \frac{p_2}{p_1} \le 0.2 \\ A_c = 96 \text{ for } 0.2 < \frac{p_2}{p_1} < 1 \end{bmatrix} \tag{2.76}$$

Similarly, the voltage loss due to change in hydrogen partial pressure at the anode is given by

$$(\Delta V)_{\text{anode}} = A_a \ln\left(\frac{p_2}{p_1}\right)_{H_2} \quad [A_a = 55] \tag{2.77}$$

The anode reaction is reversible; hence hydrogen utilisation does not have much significance on cell performance. Hydrogen is derived from coal gases/liquids petroleum products and hydrocarbons such as methane. The reformed fuel may contain CO and CO_2 as traces. CO_2 and unreacted hydrocarbons act as diluents because they are electrochemically inert. Low utilisations cause poor fuel use but yield high performance. The current density also influences the cell performance and the voltage losses can be given by:

$$(\Delta V)_{\text{CD}} = A_{\text{CD}}\, \Delta J \begin{bmatrix} A_{\text{CD}} = -0.53 \text{ for } 100 < J < 200\,\text{mA/cm}^2 \\ A_{\text{CD}} = -0.39 \text{ for } 200 < J < 650\,\text{mA/cm}^2 \end{bmatrix} \tag{2.78}$$

The value of A_{CD} are obtained at 207°C, 8.2 atm and oxidant/fuel utilisation of 70 /85%, respectively. At atmospheric conditions of 1 atm, operating temperature 204°C, and oxidant/fuel utilisation of 60 /80%, respectively, Eq. (2.78) becomes

$$(\Delta V)_{\text{atm}} = A_{\text{atm}}\, \Delta J \quad \begin{bmatrix} A_{\text{atm}} = -0.74 \text{ for } 50 < J < 120\,\text{mA/cm}^2 \\ A_{\text{atm}} = -0.45 \text{ for } 120 < J < 215\,\text{mA/cm}^2 \end{bmatrix} \tag{2.79}$$

The PAFC technology is considered to be the only commercial available fuel cell. The PAFC plants rely heavily on the quality of power produced from the stacks. The largest power of 11 MW has been achieved using the PAFC plant built by international fuel cells and Toshiba for Tokyo Electric Power. The major industrial developers of PAFC are Fuji Electric, Toshiba and Mitsubishi Electric Company (Japan) and UTC fuel cells (ONSI Corporation, United States). Mitsubishi Electric Company has demonstrated cell performance of 0.65 mV at 300 mA/cm^2. In 1991, Tokyo Electric Company demonstrated cell performance of 0.71 V at 431 mA/cm^2. Bharat Heavy Electrical Ltd. of India and Caltex Oil Corporation of South Korea are new entrants for the PAFC technology.

2.4.2 Performance of AFC

Till now, we have been familiar with the concept that the activation overvoltage on the cathode is reduced upon increase in pressure and temperature. Upon increasing the pressure, the change in OCV is given by the second term of Eq. (2.74), i.e.,

$$\Delta V = \frac{RT}{4F} \ln\left(\frac{p_2}{p_1}\right) \quad (2.80)$$

Temperature pressure and electrolyte concentration should be balanced to achieve maximum efficiency. For example, to obtain high performance for the Bacon cell, the pressure has to be reduced and temperature should be increased. Moreover, the electrolyte concentration of KOH should be increased by 75%, so that the electrolyte does not boil. Figure 2.7 depicts the variation of KOH electrolyte concentration with temperature and pressure.

When the concentration of KOH is increased, the vapour pressure is reduced. For AFC, the reactant gas pressure and electrolyte pressure differ from each other. For example, both the reactant gases in the Apollo system were about 0.7 bar above the electrolyte pressure. For the Orbitor fuel cell, the oxygen pressure was 0.35 below the hydrogen pressure. The crucial problem in AFC is the carbonation of electrodes when air is used instead of pure oxygen. There are many groups which are actively engaged in the production and development of AFC. EFFCELL GmbH, a Swiss company, has developed AFC for mobile applications. Hydrocell (Finland) developed portable AFC with an integrated CO_2 scrubber system. Scottish Fuel Cell Consortium (SFEC) of Scotland has been focused on the development of fuel cell hybrid battery electric drive trains. Gaskatel GmbH (Germany) is focusing on the electrochemical processes within AFC, particularly gas diffusion electrode kinetics. Astris Energi Inc. (Canada) has developed 1–10 kW systems for AFC. Cenergie Corporation (Atlanta, USA) has

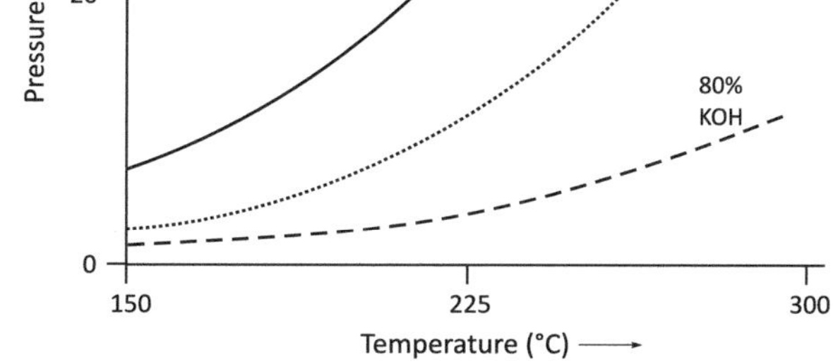

Fig. 2.7 Variation of KOH concentration with different pressure and temperature

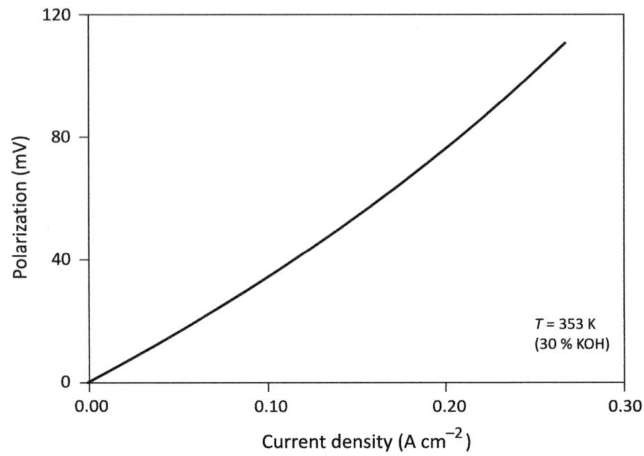

Fig. 2.8 Polarisation versus current density characteristics for the AFC

been focused on developing novel designs for SOFC. Apollo Energy Systems (Pampona Beach, USA) is developing the AFC for mobile and portable applications. Independent Power Technologies Ltd. (Moscow) are developing AFC for stationary applications. Fuel Cell Control Ltd. (UK) develops system integrators for AFC engines, which find applications in remote base transmitters and cellular communications. DLR German Aerospace Center (Stuttgart) focuses on the degradation mechanics and production of gas diffusion electrodes. The typical polarisation characteristics for AFC (Ni anode) are shown in Fig. 2.8.

2.4.3 Performance of PEMFC

The research is focused on lowering the cost of PEMFC and improving its performance. Usually, increasing the power for a system makes it more efficient. Additionally, it will also increase the cost and weight of the fuel cell system. The pressure increase raises the exchange current density and hence the OCV also increases lowering the cathode overpotential. The rate in OCV is given by

$$\Delta V = \frac{RT}{4F}\ln\left(\frac{p_2}{p_1}\right)$$

Let current *I* Amperes in flowing through *n* stacks of PEMFC; then the power gain will be given by

$$P_{\text{gain}} = \frac{RT}{4F}\ln\left(\frac{p_2}{p_1}\right)nI \tag{2.81}$$

We derive the voltage loss and hence power loss in a more detailed way as follows:

Whenever a gas expands, work is done by gas, or in other words, the work is done on the gas for compressing it. In an adiabatic reversible process (no heat loss), the pressure change from p_1 to p_2 will lead to temperature change of T_1 to T_1'; i.e.,

$$\frac{T_1'}{T_1} = \left(\frac{p_2}{p_1}\right)^{\frac{\gamma-1}{\gamma}} \tag{2.82}$$

where $\gamma = \frac{C_P}{C_V}$ or ratio of specific heat at constant pressure to specific heat at constant volume. T_1' is the isentropic temperature, anyhow practically the temperature change is higher than T_1' and is given by T_2. The work done can be given by the change in enthalpy of gas as

$$\text{Actual work} = W = C_P(T_2 - T_1)m \tag{2.83}$$

where m is the mass of gas compressed. For Eq. (2.83), the heat flow and kinetic energy of gas has been considered to be negligible. The gas is taken to be perfect gas and hence C_P is taken to be constant. For the isoentropic system, the work done can be written as

$$\text{Isoentropic work} = W' = C_P\left(T_1' - T_1\right)m \tag{2.84}$$

To obtain the efficiency, we have to divide Eq. (2.83) and (2.84) as follows:

$$\begin{aligned} \eta &= \frac{\text{Isentropic work}}{\text{Actual work}} \\ \Rightarrow \quad \eta &= \frac{C_P(T_1' - T_1)m}{C_P(T_2 - T_1)m} = \frac{T_1' - T_1}{T_2 - T_1} \end{aligned} \tag{2.85}$$

Substituting Eq. (2.82) in (2.85) we get,

$$\begin{aligned} \eta &= \frac{\left(T_1\left(\frac{p_2}{p_1}\right)^{\frac{\gamma-1}{\gamma}} - T_1\right)}{T_2 - T_1} \\ \Rightarrow \quad \eta &= \frac{T_1}{T_2 - T_1}\left(\left(\frac{p_2}{p_1}\right)^{\frac{\gamma-1}{\gamma}} - 1\right) \end{aligned} \tag{2.86}$$

$$\Rightarrow \quad \Delta T = \frac{T_1}{\eta}\left(\left(\frac{p_2}{p_1}\right)^{\frac{\gamma-1}{\gamma}} - 1\right) \tag{2.87}$$

Equation (2.87) provides the temperature change, which will be used to obtain the power lost (P_{lost}) for unit time; i.e.,

$$\text{Power lost} = W = C_P \Delta T \dot{m} \tag{2.88}$$

$\dot{m}$ indicates the rate of flow of gas in kg/s; we get

$$\text{Power lost} = C_P \frac{T_1}{\eta} \left(\left(\frac{p_2}{p_1} \right)^{\frac{\gamma-1}{\gamma}} - 1 \right) \dot{m} \tag{2.89}$$

Hence, we can make the net assessment of the system on the basis of Eqs. (2.81) and (2.89). Two models are usually considered to obtain the voltage gains: the optimistic and realistic models. The optimistic model always yields gains at high pressure, whereas the realistic model gives the net loss when pressure is increased. Figure 2.9 shows the boundary conditions for both models.

An increase in oxygen pressure also affects the performance of PEMFC as shown in Fig. 2.10. In increase in oxygen pressure from 3 to 10.2 atmospheres, and current density 215 mA/cm^2, produces an increase of 42 mV in the cell voltage (at 93°C). The current density and operating temperature value play a crucial role in determining the cell gain. For instance, the PEFMC at 80°C (431 mA/cm^2) yielded a voltage gains of 22 mV for pressure increase of 2.4 to 3.4 atm, whereas for a PEMFC at 50°C (500 mA/cm^2) demonstrates an increase of 83 mV in voltage upon pressure change from 1 to 5 atmospheres.

An increase in operating temperature lowers the internal resistance (mainly ohmic resistance) of the cell. The experimental data suggest an increase of voltage

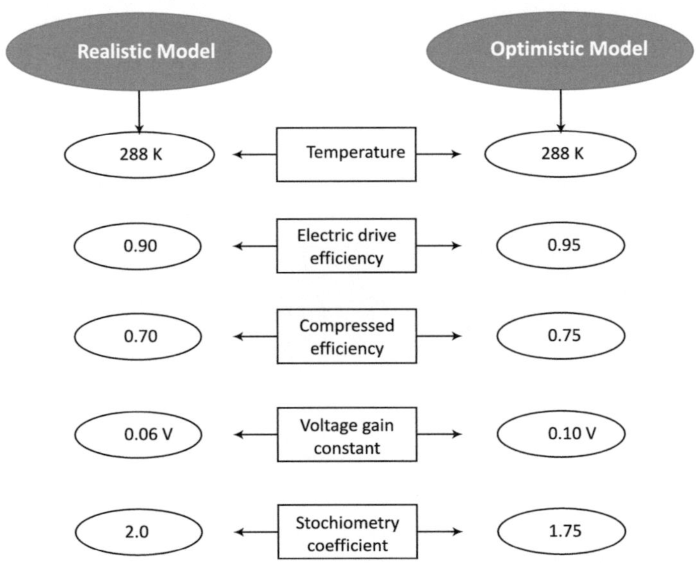

Fig. 2.9 The values of different constants used in realistic and optimistic models

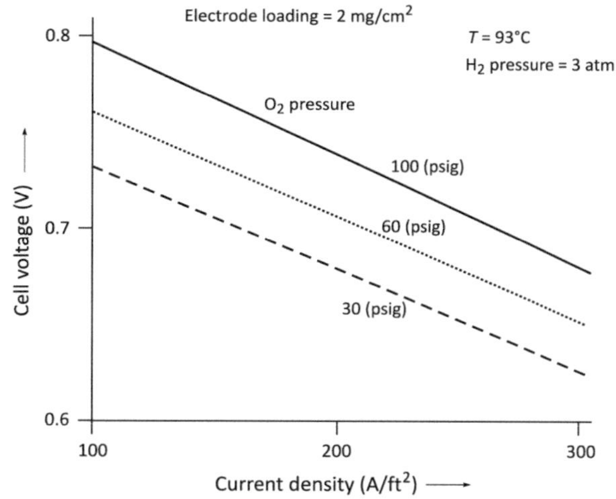

Fig. 2.10 The effect of O_2 pressure on the PEMFC performance for operating temperature 93°C (H_2 at 3 atm)

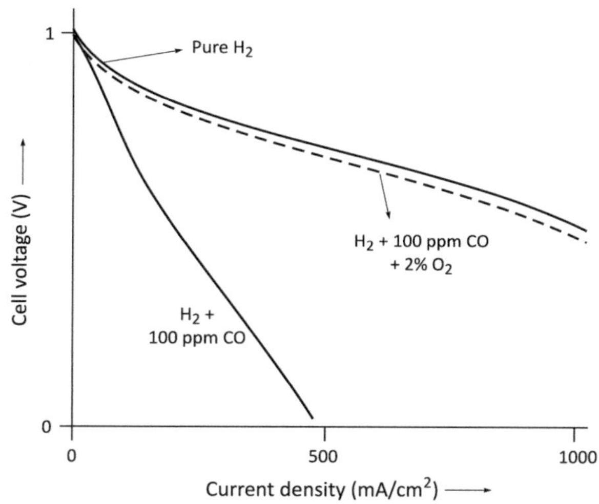

Fig. 2.11 PEMFC performance of bleeding oxygen into anode compartment

within 1.1 mV to 2.5 mV range per °C rise in operating temperature. At higher temperatures, the CO chemisorptions also decrease, hence increase the voltage gain. The stack operation has considerably affected the fuel cell performance: the CO content in methanol reformed gas can be reduced from 1% to 10 ppm using a selective oxidation process. Los Alamos National Laboratory (LANL) has addressed this problem by bleeding a small amount of oxygen into the anode chamber. Figure 2.11 shows the PEMFC performance of bleeding oxygen in the anode compartment.

The air and hydrogen partial pressures are 5 atm and 3 atm, respectively (Nafion—117 μm thick). Using the bleeding oxygen concept, the performance of PEMFC with fuel H_2 becomes equivalent to the performance of PEMFC with fuel H_2 + 100 ppm CO + 2% O_2. But the main drawback of this approach is the loss of fuel, which could be overcome using platinum|ruthenium anodes. LANL also obtained voltage gain of 0.78 V at 200 mA/cm^2 current density (operating temperature 80°C), Nafion membrane and platinum loading of 0.4 mg/cm^2. Currently, the major focus is to develop a fuel cell with low cost and high system efficiency for terrestrial applications.

2.4.4 *Performance of MFC*

For MFC, the performance can be analysed using half-cell reactions. If acetate is oxidised by bacteria at the anode, the half cell reaction is given by

$$2HCO_3^- + 9H^+ + 8e^- \rightarrow CH_3COO^- + 4H_2O \quad (2.90)$$

The Nernst equation for oxidation of acetate is given by

$$E_a = E_a^{\circ} - \frac{RT}{8F}\left(\frac{[CH_3COO^-]}{[HCO_3^-]^2[H^+]^9}\right) \quad (2.91)$$

The reaction and Nernst potential at cathode is given by

$$O_2 + 4H^+ + 4e^- \rightarrow 2H_2O \quad (2.92)$$

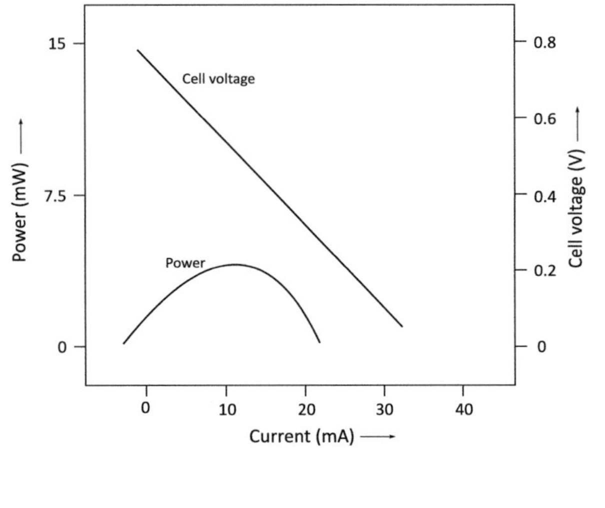

Fig. 2.12 The polarisation and power curve of MFC operating on starch

$$E_c = E_c^{\circ} - \frac{RT}{4F}\left(\frac{1}{p_{O_2}[H^+]^4}\right) \tag{2.93}$$

For MFC, ferricyanides and manganese oxides have been used as alternatives for oxygen. The main application of MFC is to treat the wastewater. The ohmic resistance plays dominant role in the polarisation curves. The polarisation and power curves for MFC are shown in Fig. 2.12 for starch substrate. The polarisation curves for MFC are usually linear in nature and hence the internal resistance value can be easily obtained from their slope. The performance of MFC depends upon the electrochemical reactions occurring between the electron acceptor with high potential and organic substrate with low potential. In MFC, it becomes uncertain to have an idea of the cell voltage because the electron transfer takes place via the organic substrate and complex chain which varies from microbe to microbe.

For MFC, pH buffer, electrolyte, electrode material, anodic chamber and cathode chamber conditions play a crucial role in determining the cell voltage. Oxygen is the most commonly used electron acceptor in MFC. When ferricyanide is used as an electron acceptor, then the power output increase (i.e., power output of 4310 and 3600 mW/m^2) can be obtained due to the higher mass transfer rate and low activation energy required for cathodic reaction while using ferricyanide. The current and power densities are also boosted when hydrogen peroxide is used as an electron accepter. In the anodic chamber, high fuel concentrations yield a higher output in a wide range of concentrations. When *S. putrefaciens* was used in MFC, high power density could be obtained for lactate fuel concentration up to 200 mM.

The pH difference occurs between the anodic and cathodic chambers if no buffer solution is used in the MFC. This pH difference causes an increase in the proton diffusion from the anode to the cathode chamber until equilibrium is achieved. Using an HCl solution to the cathode, the current density increases and if NaCl is added to the MFC, then the power density increases due to enhanced conductivity of the catholyte and analyte. Several research groups investigated electrode materials with less activation polarisation. As compared to the graphite electrode, Pt./Pt. black yields less activation polarisation due to its higher catalytic activity. A 100-fold increase in current could be obtained using a woven graphite/Mn (IV) graphite anode as compared to the graphite anode only because Mn(IV) served as the mediator for MFC. A fourfold increase in the current could be observed when a Fe(III+) graphite cathode and Mn(IV) graphite anode combination was used instead of plain graphite electrodes. The Pt./Pt. black electrode activity gets reduced by the PtO layer on their surface. Anyhow, the current density is observed to increase from 0.84 to 1.445 mA/cm^2 when polyaniline is coated on the Pt. black anode. Cobalt tetramethoxyphenylporphyrin (CoTMPP) and Iron (III) phthalocyanine (FePc)-based cathodes are also used as an alternative MFC cathodes because they deliver almost similar performance and are relatively cheap.

Nafion membranes (Du Pont, Delaware) are the most commonly used membrane in the MFC system for the selective permeability of protons, in spite of the cost issues. The surface area of the polymer exchange membrane influences the power output because the increase in PEM surface area decreases the internal resistance. The power output of 2.2 mW/m^2 was observed when the salt bridge was used in MFC inoculated with *G. metallireducens*. When graphite electrodes with a surface area of 17 cm^2 for both anode and cathode were used, then the voltage of 67 mV could be achieved for MFC inoculated with *Escherichia coli*. Though many efforts are being made to improve the cell performance, the slow metabolism of the microbes limits the performance.

2.4.5 Performance of DCFC

Scientific Applications and Research Associate (SARA) Inc. started an R&D program on DCFC (molten hydroxide) in the 1990s. SARA built prototypes for demonstrating the molten hydroxide DCFC technology and the latest prototypes include MARK II-D and MARK III-A. Figure 2.13 demonstrates the power and

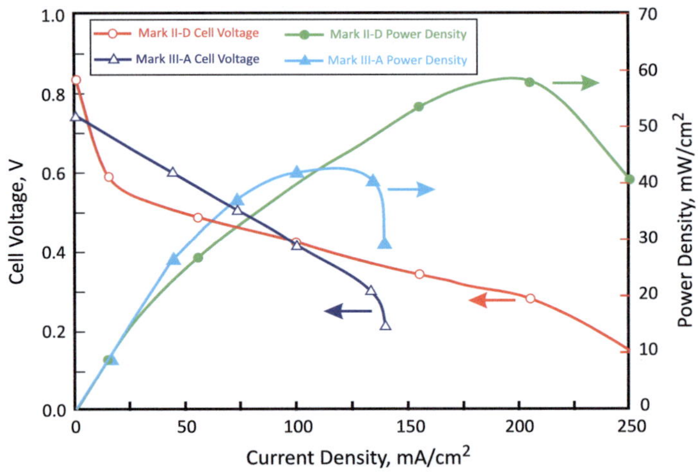

Fig. 2.13 The characteristics of MARK II-D and MARK III-A at an operating temperature of 630°C (Photo courtesy by SARA)

Table 2.1 Materials and design parameters of prototype cells

Prototype cell	Anode	Cathode	Cathode container	Electrode spacing (cm)	Anode surface area (cm^2)
MARK III-A	7.6 cm dia, 12.6 cm long graphite	Fe2Ti	Prismatic	2	300
MARK II-A	2 cm dia, 4.1 cm long graphite	Ni foam-lined C-1018 steel	Cylindrical	1.3	26

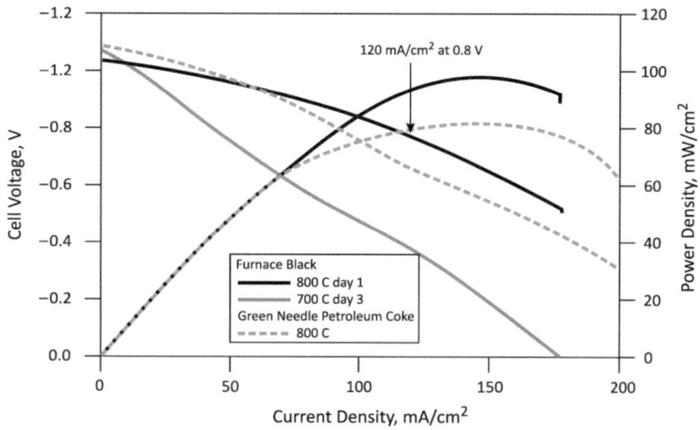

Fig. 2.14 Cell potential and power density versus current density for the DCFC (Photocourtesy by LLNL, California)

voltage density characteristics for MARK II-D and MARK III-A at the operating temperature of 630°C.

The maximum efficiency for the MARK III-A prototype was 60% at a current density of 50 mA/cm^2, whereas the peak power densities achieved by MARK III-A and MARK II-D are 42 and 58 mW/cm^2, respectively. The major losses in DCFC occurred due to ohmic losses, which are attributed to the large spacing between anode and cathode. High performance could be achieved by increasing the cathode surface area, coal derived anode, optimisation of air flow and improved current collector–anode contact. Table 2.1 demonstrate the design parameter for MARK III-A and MARK II-D.

A pure graphite rod acts as the anode and fuel which is immersed into a molten sodium hydroxide solution. The electrolyte is contained within the prismatic or cylindrical container, which also acts as the cathode. The cell operates between 400° and 650° and humidified air is fed into the cell via a gas distributor.

Fig. 2.15 Peak power density characteristics obtained for the DCFC. (Photocourtesy by SRI)

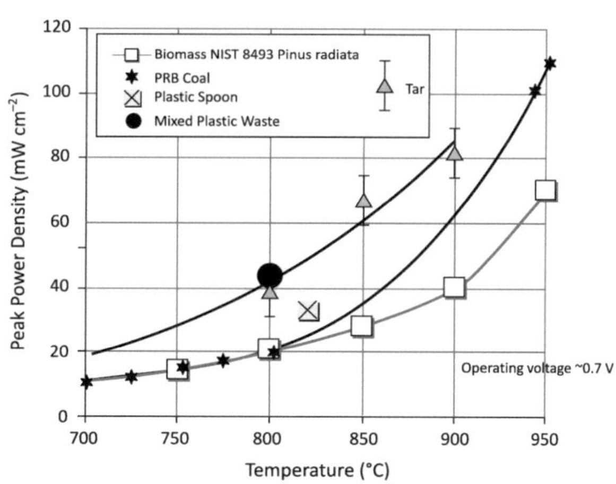

Figure 2.14 demonstrates the potential and power density characteristic obtained for molten carbonate electrolyte DCFC obtained by Lawrence Livemore National Laboratory (LLNL, California). LLNL used 32% Li_2CO_3–68% K_2CO_3 melt as an electrolyte and the Nernst potential is given by:

$$E = E^{\circ} - \frac{RT}{4F}\ln[CO_2]_a^3 + \frac{RT}{4F}\ln\left([O_2][CO_2]_c^2\right) \tag{2.94}$$

The anode and cathode cell reactions are

$$\begin{bmatrix} \text{Anode: } C + 2CO_3^{2-} \rightarrow 3CO_2 + 4e^- \\ \text{Cathode: } O_2 + 2CO_2 + 4e^- \rightarrow 2CO_3^{2-} \end{bmatrix}$$

LLNL obtained current densities from 58 to 124 mA/cm^2 for a cell voltage of 0.8 V. The performance of solid electrolyte based DCFC obtained by SRI is shown in Fig. 2.15. At 950°C, power densities of 100 mW/cm^2 could be obtained.

CellTech Power LLC obtained power obtained of 10 mW/cm^2 at 0.248 V and 50 mW/cm^2 at 0.507 V for temperatures of 800 and 1000°C, respectively. $(ZrO_2)(HfO_2)_{0.02}(Y_2O_3)_{0.08}$ was used as an electrolyte, carbon black as an anode and 0.6 mm thick $La_{0.84}Sr_{0.16}MnO_3$ as cathode for the CellTech fuel cell. The DCFC cell performance depends upon the carbon source surface area and oxygen surface groups, which increases the active sites on the carbon source.

2.4.6 *Performance of MCFC*

Significant polarisation losses occur in MCFC and its operating conditions are almost same as those of PAFC. For an MCFC, the anode and cathode compartments are at the same pressure, and the change in reversible voltage can be given by

$$\Delta V = \frac{RT}{4F}\ln\frac{p_2}{p_1}$$

If T = 650°C, then the above equation becomes

$$\Delta V = 20\ln\frac{p_2}{p_1} \tag{2.95}$$

Equation (2.95) infers that for a tenfold increase in pressure, an increase of 46 mV in reversible cell potential is observed. As the pressure is increased, the mass transport rates and gas solubilities also increase. Anyhow, the increased pressure may promote some reactions: decomposition of methane to carbon/hydrogen, methanation (methane formation) and carbon deposition and methane formation are favoured. The carbon deposition may lead to the plugging of gas passages in the anode. The steam reformation also gets inhibited due to higher pressure, which is a

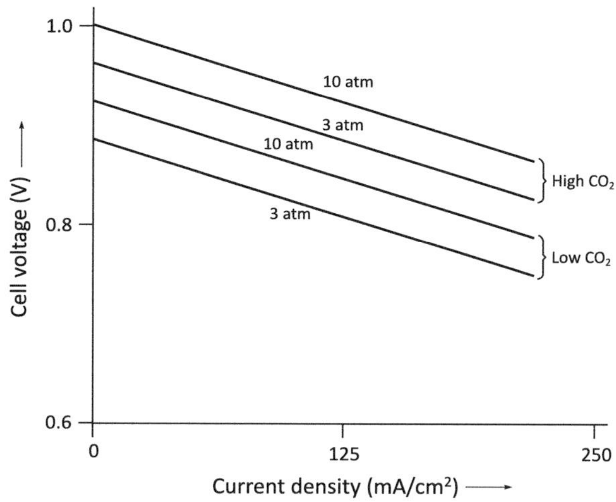

Fig. 2.16 MCFC performance with pressure and CO_2

demerit for the system using an internal reformer. If methane forms during the operation of MCFC, then 3 mol of H_2 are consumed, which is detrimental for the fuel cell because fuel loss and hence power wastage occurs. The pressure difference between the anode and cathode compartments must be kept as minimum as possible. The MCFC electrolyte is contained in the ceramic matrix, which is susceptible to cracking under stresses. For overcoming these things, the partial pressure of H_2O in gas stream is increased to avoid carbon deposition. In addition to this, water and CO_2 are added to the fuel to modify the equilibrium such that methane formation is minimised. The performance of MCFC is affected by the gas composition, i.e., the increase in CO_2 and oxygen content (Fig. 2.16). As the CO_2 content increases, the cell performance also increases. The stack size and design also affect the performance of MCFC. Figure 2.17 shows that the stack networking helps in increasing the overall performance as well as simplifying the system design.

The stack design concept was given by the ECN project and this stack is known as SMARTER™ (used both for MCFC and SOFC stacking). This geometry contains two different geometries of cells stacked together: single pass operation is applied to one cell and anode recycle to other. These systems require no preheating and glass flow rates are larger increasing the heat capacity of the system. In addition to this, the SMARTER™ stack system exhibits less pressure on the internal reformation catalyst because the fuel entering the stack has already been partially reformed.

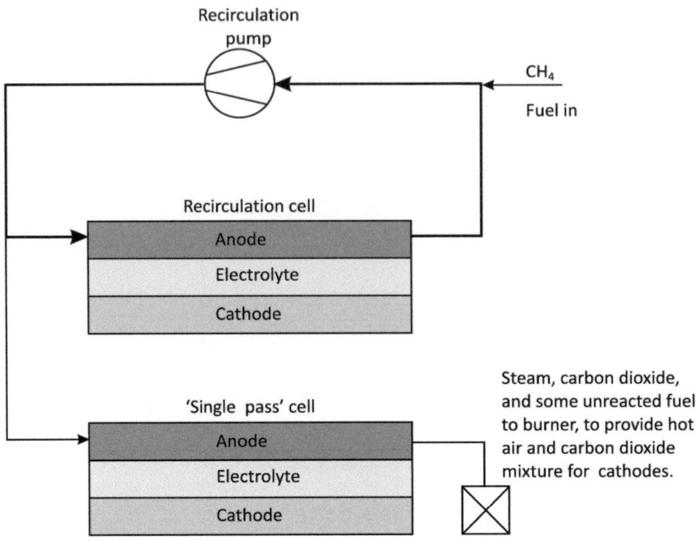

Fig. 2.17 The SMARTER stack concept for enhancing the power output (Larminie and Dicks 2003)

The increase in temperature increases the equilibrium constant (K) as the equilibrium composition increases for the product:

$$K = \frac{p_{CO}p_{H_2O}}{p_{H_2}p_{CO_2}} \tag{2.96}$$

The polarisation is also small at higher temperatures. The results obtained for the effect of temperature on the performance of 8.5 cm^2 MCFC cell are as follows:

$$\Delta V(\mathrm{mV}) = A(T_2 - T_1) \quad \begin{bmatrix} A = 2.16 & 575\ ^\circ\mathrm{C} \le T \le 600\ ^\circ\mathrm{C} \\ A = 1.40 & 600\ ^\circ\mathrm{C} \le T \le 650\ ^\circ\mathrm{C} \\ A = 0.25 & 650\ ^\circ\mathrm{C} \le T \le 700\ ^\circ\mathrm{C} \end{bmatrix} \tag{2.97}$$

Fig. 2.18 MCFC cell voltage dependence on fuel utilisation

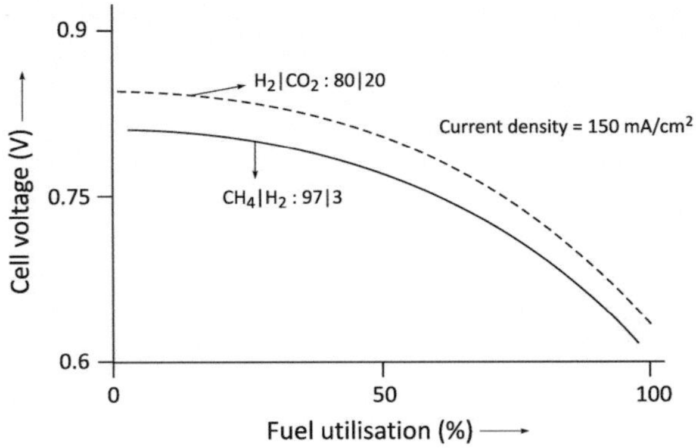

The carbonates usually melt above 500°C and cell performance is enhanced on increasing the temperature. Beyond 650°C, the gains decrease along with the electrolyte loss due to evaporation and high corrosion rates. As the reactants are consumed, the activation and concentration polarisation increase, hence decreasing the MCFC performance. The average cell voltage at 172 mA/cm^2 is found to decrease by about 30 mV for 20–50% utilisation of oxidant. Figure 2.18 depicts the cell performance for 30–60% fuel utilisation.

The major loss in MCFC occurs due to ohmic polarisation and the magnitude of ohmic losses can be given as

$$\Delta V_{\mathrm{CD}}(\mathrm{mV}) = A_{\mathrm{CD}}\Delta J \quad \begin{bmatrix} A_{\mathrm{CD}} = -1.21 & 50 \leq J \leq 100 \\ A_{\mathrm{CD}} = -1.78 & 150 \leq J \leq 200 \end{bmatrix}$$

where J is the current density in mA/cm^2.

We have discussed the current density, performance and ideal behaviour of different fuel cells. The SOFC, its working, design principles and performance variables are studied in Chap. 3. Hence, we do not discuss it in the current chapter.

Bibliography

Larminie J, Dicks A (2003) Fuel cells system explained, 2nd edn. Wiley, England

Suggested Bibliography

Advanced water-cooled phosphoric acid fuel cell development, final report, Report No. DE/MC/24221-3130, International Fuel Cells Corporation for U.S. DOE under Contract DE-AC21-88MC24221, South Windsor, CT

Balachov II, Dubois LH, Hornbostel MD, Lipilin AS (2005) Presented in fuel cell seminar, direct carbon fuel cellworkshop, Palm Springs, CA, USA, 14 Nov 2005. Proceedings online: http://www.fuelcellseminar.com/pdf/ Direct Carbon Fuel Cell Workshop/Balachov Iouri.pdf. (Photocourtesy by LLNL, California)

Appleby AJ, Electroanal J Chem 118(31)

Cairns EJ, Liebhafsky HA (1969) Energy conversion, pp 9, 63

Cao D, Sun Y, Wang G (2007) J Power Sources 167:250

Cooper JF (2003) Reactions of the carbon anode in molten carbonate electrolyte. Pittsburgh, PA: direct carbon fuel cell workshop, NETL. http://www.netl.doe.gov/publications/proceedings/03/dcfcw/Cooper%202.pdf

Fickett AP (1977) In: McIntyre JDE, Srinivasan S, Will FG (eds) Proceedings of the symposium on electrode materials and processes for energy conversion and storage. The Electrochemical Society, Inc. Pennington, NJ, p 546

Giddey S, Badwal SPS, Kulkarni A, Munnings C (2012) Prog Energy Combust Sci 38:360

Gottesfeld S (1993) Polymer electrolyte fuel cells: potential transportation and stationary applications. No. 10, An EPRI/GRI Fuel Cell Workshop on Technology Research and Development, Stonehart Associates, Madison, Connecticut

Gulzow E (2004) Fuel Cells. doi:10.1002/fuce.200400042

Haile SM (2003) Acta Materialia 51:5981

Hirschenhofer JH, Stauffer DB, Engleman RR, Klett MG (1998) Fuel cell handbook, 4th edn

Huff J (1986) Status of fuel cell technologies. In: Fuel cell seminar abstracts, 1986 national fuel cell seminar, Tucson, AZ

Jalan V, Poirier J, Desai M, Morrisean B (1990) Development of CO and H2S tolerant PAFC anode catalysts. In: Proceedings of the second annual fuel cell contractors review meeting

Kunz HR, Murphy LA (1986) In: Selman JR, Maru HC (eds) Proceedings of the symposium on electrochemical modeling of battery, fuel cell, and photoenergy conversion systems. The Electrochemical Society, Inc., Pennington, NJ, p 379

Laconti A, Smarz G, Sribnik F (1986) New membrane-catalyst for solid polymer electrolyte systems. Final report prepared by Electro-Chem Products, Hamilton Standard for Los Alamos National Laboratory under Contract No. 9-X53-D6272-1

Logan BE et al (2006) Environ Sci Technol 40(17):5181

Patton EM (2003) Presented in direct carbon fuel cell workshop, NETL, Pittsburg, PA, USA. Proceedings online: http://www.netl.doe.gov/publications/proceedings/03/dcfcw/Patton.pdf, 30 July 2003

Ralph T, Hards G, Keating J, Campbell S, Wilkinson D, Davis M, St. Pierre J, Johnson M (1997) J Electrochem Soc 144(11)

Saddawi A (2005) Carbon fuels for the direct carbon fuel cell, Master Thesis, West Virginia University, Morgantown, WV, USA

Simons SN, King RB, Prokopius PR (1982) In: Camara EH (ed) Symposium proceedings fuel cells technology status and applications, Institute of Gas Technology, Chicago, IL, vol 45, p 46

Singhal SC, Kendall K (2000) High temperature solid-oxide fuel cells: fundamentals. Elsevier, New York

Strasser K (1991) In: 26th intersociety energy conversion engineering conference proceedings, vol 3, conversion technologies/electrochemical conversion, Boston, Massachusetts. Society of Automotive Engineers, Inc., Warrendale, PA, 4–9 Aug 1991

Tanaka T et al (1989) 1989 Int Gas Res Conf 252

Zecevic S, Patton EM, Parhami P (2004) Carbon 42:1983

Zecevic S, Patton EM, Parhami P (2005) Direct carbon fuel cell with hydroxideelectrolyte—cell performance during initial stage of a long term operation. In: Proceedings of the 3rd international conference on fuel cell science, engineering, and technology, pp 507

Chapter 3
SOFC Technology: Its Working and Components

Till now, we have discussed all other types of fuel cells, i.e., their components, performance and applications. In this chapter, the prime focus is on the solid oxide fuel cell (SOFC) technology its working principle, components and configurations. Baur and Preis (1937) presented the concept of large-scale SOFC for stationary applications. The tubular design of SOFC technology developed by Siemens Westinghouse is running successfully, whereas the planar design has also achieved higher volumetric power densities. The following sections deal with the basics of SOFC technology followed by the most commonly used designs.

3.1 Fundamentals of SOFC

As the name solid oxide fuel cell implies, all components of SOFC are solid in nature. Hence, SOFC is a solid-state device, which converts chemical energy of a cell to electrical energy. There is no liquid electrolyte in the SOFC, which reduces the corrosion problem and electrolyte management. The SOFC at high operating temperature promotes kinetics, produces high quality by-product heat and allows internal reforming. The SOFC offers a great deal of advantages and the systems are relatively simple because only two phases, i.e., solid and gas, are required. The high operating temperature of SOFC also eliminates the requirement of electrocatalysts. The SOFC also offers fuel flexibility; as with MCFC, both hydrogen and carbon monoxide can be used as fuels. Figure 3.1a, b shows the working of SOFC with CO and H_2 as fuels, respectively.

The anode and cathode reactions for SOFC utilising hydrogen as a fuel are given as

At Anode:

$$2H_2 + 2O^{2-} \rightarrow 2H_2O + 2e^- \tag{3.1}$$

G. Kaur, *Solid Oxide Fuel Cell Components*,
DOI 10.1007/978-3-319-25598-9_3

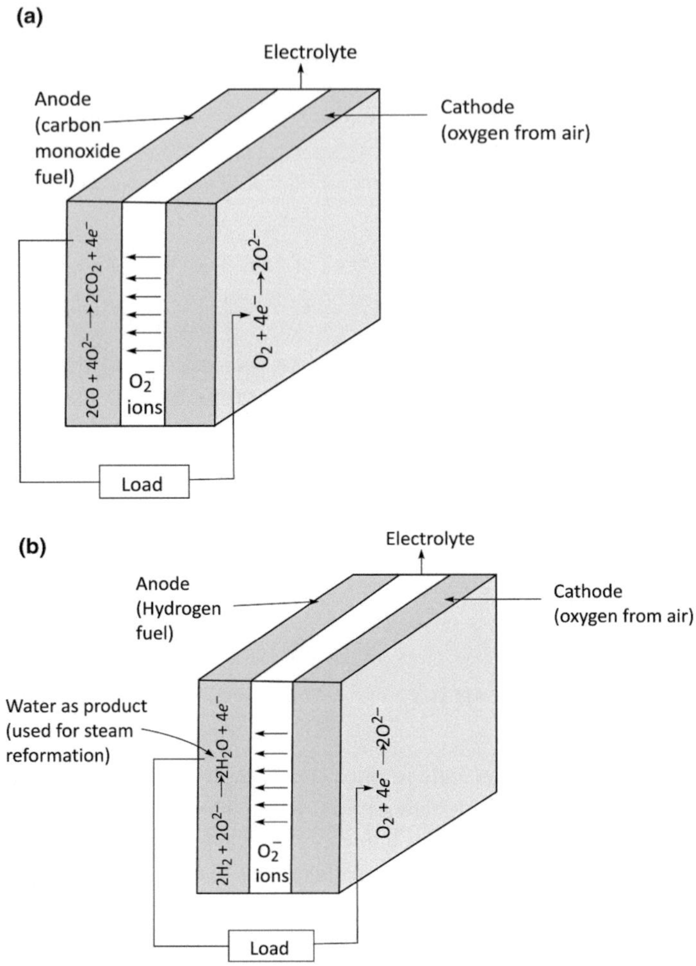

Fig. 3.1 **a** Schematic of SOFC for carbon monoxide fuel. **b** Schematic of SOFC for hydrogen fuel

At Cathode:

$$O + 4e^- \rightarrow 2O^{2-} \quad (3.2)$$

The overall cell reaction is given by

$$2H_2 + O_2 \rightarrow 2H_2O \quad (3.3)$$

The product steam can be utilised for the steam reformation process. If CO is used as fuel then no carbon dioxide recycling is required. Moreover, the absence of CO at the cathode leads to higher open cell voltage than expected. For SOFC, the electrolyte is solid and non-porous metal oxide (typically yttria stabilised zirconia). The anode is Ni–ZrO_2 cermet whereas the cathode is Sr-doped $LaMnO_3$ (see Sect. 3.4). Oxygen ions are the charge carriers for SOFC, hence, the electrolyte shall possess high ionic conduction. The SOFC do not have any liquid electrolyte, which eliminates the serious problem of electrode flooding. Therefore in SOFC, the three phase boundary is due to the solid electrolyte/electrode and gaseous fuel. The three phase boundary is imperative for the efficient electrochemical reactions within the fuel cell, hence, the electrodes shall be kept thin and porous to provide a higher interfacial region for an electrochemical reaction to take place.

The SOFC technology has undergone continuous development since the 1950s. Due to the high operating temperatures of SOFC, the fuel can be reformed within the cell. But the higher temperature of SOFC also offers certain drawbacks such as sealing issues and thermal expansion mismatch among different components of the cells. The electrolyte of SOFC exhibits high electric resistivity, which lowers the cell performance. Researchers are focused on addressing these issues to achieve the desired long-term performances and expand the applications of SOFC technology.

3.2 Early History of Solid Oxide Fuel Cell

The interest in solid ion conductors arose while developing lighting devices. In 1897, Nernst concluded that the electrolyte (thin rod shaped) could be made electrically conducting using an auxiliary heating device, which subsequently glows by maintaining the uniform passage of an electric current. One more important finding of Nernst was the use of mixed oxides because the mixtures exhibited higher conductivity as compared to conductivity of individual oxides such as lime, magnesia, etc. The conductivity of pure oxides exhibits a slow rise with temperature, whereas the mixed oxide system possesses elevated conductivity at high temperatures. Nernst obtained an almost favourable composition of 85% ZrO_2—15% Y_2O_3 (Nernst Mass) with high conductivity. Reynolds extended his work to numerous binary and ternary systems of Ce, Mg, Ca, Th, Ho, Er, Sm, Nd, Na, Le, Sc, Y, and Cr. For these systems, the conductivity was obtained for the temperature range 800–1400°C with respect to concentration changes as well. The zirconia rods developed by Nernst did not possess decomposition due to the passage of direct current, and hence, depicting their similarity to the metallic rods. Nernst also assumed the yttria to be the charge carrier and observed oxygen transport as well. He emphasised the deposition of metal cations by the direct current, which subsequently oxidises and diffuses back to the filament. Wagner in 1943 proposed the existence of vacancies in the solid solutions of mixed oxides, and hence, the conduction mechanism could be explained. "Nernst lamps" could be considered as the first commercial solid electrolyte gas cells.

In 1916, Treadwells investigated the decomposition equilibria of metal oxides in temperature range up to 1000°C with a silver/O_2 reference electrode and quartz, porcelain as solid electrolytes. In 1937, Baur concluded that the fuel cells should be made completely solid after fruitless results with many liquid electrolytes. But the search for solid electrolytes was unsuccessful till 1950 and a variety of oxide mixtures, clays, porcelains, etc. were investigated. Baur and Pries used zirconia ceramics in the fuel cells, but satisfactory results could not be obtained due to high resistance offered by the migration of cations or electrolytic shift. Other possible reasons for poor performance could be the use of Fe_3O_4 as cathode material, which forms poorly conducting Fe_2O_3 after being oxidised. Baur also used tungsten oxide and CeO_2 over the Nernst mass, however, in these ceramics, electronic conduction dominated over the ionic conduction, hence limiting their applications in fuel cells. After the development of X-ray technique, the development of solid electrolytes took a fast pace. In 1951, Hund confirmed the presence of oxide ion vacancies in the Nernst mass, using X-ray techniques and pycnometry. A model fuel cell is shown in Fig. 3.2, which uses magnesium ferrites, iron oxides and iron alumina composite electrodes. Additional research focused on the use of cerium and thorium as solid electrolytes.

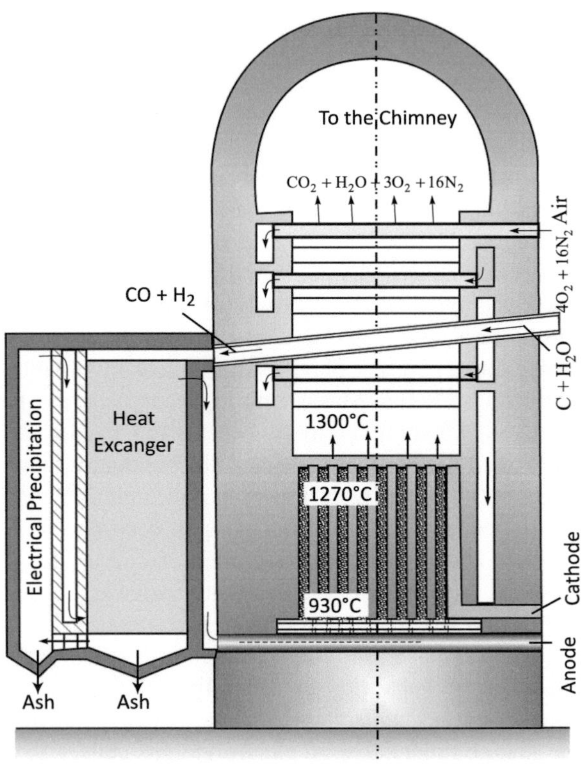

Fig. 3.2 Design of a SOFC plants as a basis for calculations

The crystallographic simplicity of thorium mixed oxides as compared to ZrO_2 was the main reason for the interest in thorium mixed oxide. With the increase in oxygen partial pressure, the electronic conductivity in thorium compounds increased. Furthermore, the signs of solubility of oxygen in the lattice could be seen while establishing the electrode potentials. Schottky showed that if the major conductivity in the electrolyte is electronic in nature, then the efficiency of the solid electrolyte fuel cell may drop down to zero with increase in lead resistance. For Th–La mixed oxides, the ion transport number 1 could be reached with 1 mol% $La_{0.5}$ with reducing atmosphere at both electrodes. Kiukkola and Wagner (1957) reported their findings for the CaO stabilised ZrO_2 system ($Zr_{0.85}$ $Ca_{0.15}$ $O_{1.85}$ systems) as solid electrolyte. For the measurements to be useful, high gas-tight solid electrolyte discs were to be used. But the results obtained by Kiukkola and Wagner were for the ferrous electrolytes sintered at 1400–1450°C. Later Peters and Mann investigated the gas-tight $Th_{0.9}La_{0.1}O_{1.95}$ solid electrolyte using a CO, CO_2 mixture at the reference electrodes. The composition $Zr_{0.85}Ca_{0.15}O_{1.85}$ was also used by Weissbart and Ruka in a high temperature galvanic cell to obtain oxygen concentration in the gas phase and they published a paper on 'A solid electrolyte cell' in 1962. In the same year, Westinghouse Corporation engaged in the development of tubular and flat plate design of SOFC. The flat plate leads to the stack and sealing issues.

In 1958, Palguyev and Volchenkova published the results on the conductivity measurements of $3ZrO_{2-\delta}CaO_2$ + 10 wt% CaO. In 1962, Sandstede gave the concept of use of hydrocarbon as a fuel for a fuel cell with 22 cm dimeter, $Zr_{0.85}Ca_{0.15}O_{1.85}$ electrolyte disc and porous platinum electrodes. Very expensive earths such as YbO_2 were also used to obtain high conductivities for the ZrO_2–Y_2O_3–Yb_2O_3 systems. In addition to this, the Al_2O_3 addition was performed for most of the electrolytes to achieve a gastight dense configuration. In 1959, Kingrey and coworkers determined the mobility of oxide ions in the $Zr_{0.85}Ca_{0.15}O_{1.85}$ electrolyte using an $^{18}O/^{16}O$ isotope exchange between the gas and solid phase. Using X-ray, Schmalzried showed that Ca and Zn occupy random sites in the $Zr_{0.85}Ca_{0.15}O_{1.85}$ cation sublattice. Subsequently decreased conductivity was observed with ageing attributed to the disorder–order transistors due to which the randomly distributed cations and oxide ions in the lattice change to an ordered state. The cation size impacted the conductivity of mixed oxides. The partial pressures of oxygen influenced the conductivity: at low oxygen pressures, excess electron conduction takes place whereas at high oxygen pressures, oxide phases show hole conduction. After developing electrolytes and the conduction phenomena associated with them, researchers focused on the electrochemical thermometry and study of polarisation phenomena. Takahashi and coworkers observed that the polarisations are much smaller at the anode as compared with the cathode, for the Ce–La mixed oxide systems. The formation of a depolarising interlayer at the anode due to partial reduction of solid mixed conductor electrolyte causes less polarisation at the anode. In the 1960s, the attention of researchers also went towards developing suitable electrode materials for the fuel cell as anode layers possessed poor adhesion. Spacil in 1964 developed the solution of Ni with solid electrolyte material. Anyhow, the suitable cathode material was relatively difficult to be found. It was found that mixed oxides with uranium exhibited stable characteristics with both electronic and ionic conductivity. Moreover, the high current densities and low polarisation could be obtained using thin conducting inter-layers between the metallic conductor and pure electrolyte. Tannenberger developed a material with polyvalent uranium ions mixed in a fluorite structure as a favourable inter-layer between the anode and cathode. Indium oxide was also proposed to be the potential cathode material when mixed with certain potential additives. The perovskites were proposed to be the suitable materials for cathodes and interconnects. The first perovskites electrode was $La_{1-x}Sr_xCaO_3$ and strontium-doped lanthanum chromite is the most used ceramic interconnect material.

In the United States, the SOFC development ran out due to the hindrance in developing electronic conducting interconnection material and suitable cathodes. Brown and Boveri extensively studied these issues along with the development of oxygen sensors. The high temperature fuel cell system studied by Brown is shown in Fig. 3.3. $LiNO_3$ doped with Bi_2O_3 and SrO doped $LaMnO_3$ were investigated as suitable cathodes.

In 1970, Isenberg developed the electrochemical vapour depositions (EVD) method which was used by Westinghouse Corporation to close the pores in the electrolyte as well as interconnect layers. The development of tubular and planar designs has been a breakthrough for SOFC technology. Though tubular designs are mechanically robust and withstand thermal stresses, the fabrication of planar designs is relatively simple and these designs offer high current density. Co-fired

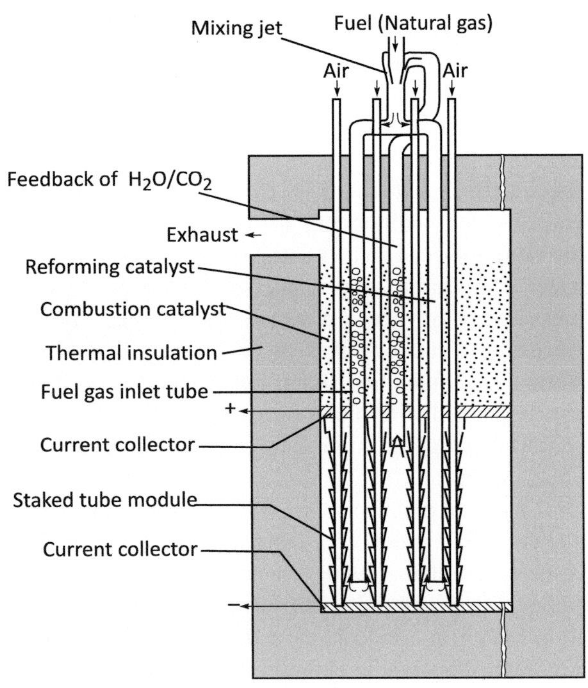

Fig. 3.3 High temperature fuel cell system for the conversion of methane by Brown (1989)

monolithic cell stacks were fabricated and investigated at Argonne National Laboratory (1983) and Ceramic Fuel Cells Ltd. (Australia) developed a 25 kW system with 3840 planar electrolyte supported cells and internal reforming anodes. These days many ideas, materials and system designs are being pursued and anode supported electrolyte SOFCs are given wide attention. The polarisation losses, low operation temperatures, materials for cathode anode and stack design are under consideration by many research groups. Moreover, the interaction among various components of SOFC has been a hot topic of research.

3.3 Operating Temperature Range of SOFC

Manufacturers are focusing on lowering cell costs by fabricating simple designs. In addition to this, efforts are also being made to lower the cell operating temperature. The high temperature, SOFC (HT-SOFC) operates within a range of 800–1000 °C. Such high temperatures require mechanical integrity of stack, interfacial layer stability and thermal expansion matching. The high operating temperature is advantageous for oxide ion conductivity through the electrolyte. In the intermediate temperature (IT) range of 600–800°C, the conductivity of electrolyte decreases considerably. Hence, the thickness of the electrolyte should also be decreased in order to maintain the desired conductivity for IT-SOFC. For the operating range of 600–800°C, material designing becomes an important criterion, so as to overcome the problems due to thermal mismatch and oxidation. The performance obtained at high temperatures is larger than those at the lower temperature. Anyhow the performance at lower temperatures can be maintained if the electrode materials possess mixed conductivity due to the electron and ions. The mixed conduction increases the probability of oxygen reduction at any point on the cathode surface rather than just being at the triple phase boundary. The thermal expansion mismatch of all the components should be addressed completely, which is necessary for the commercialisation of SOFC technology. The low operating temperatures reduce the need of expensive interconnects and the diffusion kinetics also slow down. Hence, at low temperatures, the performance degradation due to the interdiffusion of ions can be eliminated to some extent.

3.4 Components of SOFC

For a SOFC, the typical fuel gas contains 0.23 bar H_2, 0.03 bar water, 0.06 bar CO_2 and 0.2 bar CO, with N_2 yielding balance of 1 bar. Corresponding to the difference of oxygen partial pressure in fuel and air compartments, the fuel and air compartments of fuel cell have different potentials. The magnitude of current the generated in the fuel cell depends upon the rate of conversion for CO and H_2 to CO_2 and H_2O, respectively. To achieve higher powers, the cathodic and anodic

Table 3.1 SOFC cell materials

Components	Materials	Requirements
Electrolyte	Yttria stabilised Zirconia (30–40 μm thickness) by EVD technique	Oxygen ion conductivity Mechanical strength Gas tightness
Interconnect	Pt, Mn-doped cobalt chromite, doped lanthanum chromite (~ 100 μm thick by plasma spray)	High electronic and thermal conductivity High mechanical strength
Anode	Pt, Ni–ZrO_2 Cermet, Ni–YSZ Cermet (~ 150 μm thick by EVD or Slurry deposition)	High porosity Electronic conductivity Large triple phase boundary sulphur tolerance
Cathode	Pt, $La_{1-x}Sr_xMnO_3$, $La_{1-x}Sr_xFeO_3$, $La_{1-x}Sr_xCo_{1-y}Fe_yO_3$	High porosity Electronic conductivity Large triple phase boundary Activity for oxygen surface exchange

overpotentials should be minimised along with reducing the ohmic losses due to the electrolyte layer. Hence, the fast oxide ion conducting the electrolyte and highly electrochemically active electrode materials must be used. The SOFC can be regarded a laminated structure of three different functional materials: anode, electrolyte and cathode. Table 3.1 lists the materials used in SOFC technology.

3.4.1 Electrolytes

The dense electrolyte is sandwiched between the porous anode and cathode. At the anode, oxidation takes place releasing the electrons which move via external circuit towards the cathode helping in the reduction process. The electrolytes used in SOFC can be oxide (O^{2-}) conducting or proton (H^+) conducting; anyhow the research is more focused on oxide conducting electrolytes. Hence, the overall significance of the electrolyte is to provide a conduction path for O^{2-}/H^+ ions between the electrodes to complete the overall electrochemical reaction of the cell. The SOFC works in a stringent environment, i.e., high operating temperature, ceramic components, oxidation on cathode side and reduction on the anode side. Hence, the electrolyte must satisfy the following requirements:

- The electrolytes must be stable in reducing and oxidising atmospheres.
- The electrolyte must be dense and leak tight, hence free of porosity so that the gases cannot permeate from one side of electrolyte to other.
- The Electrolyte material must possess high ionic conductivity and no electronic conductivity. The transport number of ions should be close to unity whereas it should be zero (ideally) for the electrons. The oxide ion conductivity is $>10^{-2}$ S/cm for the cell operating temperature.
- The electrolyte shall exhibit excellent chemical and mechanical compatibility with its adjoining components to reduce the detrimental effects of any interfacial reaction.
- It should have high surface area to attain maximum current density.
- It should possess thermal shock resistance and hence a matching thermal expansion coefficient with the adjoining components at cell operating temperature.
- The fabrication process should be economical.

The most commonly used oxide conducting electrolytes possess a fluorite structure or perovskite structure. Apart from this, Aurivillius (BIMEVOX family), brownmillerites, proton conductors and non-cubic structured electrolytes have been investigated. For an oxide ion conductor, the flow of current is due to migration of oxide ions through the crystal lattice. The migration of oxide ion is a thermally activated process and the O^{2-} ions hop from one lattice site to the other randomly. Under the application of an external electric field, the ions drift in one particular direction instead of being random. The ionic conduction is a function of the operating temperature and oxide ion mobility. At sufficiently high temperatures, the oxide ion conductivity values approach 1 S/cm, which is comparable to the ionic conductivity for the liquid electrolytes. For the oxide conduction process to take place, the crystal lattice must contain unoccupied sites equivalent to the sites occupied by lattice oxygen ions. Moreover, the hopping/migration of O^2 ions will take place if the energy involved in the process is <1 eV.

3.4.1.1 Fluorite Structured Electrolytes

Fluorite structures are the most commonly used oxide ion conducting materials due to their open structure. The fluorite structure exhibits the general formula of AO_2, where A is usually a tetravalent cation such as Zr^{4+}, Ce^{4+}, U^{4+} and Th^{4+}. The fluorite structure is a face-centred cubic structure of cations with the anions occupying the tetrahedral sites. This structure leaves large interstitial octahedral voids causing an open structure of the lattice which is favourable for rapid ion diffusion. Oxide ion conductivity was first reported for ZrO_2 by Nernst (1899). At high temperatures, zirconica exhibit fluorite structure, but at lower temperatures the fluorite structure of ZrO_2 cannot be retained due to the small size of Zr^{4+}. Hence, at low temperatures, zirconia is partially substituted by lower valence cations such as calcium or yttrium to stabilise the structures. The dopant is introduced in the

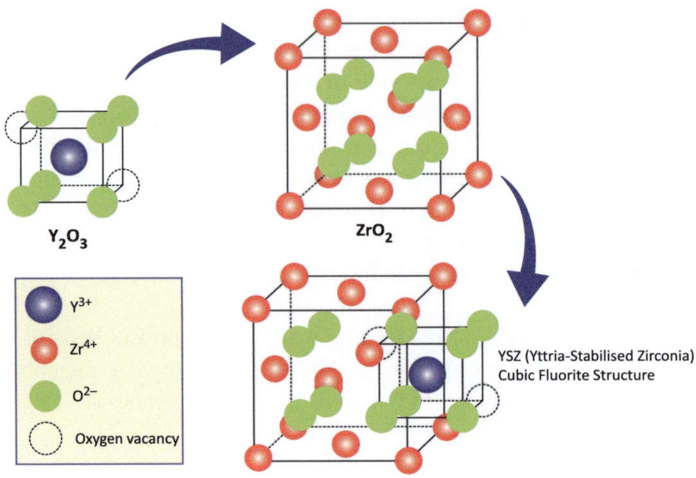

Fig. 3.4 Increase in ionic conduction of doped zirconia

zirconia structure in such a way that the charge neutrality is maintained along with the oxygen ion vacancies in the lattice for the oxide migration. CeO_2 also exhibits a fluorite structure and interstitial oxide ions and oxide ion vacancies provide the oxide ion conduction. Pure ZrO_2 and CeO_2 have very small conductivity due to low concentration of oxide ion vacancies and interstitial oxide ions. Upon adding the dopants such as Ca^{2+} and Y^{3+}, the conductivity of these fluorite oxides can be increased (Fig. 3.4). Zirconia exhibits three polymers: monoclinic structure at room temperature, tetragonal above 1170°C and cubic fluorite oxides at 2370°C. The tetragonal to monoclinic transition is associated with a large volume change of almost 3–5%, whereas the cubic phase is retained up to the melting point of 2680°C. The addition of an aliovalent dopant such as yttrium stabilises the fluorite structure and tetragonal phase to room temperature. The Kroger–Vink notation is used to represent the oxygen vacancies for a dopant M; i.e.,

$$M_2O_3(ZrO_2) \rightarrow 2M_{Zr} + 3O_o^x + V_o^{\bullet\bullet} \tag{3.4}$$

$V_o^{\bullet\bullet}$ represents the oxygen vacancy created for every 2M atoms incorporated into the lattice. The vacancy concentration follows the electrical neutrality condition:

$$2[M_{Zr}] = V_o^{\bullet\bullet} \tag{3.5}$$

The ionic conductivity can be given by

$$\sigma = en\mu \tag{3.6}$$

where μ is the mobility, e is the charge and n is the number of mobile ion vacancies. It is expected that the increase in dopant concentration would lead to an increase in conductivity. Usually at lower concentrations, the association of point defects leads to a reduction in defect mobility and hence conductivity. The ionic conductivity can be given by

$$\sigma = \frac{A}{T}\left[V_{o}^{\bullet\bullet}\right]\left[\bar{V}_{o}^{\bullet\bullet}\right]\exp\left[\frac{-E_a}{RT}\right] \tag{3.7}$$

where E_a is the activation energy, R is the universal gas constant, A is the pre-exponential factor, T is the absolute temperature and $\left[\bar{V}_{o}^{\bullet\bullet}\right]$ represents unoccupied oxide ion vacancies. The ionic conductivity of ZrO_2 can be enhanced along with extending the oxygen partial pressure range of ionic conduction by doping (and hence, forming stabilised zirconia as an electrolyte). Commonly Y_2O_3, Sc_2O_3, MgO and CaO are used as potential dopants. These dopants have high solubility in zirconia and form a stable fluorite structure. The concentration of these dopants is optimised because at higher dopant levels, the defect ordering, vacancy clustering or electrostatic interactions may decrease the ionic conductivity. In present solid oxide fuel cell systems, 3 mol% Y_2O_3(3YSZ) or 8 mol% Y_2O_3 (8YSZ) are used as potential electrolytes in spite of the fact that YSZ is not the best ion conductor. Anyhow, the cheap processing, abundance, non-toxicity and chemical stability make it a suitable candidate. For partially stabilised zirconia (PSZ), a mixture of metastable tetragonal ZrO_2 and cubic phase is obtained due to this insufficient amount of dopant. PSZ is a transformation-toughened material and is also known as tetragonal zirconia polycrystal (TZP). The main applications of PSZ involve crucible making because PSZ has a high melting temperature and low thermal conductivity. To obtain fully stabilised zirconia 8 mol% Y_2O_3 or 16 mol% CaO or 16 mol% MgO are added to the ZrO_2 structure. If 13–16% Y_2O_3 is added, then the structure becomes a cubic solid solution, which has phase stabilisation while heating from room temperature to 2500°C. It should also be noted that the content of the dopant with highest conductivity decreases with increasing dopant ion radius (Fig. 3.5).

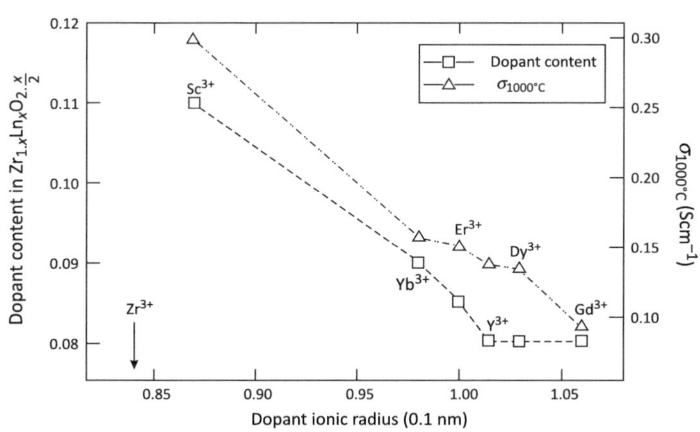

Fig. 3.5 Effect of dopant ionic radius on conductivity and dopant content

The limiting value of Dy^{3+} and Gd^{3+} concentration is 8 mol% due to their high ionic radii. Zr^{4+} and Sc^{3+} have almost the closest ion radius with each other, and possess the highest dopant concentration and highest conductivity as well. The microstructure optimisation and appropriate scandium doping has yielded high oxygen conductivity of 0.15–0.20 S/cm at 1000°C. Scandia doped zirconia is an attractive electrolyte for the intermediate temperature range of 600–800°C. Scandia doped zirconia (ScSZ) has high conductivity which is attributed to the small mismatch of Zr^{4+} and Sc^{3+} size as compared to the size difference between Zr^{4+} and Y^{3+}. The defect energy influences the oxide mobility and hence conductivity of the electrolyte. Another factor which influences the conductivity of YSZ and ScSZ is ageing during operation. 11 mol% Sc_2O_3 doped ZrO(11 ScSZ) showed no ageing effect on annealing at 1000°C even after 6000 h operation. Anyhow, phase transition from rhombohedral structure to the cubic structure at 600°C is observed for 11ScSZ accompanied by a small volume change. If a small amount of CeO_2/Al_2O_3 is added to 11ScSZ, then the cubic phase is stabilised with decreased conductivity as compared to undoped 11ScSZ. In contrast to this, 8 ScSZ showed a decrease in conductivity from 0.3 S/cm (1000°C) to 0.12 S/cm (1000°C), after 1000 h of operation. 9YSZ conductivity values are almost similar to those of 8 SCSZ conductivity after ageing. When ZrO_2 is doped with alkaline earth metal ions (A^{2+}), then the conductivity values do not increase much due to the low thermodynamic stability of cubic fluorite solutions in the ZrO_2–AO system. Addition of Bi_2O_3 to SCSZ has also resulted in a sharp increase in conductivity. Table 3.2 shows the conductivity of some zirconia-based electrolytes.

As compared to zirconia, ceria exhibits high conductivity especially at low operating temperatures. Pure ceria yields both ionic and electronic conductivity, hence, can be regarded as a mixed ionic electronic conductive material (MIEC). Due to the electronic conductivity phenomena in ceria, ceria has not received much attention as an electrolyte. In addition to this, the CeO_2 is not stable under reducing conditions and becomes CeO_{2-x}. CeO_2 exhibits an order high conductivity than the doped ZrO_2, which is attributed to the larger ionic radius of Ce^{4+} (0.87 Å in sixfold coordination)

Table 3.2 Conductivity of zirconia based electrolytes

Electrolyte	$\sigma_{1000°C}$ (S/cm)	
	As sintered	After sintering
ZrO_2—3 mol% Yb_2O_3	0.063	0.09
ZrO_2—8 mol% Yb_2O_3	0.2	0.15
ZrO_2—10 mol% Yb_2O_3	0.15	0.15
ZrO_2—2.9 mol% Sc_2O_3	0.09	0.063
ZrO_2—8 mol% Sc_2O_3	0.30	0.12
ZrO_2—11 mol% Sc_2O_3	0.30	0.30
ZrO_2—11 mol% Sc_2O_3–1 wt% Al_2O_3	0.26	0.26
ZrO_2—3 mol% Y_2O_3	0.059	0.05
ZrO_2—8 mol% Y_2O_3	0.13	0.09
ZrO_2—9 mol% Y_2O_3	0.13	0.12

than Zr^{4+} (0.72 Å) leading to a more open structure for the migration of O^{2-} ions. CeO_2 gives n-type electronic conductivity, which causes an internal short circuit of the cell. The oxygen vacancies in CeO_2 also generate non-stoichiometry, which causes mechanical failure due to expansion of the lattice. Equation (3.7) gives the ionic conductivity, whereas the electron conductivity (σ_e) is given by polaron hopping; i.e.,

$$\sigma_e = \frac{A_e}{T} \exp\left[\frac{-\Delta E}{kT}\right] p_{O_2}^{-1/4} \tag{3.8}$$

The electronic conductivity increases with increasing temperature and decreasing oxygen partial pressure, whereas oxide ion conductivity is independent of pressure. At high temperatures the electronic conductivity becomes very small, ~0.16 S/cm at 800°C. Doped ceria yields high conductivity like doped zirconia. The doping of ceria with lower valence rare earth oxides reduces enthalpy of association of defects leading to increased conductivity. As stated earlier, increased concentration of dopants may lead to formation of a secondary phase due to the solubility limit, hence decreasing the conductivity. Cerium has been doped with Sm_2O_3, Gd_2O_3 and Y_2O_3. Gadolinium doped ceria (GDC) has shown almost three times higher conductivity than the samarium doped ceria (SDC). For $Gd_{0.2}Ce_{0.8}O_2$ (GDC) the open circuit voltage of 0.89 V at 850°C is obtained for a hydrogen air fuel cell. Yahiro and co-workers obtained two conductivities for the MgO/BaO doped ceria due to insufficient solubility of alkaline oxides in the ceria. Faro and co-workers investigated the effect of preparation techniques on the characteristics of GDC. GDC prepared by oxalic-coprecipitation and sol–gel routes exhibited no electronic conductivity under air exposure. Herle and co-workers found that co-doping ceria with rare earth ion and alkaline earth ion lead to higher conductivity in air. The most commonly investigated co-doped ceria systems for IT-SOFC are $Ce_{1-x-y}Y_xLa_yO_2$, $Ce_{1-x-y}Gd_xSm_yY_zO_2$, $Ce_{1-x-y}Sm_xLa_yO_2$ and $Ce_{1-x-y}Gd_xPr_yO_2$.

3.4.1.2 Perovskite Structured Electrolytes

The second category of solid-state ion conductor is perovskites with general formula ABO_3. AB can possess different combinations to yield total charge of +6, i.e., 1 + 5, 3 + 3, 2 + 4 and in complex ways as $A'_{1/2}A''_{1/2}$ TiO_3 ($A' = \text{Li,Na}|A'' = \text{Pr, La}$) and $\text{Pb}\left[B'_{1/2}B''_{1/2}\right]O_3[B' = \text{Sc, Fe}|B'' = \text{Nd, Pa}]$. For the general structure of ABO_3, A is the large metal cation, which occupies the centre of the crystal. B is the small ion occupying the corners and oxygen is in the centre of the edges. Perovskites display a wide range of properties and applications due to their high accommodation capacity of different cations. Figure 3.6 shows a schematic of the perovskite structure.

Initial work was based on investigating the oxide ion conductivity of $LaAlO_3$ materials. Mizusaki and co-workers reported the oxide ion conductivity for $La_{1-x}Ca_xAlO_3$ single crystal sample with ($x = 0.0027$–0.008). For this material, oxygen vacancies are major defects whereas electron holes are the minor defects. The

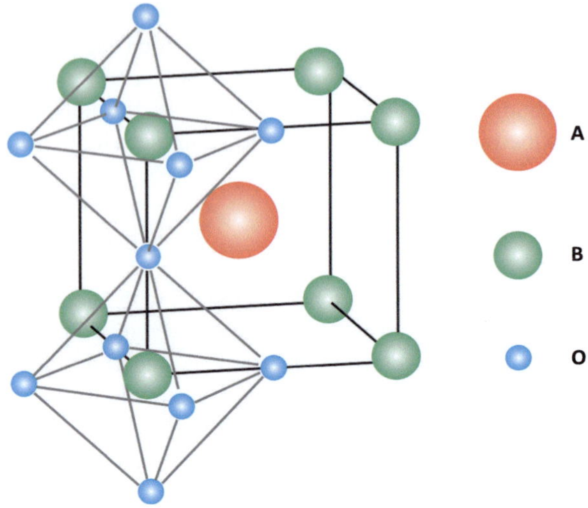

Fig. 3.6 Crystallographic depiction of perovskite ABO_3 structure

activation energy of oxide ion migration is 0.74 ± 0.05 eV, and the increase in oxygen pressure increases. Conductivity of perovskites is obtained to be smaller than the YSZ electrolytes. In ABO_3 perovskites, the migrating oxide ions have to pass through the triangular space made up of two large A and one small B cation in the crystal lattice. To improve the migration of oxide ions, the size of this triangular space should be increased, which infers that the size of the A site cation significantly determines the oxide ion conductivity. Gallate materials such as $LaGaO_3$, $PrGaO_3$, $SmGaO_3$ and $NdGaO_3$ have also been investigated as potential electrolyte systems because the conductivity of gallium-based perovskites exhibits oxygen partial pressure independent conductivity indicating oxide ion conduction to be the dominant conduction process for gallates. Furthermore, the conductivity increases when the lanthanum site is doped with barium, calcium and strontium. Particularly, the conductivity with strontium-doped gallates is higher than those of calcium/barium doped gallates. Though strontium increases the number of oxygen vacancies, the solid solubility limit for Sr in GaO_3 is low, which leads to the formation of secondary phases such as La_4SrO_7 and $SrGaO_3$ particularly for Sr concentration >10 mol%. Figure 3.7 shows the conductivity for different alkaline earth atom-doped lanthanum gallate.

For Mg dopant on the Ga-site for $LaGaO_3$, the conductivity increases due to the formation of oxygen vacancies. Furthermore, the lattice parameter also increases because the ionic radius of Mg is larger than the Ga. Another advantage of adding Mg on the Ga site is an increase in the solid solubility limit of Sr from 10 to 20 mol% due to the enlarged crystal lattice formed by Mg doping. Majewski and co-workers reported highest ionic conductivity for $La_{0.8}Sr_{0.2}Ga_{0.8}Mg_{0.2}O_3$ (LSGM) composition. In general, the highest conductivity for $La_{0.8}$ $Sr_{0.2}$ Ga_{1-y} Mg_yO_3 is obtained in the range of y = 0.15–0.2 by different groups. These doubly doped

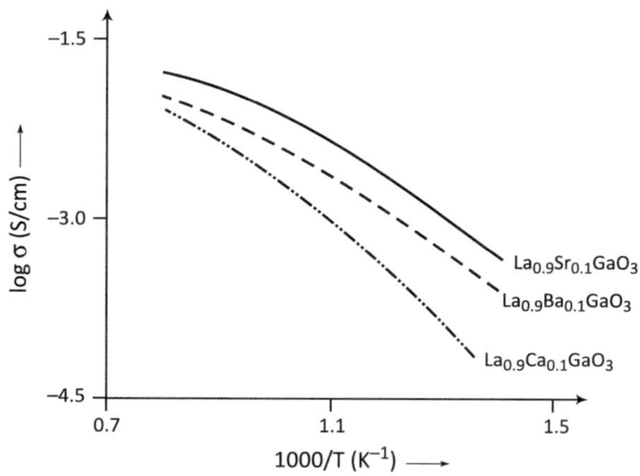

Fig. 3.7 Effect of Ca, Ba and Sr on the conductivity of lanthanum gallate

$LaGaO_3$ are promising materials as electrolytes for SOFC due to their high conductivity even more than the YSZ electrolytes. Anyhow, these are certain drawbacks for LSGM electrolytes, such as their mechanical strength and secondary phase formation. $LaGaO_3$ materials are weaker in bending strength and creep resistance than YSZ. The average strength of $La_{0.9}Sr_{0.1}Ga_{0.8}Mg_{0.2}O_3$ at room temperature and at 900°C are 162 ± 14 MPa and 55 ± 11 MPa, respectively. To overcome this problem, Al_2O_3 has been added to enhance the mechanical properties of LSGM without affecting the conductivity. Secondary phased including $LaSrGaO_4$, $SrGaO_3$, La_4SrO_7 and $LaSrGa_3O_7$ may be observed in these electrolytes. Hence, great care must be taken while synthesising them, particularly with solid-state technique which usually gives the single phase material.

Yokokawa and co-workers estimated electrolyte efficiency LSGM as a function of fuel utilisation and internal resistance. With increased electrolyte thickness, the electrolyte efficiency decreased due to the increase in internal resistance of the cell. In contrast to this, if the electrolyte is made thin, then the leakage of oxygen occurs due to electronic conduction, which leads to the decreased efficiency and increased fuel consumption. This leads to the concept of optimisation of electrolyte thickness to obtain maximum efficiency of the cell. In addition to the effect of alkaline earth metals on the conductivity of $LaGaO_3$, the Cr and Fe substitution on the Ga site was also investigated, which resulted in decreased stability against reduction. In contrast to this, it was found that the substitution of small amounts of Co or Ni increased the oxide ion conductivity for LSGM electrolytes. Cu and Mn doping decrease conductivity whereas Co, Ni and Fe increase conductivity for $LaGaO_3$. Cu increases *p*-type conduction whereas Mn and Ni enhance *n*-type conduction. Doping with cobalt has led to high power output and current density. For 10 mol% of cobalt doping, the open circuit voltage dropped, whereas power density was maximum for 8.5% of Co doping. Though the ionic conductivity increased at high concentrations of cobalt, the current leakage also became prominent.

3.4.1.3 Brownmillerite Structured Electrolytes

The brownmillerite structure has general formula of $A_2B_2O_5$ or $A_2B'B''O_5$. Brownmillerites can be regarded as perovskite structures containing oxygen vacancies along [101] direction in alternate layers. The a and c lattice parameters of the brownmillerite structure are twice those of the perovskite structure, retaining the c-axis to be the same for both of them. High oxide ion conductivity is expected in brownmillerites because sometimes the oxygen vacancies do not order in them. $Ba_2In_2O_5$ has been the most investigated brownmillerite for SOFC applications. Studies show that the cerium doped $Ba_2In_2O_5$ (i.e., $Ba_2In_{1.75}Ce_{0.25}O_{5.1}$) show almost similar conductivity as that of the YSZ. At 900°C, a large jump in the conductivity characteristics of $Ba_2In_2O_5$ is observed, attributed to the order/disorder transition of an oxygen vacancy ordered brownmillerite to the disordered pseudo-perovskite. If $Ba_2In_2O_5$ is doped with ceria, then the order/disorder transition is eliminated and the structure is stabilised at lower temperatures. Hence, no jump at 900°C is observed, but the conductivity of Ce doped $Ba_2In_2O_5$ is reduced at high temperature. Doping of La on the Ba site has also yielded promising results for the $Ba_2In_2O_5$ structures. For La doping, the lattice oxygen decreases and the oxygen disorder structure can be stablilised. Fisher and co-workers obtained the oxygen Frenkel pairs to be the dominant intrinsic defects in low temperature phases. $Ba_2In_2O_5$ is more stable in reducing conditions and the reduction removes oxygen from the octahedral layers. Zhang and co-workers also concluded that the $Ba_2In_2O_5$ exhibit proton conduction at temperatures lower than 400°C. The $Ba_2In_2O_5.H_2O$ crystal phase is mainly responsible for the proton conduction. The oxidation process in $Ba_2In_2O_5$ involves oxide ion insertion in the interstitial layers.

3.4.1.4 Aurivillius Structured Electrolytes

Bismuth vanadates are the metal ion doped vanadates, which are fast oxide ion conducting solid electrolytes and exhibit aurivillius structure. The aurivillius structure for $Bi_2VO_{5.5-\delta}$, consists of an alternate bismuth–oxygen (Bi–O) and vanadium oxygen (V–O) layers as shown in Fig. 3.8. The Bi–O layer in $(Bi_2O_2)^{2+}$ exhibits square pyramidal coordination for the bismuth. Bi contains non-bonding 6 s^2 lone pairs, which contribute to the stereo chemical activity and asymmetric coordination geometry. The V–O layer contains VO_6 octahedrons, which maintains the electroneutrality of the system. $Bi_2VO_{5.5-\delta}$ consists of three different structures depending on the temperature range as explained below:

- α-phase: Low temperature phase (<445°C) with unit cell volume $3a_m \times b_m \times c_m$. α-phase is monoclinic in nature and *n*-type semiconductor.
- β-phase: Intermediate temperature phase (445–570°C) with unit cell volume $2a_m \times b_m \times c_m$. β-phase is orthorhombic in nature.

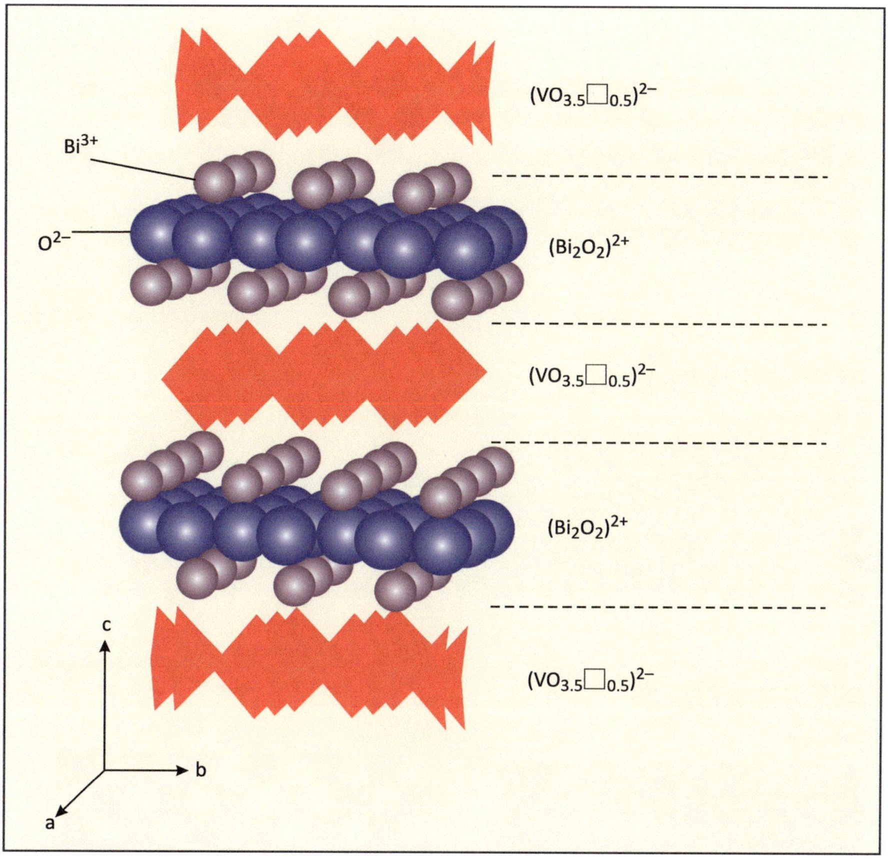

Fig. 3.8 Aurivillius structure as layered configuration

- γ-phase: High temperature phase (~700°C) with unit cell volume of $\frac{a_m}{\sqrt{2}} \times \frac{b_m}{\sqrt{2}} \times c_m$. γ-phase is tetragonal in nature.

$$[a_m = 0.55\,\text{nm}, b_m = 0.56\ \text{nm and } c_m = 1.53\,\text{nm}]$$

Figure 3.9 shows the arrangements of three main phases in bismuth vanadate aurivillius structured electrolytes.

Figure 3.10a shows the X-ray diffraction profiles for the $Bi_2VO_{5.5-\delta}$. The profiles obtained for the three phases are almost similar except the split of peaks; i.e., for α-phase superlattice diffraction peaks are observed at $24°\left(\frac{1}{3}13\right)$, for β-phase the peaks are observed at $\left(\frac{2}{3}13\right)26°$ and for γ-phase the peaks merge at 32°(200) and hence no superlattice is observed. The electrical conductivity of these three phases also varies as shown in Fig. 3.10b.

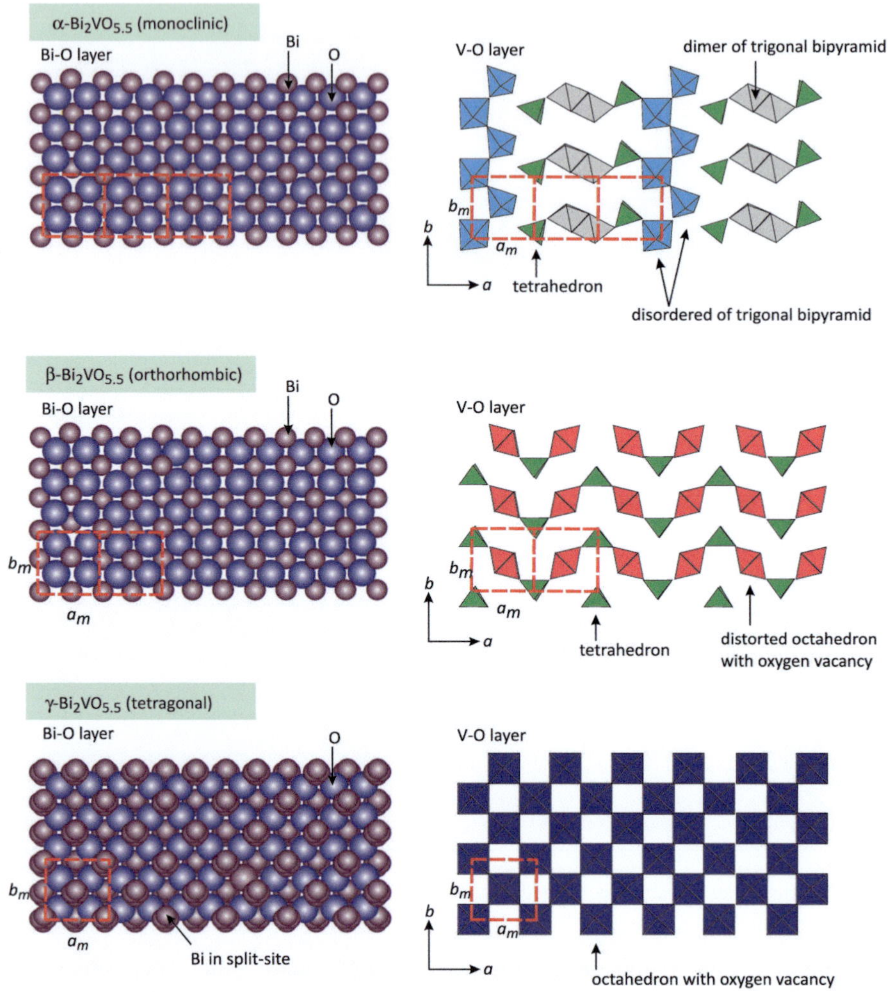

Fig. 3.9 Different structures for the bismuth vanadates

The γ-phase is a highly conducting phase with oxide ion conductivity of almost 0.6 S/cm at 700°C. Aurivillius bismuth vanadates are suitable candidates for IT-SOFC application. Abraham and co-workers explained the defect structure of BIMEVOX's using two models, i.e., the equatorial vacancy and apical vacancy model in which the vacancies are assumed to be present in the bridging and non-bridging apical sites present in bismuth and large vacancy concentration of vanadate layers make them very suitable electrolytes. Doping of the $Bi_2VO_{5.5-\delta}$

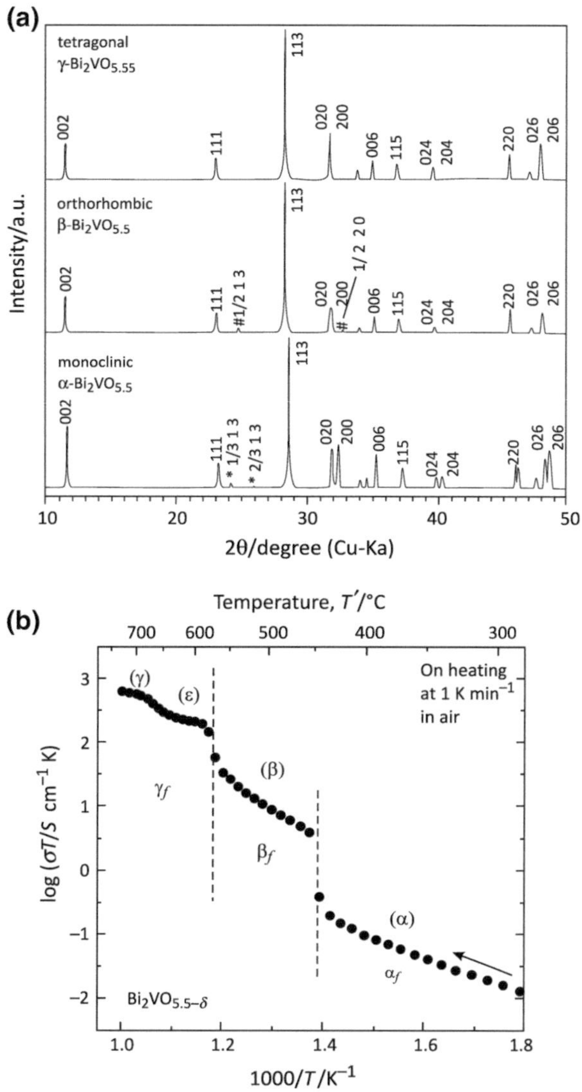

Fig. 3.10 **a** XRD and **b** electrical conductivity for all the bismuth vanadate phases [Taninouchi 2010]

system has increased their conductivity particularly for the copper doped $Bi_2VO_{5.5-\delta}$. Simmers and co-workers obtained high oxygen ion conductivity in the $Bi_2V_{0.9}Cu_{0.1}O_{5.35}$ system at 300°C. Along with obtaining high conductivity, phase stabilisation is also very important. Pernot and co-workers obtained stabilisation of the tetragonal phase in $Bi_2VO_{5.5-\delta}$ by doping up to 7% Cu whereas good enough obtained it for 15% Cu doping. Apart from copper, titanium-doped BIMEVOX is the most promising composition.

3.4.1.5 Proton Conducting Electrolytes

Proton conductors find applications for intermediate temperature fuel cells. Most of the proton conducting materials decompose at 300°C. It has been demonstrated that the H^+-SOFC system possesses higher thermodynamic efficiency than the O^{2-}-SOFC system. For the proton conducting SOFC system, water is generated at the cathode when H_2 is used as a fuel. The materials with proton conduction derive their ionic conductivity from the protonic defects which have high mobility. Usually, the perovskite materials have a general formula of $AB_{1-x}M_xO_3$ (M is trivalent doped rare earth element and $x < 1$). For the doped perovskite, the oxygen ion vacancies are formed in the structure. At elevated temperatures, these perovskites may exhibit protonic conductivity especially in the presence of water atmosphere. Therefore, for a protonic conductor, the humidity-induced conductivity is dominant over the oxide ion or electron conductivity. Protons are embedded in the electron cloud of an oxygen ion because due to the small size of the protons, they cannot occupy interstitial positions. The Kroger–Vink notation for protonic conductors can be written as

$$H_2O + V_o^{\bullet\bullet} + O_o^x \rightarrow 2(OH)_o^{\bullet} \tag{3.9}$$

Equation (3.9) indicates that the water molecule dissociates to form two hydroxide ions, which replace the oxide ions hence forming two protonic defects in the lattice structure. Earlier $BaCeO_3$ and $SrCeO_3$ have been extensively researched as the protonic conductors. Later on perovskites based on $BaSnO_3$, $SrZrO_3$, $SrTiO_3$ and complex perovskites such as $A_2B'B''O_6/A_3B'B''O_9$ also exhibited protonic conduction. $SrZrO_3$ material tends to possess low proton conductivity and doped $SrTiO_3$ also shows less formation for proton defects. Ytrium doped $BaCeO_3$ has shown protonic conductivity of 1.9×10^{-3} S/cm (800°C) in O_2/H_2O atmosphere and 1.27×10^{-2} S/cm (800°C) in H_2/H_2O atmosphere, respectively. The $BaCe_{0.8}Y_{0.2}O_3$ (BCY) electrolyte possesses conductivity of 3.2×10^{-3} S/cm (700°C), which is less than the conductivity of GDC. The smaller conductivity of $BaCe_{0.8}Y_{0.2}O_3$ could be attributed to the formation of an insulting phase such as $BaCeO_3$. The BCY-based cell exhibits power density of 183 mW/cm^2 (700°C) whereas the GDC-based cell has power density of 400 mW/cm^2 (700°C). At low current densities, the performance of a BCY-based cell is obtained to higher ionic transport number. $BaZr_{0.4}Ce_{0.4}In_{0.2}O_3$ has relatively high durability against moisture and is regarded as a good protonic conductor. $SrCeO_3$ has been doped by terbium, ytterbium and thulium and it was concluded that the Tb and Tm doped $SrCeO_3$ obeyed the Arrhenius relationship in air, N_2 and O_2 atmospheres. Tb doped $SrCeO_3$ is a pure protonic conductor, whereas for Th doped $SrCeO_3$ shows electron–proton conductivity in the presence of water vapours. $SrCe_{0.95}Yb_{0.05}O_3$ exhibits electronic conductivity of 0.01 S/cm (800°C). Upon exposure to humidified air, the protons are generated as following:

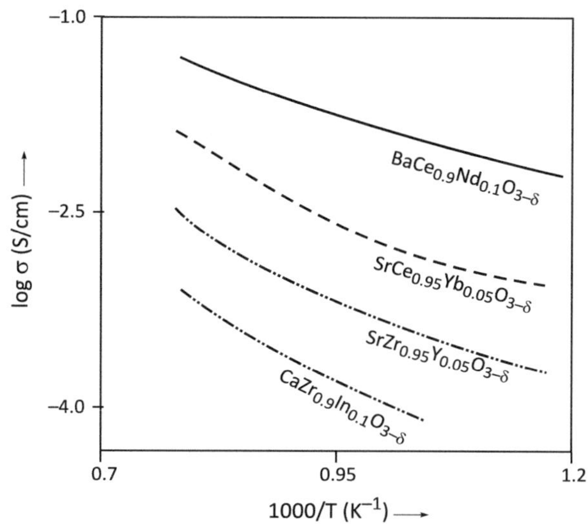

Fig. 3.11 Conductivity of different perovskite-based protonic conductors

$$H_2 + 2h \rightarrow 2H^+ \quad (3.10)$$

The formation of protons leads to the increase in conduction. The order of proton conduction is given by $CaZrO_3 < SrZrO_3 < BaCeO_3$ whereas the chemical stability followed the reverse trend for the $SrZrO_3$ as the most stable material. Figure 3.11 shows the proton conduction for different materials.

3.4.2 Cathodes

Cathodes are the air electrodes for the SOFC system and ideal cathodes must meet the following requirements.

- Matching coefficient of thermal expansion with the electrolyte and interconnect material
- High electronic conductivity >100 S/cm under the oxidising conditions
- Porosity to allow diffusion of gaseous oxygen towards the cathode/electrolyte interface
- High catalytic activity required for the oxygen reduction

During the earlier stages of SOFC development, platinum was used as a cathode material, but due to its expensive nature, researchers tried to obtain economically processable and stable cathodes. The oxygen reduction takes place at the cathode surface as follows.

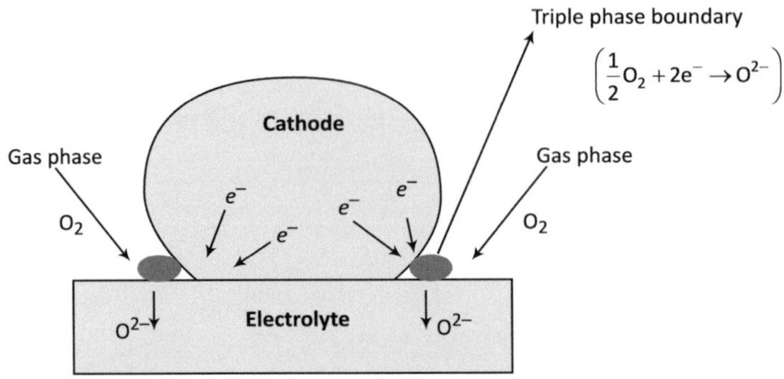

Fig. 3.12 Triple phase boundary in cathode materials

$$\frac{1}{2}O_{2(g)} + 2e^-_{\text{cathode}} \rightarrow O^{2-}_{\text{electrolyte}} \tag{3.11}$$

Usually these electrochemical reactions take place at the triple phase boundaries (TPB) as shown in Fig. 3.12.

At triple phase boundaries, the gas phase comes in contact with the electrolyte (oxygen ion conductor) and electrode (electronic conductor). If there is an obstruction to the gases, electrons, or ions at the reaction sites, then the connectivity may break and the reaction cannot proceed further. Hence, the cathodes used must possess single phase behaviour with mixed conductivities (oxide and electronic) or a composite structure consisting of electron-conducting cathode and ion-conducting electrolyte. The cathode polarisation losses may increase if the operating temperature of SOFC is reduced in spite of the fact that low operating temperatures of SOFC possess many advantages. The generation and transport of oxide ions within the porous cathode structure mostly contribute to the polarisation losses. To obtain maximum performance at intermediate temperatures, the Adler model provides the chemical reaction contribution (R_{chem}) as follows:

$$R_{\text{chem}} = \frac{RT}{2F^2}\sqrt{\frac{\tau}{(1-\varepsilon)aC_o^2DK}} \tag{3.12}$$

where ε, τ, a, and C_o are fractional porosity, tortuosity, internal surface area/volume, and surface concentration of oxygen, respectively. D is the self-diffusion coefficient (cm^2/s) and K is the oxygen surface exchange coefficient (cm/s). The polarisation losses can be decreased by optimising the microstructure or by choosing an appropriate composition with impressed oxygen exchange kinetics, and a compatible coefficient of thermal expansion (CTE). Perovskite materials (ABO_3) are the much investigated cathode materials because full or partial substitution of A/B cations with different valence cations is possible. When the overall valence of - and B-site cations $a + b < 6$, then the missing charge contributes to the oxygen vacancies. For most of the perovskites, A and B substituted atoms cause displacement and tilting of the octahedral leading to non-cubic symmetry of perovskites. The degree of distortion in ABO_3 is given by the Goldschmidt tolerance factor (t) as follows:

$$t = \frac{(r_A + r_B)}{\sqrt{2}(r_B + r_O)} \tag{3.13}$$

r_A, r_B, and r_O represent the effective ion radii of A, B, and oxygen ions, respectively. For the structure to be ideal cubic, the tolerance factor should be unity whereas bigger B/smaller A decrease the tolerance factor due to cation displacement and tilting of corner sharing BO_6 octahedra. For most of the perovskites, B-site cations provide redox catalytic because B-cations such as Mn, Fe, Co, and Ni are reducible transition metals and the octahedral symmetry around the transition metal ions promotes the electron conduction due to the formation of semiconducting band structure. Some cathode compositions with their conductivity and CTE are listed in Table 3.3. Investigations with a variety of cathode materials including $PrCoO_3$, Pr_2CoO_4, $La_{0.9}Sr_{0.1}CoO_3$, $LaMnO_3$, $La_{1-x}Sr_xMnO_3$, and $La_{1-x}Sr_xMnO_3$–YSZ (LSM-YSZ) have been done. Some of the important and commonly used cathode materials are explained in the following sections.

3.4.2.1 Lathanum Strontium Manganate (LSM) Cathodes

Lathanum strontium manganate (LSM)-based perovskites are the most investigated cathode materials. $LaMnO_3$ materials exhibit oxygen deficient or oxygen excess non-stoichiometry, which have a significant impact on the electronic and ionic transport properties. For $La_{1-x}A_xMnO_{3+\delta}$, material '−' sign indicates oxygen deficiency and the '+' sign represents oxygen excess with 'A' to be 'Sr^{2+}'/Ca^{2+}. Figure 3.13 shows the crystal structure of $La_{0.7}Sr_{0.3}MnO_3$ which indicates that the transition on metal ions occupies the octahedral B-site and large rare earth ions (such as lanthanum) occupy the 12 coordinated A sites.

Undoped $LaMnO_3$ is an orthorhombic rhombohedral transition at 600°C. The orthorhombic to rhombohedral transition temperature depends upon the content of Mn^{4+} ions formed due to oxidation of Mn^{3+}. The Mn^{4+} concentration depends on the dopant and doping concentration; that is, doping the lanthanum site with Ca2+/Sr2+ ion increases the Mn^{4+} concentration. The oxidation reaction can be given by

$$Mn_{Mn}^{*} + SrO \underset{LaMnO_3}{\longleftrightarrow} Sr_{La}' + Mn_{Mn}^{\bullet} + O_O^{*} \tag{3.14}$$

The $La_{1-x}Sr_xMnO_{3-\delta}$ can exhibit three lattice geometries depending upon the doping as follows.

- Rhombohedral for $0 \leq x \leq 0.5$
- Tetragonal for $x = 0.5$
- Cubic for $x = 0.7$

Table 3.3 CTE and conductivity of various cathode perovskite materials

Material	Conductivity (at 100°C)	CTE × 10^{-6}/K
$LaMnO_3$	83	12.5
$La_{0.84}Sr_{0.16}MnO_3$	83	11.6
$La_{0.7}Sr_{0.3}MnO_3$	178	11.7
$La_{0.6}Sr_{0.4}MnO_3$	320	12.7
$La_{0.8}Sr_{0.2}MnO_3$	190(900°C)	12.0
$La_{0.9}Sr_{0.2}MnO_3$	120	11.2
$NdMnO_3$	71	4.55
$Nd_{0.5}Sr_{0.5}MnO_3$	316	13.6
$Nd_{0.5}Sr_{0.3}MnO_3$	200	12
$CaMnO_3$		17.5
$Y_{0.5}Ca_{0.5}MnO_3$	161	11
$Y_{0.7}Ca_{0.3}MnO_3$	80	7
$Pr_{0.7}Ca_{0.3}MnO_3$	148	11.9
$Sr_{0.7}Ce_{0.3}MnO_3$	300	10.1
$Pr_{0.7}Sr_{0.3}MnO_3$	117	12.4
$Pr_{0.85}Sr_{0.15}MnO_3$	117	10.7
$Pr_{0.5}Sr_{0.5}MnO_3$	12.5	250
$Gd_{0.7}Ca_{0.3}Co_{0.4}Mn_{0.6}O_3$	37	10.5
$Gd_{0.8}Sr_{0.2}Co_{0.2}Mn_{0.8}O_3$	28	10.5
$Gd_{0.8}Sr_{0.2}MnO_3$	123	Negative
$Gd_{0.7}Sr_{0.3}MnO_3$	83	9.5
$Gd_{0.5}Sr_{0.5}MnO_3$	229.5	6.84
$Sm_{0.7}Sr_{0.3}MnO_3$	133	9.9
$La_{0.8}Sr_{0.2}Ca_{0.4}Mn_{0.6}O_3$	50(900°C)	13
$La_{0.75}Sr_{0.25}Ca_{0.5}Mn_{0.5}O_3$	25.3	12
$La_{0.82}Sr_{0.16}Mn_{0.7}Ni_{0.3}O_3$	68(900°C)	11.4
$La_{0.84}Sr_{0.16}Mn_{0.6}Fe_{0.4}O_3$	18	11.55
$La_{0.84}Sr_{0.16}Mn_{0.6}Co_{0.4}O_3$	45	13.65

The Sr^{2+} doping affects the lattice constants and cell volume more than the effect observed by the A/B ratio. Anyhow the A/B cation ratio affects the phase formation; that is, A/B > 1 leads to the formation of La_2O_3 and $La(OH)_3$ secondary phases whereas A/B < 1 leads to formation of Mn_3O_4 as minor phases. The upper limit factor δ can be explained by the charge disproportionation reaction of Mn^{3+} into Mn^{2+}/Mn^{4+}. The La/Mn ratio is kept constant on incorporation of excess oxygen; then the equation can be given as

$$LaMnO_{3+\delta} + \frac{\delta}{2}O_2 \rightarrow \frac{3+\delta}{3}\left(La^{3+}_{1-\frac{2\delta}{3+\delta}}V_{\frac{2\delta}{3+\delta}}\right) \times \left(Mn^{3+}_{\frac{3(1-2\delta)}{(3+\delta)}}Mn^{4+}_{\frac{3.2\delta}{(3+\delta)}}La^{3+}_{\frac{\delta}{3+\delta}}\right)O_3^{2-} \quad (3.15)$$

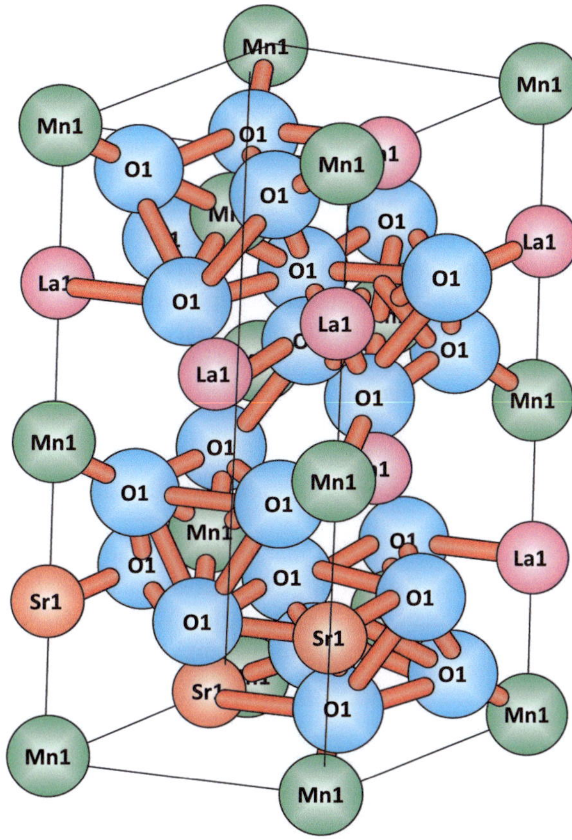

Fig. 3.13 The crystal structure for the lanthanum strontium manganite ($La_{0.7}Sr_{0.3}MnO_3$) (Photo courtesy of IHPC, Singapore)

Six Mn^{4+} ions and two vacancies (one each at Mn and La site) are formed upon dissolution of $\frac{3}{2}O_2$ molecules; i.e.,

$$\overset{Mn^{4+}}{6Mn^x_{Mn}} + \frac{3}{2}O_2 \longleftrightarrow \overset{Mn^{3+}}{6Mn^{\bullet}_{Mn}} + 3O^x_O + V'''_{La} + V'''_{Mn} \tag{3.16}$$

and

$$k_{ox}p(O_2)^{3/2} = \frac{[V'''_{La}]x + [V'''_{Mn}][Mn^{\bullet}_{Mn}]^6}{[Mn^x_{Mn}]^6} \tag{3.17}$$

In the oxygen deficient region, $V^{\bullet\bullet}_O$ acts as the predominant ionic defect and the B-site ion is reduced to a lower valence state in order to maintain charge neutrality. $LaMnO_3$ is an intrinsic *p*-type conductor and Sr^{2+} substitution is preferred on the

Lanthanum sites as the resultant perovskite structure exhibits high conductivity and stability in the oxidising conditions. To maintain the electroneutrality, a hole is created on the B-site upon substitution of Sr^{2+} ion onto the A-site:

$$LaMnO_3 \longrightarrow xSrOLa^{3+}_{1-x}Sr^{2+}_{x}Mn^{3+}_{1-x}Mn^{4+}_{x}O_3 \tag{3.18}$$

The electronic conductivity depends on the crystalline structure, preparation technique, and porosity of the material. For example, the LSM samples prepared by plasma spray coating exhibit conductivity between 50 and 200 S/cm as compared to 40–485 S/cm for the sintered samples at 1000°C. When the porosity increases from 10 to 41%, a decrease in conductivity from 148 to 45 S/cm is observed for LSM samples at 800°C. The CTE for $LaMnO_3$ (undoped) is slightly higher than that of the YSZ electrolyte.

3.4.2.2 Lanthanum Ferrite/Cobaltite Cathodes

Sr-doped $LaFeO_3$ (LSF) materials have shown promising results for SOFC applications due to their high power density and stability at 750°C. The LSF cathodes also exhibit matching CTE with the YSZ and GDC electrolyte. The Kroger–Vink notation for the LSF cathodes at high temperature is given by

$$\underset{(Fe^{4+})}{2Fe'_{Fe}} + O^{*}_{O} \longrightarrow \frac{1}{2}O_2 + \underset{(Fe^{3+})}{2Fe^{*}_{Fe}} + V^{\bullet\bullet}_{O} \tag{3.19}$$

LSF loses oxygen to form oxygen vacancies at the expense of decreasing hole concentration. When Sr^{2+} is added to the $LaFeO_3$ then charge imbalance occurs due to the substitution of strontium on the lanthanum site. The charge neutrality is maintained either by the oxygen vacancies or by formation of Fe^{4+} ions. Usually the highest conductivity is expected when the Fe^{3+} and Fe^{4+} ions are in a ratio of 1:1. The dominating conductivity mechanism in the ferrites is due to hole-conduction; that is, the mobility of holes or electrons is higher than those of oxygen ions especially at high oxygen partial pressures. Cu doped lanthanum strontium ferrites including $La_{0.7}Sr_{0.3}Cu_{0.2}Fe_{0.8}O_3$ exhibit two to three times higher current densities than the $La_{0.8}Sr_{0.2}FeO_3$ material. High power densities of about 0.95 W/cm^2 (750°C) were observed for an anode supported cell with YSZ electrolyte and $La_{0.8}Sr_{0.2}FeO_3$ cathode. $LaCoO_3$ cathodes possess high electronic density of states near Fermi level (E_F) and the catalytic properties of $LaCoO_3$ are attributed to the electron occupation of d state (near E_F). $La_{1-x}Sr_xCoO_{3-\delta}$ exhibits complex behaviour and a defect model is associated with it, which assumes the occupancy of La sites by Sr^{2+} ions leading to the formation of electron holes. For the electroneutrality condition, the following equation must hold good.

$$\overset{\curvearrowright Co^{4+}}{} \qquad\qquad \overset{\curvearrowright Co^{3+}}{}$$
$$Sr_{Sr}^{*} + Co_{Co}' + 2V_{O}^{\bullet\bullet} \longrightarrow Sr_{La}' + Co_{Co}^{*} \tag{3.20}$$

$$\overset{\curvearrowright Co^{4+}}{} \qquad\qquad \overset{\curvearrowright Co^{3+}}{}$$
$$2Co_{Co}' + O_{O}^{*} \longrightarrow 2Co_{Co}^{*} + V_{O}^{\bullet\bullet} + \frac{1}{2}O_2(g) \tag{3.21}$$

Equations (3.20) and (3.21) indicate the compensation of electron holes via the formation of Co_{Co}' and oxygen vacancies as positive charges. It should be noted that the amount of Co should be optimised as high cobalt content may result in increased CTE which leads to a cracked interface with the electrolyte. For decreasing the CTE, lanthanum can be substituted by Gd, Pr, Nd, or Sm. The Co–O bonds are weaker and Co^{2+} is associated with the spin-state transitions leading to high CTE of cobaltites. When A site is doped, the Pr^{3+}, then the highest conductivity and lowest overpotential are observed. Doping the B-site with copper leads to increased catalytic activity and ionic conductivity along with decreased electrical conductivity [particularly for (La, Sr) (Co, Cu)$O_{2-\delta}$ (LSCC)]; studies indicate that $Pr_{0.7}Sr_{0.3}Co_{1-y}Cu_yO_{3-\delta}$ for 0.05 < y < 0.3 yield low overpotentials as compared to $Pr_{0.7}Sr_{0.3}Co_{0.9}Cu_{0.1}O_{3-\delta}$ composition. Another potential material for the IT–SOFC application is strontium doped samarium cobaltite (SSC). The conductivity of SSC materials may reach 1000 S/cm (800–1000°C), hence exhibiting metallic behaviour. For the B-site manganese doped cobaltites such as $La_{0.6}Sr_{0.4}Co_{0.8}Mn_{0.2}O_{3-\delta}$, the conductivity values may reach up to 1400 S/cm (500°C). The major drawback of B-site manganese doped cobaltites is the high CTE of $\sim 18 \times 10^{-6}$/K.

3.4.2.3 Lanthanum Ferro-Cobaltite Cathodes

Other potential cathode candidates are $La_{1-x}Sr_xFe_{1-y}Co_yO_3$ (LSCF) materials due to their good electrical conductivity and high oxygen surface exchange coefficient. With the increase in doping content of strontium (Sr), the electrical conductivity decreases. Usually the B-site dopants contribute to the electronic conductivity and A-site dopants as strontium contributes to the ionic conductivity. The current density of the cells with $La_{1-x-y}Sr_xCo_{0.2}Fe_{0.8}O_{3-\delta}$ cathode is twice as high as the LSM–YSZ composite cathode. These are two main drawbacks for the LSCF electrodes; that is, LSCF possesses higher degradation rates due to the diffusion of strontium out of LSCF leading to depletion of strontium in the cathode. Secondly, LSCF exhibits poor compatibility with the electrolyte interface as detrimental products are formed at the interface. The cathode performance can be increased by doping the A-site by the Pr cation. $Pr_{0.6}Sr_{0.4}Co_{0.8}Fe_{0.2}O_{3-\delta}$ (PSCF) exhibits high conductivity of 1040 S/cm (300°C) whereas $Sr_{0.8}Ce_{0.1}Fe_{0.7}Co_{0.3}O_{3-\delta}$ (SCFC) cathode materials also provide enhanced oxygen ion diffusion as compared to LSCF cathodes at 700°C. Though LSF is thermodynamically more stable as compared to cobaltites, LSF possesses lower electrochemical performance as compared to LSCF. Usually, the cathode materials exhibit reactivity with the electrolytes in contact, but thin ceria-based barrier layers are used to prevent detrimental reactions between the cathode and YSZ surface.

$Ba_{0.5}Sr_{0.5}Co_{0.8}Fe_{0.2}O_{3-\delta}$ (BSCF) has been reported as a potential mixed conducting cathode material. Systems such as $Ln_{1-x}A_xCo_{1-y}B_yO_{3-\delta}$ [Ln–Pr, Nd, La, Sm; A = Sr, Ca, Ba; B = Cr, Fe, Ni, Cu, Co] have been investigated as cathode materials. For the $SrCo_{0.8}Fe_{0.2}O_{3-\delta}$ perovskite, the highest oxygen permeation flux was observed, which involved total substitution of lanthanum ion with strontium. Due to the presence of cobalt and iron in the composition of perovskite oxide, the Goldschmidt tolerance factor can be varied, attributed to the multiple oxidation state of iron and cobalt. The Sr^{2+} cation of $SrCo_{0.8}Fe_{0.2}O_{3-\delta}$ is too small to sustain pervoskite cubic symmetry, hence the larger cation Ba^{2+} is used to substitute Sr^{2+} ion partially, which increases the Goldschmidt tolerance of the perovskite. The most stable structure is $Ba_{0.5}Sr_{0.5}Co_{0.8}Fe_{0.2}O_{3-\delta}$ obtained by 50% doping of Sr^{2+} by Ba^{2+}. Additionally higher oxygen vacancy concentration, favourable phase stability, and higher oxygen permeation are also expected for $Ba_{0.5}Sr_{0.5}Co_{0.8}Fe_{0.2}O_{3-\delta}$. Results have shown that doping of Sr^{2+} by Ba^{2+} in $Ba_{0.5}Sr_{0.5}Co_{1-y}Fe_yO_{2.5}$ ($0 < x$; $y < 1$) resulted in increased ionic conductivity. BSCF is a mixed conducting oxide and can be regarded as a mixed ionic electronic conductor (MIEC), i.e., the material which exhibits both electronic and ionic conductivity. The total conductivity for a MIEC is given as follows.

$$\sigma = \sigma_{\text{ionic}} + \sigma_{\text{electronic}} = \sum_{\text{i}} \left|Z_{i,\text{ion}}\right| e\mu_{i,\text{ion}} C_{i,\text{ion}} + \sum_{j} \left|Z_{j,\text{elec}}\right| e\mu_{j,\text{elec}} C_{j,\text{elec}} \tag{3.22}$$

where μ_i is the mobility and C_i is the concentration. Li and co-workers obtained the increase in conductivity of $(Ba_{0.5}Sr_{0.5})_{1-x}Sm_xCo_{0.8}Fe_{0.2}O_{3-\delta}$ (BSSCF, x = 0.05–0.15); i.e., 21.2% increase in conductivity was observed at 500°C when A site of BSCF is doped with rare earth ions. Anyhow, doped BSCF material exhibits high CTE in the range of 19.1–20.3 × 10^{-6}/K, which is unacceptable for SOFC functioning. $La_{0.68}Sr_{0.3}Fe_{0.8}Co_{0.2}O_{3-\delta}$ and $Pr_{0.68}Sr_{0.3}Fe_{0.8}Co_{0.2}O_{3-\delta}$ have shown high current densities of 1.4 and 1.25 A/cm^2, respectively, at an operating temperature of 900°C. Anyhow $Ba_{0.68}Sr_{0.3}Fe_{0.8}Co_{0.2}O_{3-\delta}$ and $Sm_{0.68}Sr_{0.3}Fe_{0.8}O_{3-\delta}$ depict the lowest activation energies, which could be due to the catalytic activity of surface species involved in oxygen reduction and further incorporation into the perovskite structure. If titanium ion is incorporated for cobalt ion, then the CTE is reduced and excellent cyclic stability of cathodes is observed.

3.4.2.4 Lanthanum Nickelate and $A_2BO_{4+\delta}$ Cathodes

Nickelate perovskite, i.e., $LaNiO_{3-\delta}$ materials, are stable below 950°C. If $LaNiO_{3-\delta}$ materials are heated above this temperature, then the decomposition reaction may follow leading to the formation of $La_2NiO_{4+\delta}$ ($A_2BO_{4+\delta}$) type structure and NiO. The phase formation in lanthanum nickelates limits their application and hence they are doped with suitable ions in the nickel sublattice. The $La_2NiO_{4+\delta}$ compounds consist of stacked ABO_3 alternating layers with rock salt AO layers (along the c-axis) (Fig. 3.14).

The B contains mixed valence leading to high electronic conductivity of these compounds. $A_2BO_{4+\delta}$ cobaltites and nickelates follow a *p*-type conduction mechanism. At low oxygen partial pressures, the conductivities of these materials may decrease whereas at high oxygen partial pressures, the conductivity may reach 100 S/cm (800°C). Materials such as $La_2Sr_2NiO_{4+\delta}$ are stable materials and possess high oxygen diffusivity at lower temperatures than the LSCF cathodes. Fe-doped nickelates $LaNi_{0.6}Fe_{0.4}O_3$(LNF) possess desirable characteristic of high catalytic activity, matching CTE with YSZ and high electrical conductivity. Anyhow, LNF exhibit high reactivity towards the YSZ electrolyte and lead to formation of the $La_2Zr_2O_7$ phase. The LNF cathode also reacts with the samarium doped ceria

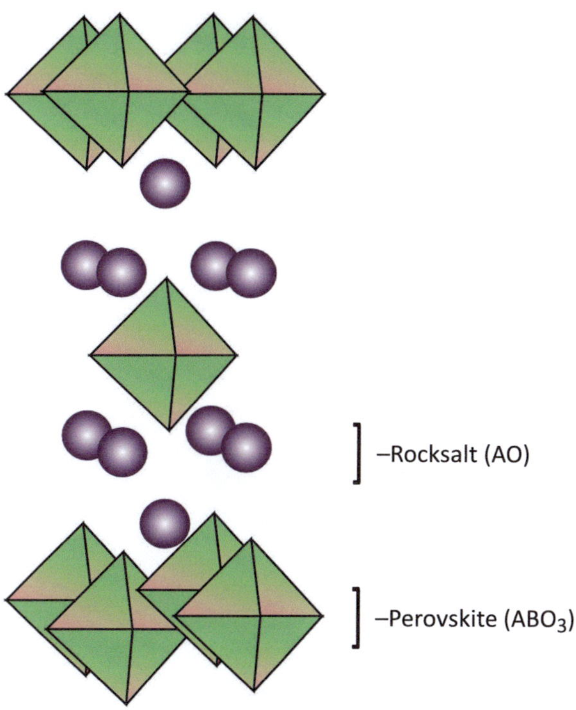

Fig. 3.14 Alternate AO and ABO_3 structure for the $A_2BO_{4+\delta}$ structure

(SDC) electrolyte at sintering temperatures of 1100°C. Strontium doped LNF (LSNF) exhibits high electronic conductivity when the concentration of Ni ≈ Sr ≈ 0.6 and temperature remains in the range of 600–800°C.

3.4.2.5 LSM Composite Cathodes

To improve the electrocatalytic activities of the LSM-based cathodes, ion conducting secondary phases such as GDC and YSZ are introduced to enhance the ionic conductivity. The electrode's polarisation resistance R_E is reduced considerably when YSZ is added to LSM because YSZ reduces grain growth of LSM and enhances the triple phase boundary regions. Figure 3.15 shows the AFM images of the YSZ surface obtained after removal of LSM using HCl treatment.

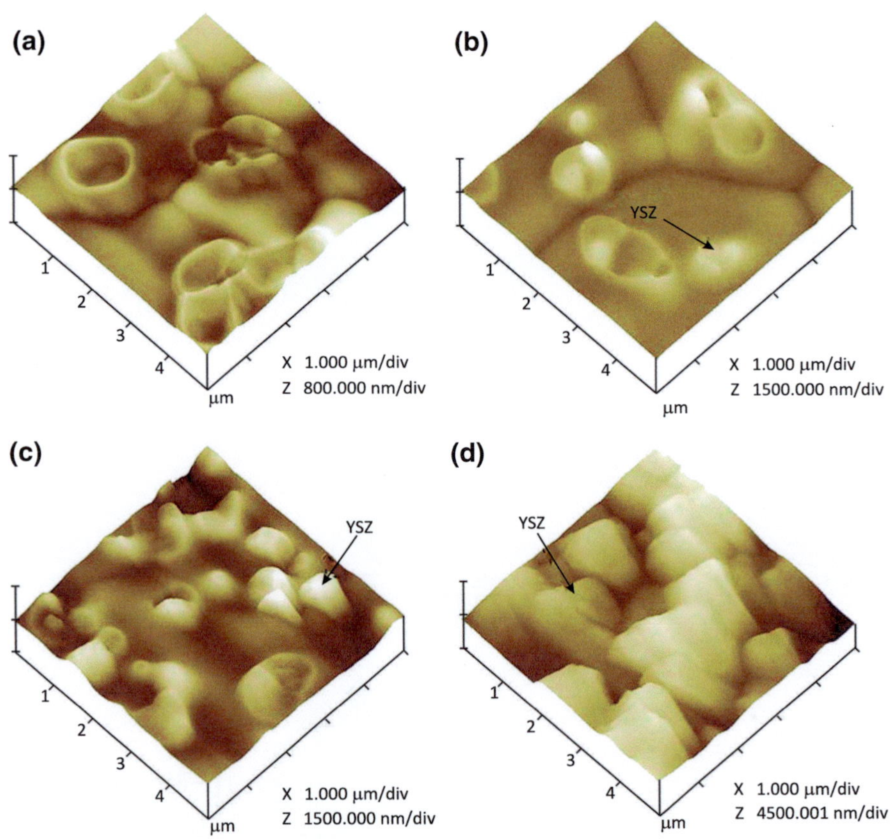

Fig. 3.15 AFM images for the composite cathodes of the YSZ electrolyte in contact with **a** LSM, **b** 90 wt% LSM/10 wt% YSZ, **c** 70 wt% LSM/30 wt% YSZ, and **d** 50 wt% LSM/50 wt% YSZ (Zhen and Jiang 2006)

When the YSZ content increases to 50% then the interconnected geometry of YSZ is observed on the YSZ electrolyte surface indicating the formation of a 3D network of YSZ phase for 50 wt% LSM. Sometimes, a $La_2Zr_2O_7$ resistive phase is formed at TPB, but the addition of YSZ phase to LSM reduces the formation of this resistive phase also. The sintering temperature and initial constituents affect the performance of LSM/YSZ cathodes. For instance, the composite cathode prepared from coarse (~210 nm) and (~8 nm) LSM powders affect grain growth. Extensive grain growth leading to increased polarisation and decreased TPB was observed in composite cathodes made up of fine LSM powders. The cathode sintered at 1300 and 1150°C posses R_E values of 0.1 and 0.03 Ω/cm^2 at 1000°C, respectively. The higher R_E value obtained at 1300°C is due to the larger grain size and denser structure of LSM/YSZ cathode. Yi and co-workers obtained (LSM/LSGM) composite cathodes with lower overpotential and smaller particle size than that of the LSM cathode. It should be noted that to avoid any interface reaction between the YSZ and LSM/LSGM composite cathode, the sintering temperature of the LSM/LSGM interlayer on YSZ should be kept below 1200°C. It has been observed that the LSM cathode exhibits a

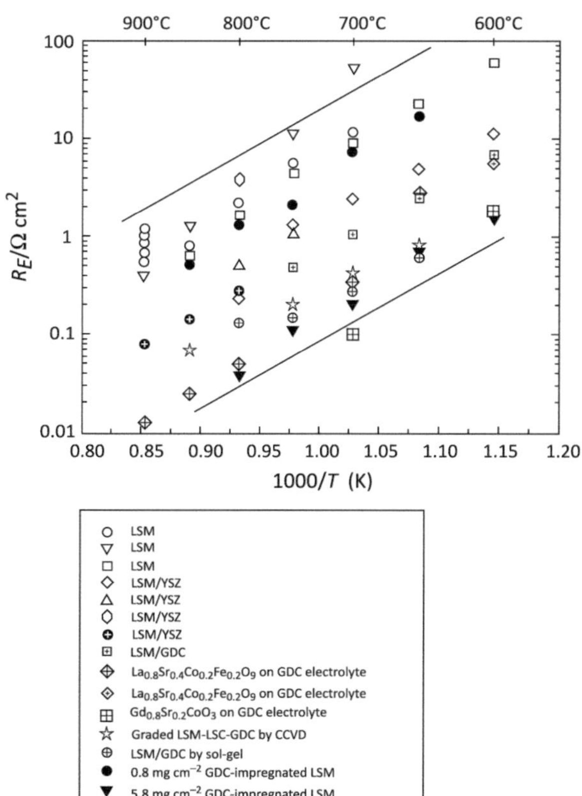

Fig. 3.16 A comparison of the R_E for various electrodes obtained by different researchers (Jiang 2008)

more porous structure, low electrode polarisation resistance, and low overpotential for the O_2 reduction as compared to LSM/ LSGM electrodes. Murray and co-workers obtained reduction in R_E values due to a decrease in electrical conductivity. The impregnated cathodes also display high performance; i.e., introducing catalytic active CeO_2 nanoparticles in LSM increases the electrochemical performance of the cathode. The R_E for pure LSM at 700°C is 11.7 Ω/cm^2, which is 56 times higher than the R_E values 0.21 Ω/cm^2 of GDC-impregnated LSM at the same temperature. Figure 3.16 gives a comparison of the R_E values obtained for different electrodes.

3.4.3 Anodes

At anodes, the electrochemical oxidation of fuels such as hydrogen (H_2) and natural gas occurs. Anodes must fulfill the following requirements.

- Should exhibit catalytic properties to enhance the fuel oxidation kinetics.
- Should possess high porosity (20–40%) for the mass transport of reactant and products gases.
- Should pass high electronic conductivity.
- Exhibit compatibility with the adjoining compounds.
- Mixed conductivity is also preferred because it extends the TPB between anode, reactant, and electrolyte.

The Kroger–Vink notation of H_2 oxidation can be given as

$$H_2 + O^{2-}_{\text{electrolyte}} \rightarrow H_2O + V_O^{\bullet\bullet} + 2e^- \tag{3.23}$$

Figure 3.17 illustrates the TPB regions for electronic and mixed conductor anode materials.

Basically, Ni–YSZ cermet materials are the most commonly used anode materials. The following sections describe the main types of anode materials, their advantages, and disadvantages.

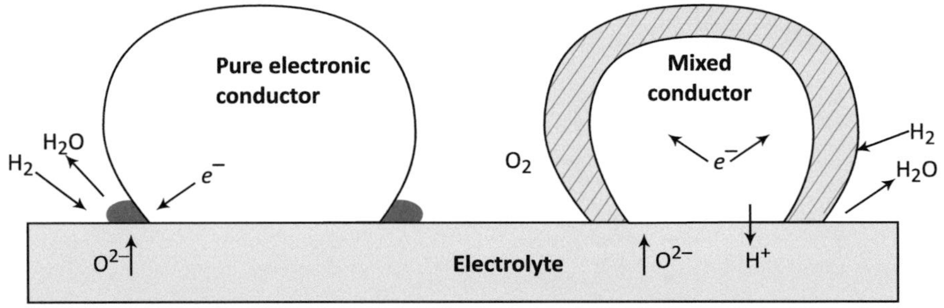

Fig. 3.17 TPB regions for electronic and mixed conductor anode materials

3.4.3.1 Ni–YSZ Cermet

Pure Ni exhibits the highest electrochemical activity for H_2 oxidation along with high CTE of $13.3 \times 10^{-6}/K^{-1}$. In addition to this, the melting point of nickel is 1453°C, hence it has a tendency to sinter at low temperature. Ni and YSZ form an immiscible solid solution with each other and YSZ also reduces the CTE of composite significantly. Moreover, YSZ creates additional reaction sites by extending the 2D reaction zone into the 3D reaction zone, hence enhancing the electrocatalyte followed by YSZ particle growth around the Ni particle using electrochemical vapour deposition techniques. To minimise the concentration and activation polarisation, a two layered anode structure is prepared using a 10–50 μm layer of NiO–YSZ by spray coating and then fabricating an outer layer by tape-casting the slurry of the NiO–YSZ mixture with another composition. The outer support layer possesses a coarse microstructure whereas the inner layer is fine. The fine interlayer of NiO–YSZ maximises the TPB area hence increasing the electrochemical reaction sites whereas the outer coarse layer facilitates quick transport of fuel gases and removal gases out of the anode, which help in reducing the concentration losses.

Though Ni is an excellent electrocatalyst for the electrochemical oxidation of hydrogen, if the fuel cell is using natural gas as the fuel, then no external fuel cell reformer is required because the reforming reactions can take place inside the Ni/YSZ cermet anode as follows.

(Steam reforming)

$$CH_4 + H_2O \rightarrow CO + 3H_2 \tag{3.24}$$

(Catalytic oxidation)

$$2CH_4 + O_2 \rightarrow 2CO + 4H_2 \tag{3.25}$$

The gas produced can be electrochemically oxidised as follows

$$H_2 + O^{2-} \rightarrow H_2O + 2e^- \tag{3.26}$$

$$CO + O^{2-} \rightarrow CO_2 + 2e^- \tag{3.27}$$

If H_2O content is small, then reaction (3.24) will happen as follows

$$CH_4 \rightarrow C + 2H_2 \tag{3.28}$$

This carbon will be deposited on the anode surface due to methane cracking. This carbon will cover the active sites on the anode surface hence leading to performance degradation of the cell. High steam/carbon ratios are used to suppress the carbon formation; anyhow the steam addition leads to a reduction in anode electrical potential due to the presence of oxygenation in the fuel. The carbon deposition can be controlled by carefully choosing the fuel composition and operating temperature.

Moreover, the addition of steam may also lead to Ni coarsening on agglomeration, which leads to reduced electrical conductivity and triple phase boundary. To avoid Ni agglomeration, some low surface energy oxides such as TiO_2, Mn_3O_4, Cr_2O_4, and MgO, etc. are added to the anode-cermet. These oxides enhance the wettability of Ni particles by anchoring them with the interface. Additionally, these oxides increase the mechanical strength of anodes and retard Ni coarsening during high temperature operation.

Usually under a high carbon activity environment, the bulk Ni metal disintegrates into metal particles (300–850°C) in the supersaturated carbon. This phenomenon is also known as metal dusting. Usually, the graphite form of carbon is observed at the anode surface. Ni is also subjected to large volume change in oxidation and reduction cycles leading to instability of the Ni/YSZ microstructure. Moreover, at low temperatures between 500 and 600°C, Ni/YSZ does not exhibit acceptable performance due to low conductivity of the YSZ electrolyte. Ni/YSZ anodes have low sulphur tolerance because H_2S strongly absorbs on active sites of nickel, which leads to the degradation of electrochemical reactions occurring at TPB sites. It should also be noted that the oxygen potential should be between −0.8 and −1.0 V for the operation of the SOFC anode. Above −0.68 V, $Ni(OH)_2$ becomes the predominant volatile compound, which increases with increasing operating temperature or oxygen potential. It should be noted that the NiO particles are reduced to Ni and react readily with Cr_2O_3 (of inter connect) to yield Ni–Cr spinals. Large Cr_2O_3 crystals evaporate in the presence of steam and hence lead to stack degradation. The formation of Ni–Cr spinals is not prominent with the ceramic $LaCrO_3$ interconnects. For the Ni/YSZ cermet, Ni is an electronic conductor and YSZ is an ionic conductor. The electrical conductivity of Ni/YSZ is dependent on the particles size and distribution of YSZ Ni phases in the composite. Other criteria for the working of Ni/YSZ cermet is the porosity for the transportation of fuel gas reactants towards the TPB interface region. When NiO is reduced to Ni, a volume reduction of 25% occurs and hence affects the porosity. Figure 3.18 shows the change in microstructure of the Ni anode after and before reduction in a humidified H_2 atmosphere, which indicates an increase in porosity after reduction. Though porosity affects Ni–Ni connectivity, the conductivity is expected to decrease with the increase in porosity of Ni/YSZ cermet.

Many alternative anode cermets have been investigated including Ni cermets prepared from Ca–CeO_2, Y–CeO_2, and CeO_2–ZrO_2. The order of activity for the Ni/oxide cermet is found to be Ni/PrO_x > Ni/SDC > Ni/CeO_2 > Ni/YSZ by Eguchi and co-workers. For Ni/SDC, the best performance was obtained for 50% Ni content with lower overpotential for H_2 oxidation at 850°C as compared to the Ni/YSZ anode. For Ni/GDC cermets, it was observed that YSZ and GDC diffuse into each other during the process of sintering at 1200°C leading to the formation of the less-conducting Gd rich phase. The ideal composition to suppress the reactivity between GDC and YSZ 40% Ni/ 60%GDC as the presence of Ni inhibits the interfacial reactions between GDC and YSZ.

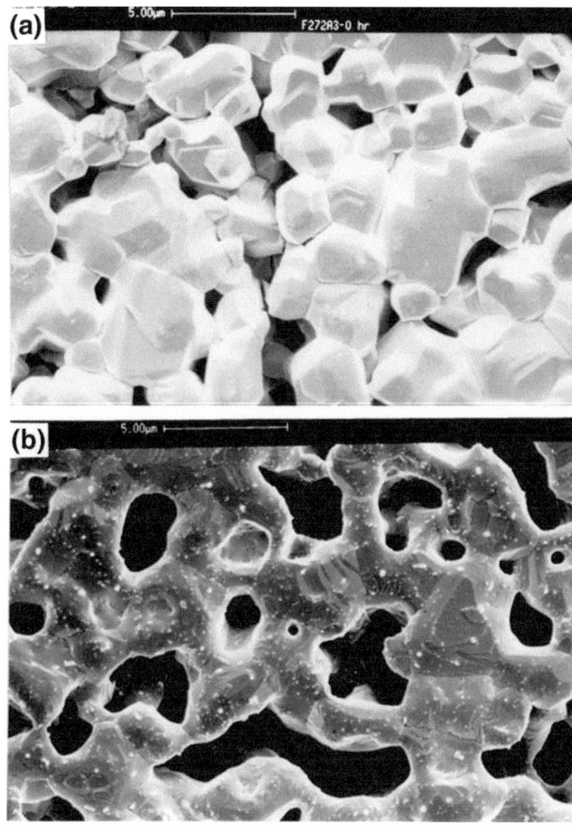

Fig. 3.18 Change in the microstructure of Ni anode **a** before and **b** after exposure in humidified 10%H2/90%N2 for 1 h at 1000°C (Jiang and Chan 2004)

3.4.3.2 Ceria-Based Anodes

The doped ceria anode was developed because ceria has been an excellent electrocatalyst for methane oxidations. The methane oxidation at TPB can be expressed as:

$$CH_4 + O^{2-} \rightarrow CO_2 + 2H_2O + 8e^- \tag{3.29}$$

Doped and undoped ceria exhibit ionic and electronic conduction at low oxygen partial pressures. Ce^{4+} gets reduced to Ce^{3+}, which is the main cause for mechanical failure due to lattice expansion (for low oxygen partial pressure). The electrolyte/electrode interface may be delaminated due to the formation of cracks. When 40–50% of Ce^{4+} ions are substituted with Gd^{3+} ions ($Ce_{0.6}Gd_{0.4}O_{1.8}$–CGO), then a balance between lattice parameters and conductivity with the YSZ electrolyte and an anchoring YSZ layer provides the ability to resist the thermal expansion mismatch. Uchida and co-workers obtained the electrocatalytic activities for the SDC and Yttria doped CeO_2 (YDC) for the electrodes sintered at 1150 and 1250°C. The current density of 0.2–0.25 A/cm^2 is observed for SDC and YDC anodes,

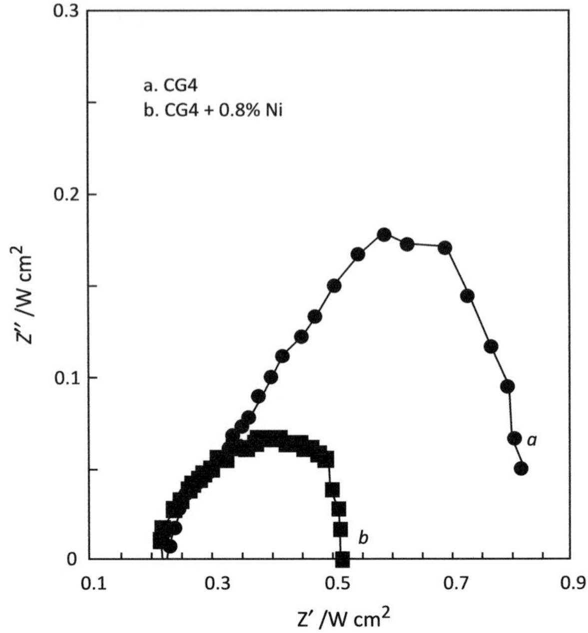

Fig. 3.19 The ac impedance for Ni containing CG4 electrode showing suppression of semicircles at low frequencies (Primdahl and Mogensen 2002)

respectively, which increased to 0.4–0.5 A/cm^2 with the addition of ruthenium particles. YDC show greater electronic conductivity than SDC and comparable ionic conductivity with YSZ. The presence of Ni in the CG4 composition leads to suppression of low frequency impedance as shown in Fig. 3.19. It should be noted that highly dispersed Ni in SDC promoted the electrocatalytic activity of H_2 oxidation more than of Ni/SDC cermet anodes.

3.4.3.3 Titanium-Based Anodes

Titanates show *n*-type conductivity at low p_{O_2} of 10^{-15} to 10^{-2} Pa and *p*-type conductivity at high p_{O_2} of 10^0–10^5 Pa. Holt and co-workers obtained the solubility unit of Ti for $NdGaO_3$ to be 20%, with electrical conductivity of 1 S/cm at 1000 °C (low pressure range). Titania doped YSZ exhibit high thermal stability, electronic conductivity, and matching thermal expansion with YSZ electrolyte. For 5 mol% titania-doped YSZ, the conductivity was reported to be 0.02 S/cm (800°C), which decreased with increasing titania content. Secondary phase $ZrTiO_4$ precipitated out for titania content >15%. Another 30% terbium doped system, i.e., yttria stabilised zirconia terbia (YSZT), showed mixed electronic and ionic conductivity of 2.3×10^{-3} S/cm and 2.8×10^{-3} S/cm, respectively. Tao and co-workers obtained $Sc_{0.15}Y_{0.05}Ti_{0.18}O_{19}$ to be the most promising anode system with electronic and ionic conductivity of 0.14 S/cm, respectively, at 900°C. Middeleton and co-workers studied $MgTi_{1.95}Nb_{0.05}O_5$ pseudobrookite, $Sm_2Ti_{1.9}Nb_{0.1}O_7$ pyrochlore, and

$Ti_{0.97}Nb_{0.03}O_2$ rutile anode materials. $Mg_{0.3}Nb_{0.1}Ti_{2.6}O_5$ yielded the highest activity for the *n*-type conducing titanates and $Ti_{0.97}Nb_{0.03}O_2$ high C_2 formation among all the oxides. Layered perovskite such as $La_2Sr_4Ti_6O_{19-\delta}$ exhibited high conductivity of 40 S/cm (900°C) in 5% H_2/Ar atmosphere but yielded high electrode polarisation resistance values of 3 and 9 Ωcm^2 in wet H_2 and CH_4 atmosphere, respectively. Yttrium doped $SrTiO_3$ ($Sr_{0.88}Y_{0.08}TiO_{3-\delta}$) exhibited high conductivity of 64 S/cm (800°C) whereas $Sr_{0.85}Y_{0.10}Ti_{0.95}Co_{0.05}O_{3-\delta}$ yielded conductivity values of 45 and 30 S/cm (30% porosity) at 800°C. Apart from these anode materials additional research is being carried out to develop an anode material which yields high power densities, low polarisation losses, and mechanical and chemical integrity.

3.4.4 Interconnects

The interconnect provides an electrical connection between two neighbouring cells and also provides gas separation within the cell. Along with maintaining the structural integrity of the SOFC cell, it separates the anode cavity from the cathode cavity. Following are the stringent requirements for the interconnect materials.

- Chemical compatibility with the fuel cell components in reducing and oxidising atmospheres is the most important requirement for the interconnect material. The interconnect should exhibit structural integrity and phase stability in both reducing and oxidising the atmosphere at the operating temperatures. If any structural or dimensional changes occur in the interconnect in the presence of oxidising and reducing atmospheres then cracking or pore formation may take place due to generation of stresses degrading the overall stack performance. The interconnect should not change with time of operation or with the exposure to different atmospheres to maintain uniform electrical conductivity.
- The interconnect must possess excellent imperviousness to oxygen and hydrogen to avoid direct combination of oxidant and fuel during SOFC operation.
- High thermal conductivity (5 W/(mK)) and mechanical strength.
- Low ionic conductivity and high electronic conductivity. Hence, the electronic transference number should be high along with high electrical conductivity.
- Ease of fabrication to gas-tight density.
- Economical and processable.
- Matching CTE with the adjoining cell components. Interconnect should exhibit maximum compatibility with the anode and cathode under operation conditions.
- Interconnect should possess excellent oxidation, sulphidation, and carburisation resistances.
- For planar SOFC, the interconnect serves as the structural support system, hence it should have adequate creep resistance and mechanical strength.

Usually two approaches are used while fabricating the interconnect materials, i.e., ceramic and metallic. Ceramic oxides possess stability in an oxidising atmosphere but possess lower conductivities as compared to the metals. Usually ceramic oxides are used at higher temperatures and metals are used at lower temperatures.

3.4.4.1 Ceramic Interconnects

Lanthanum chromites are the most commonly used ceramic interconnect material because they are stable in low oxygen partial pressures. Lanthanum chromite materials are *p*-type electronic conductors in oxidising atmospheres and this conductivity decreases with a decrease in oxygen partial pressure. With the decreasing oxygen partial pressure, lanthanum chromite becomes oxygen deficient, i.e., $LaCrO_{3-\delta}$; the conductivity of lanthanum chromite can be increased by doping with a suitable ion. For an interconnect material to exhibit high performance, its conductivity should be >1 S/cm (1000°C). Table 3.4 shows the data for conductivity of doped lanthanum and yttrium chromite material.

At the cathode, the oxygen partial pressure is relatively high (10^{-7} to 10^{-4} atm), hence Cr^{3+} to Cr^{4+} transition via the electronic compensation mechanism promotes conductivity. In contrast to this, at the anode, the oxygen partial pressure is low (10^{-8} to 10^{-18} atm), and hence the conductivity decreases due to the occurrence of oxygen vacancies via the ionic compensation mechanism. Above 800°C, the overall conductivity of the doped $LaCrO_3$ is acceptable whereas below 800°C, the electrical conductivity of the doped $LaCrO_3$ becomes very low, which restricts its use for the intermediate temperature SOFCs. Cobalt doping increases the electrical conductivity and CTE significantly whereas iron doping slightly improves the electrical conductivity and lowers CTE. Nickel doping results in a lower solid solution limit than cobalt and iron and higher electronic conductivity due to transition to a $LaNiO_3$ metallic conductor. Vanadium decreases both electrical conductivity and CTE whereas it increases donor doping at the chromium site to reduce overall acceptor concentration. Doping at both lanthanum and chromium sites with a double acceptor causes a significant increase in the electrical conductivity, but the

Table 3.4 Electrical conductivity data for doped $LaCrO_3$ and $YCrO_3$ interconnects (in air)

Composition	Electrical activation (1000°C S/cm)	Activation energy (eV)
$LaMg_{0.1}Cr_{0.9}O_{3-\delta}$	3	0.20
$La_{0.9}Sr_{0.1}CrO_{3-\delta}$	14	0.12
$La_{0.8}Ca_{0.2}CrO_{3-\delta}$	35	0.14
$LaCrO_{3-\delta}$	1	0.19
$Y_{0.95}Ca_{0.05}CrO_{3-\delta}$	4.5	0.17
$Y_{0.9}Ca_{0.1}CrO_{3-\delta}$	7.7	0.18
$Y_{0.85}Ca_{0.15}CrO_{3-\delta}$	13.0	0.18
$Y_{0.80}Ca_{0.20}CrO_{3-\delta}$	15.5	0.18

increase in CTE is unacceptably large. $LaCrO_3$ or doped $LaCrO_3$ display poor sintering behaviour in air due to the easy volatilisation of Cr (VI) species leading to the development of a thin layer of Cr_2O_3 at the interparticle neck during initial stages of firing. Usually $LaCrO_{3-\delta}$ materials are doped with calcium and strontium dopant ions because they act as electron acceptors on the chromium or lanthanum sites, hence increasing the *p*-type conduction by a polaron mechanism. Calcium-doped lanthanum chromites possess higher conductivity as compared to strontium-doped lanthanum chromites because the radius of Ca^{2+} (0.134 nm) is closer to the radius of La^{3+} (0.136 nm) whereas the radius of Sr^{2+} is 0.144 nm. Similar radii lead to small lattice distortion and hence higher conductivity as well as higher mobility. Calcium-doped lanthanum chromites offer more advantages than the strontium-doped chromites; because e.g., the solid solubility limit of calcium is higher than strontium for the lanthanum chromites due to which calcium-doped lanthanum chromites are single phase materials. Anyhow for low oxygen partial pressures, strontium possesses higher solubility than the chromium leading to higher stability of strontium lanthanum chromite. Other dopants such as magnesium, copper, cobalt manganese, and nickel have also been used for the lanthanum chromites. Nickel, cobalt, and copper do not enhance the conductivity much, anyhow at lower temperatures Ni addition results in enhanced conductivity. Magnesium being smaller in size occupies the chromium site. Manganese addition shows dual behaviour; i.e., when added in high concentration, manganese enhances the conductivity, but when added along with strontium, the conductivity decreases due to inhibition of Cr^{3+} to Cr^{4+}. Titanium additions also enhance the conductivity because Ti^{4+} compensates for the calcium defect. At low oxygen partial pressures, the conductivity of $La_{0.7}Ca_{0.3}Cr_{0.2}Ti_{0.8}O_{3-\delta}$ is higher than that of $La_{0.7}Ca_{0.3}Cr_{0.8}Ti_{0.2}O_{3-\delta}$. Figure 3.20 shows the conductivity of lanthanum chromite doped with different cations. Though magnesium is less effective in increasing

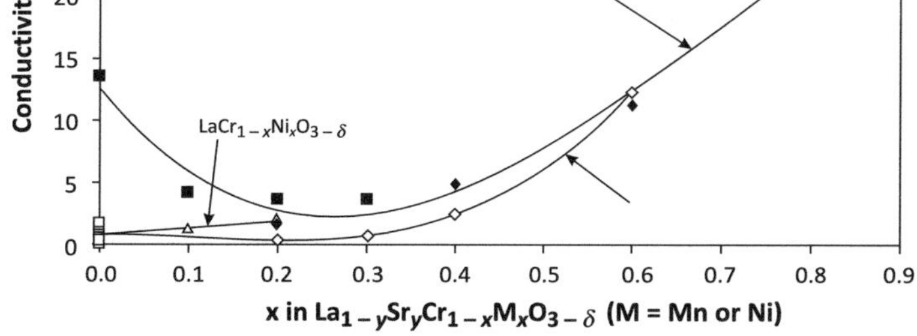

Fig. 3.20 Conductivity of Ni-and Mn-doped lanthanum chromate (Fergus 2004)

the conductivity, manganese-doped chromites exhibit high strengths. Even strontium-doped lanthanum chromites exhibit high strength as compared to the calcium-doped chromite due to the maintenance of a rhombohedral phase at high temperature for Sr-doped chromites.

3.4.4.2 Metallic Interconnects

Metallic interconnects have low fabrication costs, better thermal and electrical conductivity, and no failure due to deformation. Metallic interconnects can be classified as discussed in this section.

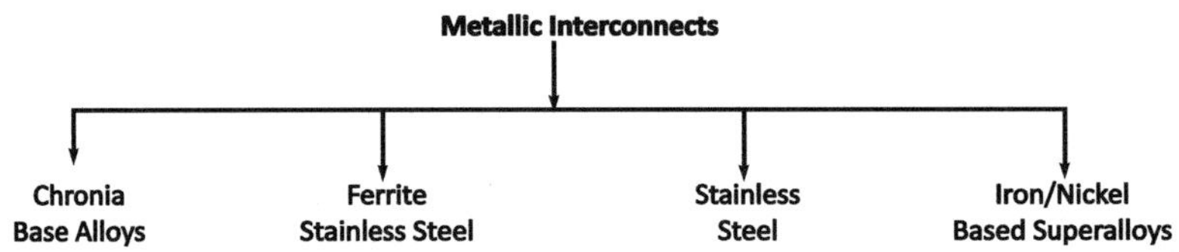

(a) **Chromia-Based Alloys**

Earlier chromia-based alloys include $Cr5Fe1Y_2O_3$ (Ducrolloy) developed by Metallwerke Plansee AG which contained 5 wt% iron and 1 wt% yttria. In Ducrolloy, the chromia scale formation is very thin even after long-term exposure in air. Chromia-based alloys are not considered as superalloys because they crystallise in body centred structures. Anyhow, Cr_2O_3 species provide moderate oxidation and corrosion resistance in addition to the large electronic conductivity. In the temperature range of 25–1000°C, these alloys exhibit matching CTE with other ceramic components. The major drawback of chromia-based alloys is the poisoning effect of Cr_2O_3 at the electrode/electrolyte interface along with evaporation of Cr_2O_3 species. These high valence Cr species such as CrO_3(g) and $Cr(OH)_2O_2$(g) develop simultaneously with the formation of the Cr_2O_3 scale at the cathode side of the interconnect. Cr_2O_3 species usually oxidise at the higher oxygen partial pressure of the cathode prior to the electrochemical reduction of oxygen. Hence, these species diffuse into and interact with electrodes, ultimately leading to change in the cathode composition and formation of new phases. Different coatings are being developed to avoid degradation of cell performance and to improve the contact between electrodes and interconnect material.

(b) **Ferritic Stainless Steel**

Crofer 22APU, AISI430, and AL453 are among the most common ferritic stainless steels which possess high advantages as compared to the $Cr5Fe1Y_2O_3$ such as easy processing and matching CTE with increase of chromium content up to 20 wt%. The oxidation of steels depends on aluminium and silicon content along with the chromium content because instead of chromia layers, alumina and silica layers are formed leading to a reduction in oxidation rate. But aluminium and silicon layers are insulating in nature and hence affect performance when the interconnect is in contact with electrodes. To enhance the performance of ferrite stainless steel by reducing the release of volatile Cr species, various alloying elements have been added in 0.1–2.5 wt% to alloys. Addition of titanium causes a higher oxidation rate by enhancing chromia scale growth and formation of internal titanium oxides. Lanthanum promotes thin oxide scale growth. In contrast to this, Ni does not form protective scale formation. To increase the power density of interconnects, they are fabricated as thin foils; the thin foils are highly affected by the selective oxidation and depletion of alloyed elements leading to the change in composition of the foil. The interconnect composition is not affected if the metallic interconnect is thick enough to serve as a reservoir of chromium.

(c) **Stainless Steel**

The stainless steel is composed of chromium and Fe–C base alloys which are used due to their pitting resistance. To maintain a body centred cubic ferritic structure, a minimum of 13% chromium is required. Stainless steels can be categorised into duplex, austenitic, ferritic, and martensitic. Carbon and nitrogen are austenitic formers and characterised by *a* γ-austenitic face-centred cubic. For planar SOFC, the stainless steels should contain Cr content ~17% considering the fact that chromium may be depleted by interaction with adjoining components.

(d) **Fe/Ni-Based Superalloys**

For Fe/Ni-based superalloys, a minimum of 20–25% chromium is required to ensure the formation of a protective Cr_2O_3 layer. Some cobalt-based alloys require 25–30% Cr_2O_3 and do not contain any aluminium. These alloys are resistant to high temperature and possess slower oxidation kinetics. Anyhow, their disadvantage is larger thermal CTE as compared to the iron-based alloys.

Bibliography

Baur E, Preis H (1937) ZElektrochem 43:727

Brown JT (1989) Solid oxide fuel cells. In: Takahashi T (ed) High conductivity solid ionic conductors. World Scientific, Singapore

Fergus JW (2004) Solid State Ionics 171:1

Jiang SP, Chan SH (2004) J Mater Sci 39:4405

Jiang SP (2008) J Mater Sci 43:6799

Kiukkola K, Wagner C (1957) Electrockem Soc 104:379

Primdahl S, Mogensen M (2002) Solid State Ionics 152–153:597

Taninouchi Y (2010) High Oxide-Ion conductivity and phase transition of doped bismuth vanadate. Kyoto University

Nernst W (1899) US Patent 685 730,1899

Zhen YD, Jiang SP (2006) J Electrochem Soc 153:A2245

Suggested Bibliography

Abrahmas I, Krok F (2002) J Mater Chem 12:3351
Abrahmas I, Krok F (2003) Solid State Ionics 157:139
Bastidas DM (2006) Rev Metal 42(6):425
Baur E, Preis H (1938) 2. Elektrochem 44:695
Berger D, Jitaru I, Stanica N, Perego R, Schoonman J (2001) J Mater Synth Process 9(3):137
Binder H, Kohling A, Krupp H, Richter IC, Sandstede G (1963) Electrochim Acta 8:781
Chen XJ, Chan SH, Khor KA (2004) Electrochem Solid-State Lett 7:A144
Cook RL, MacDuff RC, Sammells AF (1990) J Electrochem Soc 137:3309
Demin A, Tsiakaras P (2001) Int J Hydrogen Energy 26:1103
Dygas JR, Krok F, Bogusz W, Kurek P, Reiselhuber K, Brietter MW (1994) Solid State Ionics 70–71:239
Eguchi K, Setoguchi T, Okamoto K, Arai H (1992) In: Proceedings of the international fuel cell conference, Makihari, Japan, p 373
Faro ML, La Rosa D, Antonucci V, Arico A (2009) J Indian Inst Sci 89, 4 Oct–Dec 2009
Fisher CAJ, Islam MS (1999) Solid State Ionics 118:355
Goodenough JB, Manthiram A, Paranthaman M, Zhen YS (1992) Solid State Ionics 52:105
Haile SM (2003) Acta Mater 51:5981
Hammouche A, Siebert E, Hammou A (1989) Mater Res Bull 24:367
Herle JV, Seneviratne D, McEvoy AJ (1999) J Eur Ceram Soc 19:837
Hirschenhofer JH, Stauffer DB, Engleman RR, Klett MG (1998) Fuel cell handbook, 4th edn
Holt A, Ahlgren E, Poulsen FW (1993) In: Singhal SC, Iwahara H (eds) SOFC-III, vol. 93–104. Electrochemical Society, Pennington, p 562
Isenberg A, Pabst W, Sandstede G (1966) Oxydisches Kathodenmaterial fur galvanische Brennstoffzellen fur hohe Temperaturen. DE-P 1 5 71 991. Filed 22 Oct 1966
Ishihara T, Matsuda H, Takita Y (1995) Solid State Ionics 79:147
Iwahara H (1992) Bull Cerum Soc Jpn 27:112
Iwahara H, Uchida H, Ono K, Ogaki K (1988) J Electrochem Soc 135:529
Jiang SP, Wang W (2005) J Electrochem Soc 152:A1398
Kamata H, Yonemura Y, Mizusaki J, Tagawa H, Naraya K, Sasamoto T (1995) J Phys Chem Solids 56:943
Karim DP, Aldred AT (1979) Phys Rev B 20(6):2255
Kingery WD, Pappis J, Doty ME, Hill DC (1959) J Am Ceram Soc 42:393
Koc R, Anderson HU (1992) J Mater Sci 27:5477
Koc R, Anderson HU, Howard SA (1989) Structural, sintering and electrical properties of the perovskite-type (La,Sr)(Cr,Mn)O_3. In: Singhal SC (ed) Solid oxide fuel cells—The electrochemical proceeding, vol PV 89-11. The Electrochemical Society, Pennington, pp 220–241
Kuo JH, Anderson HU, Sparlin DM (1990) J Solid State Chem 87:55
Larminie J, Dicks A (2003) Fuel cells system explained, second edition. Wiley, England
Lee HY, Cho WS, Oh SM, Wiemhofer HD, Gopel W (1995) J Electrochem Soc 142:2659
Li S, Lu¨Z, Wei B, Huang X, Miao J, Liu Z, Su W (2008) J Alloys Compd 448:116
Lubke S, Wiemhofer H-D (1999) Solid State Ionics 117:229
Majewski P, Rozumek M, Aldinger F (2001) 1. Alloys Compds 329:253
Middleton PH, Steiner HJ, Christie GM, Baker R, Metcalfe IS, Steele BCH (1993) In: Singhal SC, Iwahara H (eds) SOFC-III, vol 93–104. Electrochemical Society, Pennington, p 542
Minh NQ, Horne CR (1993) High temperature electrochemical behaviour of fast iron and mixed conductors. In: Poulsen FW, Benzen JJ, Jacobsen T, Sltou E, Ostergard MJL (eds) Proceedings of the 24th R i s international symposium on materials science. Riser National Laboratory, Roskilde, Denmark, p 337
Mizusaki J, Tagawa H, Naraya K, Sasamoto T (1991) Solid State Ionics 49:111
Mizusaki J, Yasuda I, Shimoyama J, Yamauchi S, Fueki K (1993) J Electrochem Soc 140:467
Mori T, Drennan J, Lee J-H, Li J-G, Ikegami T (2002) Solid State Ionics 154–155:461

Murray EP, Barnett SA (2001) Solid State Ionics 143:265
Palguyev SF, Volchenkova ZS (1958) Tr Inst Khim Akad Nauk SSSR, UralFilial 2:183–200; C. A. 54:9542 i (1960)
Pernot E, Anne M, Bacmann M, Stroble P, Fouletier J, Vannier RN, Mairesse G, Abraham F, Nowogrocki G (1994) Solid State Ionics 70–71:259
Peters H, Mobius H-H (1958a) Z Physik Chern 209:298
Peters H, Mobius H-H (1958b) Z Physik Chem (Leipig) 209:298
Peters H, Mobius H-H (1958c) Verfahren zur Gasanalyse bei erhohten Temperaturen mit Hilfe galvanischer Festelektrolytelemente, DDR-P 21673, filedMay 20
Quadakkers WJ, Greiner H, Kock W (1994) In: Bossel U (ed) Proceedings of the 1st European SOFC forum, Switzerland, p 525
Reynolds H (1902) Uber die Leitfahigkeit fester Mischungen bei hohen Temperaturen. Ein spezieller Fall der festen Losungen. Thesis, University of Gottingen
Sammells AF, Cook RL, White JH, Osborne JJ, MacDuff RC (1992) Solid State Ionics 52:111
Sarantaridis D, Atkinson A (2007) Fuel Cells. 7:246
Schmalzried H (1962) Z Elektrochem Ber Bunsenges Physik Chem 66:572
Schottky W, Veroff W (1935) Z Elektrockem, Bey Bunsenges Physik Chem 63:244
Shannon RD (1976) Acta Crystallogr A 32:751
Simner SP, Sandoval DS, Mackenzie JD, Dunn B (1997) J Am Cera Soc 80:2563
Simner SP, Hardy JS, Stevenson JW (2001) J Electrochem Soc 148(4):A351
Simner SP, Bonnett JF, Canfield NL, Meinhardt KD, Shelton JP, Sprenkle VL, Stevenson JW (2003) J Power Sources 113:1
Sin A, Kopnin E, Dubitsky Y, Zaopo A, Aricó AS, Gullo LR, La Rosa D, Siracusano S, Antonucci V, Oliva C, Ballabio O (2004) Solid State Ionics 175:361
Singhal SC, Kendall K (2000) High temperature solid-oxide fuel cells: fundamentals. Elsevier, New York
Spacil HS, Tedmon CS (1969) J Electrochem Soc 116:1618
Sun C, Hui R, Roller J (2010) J Solid State Electrochem 14:1125
Takahashi T (1966) 1. Electrochem Soc 34:60
Takahashi T, Iwahara H (1971) Energy Convers 11:105
Takahashi T, Ito K, Iwahara M (1965) Rev Energ Primaire Journies Int d'Etude Pilesa Combustible, Bruxelles, vol 3, pp 42–48
Tannenberger H (1962) Electrolyte solide pour piles a combustible. Swiss Patent 400 264, filed 23 Nov 1962
Tannenberger H, Schachner H, Simm W (1962) Festelektrolytbrennstoffelement, DE-P 1 471 768, filed 22 May 1963; Swiss priority 23 May 1962
Tao S, Irvine JTS (2002) J Solid State Chem 165:12
Treadwell WD (1916) 2. Elektrochem 22:414
Uchida H, Yoshida M, Watanabe M (1995) J Phys Chem 99:3282
Uchida H, Suzuki H, Watanabe M (1998) J Electrochem Soc 145:615
Webb JB, Sayer M, Mansingh A (1977) Can J Phys 55:1725
Weber WJ, Griffin CW, Bates JL (1987) J Am Ceram Soc 70(4):265
Weissbart J, Ruka RJ (1961) Rev Sci Instr 32:593
Weissbart J, Ruka RJ (1962) 1. Electrochem Soc 109:723
Yahiro H, Eguchi Y, Eguchi K, Arai H (1988a) J Appl Electrochem 18:527
Yahiro H, Ohuchi T, Eguchi K, Arai H (1988b) J Mater Sci 23:1036
Yamaji K, Horita T, Ishikawa M, Salrai N, Yokokawa H, Dolriya M (1997) Solid oxide fuel cells V. In: Stimming U, Singhal SC, Tagawa H, Lehnert W (eds) The Electrochemical Society proceedings, Pennington, NJ, PV97–401, 301
Yang YJ, Wen TL, Tu H, Wang DQ, Yang J (2000) Solid State Ionics 135(1–4):475
Yasumoto K, Mori N, Mizusaki J, Tagawa H, Dokiya M (2001) J Electrochem Soc 148:A105

Yasumoto K, Shiono M, Tagawa H, Dokiya M, Hirano K, Mizusaki J (2002) J Electrochem Soc 149:A531
Yi JY, Choi GM (2004) J Eur Ceram Soc 24:1359
Yokokawa H, Sakai N, Horita T, Yamaji K (2001) Fuel Cell 1:117
Yoshida H, Deguchi H, Miura K, Horiuchi M (2001) Solid State Ionics 140:191
Zhang GB, Smyth DM (1995) Solid State Ionics 82:153
Zhou W, Ran R, Shao Z (2009) J Power Sources 192:231
Zhu WZ, Deevi SC (2002) Mater Sci Eng A 348(1–2):227
Zhu WZ, Deevi SC (2003) Mater Sci Eng A 362:228

Chapter 4
Thermodynamics, Performance, and Configurations of SOFC

Till now, we have discussed the basic concepts of SOFC technology and the components used. As SOFC converts chemical energy directly into electricity without irreversible oxidation, we can calculate reversible work at equilibrium for the reversible reaction by giving thermodynamic treatment in terms of free enthalpy of reaction between the fuel and the oxidant. This process involves the reversible transfer of heat to the surroundings. There are many losses in SOFC and predominant ones are voltage fall due to the irreversible mixing of gases and generation of heat due to ohmic resistance. In this chapter, we discuss both types of systems, i.e. a reversible and a fuel gas mixture.

4.1 Ideal Reversible SOFC Thermodynamics

For a fuel cell, there is an inlet through which the fuel enters the system and outlet through which the gases escape the fuel cell system. Figure 4.1 shows the schematic of a fuel cell as a reversible system. Fuel and air enter the fuel cell separately and leave the fuel non-mixed provided the SOFC system is considered to be reversible. The total enthalpy of the system is n_iH_i while entering the system and is n_jH_j while leaving the cell. To attain equilibrium, heat Q must be extracted reversibly from the fuel cell which is subsequently transferred to the environment. Q is positive if heat is transferred to the system and negative if heat is given by the fuel cell to the surroundings. Similarly work done W is taken to be negative if work is done by the system and positive if work is done on the system. According to the first law of thermodynamics,

$$\Delta H = \Delta Q + \Delta W \tag{4.1}$$

where ΔH is the enthalpy of reaction, according to the second law of thermodynamics, for a reversible process of $\int \mathrm{d}S = 0$, and the entropy of reaction (ΔS) is given by

G. Kaur, *Solid Oxide Fuel Cell Components*,
DOI 10.1007/978-3-319-25598-9_4

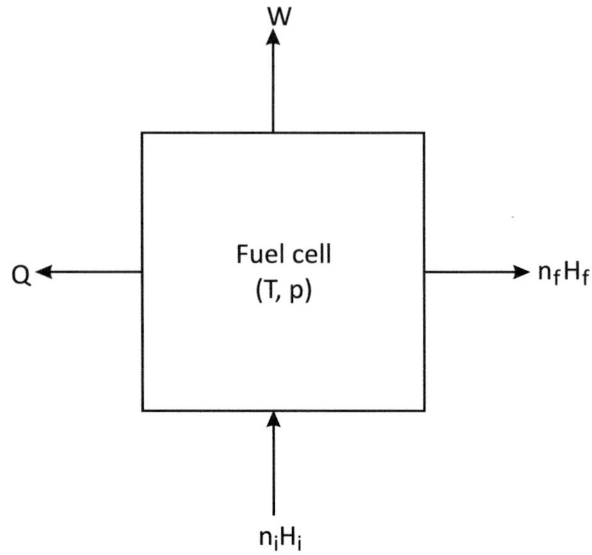

Fig. 4.1 SOFC as a reversible system

$$\Delta S = \frac{\Delta Q}{T} \tag{4.2}$$

Usually, the Gibbs free energy ΔG is equal to the reversible work done, hence Eq. (4.1) can be written as

$$\begin{aligned} &\Delta H = \Delta Q + \Delta G \\ \Rightarrow\ &\Delta G = \Delta H - \Delta Q \\ \Rightarrow\ &\Delta G = \Delta H - T\Delta S \end{aligned} \tag{4.3}$$

Moreover, to define the efficiency, we require Gibbs free energy and enthalpy of reaction, i.e. efficiency of a fuel cell, which is defined as the ratio of Gibbs free energy and enthalpy change.

$$\Rightarrow \eta = \frac{\Delta G}{\Delta H} \tag{4.4}$$

$$\Rightarrow \eta = \frac{\Delta H - T\Delta S}{\Delta H} = 1 - \frac{T\Delta S}{\Delta H} \tag{4.5}$$

As defined in Chap. 2, the fuel utilisation factor (U_f) describes the efficiency of the fuel cell (Eq. 2.78), and U_f is given by

$$U_f = \frac{\text{Fuel}_{\text{in}} - \text{Fuel}_{\text{out}}}{\text{Fuel}_{\text{in}}}$$

Hence, at the anode, the concentration of product H_2O increases with the increasing fuel utilisation factor. As discussed in Chap. 3,

$$\text{Anode: } H_2 \rightarrow 2H^+ + 2e^-$$

$$\text{Cathode: } \frac{1}{2}O_2 + 2e^- \rightarrow O^{2-}$$

O^{2-} migrates through the electrolyte and reaches anodes where it combines with hydrogen ions to form water; i.e. $2H^+ + O^{2-} \rightarrow H_2O$.

We can also consider molar flow of electrons to further obtain the voltage gain. It is apparent that the molar flow of electrons ($\bar{n}_e$) is twice that of molar flow of hydrogen ($\bar{n}_{H_2}$); i.e.,

$$\bar{n}_e = 2\bar{n}_{H_2} \tag{4.6}$$

The electric current (I) can be given in terms of molar flow of electrons, and Avogadro's number (N)

$$\Rightarrow I = \bar{n}_e(-e)N \tag{4.7}$$

$$\Rightarrow I = -2\bar{n}_{H_2}eN \tag{4.8}$$

where e is the charge of an electron with magnitude 1.6×10^{-19} C.

$$\Rightarrow I = -2\bar{n}_{H_2}F \tag{4.9}$$

where F is Faraday constant and is defined by the product of charge and Avogadro's number; i.e.,

$$F = Ne = 6.023 \times 10^{23} \times 1.6 \times 10^{-19} = 96{,}485\ \text{k/mol}$$

The current I measures the amount of fuel spent. To define the reversible power (P), we need to obtain the product of voltage and current

$$\Rightarrow P = VI \tag{4.10}$$

Additionally, the power can also be given as the product of Gibbs free energy and molar flow of hydrogen flow ($\bar{n}_{H_2}$); i.e.,

$$\Rightarrow P = \Delta G \cdot \bar{n}_{H_2} \tag{4.11}$$

From (4.7, 4.10) and (4.11), we obtain

$$\begin{aligned} V &= \frac{\Delta G \cdot \bar{n}_{\mathrm{H}_2}}{-n_{\mathrm{e}} e N} \\ \Rightarrow V &= \frac{-\Delta G \cdot \bar{n}_{\mathrm{H}_2}}{\bar{n}_{\mathrm{e}} F} \end{aligned} \tag{4.12}$$

When the fuel mixing occurs (during fuel utilisation) within the SOFC, then the system no longer remains reversible and then the voltage reduction has to be calculated, which takes partial pressures into account; i.e., Eq. (4.3) can be written as

$$\Rightarrow \Delta G(T,p) = \Delta H(T,p) - T\Delta S(T,p) \tag{4.13}$$

For an ideal gas, Eq. (4.13) can be written as

$$\begin{aligned} \Delta G(T,p) &= \Delta H(T) - T\Delta S(T,p) \\ \Delta S(T,p) &= \frac{\varDelta H(T) - \varDelta G(T,p)}{T} \end{aligned} \tag{4.14}$$

$$\left[\begin{array}{ll} \text{As} & \begin{array}{l} W = pV, \text{hence} \\ W = p\mathrm{d}V + V\mathrm{d}p \end{array} \\ \text{For constant volume} & \mathrm{d}W = V\mathrm{d}P \end{array}\right]$$

The Gibbs free energy can be treated equivalent to the work done and Eq. (4.14) becomes (for infinitesimal changes)

$$\Rightarrow \mathrm{d}S = \frac{\mathrm{d}H - V\mathrm{d}p}{T} \tag{4.15}$$

In terms of gas constant (R), the entropy can be written as

$$\Delta S(T,p) = \Delta S(T) - R\ln K \tag{4.16}$$

where K represents the equilibrium constant, and hence the Gibbs free energy can be written as

$$\Delta G(T,p) = \Delta G(T) + RT\ln K \tag{4.17}$$

Using (4.17) in (4.12), we obtain

$$V = -\frac{\{\Delta G(T) + RT\ln K\} n_{\mathrm{H}_2}}{\bar{n}_{\mathrm{e}} F} \tag{4.18}$$

It should be noted that the reaction enthalpy and entropy are assumed to vary slightly with the temperature, though the reversible cell voltage increases with increasing system pressure and decreases with increasing system temperature.

4.2 Performance of SOFC

The ongoing development of SOFC includes the fabrication of material components which can sustain mechanical pressure, address cost issues, increase cell efficiency, and deliver the required power at low and intermediate temperatures. By lowering the operating temperatures, the cell costs can be reduced drastically and within the intermediate operating range of 600–800°C, the precious metals can be substituted by the ceramic materials. The efficiency of SOFC also depends on the fabrication method; e.g., the process of PMSS (pyrolysis of metallic soap slurry), in which NiO particles are surrounded by thin films of YSZ, has highly improved the nickel adhesion leading to strong adhesion of anode and YSZ electrolyte. Similarly, the use of mixed lanthanides as electrodes has reduced the cost of cell by 70%, whereas the efficiency gets reduced by 8%. Perovskite materials are being synthesised as electrolytes due to their low activation energy and high ionic charge carriers. To maintain the performance at low temperatures, the materials with mixed conduction (electronic + ionic) are used as electrodes instead of purely electronic conductors. Hence, the performance of SOFC is dependent on a number of factors including pressure, temperature, impurities, current densities, etc. The high operating temperatures of SOFC offer low open circuit voltage but the polarisation is reduced. This section describes the performance of SOFC with temperature, pressure, and other factors.

4.2.1 *Effect of Gas Composition and Utilisation*

SOFC does not require reforming the catalyst as the anode acts as the catalyst for the process of fuel reformation. Moreover, the high temperature of SOFC allows the internal reformation of fuel gases. The reformation of CO_2 from the fuel stream is not required because the cathode utilises oxygen only for SOFC. The fuel gas consists of H_2, CO, and H_2O; the pressure and concentration of H_2 highly influence the SOFC performance; i.e., increasing the hydrogen content increases the potential and the complete oxidation range extends to the higher value. In addition to this, the water gas shift reaction [$CO + H_2O \rightarrow CO_2 + H_2$] is not favoured at high temperatures. At high temperatures, the theoretical potential of CO/O_2 reaction decreases whereas the H_2/O_2 reaction potential increases. Usually, the fuel composition is defined by the H/C and O/C atom ratio; i.e., for pure CO, O/C = 1, pure CO_2 yields O/C = 2 and for no hydrogen H/C = 0. Figure 4.2 represents the variation of open circuit voltage with the O/C ratio corresponding to difficult H/C ratios.

The fuel gas composition describes the cell voltages: for a fuel gas composition of 97% CO/3% H_2O, the current density is 170 mA/cm^2, whereas for 1.5% H_2/3% CO/75.5% CO_2/20% H_2O, the current density is 100 mA/cm^2. In contrast to this, the composition with higher hydrogen (97% H_2/3%H_2O) content gives the current

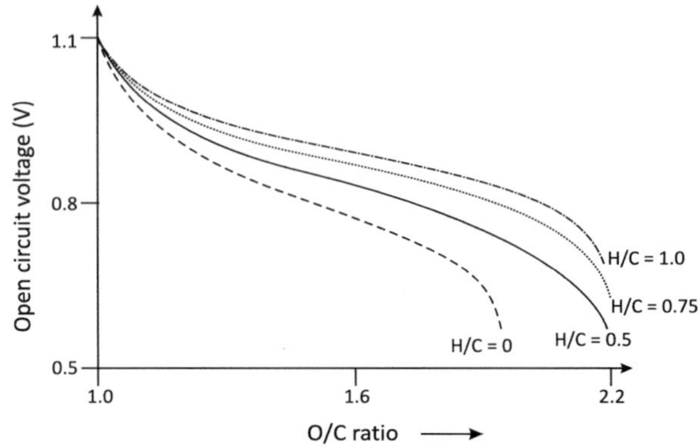

Fig. 4.2 Open cell voltage versus the O/C ratio for SOFC at 1000°C

density of 220 mA/cm^2. For the composition with less CO and H_2 content, the performance is lowered due to higher concentration polarisation. Similarly, for a cathode which uses pure oxygen as fuel, the cell performance increases. When current density increases, the difference between open circuit voltage with pure oxygen and air increases, which can be attributed to the increased concentration polarisation. At T = 100°C, the voltage gain calculated can be given by

$$\Delta V_{\text{cathode}} = 92 \log \frac{(p_{O_2})_2}{(p_{O_2})_1} \tag{4.19}$$

Figure 4.3 gives the effect of fuel utilisation on the cell voltage for pure O_2 as oxidant and 67% H_2/22% CO/11% H_2O as reference fuel. It has been observed that the cell voltage increases with pure O_2 as oxidant provided the temperature is constant. The cell voltage decreases with increasing fuel utilisation. Furthermore,

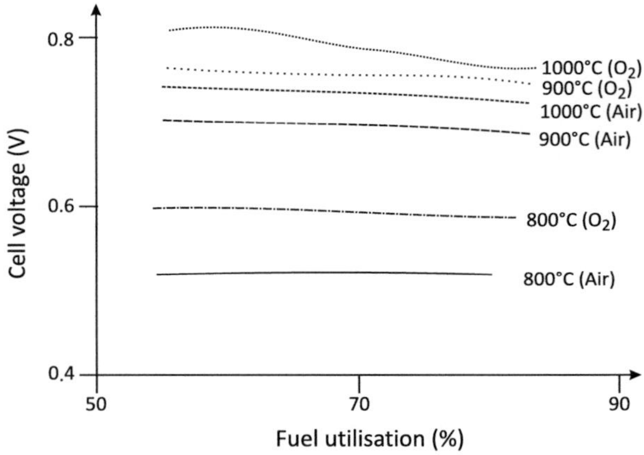

Fig. 4.3 Cell voltage as function of temperature and fuel utilisation

the decrease in cell voltage is steeper at the higher temperature of 1000°C than at 900/800°C.

4.2.2 *Effect of Temperature and Pressure*

Upon increasing the temperature, the current density increases because the ohmic polarisation decreases. Though the cell voltage decreases with increasing current density, this decrease is sharper at low temperatures than at high temperatures (Fig. 4.4).

Like PAFC and MCFC, SOFC also depicts enhanced performance upon increase in pressure. At 1000°C, the cell performance can be approximated as

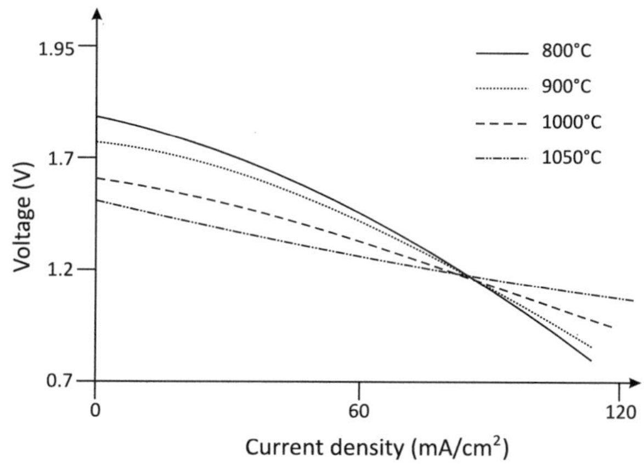

Fig. 4.4 Cell voltage versus current density for two cell stack

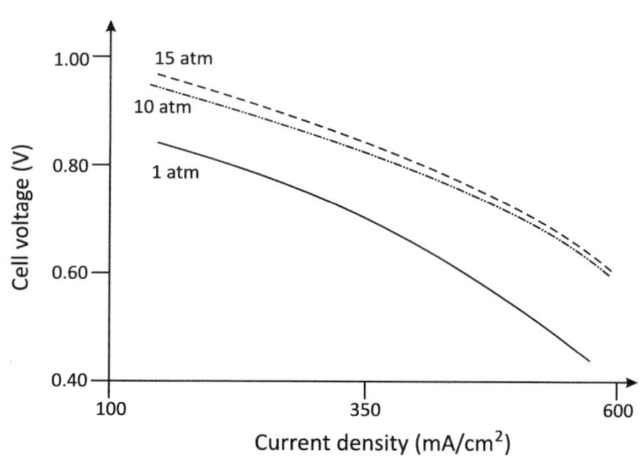

Fig. 4.5 AES cell performance at different pressures (1000°C) (89% H_2/ 11% H_2O, O_2^-)

$$\Delta V(\mathrm{mV}) = 59 \log \frac{p_2}{p_1} \tag{4.20}$$

The variation of cell voltage with different pressures and current density is shown in Fig. 4.5 (listed for AES cells, Siemens Westinghouse Corporation).

4.2.3 *Effect of Current Density, Cell Life, and Impurities*

The effect of current density on the cell voltage is given by the following equations.

$$V_{\mathrm{CD}} = -0.73\,\Delta J \quad \text{at } T = 1000\,^{\circ}\mathrm{C} \tag{4.21}$$

Figure 4.6 shows the performance of single SOFC single cell performance at reducing temperatures. With increasing current density, ohmic activation and concentration losses also increase. In addition to this, the tubular design with air electrode supported (AES) has shown a 33% increase in power density as compared to the calcia-stabilised supported porous support tube design. As far as the impurities are concerned, ammonia, HCl, and hydrogen sulphide are the most commonly found impurities in the coal gas. Up to 5000 ppm of ammonia and 1 ppm HCl, no

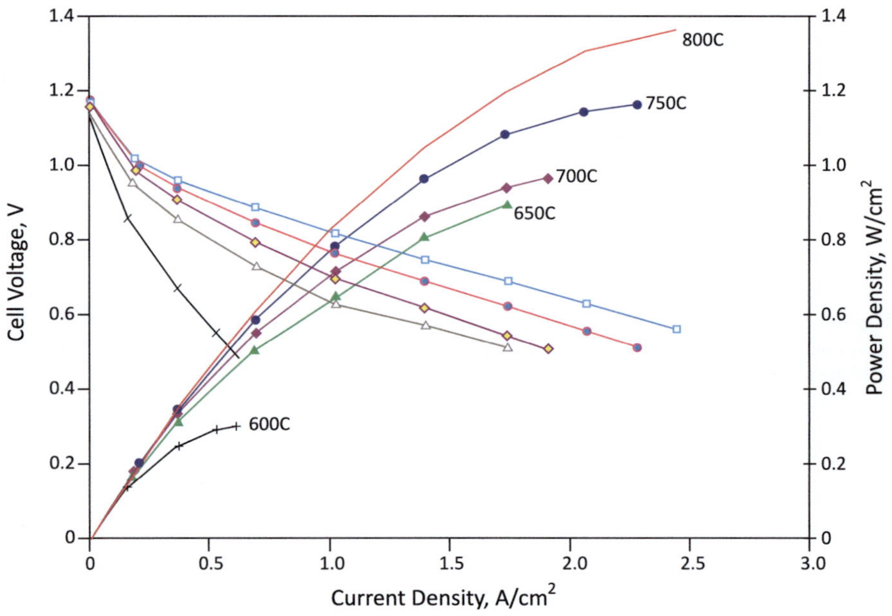

Fig. 4.6 Single cell SOFC performance at reduced temperatures (Minh 2004)

detectable cell degradation is observed whereas even 1 ppm of H_2S leads to performance degradation. If H_2S is removed from the fuel stream, the cell resumes its original performance.

4.3 Designs of SOFC

The electrolyte is sandwiched between the anode and cathode for the SOFC. The anode/electrolyte/cathode sandwich structure is most commonly known as a single cell or PEN (positive electrolyte-negative) electrode cell. The fuel is fed at the anode which undergoes oxidation to release the electrons in the external circuit. At the cathode, the oxidant accepts these electrons from the external circuit and is reduced. Furthermore, the reduced oxygen ion migrates via electrolyte towards the anode and reacts with fuel to produce steam. The power generated by a single cell is not enough (<1 V) to sustain stationary application, hence many single cells are joined together by an interconnect to form a cell stack. This section describes the different stack designs, their advantages and disadvantages along with the requirements which should be kept in consideration while designing the SOFC.

4.3.1 Requirements for SOFC Designs

The SOFC must deliver the desired electrochemical performance along with the maintenance of mechanical integrity of the stack for achieving long-term performance goals. Hence, for choosing a specific design, following the requirements must be fulfilled.

- **Minimum Ohmic Losses**: To minimise the ohmic losses in the stack, the current parts in all the components must be kept as short as possible.
- **Good Electrochemical Contacts**: Among the components of SOFC, the contact surface area should be high and the current collector must regulate the desired current flow in the stack.
- **Uniform Temperature Distribution**: The SOFC design should exhibit uniform temperature distribution during its operation along with proper stack cooling.
- **Minimum Polarisation Losses**: Usually polarisation losses in SOFC cause a major drop of open circuit voltage. Hence, the stack should be designed in a way to minimise any polarisation losses, gas leakage, short-circuiting, and cross-leakage of gases.
- **Uniform Distribution of Gases**: The fuel gas and oxidant should be distributed uniformly over the entire cell stack so that the gases can reach reaction sites and efficiently reduce the limitation of mass transportation.

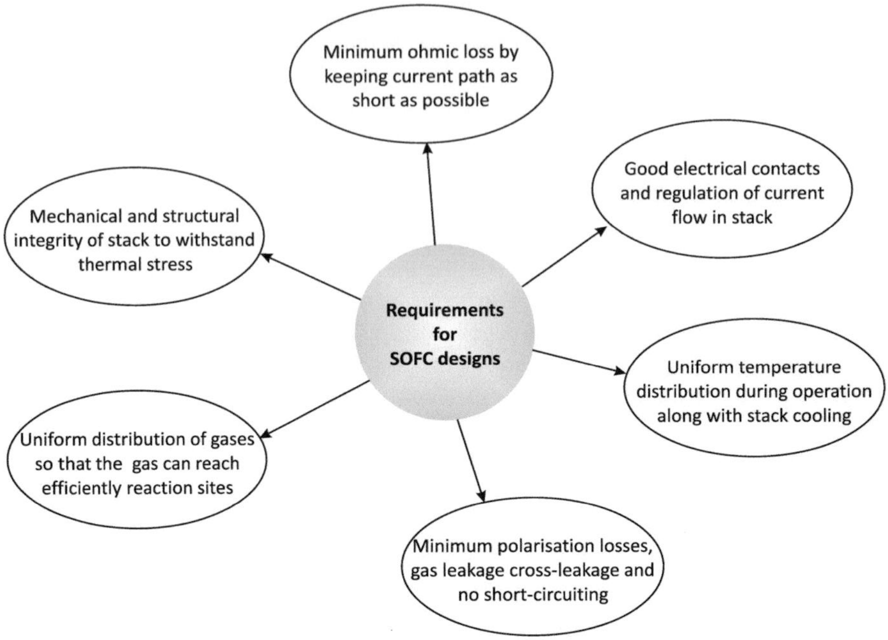

Fig. 4.7 Requirements for the SOFC designs

- **Mechanical and Structural Integrity of Stack**: The SOFC design should be robust so as to withstand the thermal stresses or pressure of gases.

Basically, the stack should be tolerable to thermal shock conditions due to cold start-up and sudden power change in addition to the conditions faced during installation and vibration loading, as well as off-design temperature gradients. Figure 4.7 shows the diagrammatic depiction of the basic requirements while designing SOFC.

4.3.2 Single Cell Configuration

Owing to the solid nature of the components of SOFC, the cell designs offer flexibility. The individual cells are stacked together to obtain a particular stack configuration. The individual cell configuration can be of two types: self-support and external support. For the self-support cell configuration, the structural support of the cell is provided by one of the thickest layers of the cell component. Anode-supported, cathode-supported, and electrolyte-supported cell configurations belong to self-support systems. For the external-support configuration, the thin

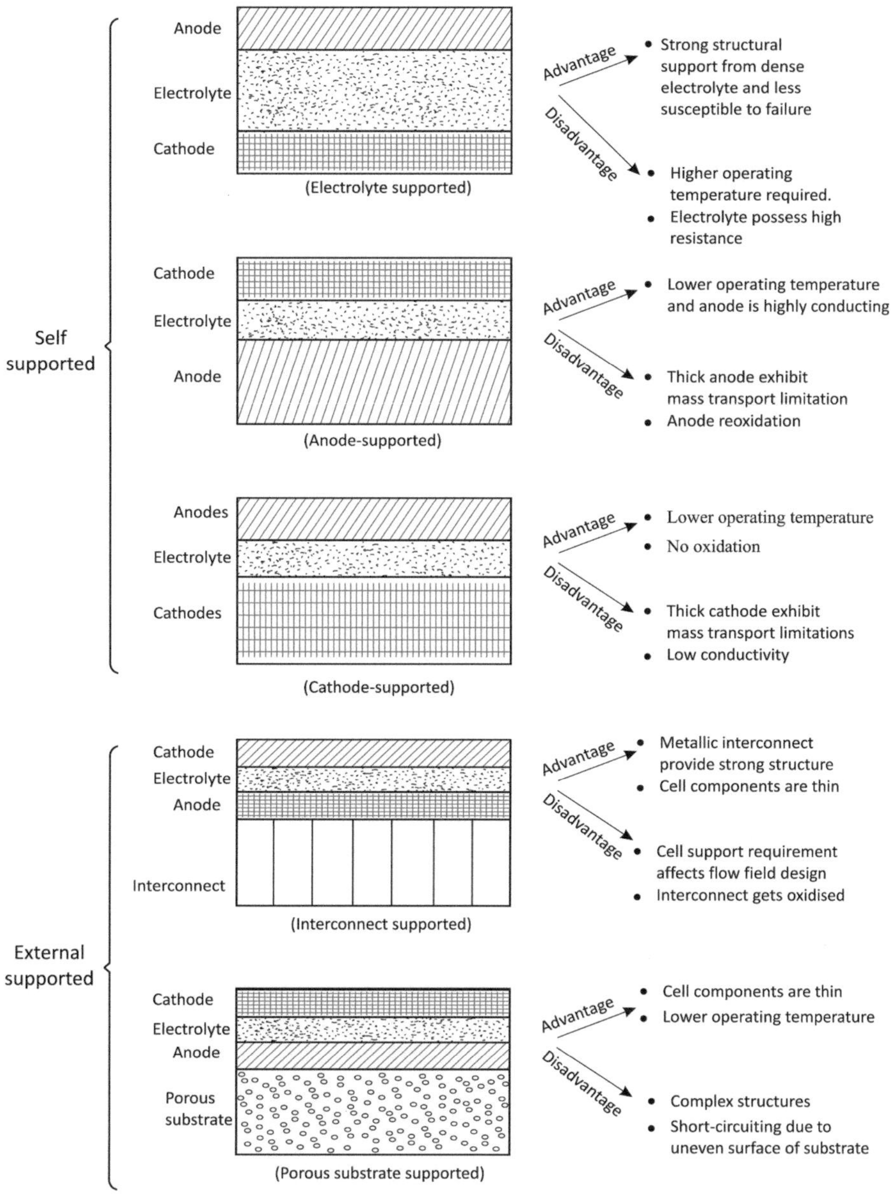

Fig. 4.8 Single cell configurations for SOFC

layers of cells are configured on a porous substrate or interconnect. Figure 4.8 shows the cell configurations along with their advantages and disadvantages. YSZ is the most commonly used electrolyte and for the electrolyte-supported cell, the electrolyte possesses typical thickness >100 μm. Anyhow, the thick electrolytes exhibit high ohmic losses, hence the operating temperature should be kept high (900–1000°C). But when the electrolyte of SOFC is 5–20 μm thick, the cell operating temperature is reduced to <800°C, which provides a wider choice of materials. At 1000°C, the ionic conductivity of the solid electrolyte is $0.1\ \Omega^{-1}\ cm^{-1}$, which is reduced to $0.02\ \Omega^{-1}\ cm^{-1}$ at 800°C. The typical resistivities of cell materials at 1000°C for a fuel gaseous environment are:

- Doped $LaCrO_3$ (Interconnect)—1 Ω cm (electronic)
- Ni/ZrO_2 cermet (Anode)—3×10^{-6} Ω cm (electronic)
- Doped $LaMnO_3$ (Cathode)—0.01 Ω cm (electronic)
- 8% YSZ (Electrolyte)—10 Ω cm (ionic)

Electrode-supported cells pass low ohmic polarisation losses as compared to the electrolyte-supported cells. In addition to this, anodic polarisation is quite less than the cathodic polarisation, hence, anode-supported SOFC is the preferred configuration.

4.3.3 Designs of SOFC Stacks

Many designs have been proposed for SOFC stacks as listed in Fig. 4.9, but two of them have been running successfully, viz. tubular and planar design.

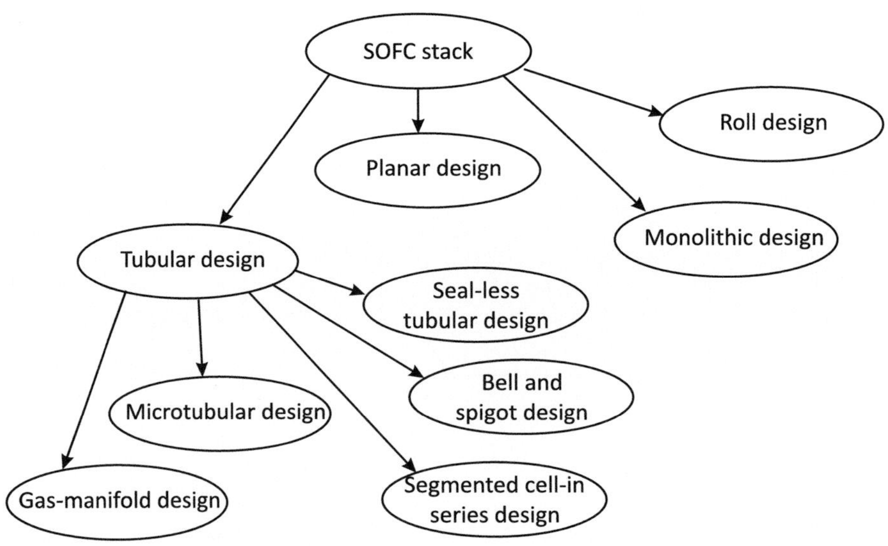

Fig. 4.9 Different stack designs for SOFC

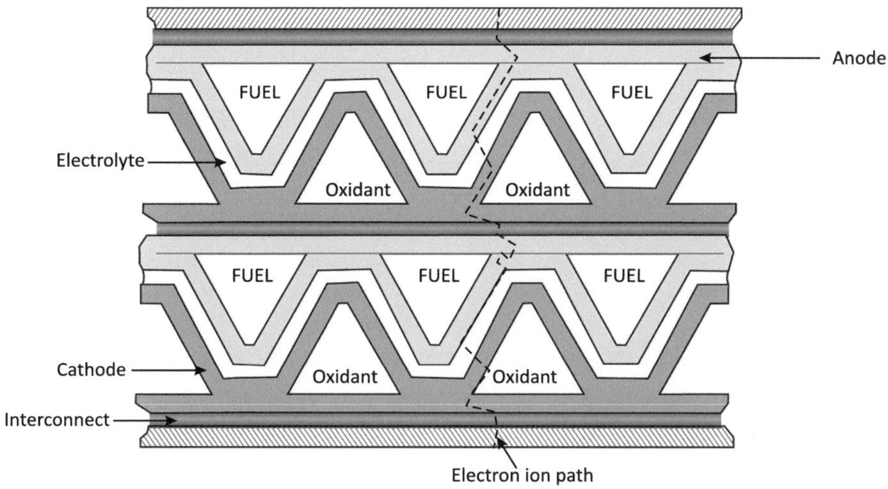

Fig. 4.10 Demonstration of current path in monolithic SOFC design (Minh 2004)

Each design may have variants, but the seal-less tubular designs, microtubular designs, and bipolar flat-plate (planar) designs have received wide attention. Hence, the discussion in the present section is limited to tubular (seal-less), microtubular, and planar designs only. The design of SOFC usually depends on three criteria as follows.

(a) **Current path**: The current path varies as per design of the SOFC stack. For example, the current follows a multicell monolithic structure for the monolithic designs. The monolithic designs are basically electrolyte-supported and cell components form a corrugated structure of gas cross-flow or co-flow configuration (Fig. 4.10). For the tubular cell, the current traverses across the thickness of the electrolyte/interconnect and along the circumference of the cathode and anode.

(b) **Cell-to-cell connections**: Either parallel or series connection is provided to the SOFC stack, but in certain cases both parallel and series connection are combined to provide cohesive voltage output of the stack. When cells are joined in series, then the output is high as compared to the parallel arrangements.

(c) **Gas flow configurations**: The fuel and oxidant flow in SOFC stacks can be co-flow, counter-flow, or cross-flow. Different flow patterns can be implied for different flow configurations, such as serpentine, radial, spiral, or 2-flow patterns (especially for planar configuration) as shown in Fig. 4.11. For planar configuration, the flow channels promote uniform gas distribution along with mass and heat transport of each cell. Usually flow field channels are designed as a part of the interconnect, whereas some SOFC designs integrate them into the electrode. The flow field patterns connect the interconnect with the

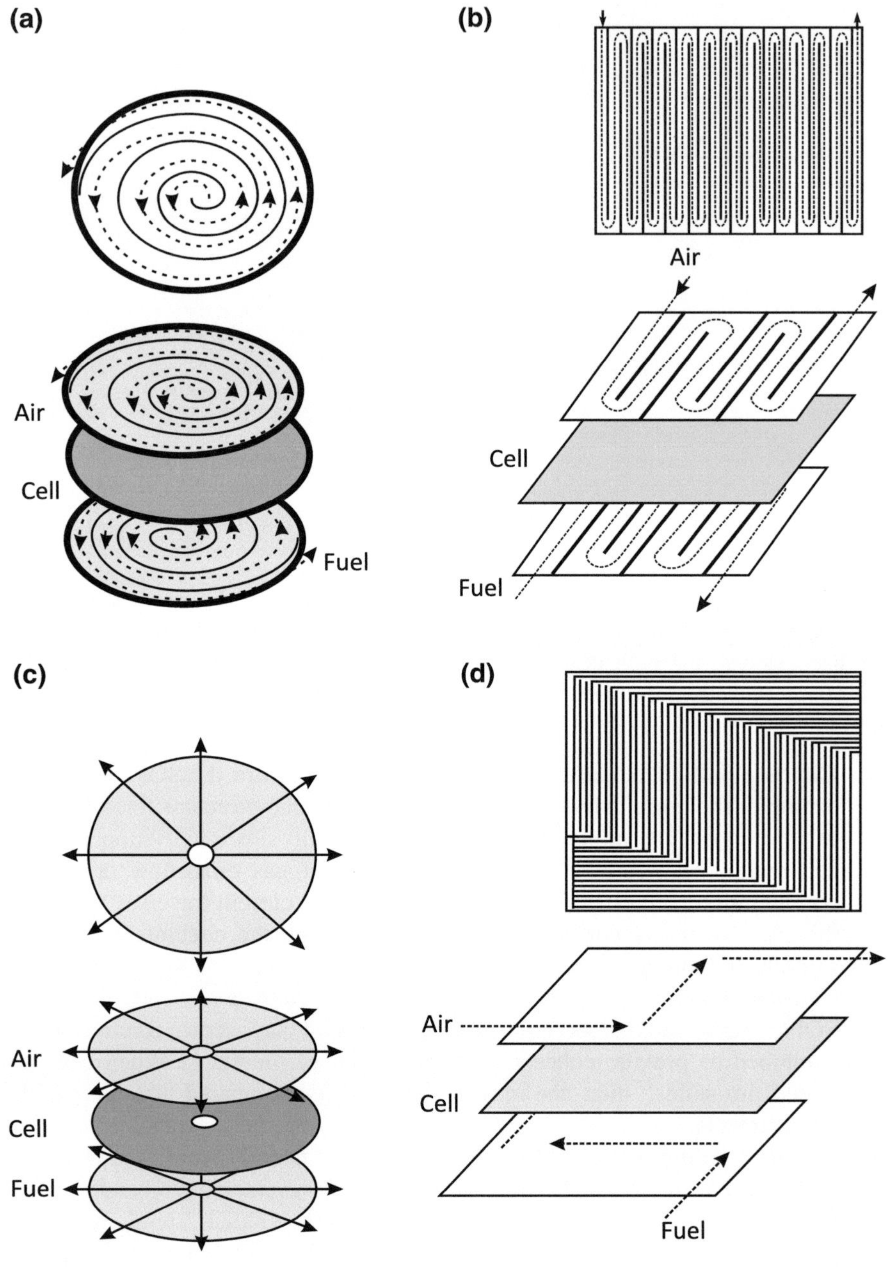

Fig. 4.11 Different flow field patterns for SOFC (Singhal and Kendall 2000). **a** Spinal-flow. **b** Serpentine-flow. **c** Radial-flow. **d** Z-flow

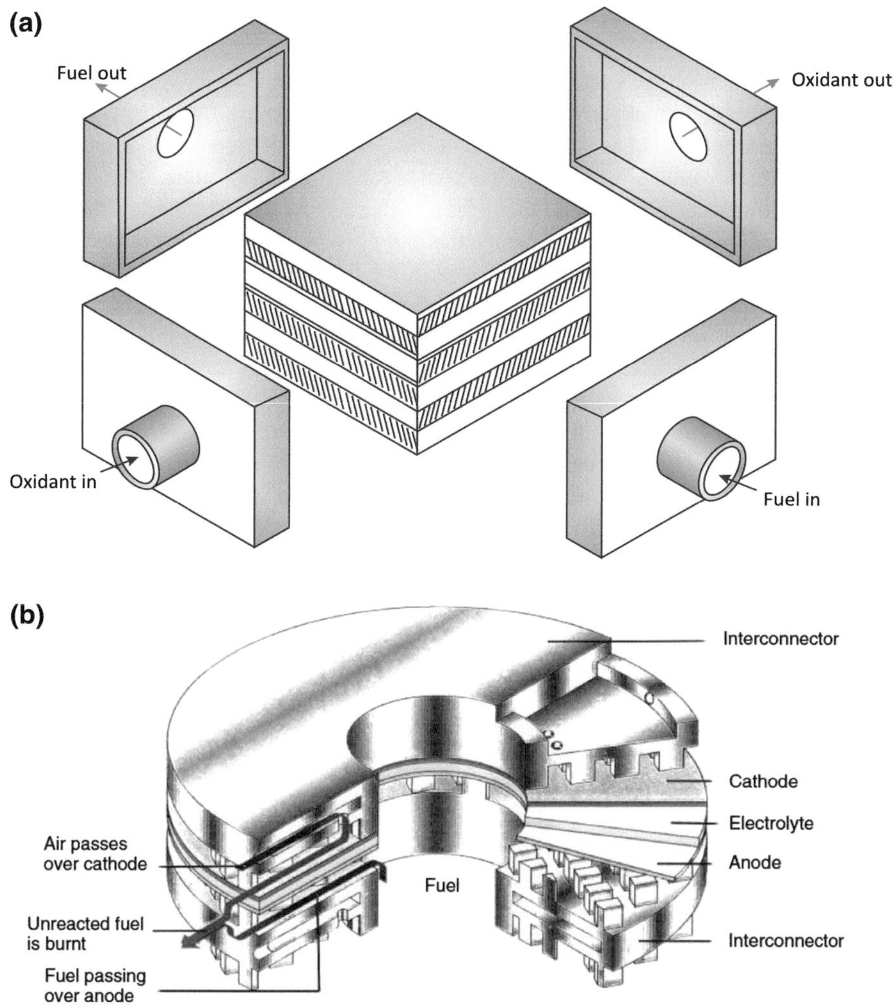

Fig. 4.12 **a** External and **b** internal manifolding in SOFC (Singhal and Kendall 2000)

electrodes, hence the contact area between electrodes and interconnect plays an important role in deciding the contact resistance losses.

(d) **Gas Manifolds**: The gas manifold provides gas routing common supply points for each cell and then removal of reaction products as well as unreacted gases. The gas manifolds can be internal or external. Internal manifolds are designed as part of the interconnect or cell whereas external manifolds are constructed separately (Fig. 4.12). The gas leakage may occur through the manifolds, hence sealing is required to overcome any leakage losses and drop in cell performance. Additionally, the manifold must provide low pressure drop and uniform flow distribution to the stack.

4.3.3.1 Tubular Cell Design (Seal-Less)

Tubular design of SOFC was initially made by Siemens Westinghouse, USA. Each tube in the tubular stack is made of a large test tube closed at one end which eliminates seal requirement. The tubular SOFC consists of two tubes, i.e. an outer tube and an inner tube (Fig. 4.13a). The outer tube (cell tube) consists of an inner surface and outer surface. The outer surface of the cell tube is regarded as the anode side of the cell and the inner surface is considered to be the cathode side of the cell tube. The solid electrolyte is fabricated within the inner and outer surface of the cell tube. The inner tube is also known as the injection/guidance tube. The injection tube is made up of alumina through which preheated air is injected towards the bottom of the cell tube. This preheated air then flows over the cathode surface of the cell through the gap between the cell and injection tubes.

Oxygen ions migrate through the electrolyte to react with fuel producing an electric current. Earlier the cell tube was made up of porous calcia-stabilised zirconia. The porous support tube served as a structural member upon which the active cell components were fabricated. The porous nature of this tube allowed the air to pass towards the cathode during cell operation. The major advantage of this design is 'no requirement of sealants' as demonstrated in Fig. 4.13b. Anyhow, air flow undergoes inherent impedance due to the tube, which decreases the power output of the cell. Therefore the porous support tube is replaced by the air electrode-supported cell assembly/$LaMnO_3$ tube. To increase the power of the cell, the length of the tube has been actively increased so that the number of cells required to produce a specific output can be reduced. To accommodate larger pressure drops, the diameter of the cell has been increased. An actual tubular cell used these days may consist of a 2.2 cm diameter air electrode tube with 2.2 mm wall thickness with length of 180 cm as shown in Fig. 4.14. The tubular structure is fabricated using the extrusion method followed by sintering such that a porosity of 30–35% can be obtained. There is 10% YSZ deposited in the form of a 40 μm thick layer using the electrochemical vapour deposition (EVD) technique. To form this layer, yttrium and zirconium chlorides are volatilised in a predetermined ratio and passed along with argon and hydrogen over the outer surface of hydrogen, steam, metal chlorides, and oxygen occurs through the porous cathode and subsequently reacts to fill the cathode pores with YSZ. These reaction kinetics are followed.

For yttrium

$$2YCl_3 + 3H_2O \rightarrow 2YO_{3/2} + 6HCl \tag{4.22}$$

$$4YCl_3 + 3O_2 + 6H_2 \rightarrow 4YO_{3/2} + 12HCl \tag{4.23}$$

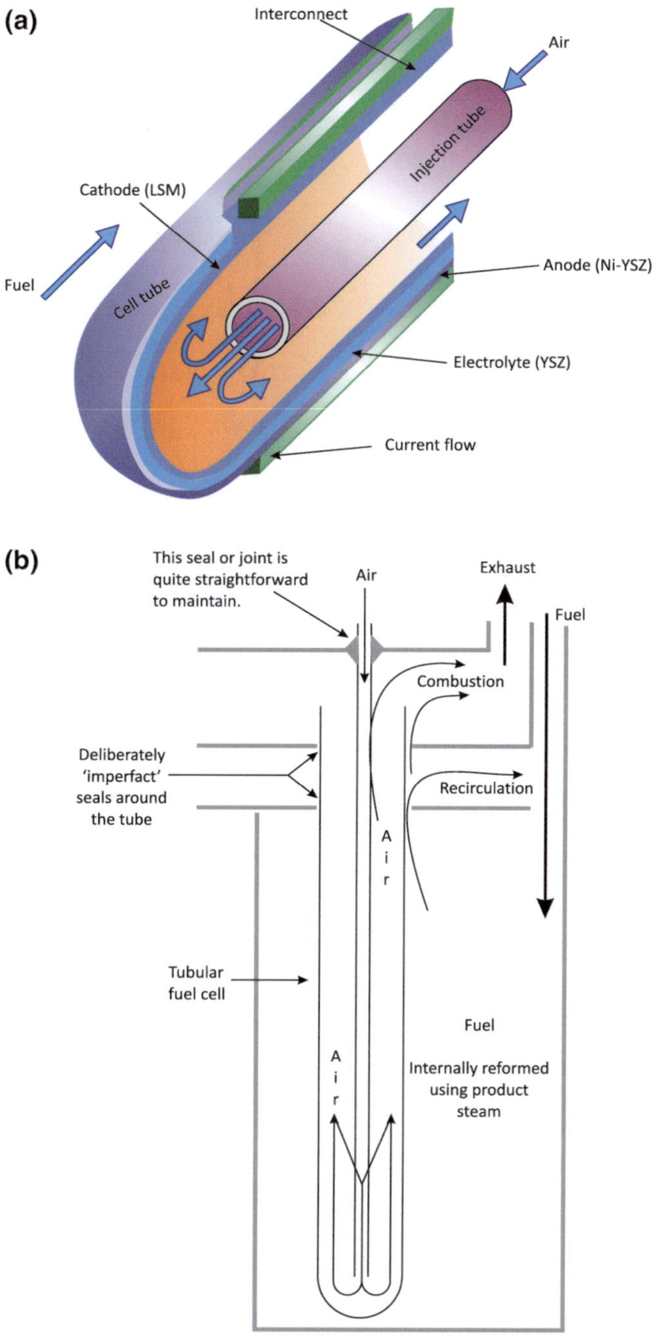

Fig. 4.13 **a** Design of tubular SOFC (Hajimolana and coworkers 2011) and **b** demonstration of no sealing requirements for Tubular SOFC (Larminie and Dicks 2003)

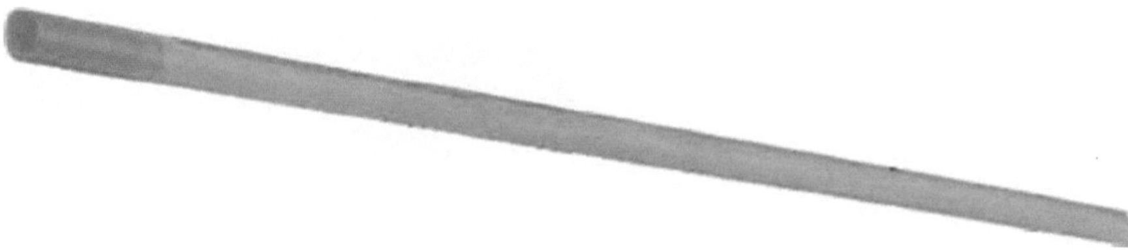

Fig. 4.14 Photograph of Single Tubular SOFC Cell (Photo courtesy of Siemens Westinghouse)

For zirconium

$$2ZrCl_2 + 2H_2O \rightarrow 2ZrO + 4HCl \quad (4.24)$$

$$4ZrCl_2 + 2O_2 + 4H_2 \rightarrow 4ZrO + 8HCl \quad (4.25)$$

When the pores of the cathode are closed, the oxide ions migrate through the electrolyte (already deposited YSZ) from the high oxygen partial pressure side to the low oxygen partial pressure side, i.e. from the oxygen/steam side to the chloride side. The oxide ions react with metal chlorides and the electrolyte layer thickens parabolically with time. Then the Ni/YSZ anode (100–150 μm thickness) is deposited onto the electrolyte layer. First, Ni powder slurry is applied over the electrolyte followed by the growth of YSZ around Ni particles using a similar EVD technique. Finally, a doped lanthanum chromite interconnection layer is deposited along the air electrode lengths using plasma spraying followed by densification sintering. The major advantages of tubular design are listed as follows.

- Higher energy efficiencies.
- Potential to replace conventional heat engineers.
- A variety of hydrocarbon fuels or their synthesis derivatives such as biomass and natural gas can be used as the fuel sources.
- Can oxidise CO to CO_2.
- No sealing requirement to isolate fuel from oxidant.
- Suitable for mid- to large-scale applications up to 2 MW.

Anyhow, tubular cells possess higher ohmic losses and lower power density due to the large current paths; i.e., when stacked together, the circumference increases which results in relatively long current paths which limits cell performance (Fig. 4.15). Tubular designs can also have many variants for example, a bell and spigot configuration, segmented in series tubular cell design, gas manifolding, and microtubular design.

For the bell and spigot design, electrolytes are developed in the form of short cylindrical segments in order to fit them onto each other and be connected to form a long tube by the bell and spigot joints as shown in Fig. 4.16.

Another SOFC design known as the SOFC roll design has been fabricated using tape-casting and co-firing as a single unit (Fig. 4.17). The roll geometry eliminates

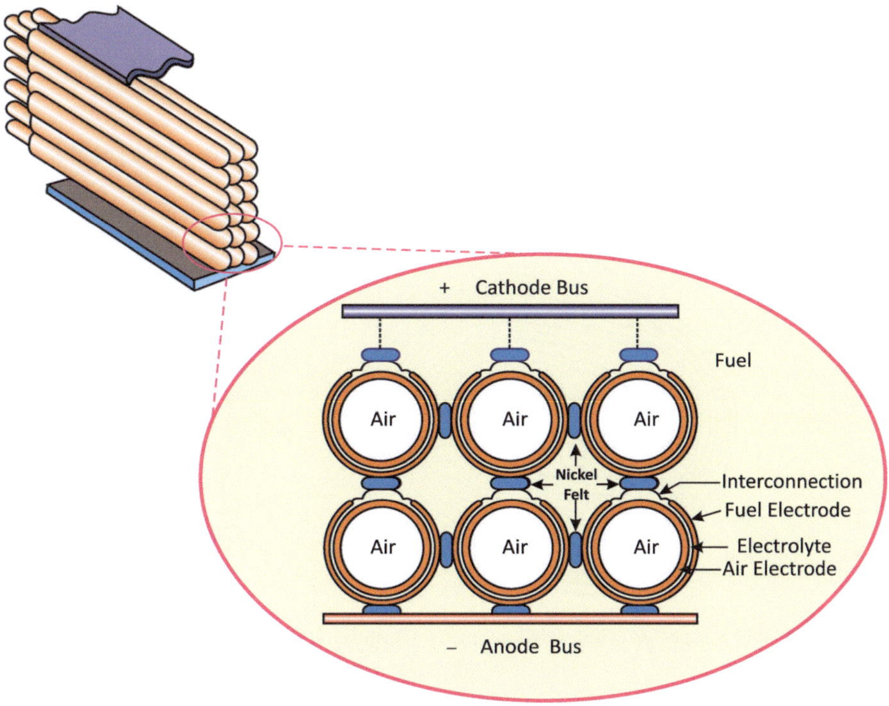

Fig. 4.15 Demonstration of large current paths due to cell-to-cell connections in tubular SOFC (Photo courtesy of Siemens Westinghouse)

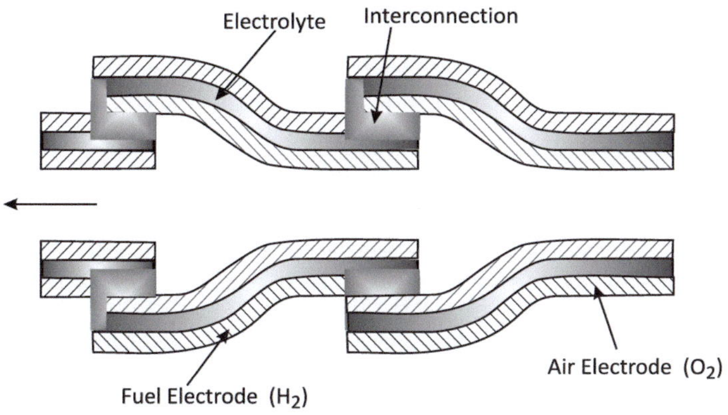

Fig. 4.16 Bell and Spigot configuration for tubular SOFC configuration (DOE/METC-86/0241, Technology Status Report, Morgantown, WV)

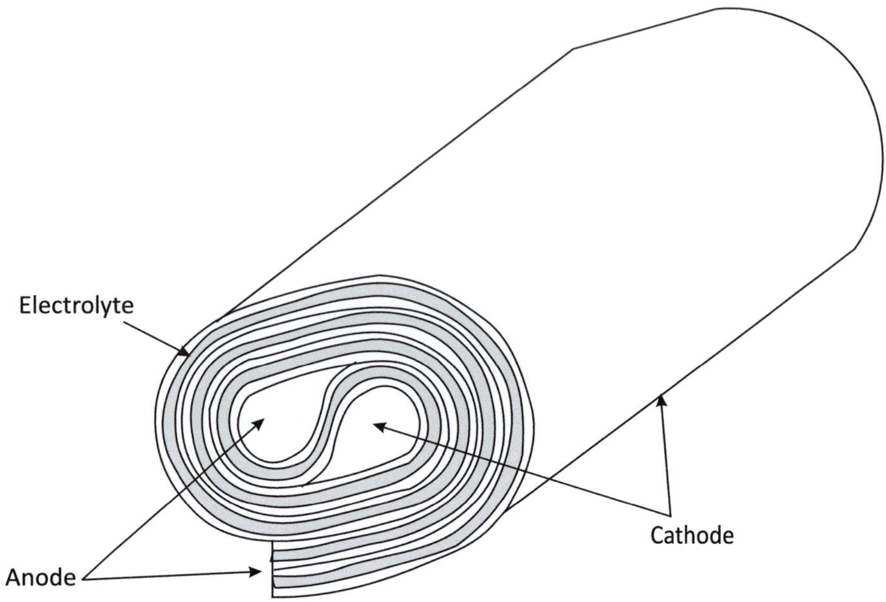

Fig. 4.17 SOFC roll design with curled geometry (Bastidas 2006)

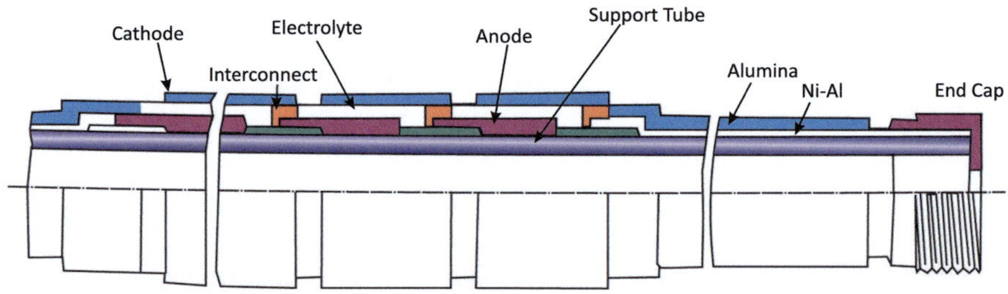

Fig. 4.18 Segmented in series design for the SOFC (Singhal and Kendall 2000)

the need of thicker support components which reduces space requirements and yields high power densities.

The segmented cell in series design is being investigated by Mitsubishi Heavy Industries (Japan), ABB, and Rolls Royce. This cell contains segmented cells arranged in a thin, banded structure on a porous support alumina tube. The sealing is provided by the interconnect, which also serves as an electrical contact between the cathode of one cell and anode of the next cell (Fig. 4.18). The oxidant flows outside and the fuel flows from one cell to the next inside the tubular cell stack.

4.3.3.2 Microtubular Cells Design

Microtubular cells were first made in 1990 by Michaela Kendall which were 1–5 mm in diameter, possessed 100–200 μm wall thickness and were made from extruded yttria-stabilised zirconia. After this achievement, Kendall and his co-workers demonstrated 20 cell, 200 cell, and 1000 cell reactors. Only a hot-air-fed reactor could heat up to the desired operating temperature and the cell could withstand an increase of temperature up to 200°C/min. Professor Kendall set up the first company in 1996 for microtubular (MT) cells known as Adelan (UK) Ltd. It was demonstrated by Bujalski that the operational temperature of these MT cells could be ramped at a rate of 4000°C/min, which indicates the rapid start-up time of 12 s for SOFC without any operational or structural damage to these cells. Second, these cells could be cooled extremely fast without any damage. The typical design of MT-SOFC and a 50 W portable MT is shown in Fig. 4.19a, b. The YSZ tube of thickness 150 μm and

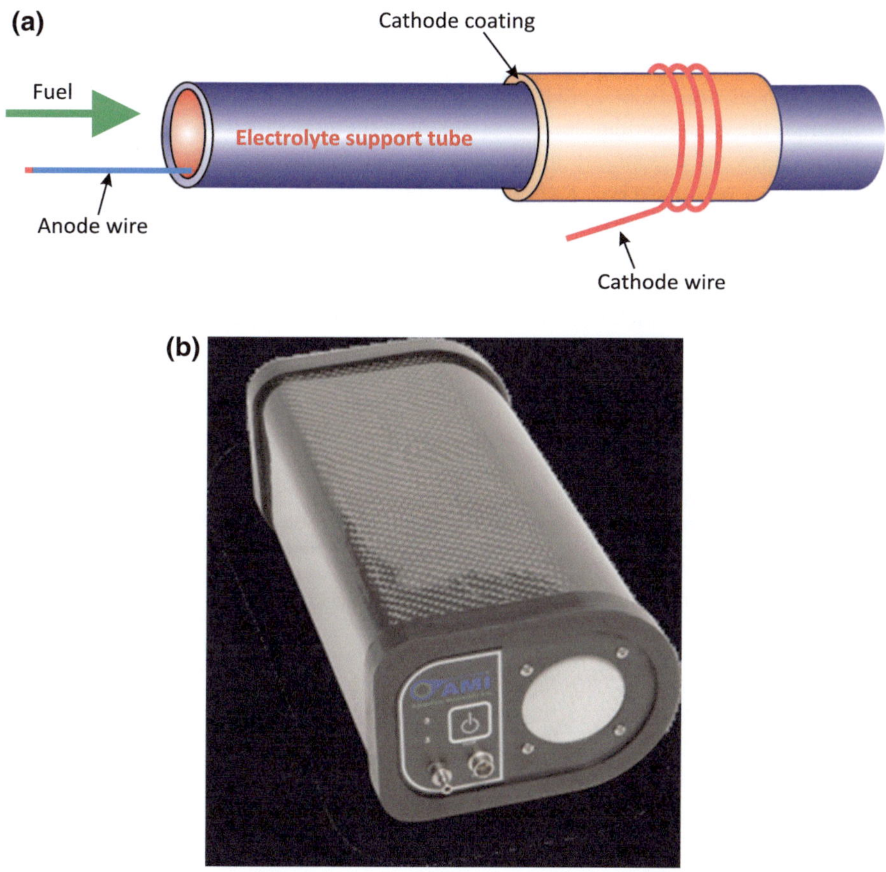

Fig. 4.19 **a** Schematic of microtubular cell (Singhal and Kendall 2000) and **b** 50 W portable microtubular cell (Kendall 2010)

2 mm diameter is a support for the electrodes. In addition to this, the electrolyte tube also serves as a gas inlet tube and combustor outlet tube. The length of the tube may vary between 100–200 mm whereas the occupancy of the cell region is about 30 mm towards the outlet end of the tube.

On the inner wall of the tube, 30 mm long and 50 μm thick Ni+YSZ anode is coated. Additionally a Ni current collector wire is inserted which comes out from the fuel inlet side of the tube. The cathode is fabricated by depositing a 100 μm thick lanthanum strontium manganite layer with a silver wire wound around it to obtain the cathode current. The microtubular cell is obtained by the extrusion process, which produces several layers in one step. For a typical SOFC, five different layers (1 electrolyte + 2 electrodes + 2 metal connections) are fabricated separately to yield one structure. Bi cermet anode tubes with a 15 μm layer of YSZ using the thermoplastic co-extrusion process could be developed. Microtubular cells exhibit two major benefits, viz. high power density and thermal shock resistance. Power density is reciprocal of the tube diameter. Hence, a 1 mm diameter MT-SOFC provides 10 times more power output of stack per stack volume and it depends on tube diameter and a real power density $\boldsymbol{p}$ of the stack. If l is the tube length, then the power output of a single tube is $\pi Dl\boldsymbol{p}$ and the stack volume is πD^2L, giving a power density of $\pi\boldsymbol{p}/4D$.

Second, the microtubular cell possesses high thermal shock resistance. The large diameter SOFC tube is susceptible to damage due to cracking, but in MT-SOFC no such cracking was observed. Experiments show that the strongest plates cracked at a flow of cold air of 1 L/min for planar 8YSZ electrolyte (2 mm diameter). But the 8YSZ injector tube for MT-SOFC never broke even at the highest cold air flows, and hence MT-SOFC was found to be 10,000 times more resistant than planar cells. Table 4.1 shows the microtubular cells stack designs researched by various firms.

Table 4.1 Microtubular cells stack designs by various firms

Group	Cell diameter (mm)	Operational temp. (°C)	Heat up time (°C/min)	Support	Stack power (W)
Adelan	~2	850	200	Electrolyte/Anode	1000 cell—500 W
Alberta Research Council	~2	800	–	Anode	30 cells—20 W
AIST	0.8	500	83	Cathode	20 cells—20 W
Nanodynamics	–	800	550	Cathode	250 W
Upper Austria University of Applied Science	~2	850	–	Anode	7 cells—3.5 W
Korean Institute of Technology	10	750	6	Anode	36 cells—700 W
University of Connecticut	13.2	850	40	Anode	40 cells—1000 W

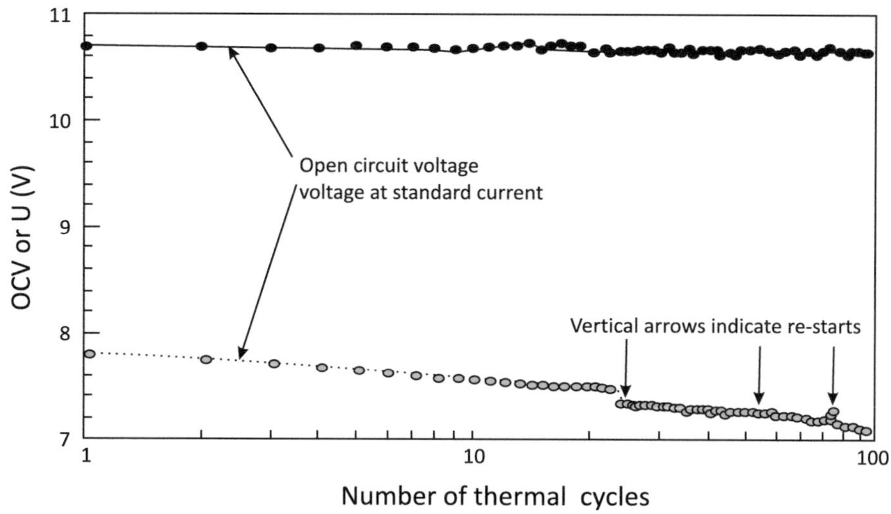

Fig. 4.20 Thermal cycles for RRFCS fuel cells (Kendall 2010)

Investigations showed that rapid heating and cooling were possible in the microtubular cell system. For a RRFCS tubular plate containing 20 cells, 10 in series on one side and other 10 on opposite side, the open circuit voltage of 1.07 V/cell at 900°C on hydrogen fuel was obtained which was more or less constant throughout the cycles (Fig. 4.20). The lower curve for the 10 series-connected cell showed a voltage fall of 0.8 V after 100 cycles. The temperature was brought down to 200°C and then raised to 900°C. The fatigue relationship could be given for number of cycles (N) as

$$\mathrm{d}V = -k \log N \tag{4.26}$$

where k depends on temperature ramp rate. During each cycle, the microcracks within the nickel cermet anode separates nickel particles causing extra resistance.

Anyhow, these cracks do not enter equally during each cycle due to the relaxation of the crack driving force by the expansion process. The test obtained from redox cycling showed more prominent damage as compared to the thermal cycling. In the redox cycle, hydrogen was switched off periodically allowing the Ni anode to reduce and oxidise leading to its repeated contraction and expansion. Figure 4.21 shows a 10-s redox cycle which yields a 0.4% degradation/cycle. In the redox cycling, the crack driving force was almost constant unlike thermal cycling where the crack driving force followed an inverse relation with the cycle number. Hence, redox cycling is more damaging in nature. But if the cycling rate increases, then the damage due to redox cycling may fall.

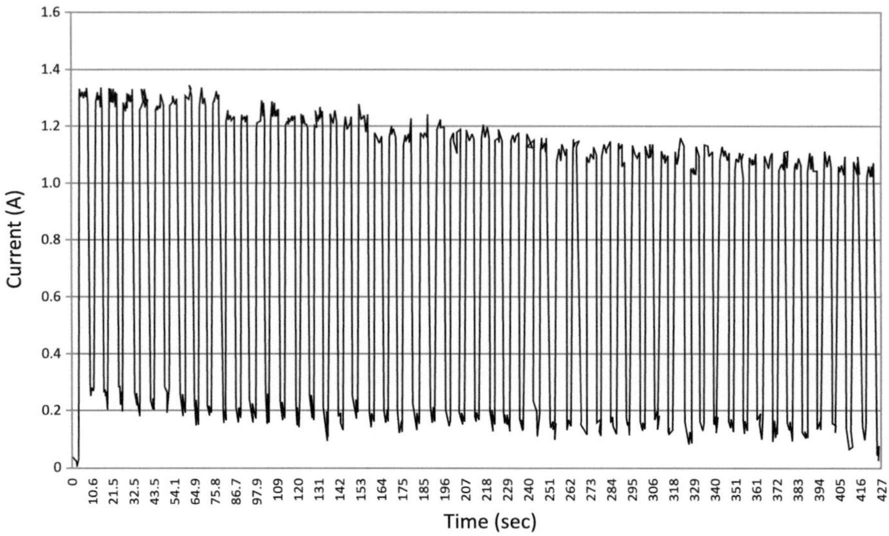

Fig. 4.21 Redox cycling for 10 s in microtubular fuel cell (Kendall 2010)

4.3.3.3 Planar Cell Design (Bipolar/Flat-Plate Design)

Planar design is usually constructed using a positive electrolyte-negative electrode (PEN) cell sandwiched between the interconnect plates (at top and bottom) (Fig. 4.22). The air channels are located between the interconnect plate electrodes; i.e., an air channel exists between the cathode and interconnect (separate plate), whereas the fuel channel exists between the anode and interconnect.

The planar fuel design offers many attractive features including:

- Low fabrication costs and simple manufacturing process
- High volumetric power densities; simple series electrical connections
- Lower ohmic polarisation due to perpendicular current collection in cell stack
- Ease of heat removal

Planar SOFC also uses the same common materials for component fabrication as used for other designs of SOFC: YSZ is used as an electrolyte, Ni/YSZ cermet as an anode, and lanthanum strontium manganite as the cathode. If the electrolyte thickness is 5–20 μm instead of 100 μm then the cell can operate at reduced temperatures, which allows a wide choice of materials along with the reduction in thermal stress. Anyhow, the planar geometries offer some drawbacks including:

- Difficulty in obtaining mechanical stable structure.
- Moderate stress can also lead to the failure of cell stack.
- Start-up time is high.
- Requirement of gas-tight high temperature seals to prevent gas leaks in bipolar configuration.

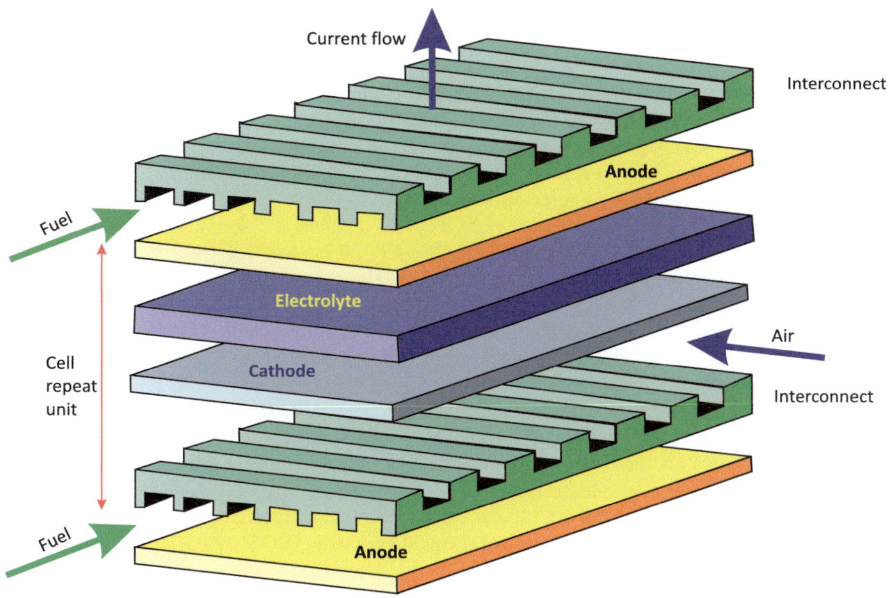

Fig. 4.22 Planar configuration of SOFC with ribbed interconnects

As shown in Fig. 4.22, the interconnect has ribs on both sides which act as gas flow channels along with acting as gas separation. The cell components for planar SOFC can be fabricated using a tape-casting method or tape calendaring. Using casting, thin sheets of ceramics can be obtained; i.e., a layer of powder suspended in liquid is made using a doctor blade followed by drying. Then the dried layer is stripped from the support and fired to obtain the ceramic layer. In contrast to this, in tape calendaring, organic binder and ceramic powder are mixed and then rolled and fixed at elevated temperature to obtain the sheet of desired thickness. Other methods to obtain the components of SOFC are deposition techniques such as spin coating, spray pyrolysis, sputtering, dip-coating, plasma spraying, electrophoretic deposition, and slip casting. These techniques are used to obtain thin layers of SOFC electrolyte. Anyhow, not many details have been discussed for these techniques because they are beyond the scope of this book. In the next chapter, the main focus is on the different seals for planar designs.

Bibliography

Bastidas DM (2006) Rev Metal 42(6):425

Hajimolana SA, Azlan Hussain M, Ashri Wan Daud WM, Soroush M, Shamiri A (2011) Renew Sustain Energy Rev 15:1893

Kendall K (2010) Int J Appl Ceram Technol 7(1):1

Larminie J, Dicks A (2003) Fuel cells system explained, 2nd edn. Wiley, England

Minh NQ (2004) Solid State Ionics 174:271

Singhal SC, Kendall K (2000) High temperature solid-oxide fuel cells: fundamentals. Elsevier, New York

Suggested Bibliography

Bujalski W, Dikwal CM, Kendall K (2007) J Power Sources 171:96
Data from Allied-Signal Aerospace Company, 1992
Fuel Cells, DOE/METC-86/0241, Technology Status Report, Morgantown Energy Technology Center, Morgantown, WV, 1986
Hirschenhofer JH, Stauffer DB, Engleman RR, KlettM G (1998) Fuel cell handbook, 4th edn
Matsuzaki Y et al (1992) High power density SOFC development at Tokyo gas. Fuel Cell Program and Abstracts, 1992 Fuel Cell Seminar, Tucson, AZ, 29 Nov–2 Dec 1992
Minh NQ (1991) High-temperature fuel cells, Part 2: The solid oxide cell. Chem Tech 21, Feb 1991
Minh NQ (1993) J Am Ceram Soc 76(3):563
Minh N (2003) In: Singhal SC, Dokiya M (eds) SOFC VIII. The Electrochemical Society, Pennington, p 43
Minh NQ, Horne CR (1993) High temperature electrochemical behaviour of fast ion and mixed conductors. In: Poulsen FW, Bentzen JJ, Jacobsen T, Skou E, Ostergard MJL (eds) Proceedings of the 14th RISO international symposium on materials science. Riso National Laboratory, Roskilde, p 337
Singhal SC (1997) Recent progress in tubular solid oxide fuel cell technology. In: Proceedings of the fifth international symposium on solid oxide fuel cells (SOFC-V). The Electrochemical Society, Inc., Pennington
Singhal SC (2002) Solid State Ionics 152–153:405
Sverdrup EF, Warde CJ, Glasser AD (1972) In: Sandstede G (ed) From electrocatalysis to fuel cells. University of Washington Press, Seattle, WA, p 255
www.doitpoms.ac.uk
Yoshida Y et al, Development of solid oxide fuel cell. Paper provided by Mitsubishi Heavy Industries Ltd.
Yamamoto O (2000) Electrochimica Acta 45:2423
Zeh C (1987) Handbook, 2nd edn (Private communication), 29 April 1987

Chapter 5
Sealing Concepts: Glasses as Sealants

With the ever-increasing needs of energy, fuel cells have emerged as tantalising alternatives due to their high efficiency (higher than diesel combustion engines and gas turbines), low maintenance, and ability directly to convert chemical energy into electrical energy via an electrochemical reaction. The solid oxide fuel cell is unique as it is fabricated entirely from the solid components, which eradicates any possibility of liquid corrosion and hence performance degradation. SOFC also offers fuel flexibility and internal reformation because it operates at high temperature. No external fuel reformer is required, and hence a wide range of hydrocarbon materials can be used as fuels. During the operation of SOFC, a large amount of heat is generated, which can be used in combined heat power (CHP) units to increase efficiency as well as to carry out internal fuel reformation. Among the different geometries of SOFC, the planar design has received wide attention due to its high volumetric density. Anyhow, the planar configuration needs an additional requirement of sealants. The following sections are focused on the requirement of sealants and different sealing concepts given for the SOFC.

5.1 Why Sealants Are Required?

The planar SOFC stack contains yttria-stabilised zirconia electrolytes, which are solid in nature. The temperature and thickness of electrolytes are important criteria, which determine the oxide ion conductivity of the SOFC. SOFC operates under stringent conditions such as the oxygen ion gradient across the electrolytes. This gradient makes it imperative that the YSZ layer be dense with minimum interconnected porosities, and hence must be in contact with the device via high-temperature gas-tight seals. During stack manufacturing or component degradation of SOFC, some leaks may develop that could lead to deterioration of SOFC performance, poor fuel utilisation, and reduced efficiency. These leakage losses may further add to the local hot spots or internal combustion within the cell. Hence, gas-tight seals are required to prevent the leakage losses, and hence stack degradation.

G. Kaur, *Solid Oxide Fuel Cell Components*,
DOI 10.1007/978-3-319-25598-9_5

Hence sealants are required to:

- Electrically isolate cells in the stack.
- Provide mechanical bonding among components.
- Avoid mixing of fuel and oxidant within the stack.
- Prevent any fuel/oxidant leakage from stack.

5.2 Criteria for the Sealants

While in operation the sealants are subjected to high temperature conditions (600–1000°C). In addition to this, the seals are continuously exposed to oxidising and reducing conditions at the cathode and anode, respectively, for an anticipated lifetime of >10,000 h. The sealing material selection heavily relies on the applications (whether portable or stationary) and the geometry of the fuel cell. For stationary applications, the fuel cell seal operates for higher durations, whereas for mobile applications, the seal undergoes the thermal cycling, dynamics mechanical loading, and thermal shocks. To sum up with all requirements for the seals, Fig. 5.1 shows an insight for the generic criteria for acceptable sealants.

The stack assembly procedure also affects the seal selection process. For instance, a laminate structure of anode, electrolytes, and cathode is formed by the plasma spraying technique. The microstructuring processing steps must be controlled so that the interfacial reactions and interdiffusion between those layers can

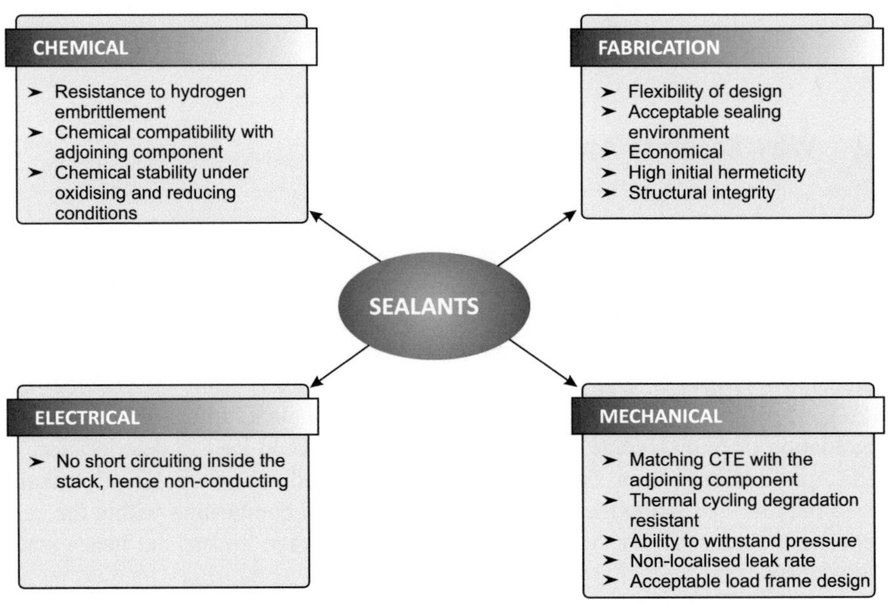

Fig. 5.1 Selection requirement for sealants

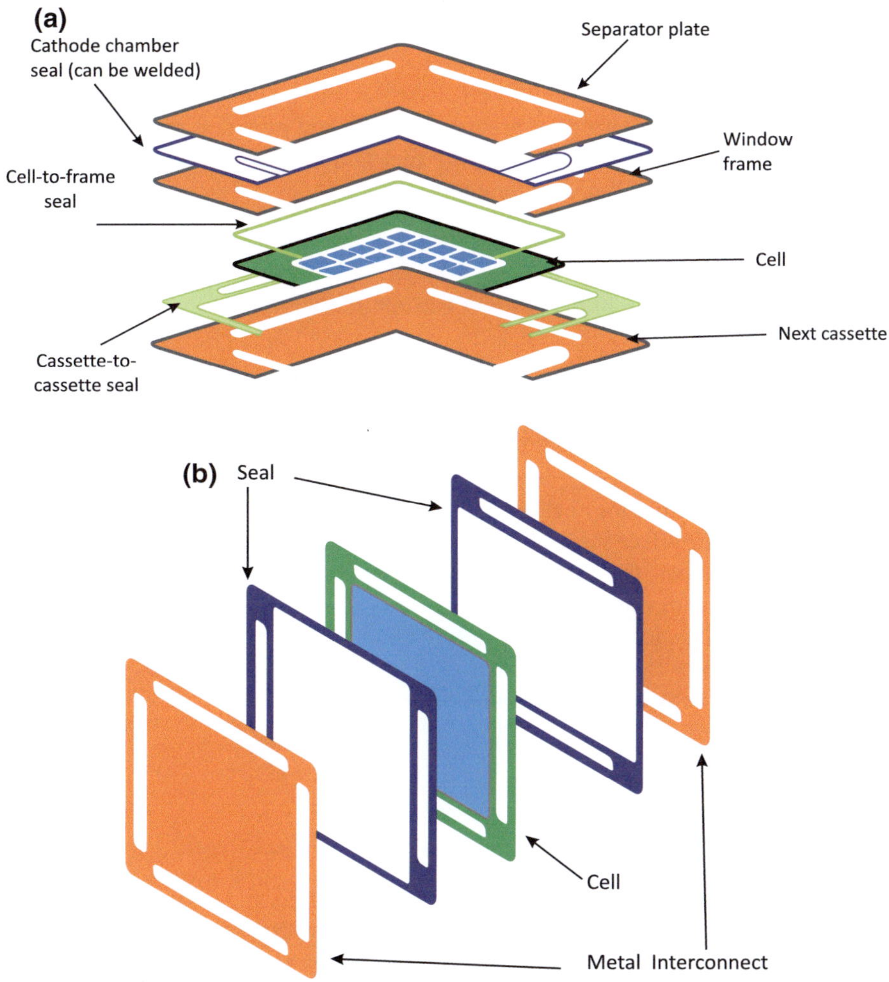

Fig. 5.2 **a** Cell-to-frame design and **b** cell–to-edge design for sealing in SOFC (Weil 2006)

be minimised. When a cell is exposed to the high temperature or inert gas environment, the property degradation and chemical reduction may take place. Hence, in order to maintain the proper oxidation state of cell materials, the air-fired sealing process is done. For particular geometries of planar SOFC, the sealant considerations must be taken into account. In window frame design (cell-to-frame design), the cell size is smaller than the separator plate and joined to metallic window frame, which provides the mandatory gas porting as shown in Fig. 5.2a.

Three seals are used for the cell-to-frame design:

- Between the cell and window separator plate forming a cassette unit
- Between every cassette of the stack

- Between the system gas manifold and stack (gas manifold supplies fresh fuel and air)

The second design is the cell-to-edge design as shown in Fig. 5.2b. In this design, the cell and separator plate are the same size and provided with the same gas manifold pattern holes through which the fuel and air transports take place. The sealing for cell–to-edge design is required:

- Along the edges of manifold holes, such that the fuel and air gases mixing can be prevented
- At the exposed edges to prevent the leakage of fuel outward

5.3 Different Types of Sealants

It has become clear that the seal quality must be high to prevent even the smallest leaks, which could affect the cell potential and thus deteriorate cell performance. For rapid startup times, the seal design is important because high thermal stress may develop in glass with rapid heating and cooling. For a typical positive electrode–electrolyte–negative electrode (PEN) cell, pSOFC cell stack, the sealant sections are shown in Fig. 5.3.

The repeating cells of SOFC consist of

- Metallic or ceramic interconnect
- Metal or ceramic window frame
- PEN cell

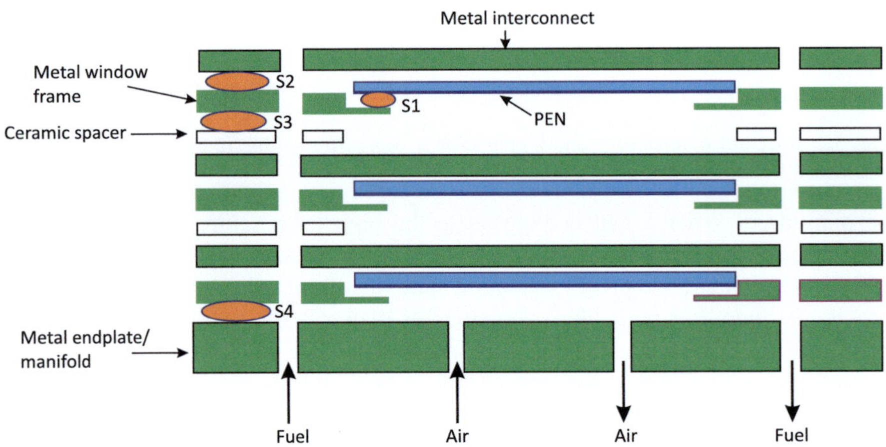

Fig. 5.3 The seal positions for the SOFC (Chou et al. 2007a, b)

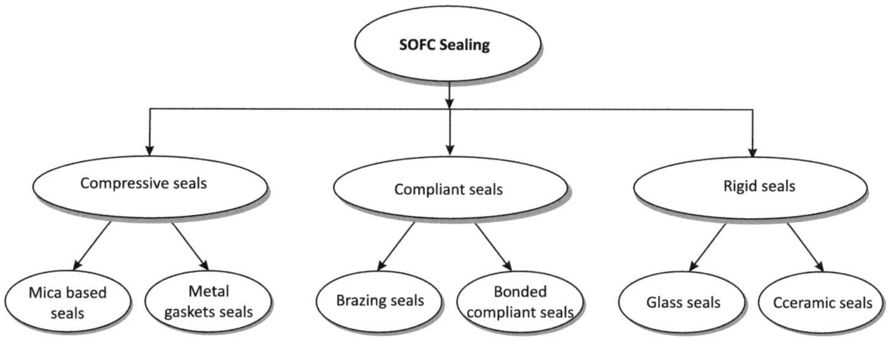

Fig. 5.4 Different sealing concepts for SOFC

For planar SOFC stacks, four different sealing sections are given:

- S4 seal between the base manifold and stack
- S3 seal between the metallic window frame/interconnect and the ceramic spacer
- S2 seal between the metallic/ceramic window frame and metallic/ceramic interconnect
- S1 seal between the metallic/ceramic window frame and PEN plate

The seals should be reliable to obtain long-term performance goals. The SOFC consists of 10–15-μm thin electrode–electrolyte delicate assembly, which requires the seal separating the cathode from the anode with minimum stress transmission. Basically, three different sealing approaches are given for SOFC which are shown in Fig. 5.4 and explained in the following section.

All sealing techniques have offered wide advantages along with certain disadvantages.

5.3.1 Compressive Seals

Compressive seals are the non-bonding deformable materials. The non-bonding nature of these seals allows the stack components to contract and expand during the thermal cycling. The sealing occurs when the entire stack is compressively loaded, and hence a dynamic seal is formed. A dynamic seal refers to the fact that the seals do not require any criteria of thermal expansion matching and sealing surfaces can slide past one another. Due to the requirement of load, compressive seals require a load frame to maintain the desired compressive load during operation. Several complexities are introduced in pSOFCstacks due to these load frames such as oxidation of cell frame, increased thermal mass load relaxation due to creep, and increased system cost. It is mandatory that the stack should withstand the compressive load. Anyhow, as CTE matching is not required, the stack designers can use either alloys or ferritic stainless steels for metal components. Mostly, two types of seal materials are used as compressive seals, i.e., metallic gaskets and mica-based materials, and are discussed in detail in Sects. 5.3.1.1 and 5.3.1.2.

5.3.1.1 Metallic Gaskets

Usually, ductile metals are used as seals so that they can deform on applying stress. In addition to this, metals such as gold/silver are also used because they are resistant to the formation of oxide scales. The metal gaskets should form hermetic sealing at higher pressures of >25 MPa. Therefore, the metal gasket should possess high strength. Chou and co-workers obtained silver metal seals degraded with thermal cycling due to the formation of cracks along the grain boundaries. Singh and co-workers exposed the silver metal seals to both oxidising and reducing atmospheres and found that the dissolved oxygen and hydrogen reacted with silver to form water, hence leading to seal failure. Furthermore, the metal gaskets are C-shaped, corrugated, and hollow tube cross-sections, and they are used to overcome the need of ductile metals. Usually, these different metal gaskets can be oxidation-resistant alloys such as stainless steels and Ni-based superalloys. When these gaskets are coated with a noble ductile metal, hermeticity of seals can be improved by increasing the tightness of the interface. Anyhow, the metal gaskets pose a serious problem of short circuiting due to electrical conduction.

5.3.1.2 Mica-Based Seals

Mica-based materials are used as an alternative to metallic gaskets. Micas are composed of sheets of silica tetrahedrons and belong to the class of layer mineral phyllosilicates. The most commonly investigated mica for SOFC applications are

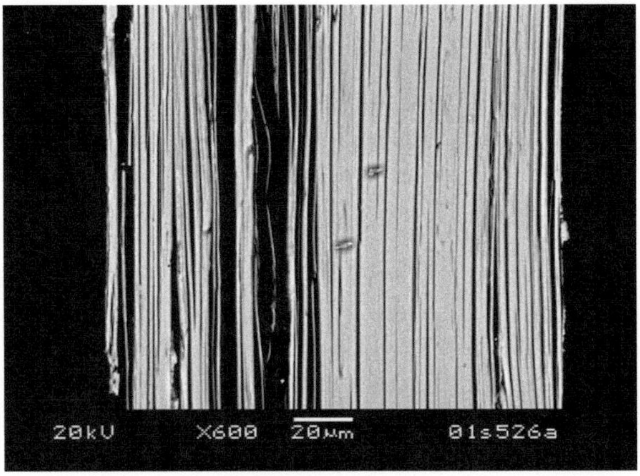

Fig. 5.5 Microstructure of Muscovite mica single crystal after heat treatment at 500°C (Chou 2002)

muscovite and phlogopite as they exhibit the lowest leak rates. Usually, naturally occurring mica has certain defects including fine scratches. Upon heat treatment, the muscovite mica sheets lose water and cleave into multiple layers (Fig. 5.5), whereas phlogopite mica shows overlapped mica flakes. Phlogopite ($KMg_3(AlSi_3O_{10})(OH)_2$) exhibits higher CTE than the muscovite ($KAl_2(AlSi_3O_{10})(F,OH)_2$).

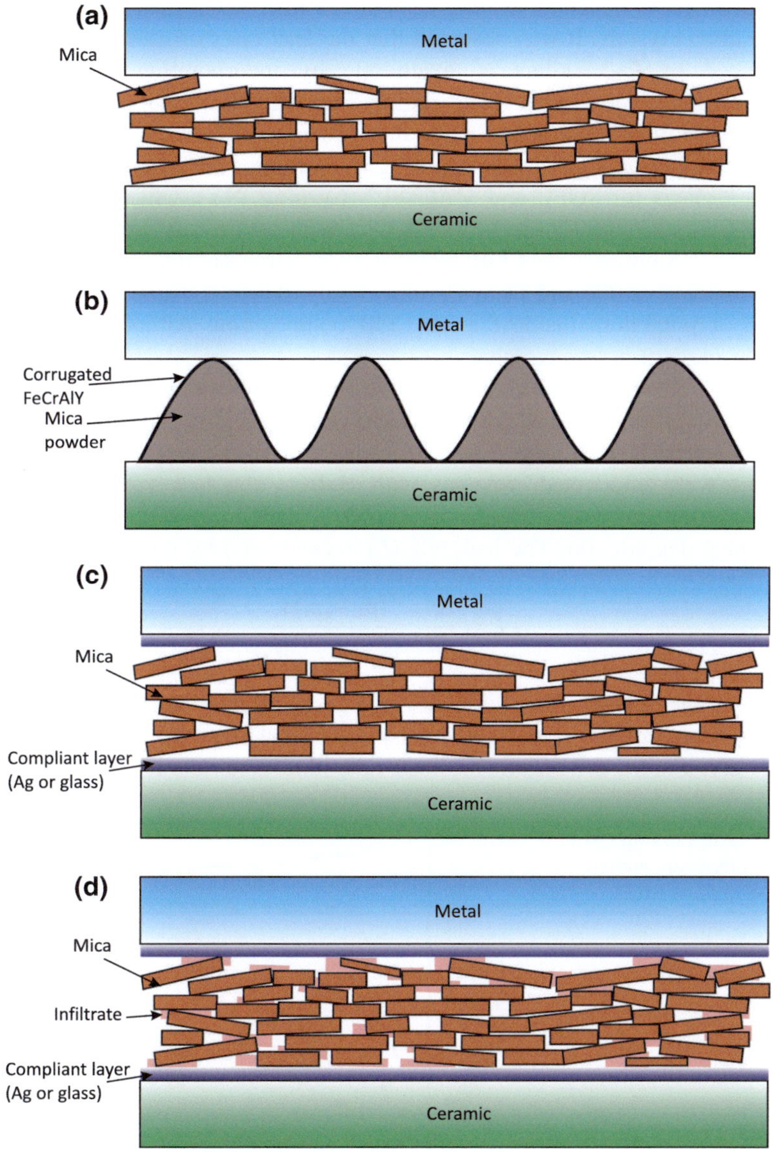

Fig. 5.6 **a** Mica seal, **b** mica powder with corrugated alloy (glass or metal), **c** hybrid mica seal with compliant layer and **d** infiltrated mica seal (Fergus 2005)

All these mica materials exhibit a uniform dielectric constant along with high resistivity. Usually, primary leak paths in compressed mica seals are the interfaces between mica sealants and the ceramic/metal component. Figure 5.6a shows the gas-tight seal formed by mica crystals and the predominant leakage routes. To improve the performance of mica seals, mica powder can be placed inside the gaps in corrugated metal seals (Fig. 5.6b).

The glass provides better sealing capabilities than silver metal when combined with the mica crystals. It was observed that the leak rates of muscovite mica seals are reduced by 2–3 orders of magnitude by incorporating compliant seals (Fig. 5.6c). However, it was found during microstructural analysis that the glass components diffused inside the mica sheet and the leak rates increased rapidly during initial thermal cycles. Mica-based glass/ceramics have been proposed as the potential sealants for SOFC and MACOR® (Corning, NY), which is one of them. MACOR® is successfully used to bind YSZ with ceramic as well as the metallic interconnect. Anyhow, MACOR showed discolouration and reacted with lanthanum chromite at 900 and 1000°C, respectively, which is usually attributed to the presence of potassium inside the MACOR composition. Although potassium is present in mica, it is less reactive.

Another concept of infiltration was proposed by which the seal could be improved by infiltrating mica with a phase or wetting and melt-forming agent such as $Bi(NO_3)_3$ and H_3BO_3 (Fig. 5.6d). It has been observed that with $Bi(NO_3)_3$, the leak rates were unaffected, whereas with H_3BO_3, the leak rates decreased for thermal cycling. Anyhow, Bi_2O_3 of $Bi(NO_3)_3$ may react with steel leading to performance degradation. The leak rate (L) is calculated by the formula:

$$L = \frac{\Delta n}{\Delta t} = \frac{(p_f - p_i)V}{RT(t_f - t_i)} \tag{5.1}$$

where V stands for reservoir volume, R is the gas constant, T is the temperature, and p_i and p_f represent the initial and final pressures at time (t_i) and final time (t_f), respectively. Figure 5.7 shows the leak rates due to different interlayers, which

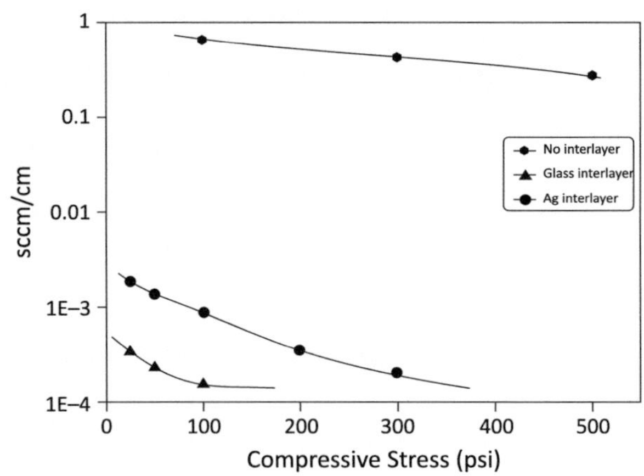

Fig. 5.7 Leak rates due to different interlayers with respect to applied stress

shows that glass layer is more effective than the silver interlayer in controlling leaks. Although promising results have been obtained with mica/glass seals, the compression load is the mandatory requirement to achieve gas tightness, which leads to the increased complexity of the system. Also, the compressive sealing stresses are not uniform over a large sealing area.

5.3.2 Compliant Seals

The joint region in compliant seals can be plastically deformed above room temperatures, which lessens the thermal stresses caused due to CTE mismatch. This category of sealants includes metallic components and hence is electrically conducting. In addition to this, the cell bowing and non-uniformities in the gas distribution are potential problems. Brazing and bonded compliant seals fall in this category and are discussed in Sects. 5.3.2.1 and 5.3.2.2.

5.3.2.1 Brazing

Metals undergo plastic deformation easily, hence providing a vent for the thermal and mechanical stresses. Braze materials are usually made up of metals which are stable in air such as silver (Ag), gold (Au), and platinum (Pt). Anyhow, other metals such as copper, nickel, titanium, palladium, and chromium can be used as potential metal brazes. The inadequate melting of ceramic materials by braze materials remains the potential problem. Usually, in brazing materials, a filler material with liquidus below the materials to be joined is heated so that it melts and fills the pores under capillary action. This melt will solidify upon cooling, hence solidifying and forming a solid joint. The addition of some reactive materials including hafnium, zirconium, and titanium along with metal oxides such as Al_2O_3, CuO, and Al_2TiO_5 is done to the brazes in order to overcome wetting issues. The addition of reactive materials will overcome the oxidation problem and also CTE can be adjusted so as to match that of electrolyte YSZ and interconnects. The brazing can be an active metal brazing or air brazing. In active metal brazing, a reactive material such as titanium is employed which facilitates the wetting of the sealing surface and filler material. The following problems are to be faced during active metal brazing.

- Joining process should take place in an inert environment or reducing gas atmosphere, which requires high maintenance of partial pressures.
- Joining should be done under high temperature.
- The joints degrade at elevated temperatures and are not resistant to the oxidation process.

The air brazing technique has been developed to overcome these shortcomings. Air brazing offers the following advantages.

- Promotes the wetting of ceramic sealing surface by employing a partial soluble molten oxide in noble metal solvent.
- Exhibits high temperature oxidation resistance.
- No vacuum or inert atmosphere is required.

The most promising air braze composition is Ag–CuO which exhibits good surface wettability, metallic ductility, and high temperature oxidation resistance. The Ag–CuO system contains a eutectic and is used to braze the perovskites and alumina. Figure 5.8 shows the variation of wetting between Ag–CuO (with the % CuO) when in contact with SOFC components [LSCF-6428 ($La_{0.6}Sr_{0.4}Co_{0.2}Fe_{0.8}O_{3-\delta}$) 5% YSZ and FeCrAlY (Fe-22Cr–5Al–0.2Y)]. The braze paste of Ag–CuO is formed by mixing them in screen printing binder. Even 1.4% CuO in silver yields good adhesion and wettability on ceramic sealing surface.

Within 10–20% CuO concentration, maximum decrease in wetting angles could be observed. The cross-section of a brazed joint with 90 mm thickness is shown in Fig. 5.9. The concentration of CuO is kept at 4% and the composition showed no reaction zone with YSZ, though dispersed CuO crystals are observed at the interface surrounded by silver matrix. Between the FeCrAlY alloy and braze, an interaction layer of ~2 mm thickness is observed. The interaction layer comprises $2CuO \cdot Al_2O_3$ and $CuO \cdot Al_2O_3$ phases. A thick alumina (Al_2O_3) scale of ~0.5 mm is also formed adjacent to the reaction zone. It has been observed that on increasing CuO concentration, increase in the interaction zone occurs due to the formation of a thick CuO layer. If alloying elements such as TiO_2 are added to the Ag–CuO system, then the temperature, wetting behaviour, and oxygen solubility are modified. If Ag–CuO brazes are used to join two components directly in air,

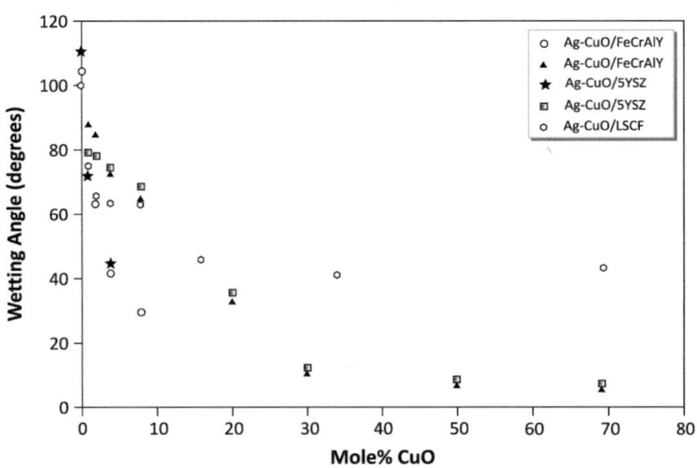

Fig. 5.8 Variation of wetting angle between Ag–CuO and YSZ/FeCrAlY/LSCF interface with respect to the concentration of CuO (Fergus 2005)

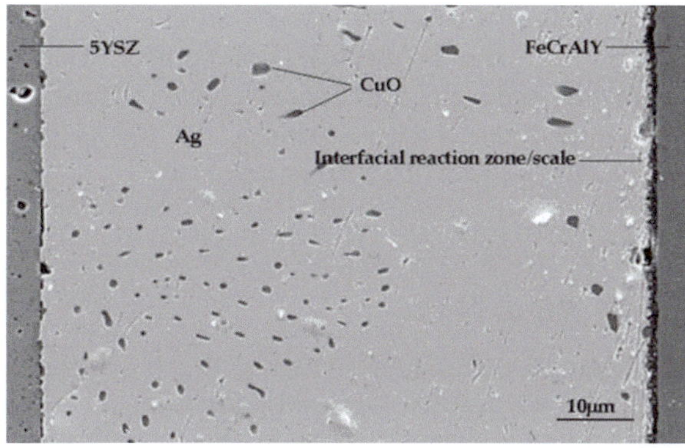

Fig. 5.9 Cross section between Ag-4% CuO braze/FeCrAlY alloy formed at 1050°C (Weil 2006)

then the joint becomes highly ductile and can withstand thermal cycling without degradation. Anyhow, the Ag-based alloys possess degradation under dual atmospheres with volatility of Ag long out of the braze. For short-term stack operation, i.e., ~ 120 h, the microstructures were intact without loss of hermeticity. In addition to this, no substantial changes in the microstructure were observed. Anyhow, extensive work is required to get an idea about the effect of extensive thermal cycles and dual atmosphere on the brazes.

5.3.2.2 Bonded Compliant Seals

In bonded compliant seals, a thin stamped foil is bonded to both sealing surfaces as shown in Fig. 5.10. A number of high-temperature alloys can be potential candidates for the foil membrane if they satisfy the following criteria.

- Low cost and ease of fabrication
- Low stiffness and high ductility
- High oxidation resistance

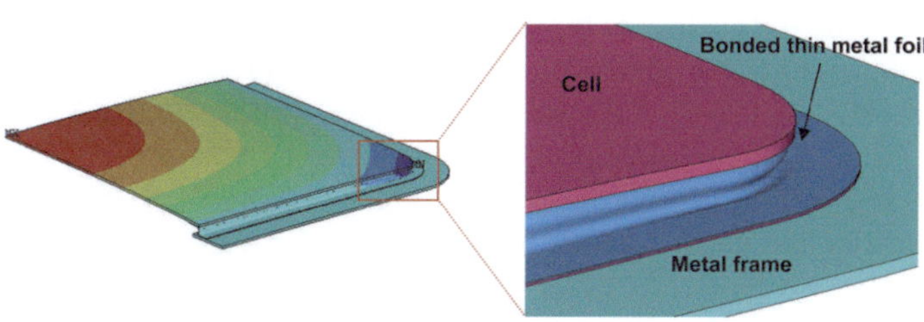

Fig. 5.10 Foil-based bonded compliant seal (Weil 2006)

Durafoil (22-Cr/7-Al/0.1-La + Ce/Fe) is a commercially used bonded foil as it satisfies the above-mentioned properties. The bonded compliant seals are non-sliding and metal foils endow a higher degree of geometric deformation. Under modest thermomechanical loading, the foil is deformed and may return to its original shape when the load is removed.

5.3.3 *Rigid Seals*

In rigid bonded seals, the joint formed is non-deformable at room temperature. During thermal events or due to the CTE mismatch, the joint is susceptible to cracks/fracture. It becomes imperative that the sealant should have matched CTE as that of the component with which it is in contact. If the CTE is not matched, then there is air maldistribution in the stack and cell bowing may occur. The CTE of ferrite steels lies in the range ($12–13 \times 10^{-6}$/K), whereas the ceramic components exhibit CTE of $(10–12) \times 10^{-6}$/K. Glasses and ceramics are the potential rigid seals and are attracting the attention of many researchers across the globe.

5.3.3.1 Glasses/Glass–Ceramics Seals

The glass materials are inexpensive, display acceptable stability under oxidising and reducing atmospheres, and exhibit good bonding and wetting behaviour along with electric insulation. Glasses are characterised by the glass transition temperature (T_g) above, which the glasses change from brittle to elastic. During operation at high temperatures, the glasses change to glass-ceramics because of the occurrence of crystallisation in them. Usually, crystallisation increases the mechanical strength of the glass and the properties of glass-ceramics are tailored w.r.t. the original glass composition. The glasses can be borate, borosilicates, aluminosilicates, and silicates. Boron is volatile in nature and may undergo a corrosion problem when in contact with adjacent components of the fuel cell. Crystallisation modifies the viscosity because the viscosity should be enough to wet the surfaces but the glass should not squeeze out of the stack. Hence, during the designing of the glass composition, glass transition temperature (T_g), glass softening temperature (T_s), and glass crystallisation temperature (T_c) should be kept in consideration. Before going deeper into the behaviour of glasses as sealants, we discuss the glass structure, its components, nucleation, crystallisation, and the phase formation in them. Sections 5.4–5.9 are entirely based on the glass structure and their role as sealants in SOFC.

5.3.3.2 Ceramic Seals

Ceramic sealants have been fabricated to join the components of SOFC with reaction bonding. Anyhow, the ceramic may not exhibit CTE matching, and hence is prone to thermal cracking. Although the bonding treatment may take place at higher temperatures, by adding organosilane components such as SiC/SiO_xC_y, the joining temperature can be lowered. The pyrolysis of polymers may lead to the formation of gaseous reaction products, and hence the volume shrinkage occurs accompanied with pore and crack formation during the processing. This technique is still under development and fundamental aspects need to be evaluated completely before ceramics can be regarded as suitable candidates for pSOFC applications.

5.4 Introduction to Glasses

Glass is an amorphous material with no long-range order of atoms, lacking periodic atomic structure, and exhibiting a glass transformation region (Fig. 5.11). Glass is usually brittle and optically transparent. In glass, the atoms are arranged in a random manner like a liquid and behave like a super-stiff liquid. A broader distribution of bond angles is observed for the glasses. Usually, the dimension of glasses may extend to three scales:

- 2–10 Å—Local atomic arrangements
- 30-few thousand Å—Submicrostructure scale
- Microns–millimetres—Microstructure/Macrostructure

Any material that may be metallic, organic, or inorganic is considered as a glass if it exhibits a glass transformation.

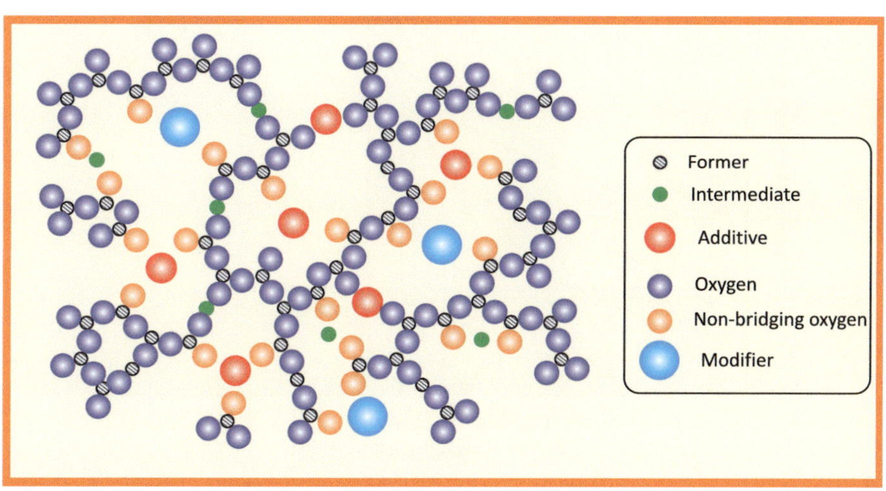

Fig. 5.11 Schematic of glass network structure

5.4.1 Enthalpy/Temperature Curve of Glass

The glass transformation behaviour can be discussed in terms of enthalpy or specific volume (**V**) w.r.t. the temperature diagram as shown in Fig. 5.12. If the crystal is cooled to any temperature below the melting temperature, then the crystalline state of the material is produced which exhibits long-range structural periodic arrangement. Upon cooling, the enthalpy will keep on decreasing due to the heat capacity of the crystals. A liquid then can be cooled below melting temperature (T_m) without any crystallisation and is known as supercooled liquid. Furthermore, when a supercooled liquid is cooled to lower temperatures, then structural rearrangement takes place and leads to an increase in viscosity. No abrupt decrease in enthalpy is observed due to the movement of molecules, leading to structural rearrangement of the atoms. Upon cooling, eventually the viscosity becomes so high that no more structural arrangements of atoms may take place. The structural arrangements are required to find equilibrium-specific volume/enthalpy of the liquid. A structural lag occurs and experimental enthalpy deviates from the equilibrium curve and follows a curve of decreasing slope. At very low temperature, the time scales become quite large compared to the time scale of experimental observations for the molecular arrangements. Hence, the high viscosity at this temperature makes the liquid structure fixed and can be considered ‘frozen’. This frozen structure is temperature independent and known as ‘glass’.

The temperature region between the enthalpy of the equilibrium liquid and that of frozen solid is known as the ‘region of glass transformation’. Now, from here we

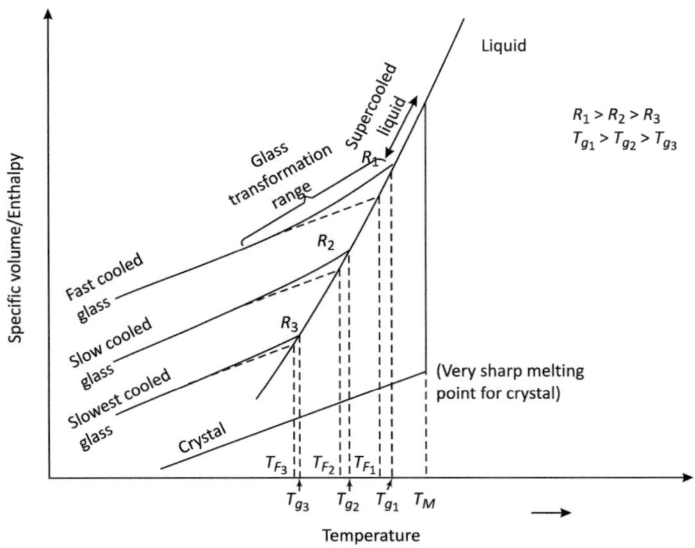

Fig. 5.12 Volume/enthalpy and temperature relation for glasses formed at different cooling rates R_1, R_2 and R_3

need to define 'glass transition temperature' which can be interpreted in different ways. The glass transformation occurs over a region as shown in Fig. 5.12, and hence cannot be characterised by a single temperature. We can define the glass transition temperature using the coefficient of thermal expansion. If a liquid crystallises, then a discontinuous change in enthalpy is observed at melting temperature but if no crystallisation occurs, then the volume (V) of liquid decreases almost at same rate as that occurring above the melting temperature until a decrease in the coefficient of thermal expansion occurs. This decrease in CTE usually occurs within a range of temperatures and is known as the glass transformation region. Usually, the crystalline and glassy states exhibit the same CTE. A higher density of glass is obtained if the slow cooling rates are used such that the supercooled liquid has more time for the structural relaxation. The CTE can be given as

$$\mathrm{CTE} = \alpha = \frac{\mathrm{d}(\ln \mathbf{V})}{\mathrm{d}T} \tag{5.2}$$

Anyhow, the CTE for glass is smaller than the liquid or supercooled liquid states because the atomic vibrations are different in both cases. The glass transition/transformation temperature can be defined as 'the intersection temperature between the curve of supercooled liquid and glassy state' (Fig. 5.12). As different cooling rates exhibit different structural relaxations, hence, we get different T_g. It is a function of heating rate (β) and is a useful indicator about the transition of supercooled liquid to glassy solid upon cooling and the solid behaves as a visco-lastic material. Figure 5.13 depicts the behaviour of selenium glasses cooled from melting temperature ($T_m = 490$ K) to 200 K for different heating rates. Another temperature known as fictive temperature (T_F) is defined as the intersection temperature between the extrapolated supercooled liquid temperature and the glass line.

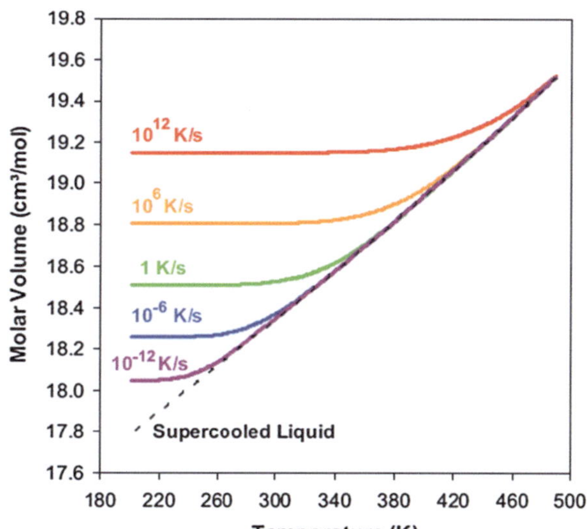

Fig. 5.13 Volume temperature diagrams for the selenium glasses cooled at different rates (Mauro and coworkers 2009a, b)

5.4.2 *Viscosity, Fragility, and Specific Heat of Glasses*

Viscosity (η) indicates the resistance to flow or shear deformation of a liquid. The ease of glass formation is dependent on the temperature and the viscosity of the glass. Viscosity is the determining factor whether the glass will be homogeneous, bubble free, and devitrified. Mathematically, viscosity may be defined as the ratio of applied shearing force to the rate of flow of liquid, i.e.,

$$\eta = \frac{Fd}{Av} \tag{5.3}$$

where F is the tangential force difference applied between two parallel planes of area A separated by distance d and relative velocity of planes v. The liquid is said to be 'Newtonian' if the viscosity and applied force share a direct relationship with each other. On the other hand, the liquid is said to be 'non-Newtonian' with pseudoplastic flow if an apparent decrease in viscosity is observed with increasing shear stress. The SI unit of viscosity is N s m^{-2} (Pascal × second = Pa s = 10 Poise). The reciprocal of viscosity is known as fluidity; i.e., a liquid with high fluidity exhibits small viscosity. Every characteristic temperature bears a special relationship to the viscosity. Usually, two expressions are given to express the temperature-dependent viscosity: Arrhenian equation and Vogel–Fulcher–Tamman (VFT) equation. According to the VFT equation, viscosity is given by

$$\eta = A + \frac{B}{T - T_0} \tag{5.4}$$

The constant A is η_0, and T and T_0 are temperatures (°C). The VFT equation provides a good fitting of data over a wide range of temperatures. Anyhow, near the lower end of the transformation region, it should be used with caution because the activation energy of viscosity (ΔE_η) is almost constant and hence viscosity values may be overestimated. When $T_0 \approx 0$, then the VFT equation changes to the Arrhenian equation, i.e.,

$$\eta = \eta_0 \exp\left(\frac{\Delta E_\eta}{RT}\right) \tag{5.5}$$

and the constant B becomes $\frac{\Delta E_\eta}{RT}$. Viscosity determines whether a melt will be fragile or strong; i.e., the melts, which exhibit non-Arrhenian behaviour over the entire viscosity range, are known as strong melts and the one, which display large curvature are known as fragile melts (Fig. 5.14).

Fragile melts exhibit short-range order and have ionic bonds, whereas the strong melts have high degree short-range order with a partially covalent bond. The fragile melts show high disintegration with increased T_g, whereas the strong melts show gradual degradation with increase in T_g. One of the major differences in strong and

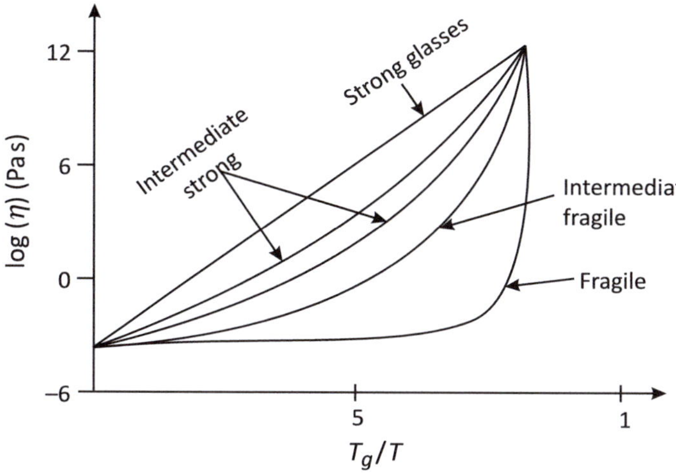

Fig. 5.14 Strong and fragile melts

fragile melt lies in the fact that the strong melts show small changes in heat capacity compared to the fragile melts upon passing through the transformation range. Figure 5.15 shows the fragility of selenium glass prepared using different cooling rates (while computing $\eta = 10^{12}$ Pa s at the T_g). The faster cooling has resulted in earlier glass transition and glass with higher molar volume. The structures with higher molar volume possess high degeneracy and hence a large number of transitions. This implies that a faster cooled glass will exhibit lower viscosity corresponding to the lower effective free energy barrier for structural relaxation.

Fig. 5.15 Variation of viscosity with the T_g/T for the selenium glasses prepared at different rates (Mauro and coworkers 2009a, b)

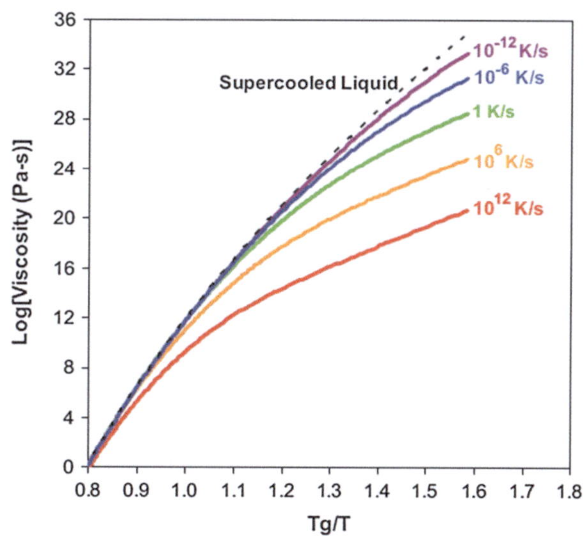

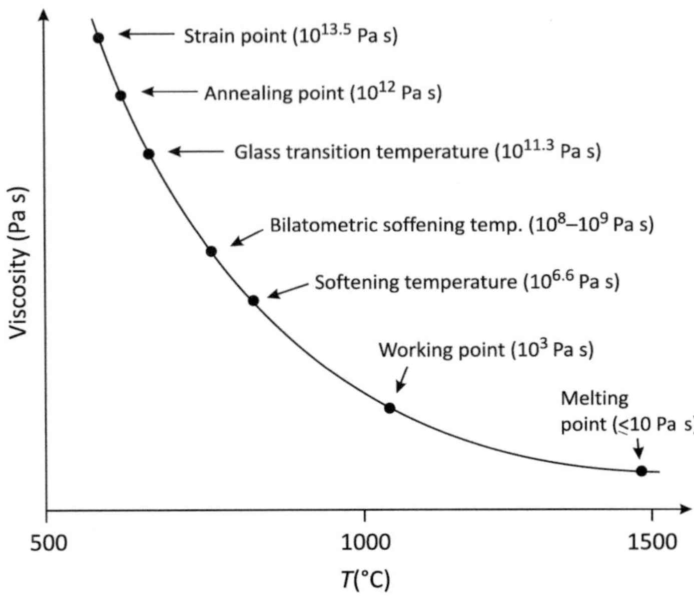

Fig. 5.16 Characteristic viscosities corresponding to different temperatures

Viscosity is different for every characteristic temperature of the glass (Fig. 5.16). The melting temperature is obtained at a typical viscosity of ≤10 Pa s for most of the glasses. The melt should be viscous enough to retain its shape after forming but fluid enough to flow under acceptable stresses. At a viscosity of 10^3 Pa s (working point), the melt is migrated to a processing chamber. At $10^{6.6}$ Pa s, the melt achieves viscosity, which is high enough to prevent the deformation under its own weight. The difference between the softening temperature and the working temperature defines the working range of the melt. The glasses with large and small working range are known as long and short glasses, respectively. For the glass, glasses which possess a working range at elevated temperatures are known as hard glasses (soda-lime silica glass melts), whereas those which work below the working range are referred to as soft glasses.

After the glass is formed, annealing of material is performed in order to reduce its internal thermal stresses. The viscosity at annealing is 10^{12} Pa s, whereas for the strain point it can be $10^{13.5}$ Pa s. Annealing reduces the stress within few minutes, whereas straining relieves stress gradually over several hours. The T_g can be determined from the heat capacity and CTE measurements, and Moynihan determined the viscosity at T_g to be $\sim 10^{11.3}$ Pa s. From the dilatometric curve, the dilatometric softening temperature (T_d) is obtained at which $\eta \approx 10^8$–10^9 Pa s. At the dilatometric softening temperature, the sample reaches maximum length during heating. To calculate the entropy of the melt, we need to perform integration over the specific heat at constant pressure (C_p; Fig. 5.17). From Fig. 5.17a, it can be seen that the specific heat C_p is highest for supercooled liquid and drops to a lower value close to that found in the crystal phase.

The thermodynamic relation between the entropy and C_p is given by

$$S_2 - S_1 = \int_{T_1}^{T_2} \frac{C_p(T)}{T} dT \tag{5.6}$$

The rate of cooling determines the variation of C_p as demonstrated in Fig. 5.17a; i.e., glass 1 possesses a higher cooling rate R_1 than glass 2 with cooling rate R_2. Entropy S versus T exhibits higher slope for the supercooled and liquid phase (C_p is also largest) (Fig. 5.17b). The entropies of supercooled liquid and crystal approach one another when the temperature decreases. There is another extraoperated temperature, 'Kauzmann temperature' (T_k). If the decrease in entropy continues and C_p remains constant at T_g, then the liquid and crystal possess almost the same entropy at T_k. Below T_k, the crystal should have higher entropy. Anyhow, if the liquid entropy keeps on decreasing below T_k without any change in slope, then entropy

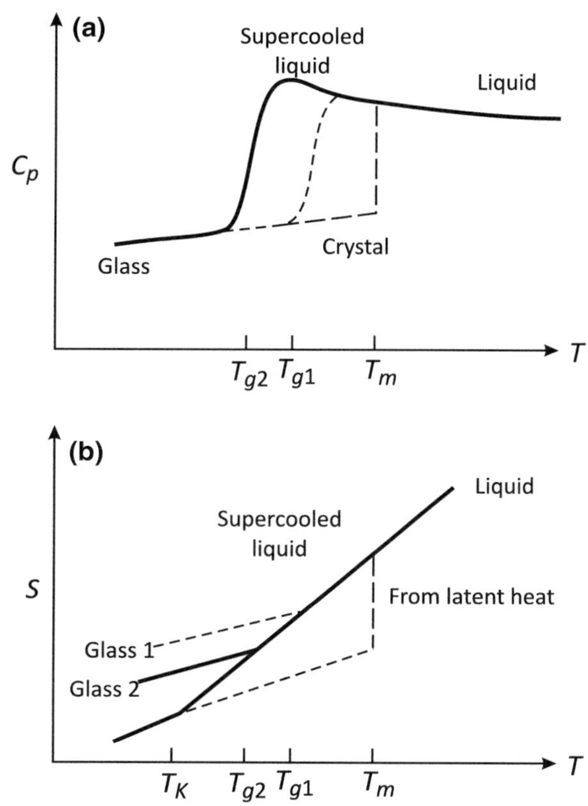

Fig. 5.17 Variation of **a** specific heat and **b** entropy with the temperature for glass 1(faster cooling rate) and glass 2 (slow cooling rate) (Ediger and coworkers 1996)

will be negative even above $T = 0$. Hence, there should be a limit on T_g such that entropy remains positive. Therefore, $T_g \geq \approx T_k$, and a drop in C_p is observed at temperature above T_k.

5.4.3 Density and Thermal Expansion Behaviour

The density of material (ρ) is defined as the mass of substance per unit volume; i.e.,

$$\rho = \frac{M}{\mathbf{V}} \tag{5.7}$$

where M represents the mass of substance, $\mathbf{V}$ is the volume, and ρ is the density. The density can be a true density, if the material is free from defects, bubbles, or voids. Anyhow, if the material contains these defects, then the measured density turns out to be higher and is referred to as the apparent density of the material. Usually, for glass, the volume is largely determined by the vitreous network. An alternative expression for density is given as

$$\rho = \frac{M_w}{\mathbf{V}_m} \tag{5.8}$$

where M_w represents the molecular weight of the substance and $\mathbf{V}_m$ is the molar volume. The density of glass-forming oxides is less than their corresponding crystalline phases, and the free volume of glass ($\mathbf{V}_f$) can be calculated using the relation:

$$\mathbf{V}_f = 1 - \frac{\mathbf{V}_c}{\mathbf{V}_g} \tag{5.9}$$

where $\mathbf{V}_c$ and $\mathbf{V}_g$ represent the molar volume of the crystalline form and glass, respectively. The density of glass is determined by the cooling rate; i.e., for fast cooling rates, the density becomes independent of the cooling rates. If the melt is cooled at a slower rate, then the equilibrium is maintained at low temperatures and a dense structure is obtained. Similarly, heat treatment of glass at a particular temperature influences the density of the melt. A glass with low density and high T_F is obtained if the melt is quenched rapidly. Crystallisation leads to large changes in the density of glasses. Upon heat treatment of glass (at crystallisation temperature), some crystallised phases may develop in glass and the glass converted into glass-ceramic whose density is approximated as

$$\rho = \rho_c \mathbf{V}_c + \rho_g \mathbf{V}_g \tag{5.10}$$

where ρ_c and ρ_g are the densities of the crystalline and glassy phases, respectively; $\mathbf{V}_c$ and $\mathbf{V}_g$ are the volume fractions of the crystalline and glassy phases, respectively. In the case of the presence of some third phase in minute concentrations or microcracks, the density can yield low values but these values are acceptable within $\sim 1\%$. The coefficient of thermal expansion (α) is an indispensable tool in determining the thermal behaviour of glass. The CTE of a material is the measure of rate of change in volume with temperature and is given as

$$\alpha_V = \frac{1}{\mathbf{V}}\left(\frac{\partial \mathbf{V}}{\partial T}\right)_\rho \tag{5.11}$$

where $\mathbf{V}$ is the volume of the sample and $\left(\frac{\partial \mathbf{V}}{\partial T}\right)_\rho$ is the slope of curve. Equation (5.11) requires the volume of a substance, which is somewhat difficult to measure. Hence, only the linear coefficient of thermal expansion is considered; i.e., ($\alpha_V = 3\alpha_L$)

$$\alpha_L = \frac{1}{L_0}\left(\frac{dL}{dT}\right)_\rho \tag{5.12}$$

where L_0 is the original length of sample and dL is the change in length. Usually, the thermal expansion coefficients are measured over a specified temperature range and their values lie in between 1 and 50×10^{-6}/K. The typical thermal expansion behaviour of glasses is shown in Fig. 5.18.

From graph 5.18, we obtain three important parameters:

- CTE = α, which gives the expansivity of the glass.
- T_d = Dilatometric softening temperature where the sample reaches maximum expansion.
- T_g = glass transition in temperature indicating onset of viscoelastic behaviour.

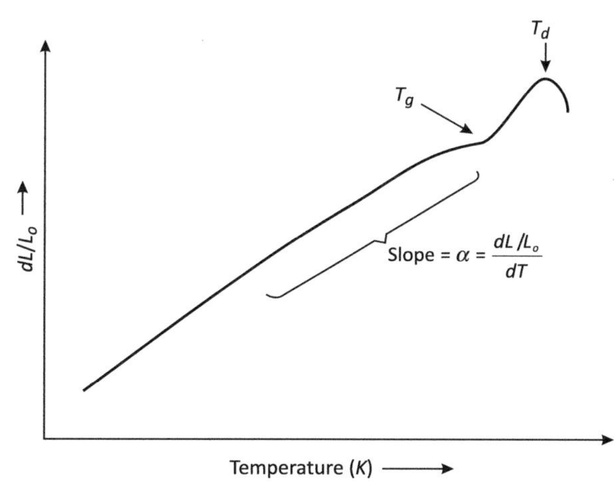

Fig. 5.18 Change in length of glass with temperature

The expansion of glass is related to the non-harmonic vibration of particles. Particularly, for oxide glasses, the expansion can be determined by taking attraction between ions into account:

$$F = \frac{2Z}{d^2} \tag{5.13}$$

where Z is the equivalent of positive ions, F is the attractive force, and d is the distance between positive and negative ions. If the interatomic potential well of an atom is parabolic, then the driving force and atomic displacement are linearly proportional to each other. For such cases, the atoms undergo harmonic oscillations without change in their position, hence yielding no thermal expansion during the heating process.

Usually, the potential well is not parabolic as Lennard Jones potential shows more resistance to the compression than the tension. Therefore, thermal vibrations tend to drive atoms apart causing thermal expansion. As the time of heat treatment increases, the thermal vibration amplitude within the glass network increases. The general increase in volume with temperature is mainly attributed to the increased amplitude of atomic vibrations about a mean position (Fig. 5.19); i.e.,

$$\text{Energy} = E = \frac{-e^2}{4\pi\varepsilon_0 r} + \frac{B}{r^n} \tag{5.14}$$

where e is the charge of ion, r is the interatomic distance between ions and B is the constant. The first and second terms indicate the attraction and repulsion between ions, respectively. The atomic separation influences the repulsive term more than the attractive term due to the presence of factor n; i.e., increase in n leads to a symmetric curve and hence CTE decreases. In addition to this, if the bonds are between the cations of equal field strength, then this leads to a symmetric curve.

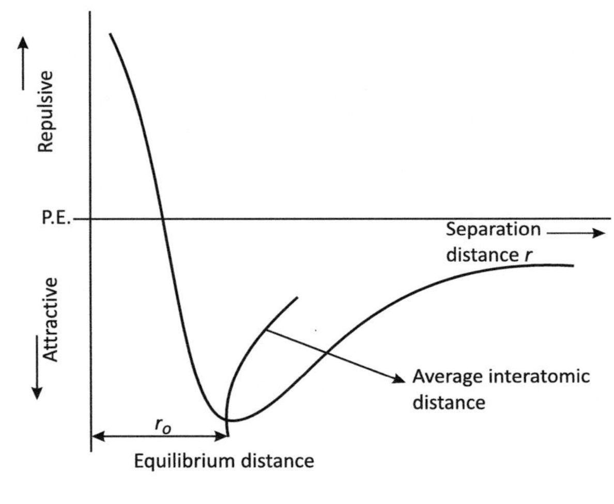

Fig. 5.19 Potential distance curve showing the cause of thermal expansion

Bond bending and filling of interstitial positions lead to change in the CTE. For crystalline materials, the bonds break at the sharp melting point and a sharp increase in CTE is observed, whereas for the glass a gradual increase in CTE is observed in the glass transformation range. In an asymmetric potential well, the atoms spend more time at $r > r_0$, where r_0 is the equilibrium distance at absolute zero. This subsequently decreases the attraction between ions and increases bond length leading to lattice expansion of the glass network. In a dilatometer, the sample is under a small load provided by push rods. During deformation, the push rod sinks into the sample bending which may occur and a temperature of maximum expansion T_d is obtained. The CTE for an elastic material can be expressed as

$$\alpha = \alpha_0 + \alpha_1 T + \alpha_2 T^2 + \alpha_2 T^3 \tag{5.15}$$

where α_0, α_1, α_2, and α_3 are the experimental constants. Some glasses display negative CTE over a limited temperature range indicating their high resistance to thermal shock and ability to absorb lattice expansion through bond bending. If some modifier ions are added to the network, then the CTE may decrease due to small bond bending. The phase separation and fictive temperature virtually exhibit no effect on the CTE. The glass relaxes when heated up to the transformation region. In fact, reheating a quenched glass at a slower rate relaxes the glass (before T_g is reached leading to contraction (small) of glasses below T_g). Anyhow, the crystallisation highly influences the CTE and formation of crystals alters T_g/T. We can sum up the important conclusions as follows:

- Every glass possesses time-dependent transformation behaviour over a range of temperature known as the 'glass transformation region'. Hence, glass is an amorphous solid, which exhibits a glass transformation region.
- Glass is a represented liquid which implies that it exhibits rigid and static behaviour. The term 'supercool' implies that the liquid continues to rearrange with the decrease in temperature without abrupt change in enthalpy.
- Glass is non-crystalline and cannot be defined in terms of unit cell.
- The macroscopic properties of glass are dependent on the thermal history and composition. In addition to this, the glass cannot be explained using statistical/equilibrium mechanics, and hence, glass is 'non-equilibrium'.
- We observe glass on a time scale, which is much shorter than the structural relaxation times. Hence, glass is non-ergodic and with the elapse of time, the ergodicity is restored and the glass properties reach an equilibrium value.

5.5 Glass Constituents

For glass making, raw materials should be chemically pure. For a glass, the batch materials can be classified into the following categories and listed in Table 5.1.

Table 5.1 Some glass network formers, glass modifier, and intermediate oxides with their properties

Role of oxide	M in MO_x	Valence electrons	Coordination number	Dissociation energy E_d per MO_x (kcal)	Single bond strength (kcal)
Glass former	Si	4	4	424	106
	B	3	3	356	119
	Al	3	4	402–317	101–79
	Ge	4	4	431	108
	As	5	4	349	87–70
	V	5	4	449	112–90
	P	5	4	442	111–88
	Zr	4	6	485	81
	Sb	5	4	339	85–68
Modifier	Na	1	6	120	20
	K	1	9	15	13
	Li	1	4	144	36
	Cs	1	12	114	10
	Mg	2	6	222	37
	Ca	2	8	257	32
	Sr	2	8	256	32
	Ba	2	8	260	33
	Pb	4	6	232	39
	Sn	4	6	278	46
	Pb	2	4	145	36
	Zn	2	4	144	36
	Ga	3	6	267	45
	In	3	6	259	43
	Th	4	12	516	43
	Cd	2	4	119	30
	La	3	7	406	58
	Y	3	8	399	50
	Rb	1	10	115	12
	Hg	2	6	68	11
Intermediate	Al	3	6	317–402	53–67
	Zn	2	2	144	72
	Pb	2	2	145	73
	Ti	4	6	435	73
	Cd	2	2	119	60
	Zr	4	8	485	61
	Be	2	4	250	63
	Th	4	8	516	64

(a) **Glass network formers**: The glass formers form the backbone of the glass network. A glass may contain one ore more glass formers and commonly used oxide glass formers are SiO_2, B_2O_3, and P_2O_5; and sulphur, tellurium, and selenium are the glass formers for chalcogenide glass. ZrF_4 and BeF_4 are the common halide glass formers. Oxides such as Sb_2O_3, Al_2O_3, Ga_2O_3, V_2O_5, and Bi_2O_3 act as glass formers when mixed with other oxides. These oxides do not form glass on their own until and unless they are vapour deposited or rapidly quenched. The bond strength for glass formers lies in the range of 60–80 kcal/mol. Glass formers exhibit high field strength (1–2) and low coordination number (CN) between 3 and 4. Usually, the glass gets its generic names from the formers. For example, if a glass has a significant amount of B_2O_3 and SiO_2, then the glass is known as borosilicate glass, whereas if the glass has significant P_2O_5 and SiO_2, then the glass is phosphosilicate glass. For making pure silica glasses, high melting temperature is required. Hence, to reduce the melting temperature, fluxes such as $Na_2O{\cdot}PbO$, Li_2O, and K_2O are used. Alkali oxide fluxes may lead to the property degradation.

(b) **Glass Modifiers**: Glass modifiers occupy random positions in the glass network and modify the network by providing additional oxygen ions. These additional oxygens are known as non-bridging oxygens (NBO) and they are not a part of polyhedral units (Fig. 5.11). Modifiers usually locate themselves in the glass structure. The glass modifiers such as alkaline earth oxides help in reducing the melting temperature of glass and improve the glass properties. The concentration of modifiers is less than that of the fluxes. The bond strength for modifiers lies in the range of 30–40 kcal/mol. The field strength of modifiers is low $\approx$0.35. Modifiers possess high CN between 6 and 8.

(c) **Glass Intermediates**: Glass intermediate exhibits a dual nature as it can either work as a network former or a network modifier. These oxides have intermediate coordination and field strength, i.e., CN between 4 and 8 and field strength between 0.84 and 1.04. Their bond strength lies in the intermediate range of formers and modifiers, i.e., 50–60 kcal/mol.

(d) **Colourants**: Colourant oxides usually belong to 4f rare earth or 3d transition metal series. Colourants endow the glass with different colours, e.g., copper, vanadium, iron, manganese, gold, and silver. Gold and silver produce colour in glass by the formation of colloids. Sometimes, glasses are required to be free of any unintentional colourants (such as iron oxides present in sand), and other colourants are used to counteract their effect and are known as decolourants.

(e) **Fining Agents**: Fining agents are present in very small amounts such that they do not affect other properties of glass. Fining agents release large bubbles which carry smaller bubbles to the surface and seed over there. Arsenic and antimony are very well-known fining agents. NaCl, CaF_2, NaF, and Na_3AlF_6 are the other fining agents. Fining agents may absorb O_2 from the bubbles and reduce the size of the seed such that the seed shrinks below the critical radius, and hence the bubble disappears. Halides are very useful, and they change the viscosity of the melt.

5.6 Preparation Techniques for Glasses

Usually, the glasses are produced by the two common techniques, i.e., sol-gel and melt quench technique, and are discussed as follows.

(a) **Melt-Derived Technique**

The most common way of obtaining a glass is by fusion of two or more component oxides followed by their quenching. The batch of glasses is prepared by mixing an appropriate mole/weight fraction of well-desired initial ingredients. Thereafter, the mixture is ground (sometimes in the presence of acetone) in order to obtain uniform powder. The resulting mixture is dried in air so that the acetone evaporates. The ground powder can be transferred into a recrystallised alumina crucible or platinum crucible and melted in a high-temperature furnace. The glass mixture should be calcined at 500°C for 2 h so that the gaseous substances are released from the composition. The temperature of the furnace can go up to 1500°C for certain aluminosilicate compositions, whereas borate and phosphate compositions melt at lower temperatures of 1200–1300°C. The glass mixture must be held at the melting temperature for at least an hour to facilitate uniform fusion and to achieve homogeneous molten materials. The molten glass is poured into a preheated graphite mould and the remaining melt is poured on the flat copper plate and quenched by an other copper plate in air to obtain flakes. Glasses can be made of desirable sizes and shapes by pouring them into moulds of particular shapes. Figure 5.20a shows the schematic of the melt-quenching process. Sometimes, the glasses are air quenched to avoid any crystallisation. The glasses can be quenched in liquid such as water, liquid nitrogen, mercury, etc. when the higher cooling rate is required. By quenching in cold water, granules and pieces of different sizes collectively known as frits can be obtained and they can be easily powdered. The calefaction effect limits the cooling rate to 100°C/s; i.e., an insulating vapour layer at the glass–liquid interface is formed. For achieving ultrafast cooling rates, the glass should be splat quenched by keeping the glass in contact with some solid material with high thermal conductivity such as copper. To remove the internal stresses generated during quenching process, the obtained glass frits are annealed at 500°C (below glass transition temperature) for 12 h in air in a calibrated pit furnace.

(b) **Sol-Gel Technique**

The dispersions of colloidal particles (diameter of 1–100 nm) in a liquid are known as sols, whereas a gel is a rigid network comprising interconnected pores (in submicrometre) and polymeric chains (in micrometres). The hydrolysis and polycondensation of an organometallic precursor leads to the formation of an interconnected 3-D network, which subsequently forms a gel. The schematic illustration of the sol-gel process is shown in Fig. 5.20b. The first step of the sol-gel process involves mixing all the alkoxide or organometallic precursors together. The second step involves the hydrolysis of liquid alkoxide precursors with the de-ionised water. When silicon alkoxide is hydrolysed by the water, it gives rise to the silanol groups

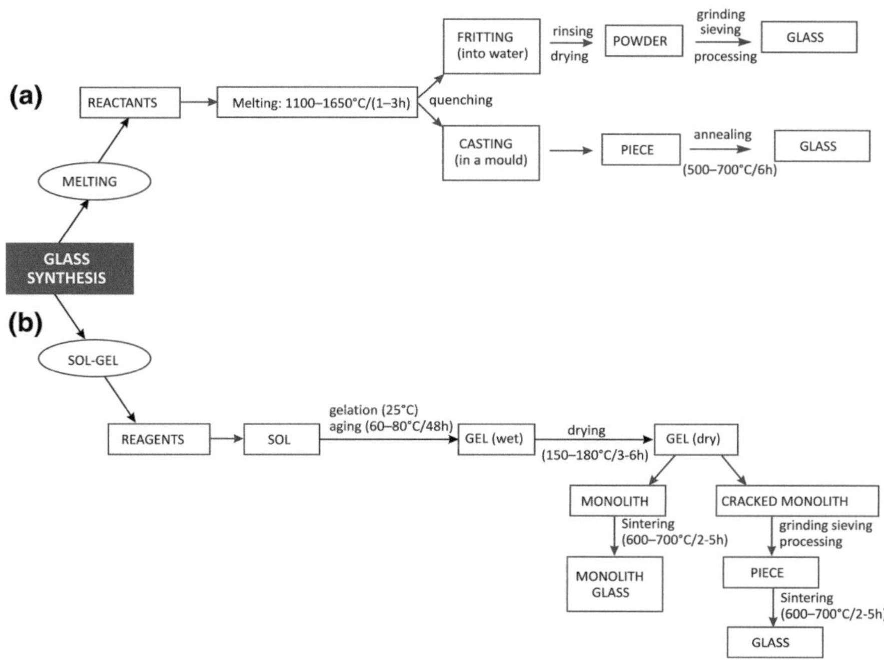

Fig. 5.20 Schematic of **a** melt quench and **b** sol-gel glasses

($Si(OH)_4$). These silanol groups interact with each other forming Si–O–Si bonds eventually giving rise to the silica network (SiO_2) via the polycondensation process during which the water is given out as a by-product. The hydrolysis (Eq. 5.16) and the polycondensation (Eq. 5.17) processes are shown as follows (where R = alkyl group).

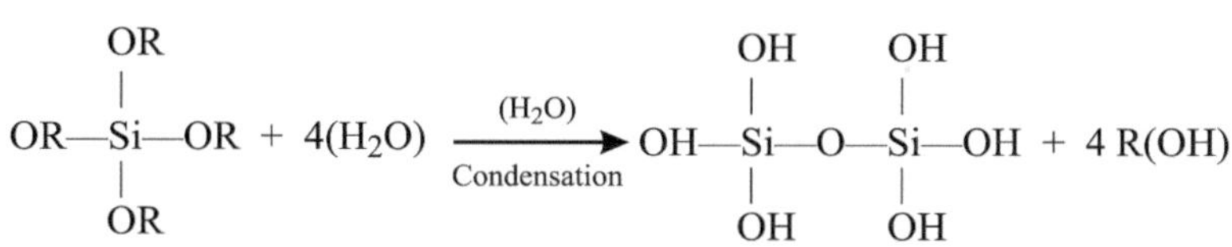

(5.16)

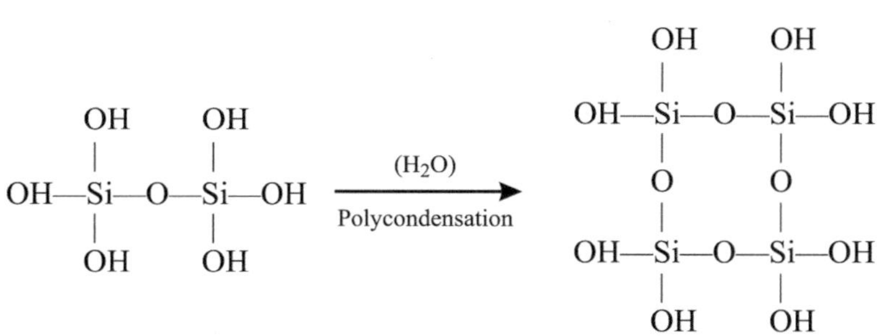

(5.17)

Due to the high field strength of Si^{4+}, the highly nucleophilic hydroxyl ions (OH^-) tend to attack silicon ion. The hydrolysis and polycondensation are simultaneous process and their kinetics are affected by several factors including pH, composition, temperature, alkoxide precursor, pressure, and concentration of different ion species. The hydrolysis and condensation rate constants are small for the bulkier alkoxide groups.

After the sol is formed, the process of gelation follows in which a three-dimensional network is formed with time by condensation and cross-linking of silica particles and other colloids. Sol is a very low viscosity liquid, and during the process of gelation, an abrupt increase in the viscosity is observed. The gelation point (t_g) is the point during which the gel behaves as an elastic solid as more and more particles become interconnected. The gelation time depends on the solvent concentration, nature of the alkoxide group, and the amount of water used for the hydrolysis. The gelation time is relatively larger for heavy alkoxy groups.

The next step is the syneresis/ageing of a gel, during which the decrease in porosity and hence increase in the strength are observed due to the continued process of polycondensation and reprecipitation of the gel network. In addition to this, the phase transformation occurs during the process of ageing. The aged gel must be dried by removing the pore liquid from the interconnected rigid 3-D network. During the process of drying, the gel should be strong enough to resist any cracking due to the capillary stresses developed within the aged gel network. When the resulting gel is heated at high temperatures, the elimination of pores results in the densification of network.

For making a glass, we need to make batch calculations. Two examples are given to illustrate the batch calculations for the melt quench technique.

Example 1 When the precursors are oxides,

40% SiO_2—30% Na_2O—20% Al_2O_3—10% CaO
Molecular weight of SiO_2 = 60.09 g
Molecular weight of NaO_2 = 61.98 g
Molecular weight of Al_2O_3 = 101.96 g
Molecular weight of CaO = 56.08 g

$$\begin{aligned}\text{Molecular weight of glass} &= 0.04(60.09) + 0.3(61.98) + 0.2(101.96) + 0.1(56.08)\\ &= 24.036 + 18.594 + 20.392 + 5.608\\ &= 68.63\end{aligned}$$

Weight fraction of each component:

$$\Rightarrow SiO_2 = \frac{24.036}{68.63} = 0.3502$$

$$\Rightarrow Na_2O = \frac{18.594}{68.594} = 0.271$$

$$\Rightarrow Al_2O_3 = \frac{20.392}{68.594} = 0.297$$

$$\Rightarrow CaO = \frac{5.608}{68.594} = 0.082$$

For 100 g of batch, we need to multiply the weight fraction of every component by 100

$$\boxed{\begin{array}{c} SiO_2 = 35.02\ g \\ Na_2O = 27.01\ g \\ Al_2O_3 = 29.7\ g \\ Cao = 8.2\ g \end{array}}$$

Example 2 When the precursors are carbonates, e.g., Na_2O is obtained from Na_2CO_3, the following decomposition reaction holds good:

$$Na_2CO_3 \rightarrow Na_2O + CO_2 \uparrow$$

Then the 100 g of batch calculations for the above series (mentioned in Example 1) would be the same except the Na_2O calculations; i.e.,

61.98 g of Na_2O is obtained from 105.98 g Na_2CO_3.

1 g of Na_2O is obtained from $\frac{105.98}{61.98}$ g Na_2CO_3.

27.1 g of Na_2O is obtained from $\frac{105.98 \times 27.1}{61.98}$ g Na_2CO_3 = 46.338 g.

Hence, for 100 g of batch, we require the following measurements.

$$\boxed{\begin{array}{c} SiO_2 = 35.02\,g \\ Na_2CO_3 = 46.338\,g \text{ to obtain } 27.1\,g \text{ of } Na_2O \\ Al_2O_3 = 29.7\,g \\ CaO = 8.2\,g \end{array}}$$

5.7 Theories of Glass Formation

Many theories have been given to explain the structural aspects and formation of glasses. In the light of these theories, we can estimate why certain compositions can form a glass melt whereas others cannot. The theories which investigate the structural aspect of glasses are 'structural theories of glass formation', whereas the theories which exploit the processing parameters, reaction kinetics, and thermodynamics behind the glass structure are known as 'kinetic theories of glass formation'. Originally, the glass models were focused on silicate crystals. The absence of crystal peaks in the X-ray diffraction pattern was explained on the basis of the presence of very small microcrystals in them. A crystallite model was proposed by Lebedew and other Russian scientists explaining that the glass structure is composed of crystallites with deformed versions instead of merely small-sized crystals. Furthermore, these crystallites are assumed to be connected by the amorphous regions. In the next section, we discuss the theories of formation of a glass network in detail.

5.7.1 Structural Theories of Glass Formation

5.7.1.1 Goldschmidt's Criterion

According to Goldschmidt criterion, the ratio of cation to anion radii for simple oxides A_mO_n should be between 0.2 and 0.4 for the glasses to vitrify. This radius ratio corresponds to the tetrahedral coordination; i.e., the cation is surrounded by 4 oxygens. Although Goldschmidt proposed that the melts, which contain tetrahedral coordination of cations are favourable for making glasses, the technical explanation for this assumption was still lacking.

5.7.1.2 Zachariasen Random Network Theory

Zachariasen could successfully explain why certain CNs are favoured for glass formation. Zachariasen random network theory formed the basis of glass formation. During his studies, he observed that the silicate crystals have a tetrahedral network and readily form glass. These tetrahedral networks are symmetrical, non-periodic, and connected with each other at the corners. This leads to the 3-D network extended over all directions with the isotropic property attributed to the amorphous nature of glasses. He elucidated the following points during his studies.

- An oxygen atom can be linked to only two network atoms. To form a non-periodic network, the oxygen should not have higher CN; otherwise, the variations in the oxygen–cation–oxide bond angles get diminished.

- The cation CN should be small, i.e., 3–4. For silicate (SiO_2) and phosphate (P_2O_5) networks, tetrahedral coordination is observed, whereas for borate (B_2O_3) network, the coordination is 3. Additionally, sufficient network cations should be present to sustain a continuous network structure.
- The oxygen polyhedra share the corners and cannot be present at edges or corners. This makes the glass structure more open and the cations are located far apart from each other.
- The network can be 3-D if at least three corner oxygens of a polyhedron are shared. This requires sufficient bonds linking the network polyhedra.

These postulates given by Zachariasen are to be satisfied by a melt for glass formation.

5.7.1.3 Smekal Theory

This theory was based on the bond nature of material. According to Smekel, glass-forming melts fall into two categories, i.e.,

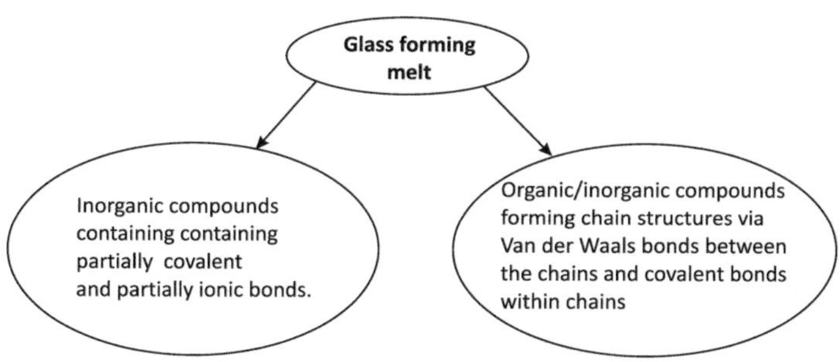

Smekal considered that pure ionic or pure covalent materials cannot form glasses because a covalent bond exhibits sharply defined bond angles, hence limiting the formation of a non-periodic network. Highly ionic bonds do not form network structures due to lack of directional characteristics.

5.7.1.4 Dietzel Theory

Dietzel suggested that the cation oxygen bond strength can be regarded as an effective tool to determine cation behaviour. Hence, the cations in oxide glasses have been classified on the basis of field strength. The field strength for a cation with Z valence and 'a' as the distance between the cation and oxygen is given as

$$F_s = \text{Field strength} = \frac{Z}{a^2} \quad (5.18)$$

Low field strength implies low cation–oxygen bond energy and vice versa. On the basis of field strength, the role of cations can be mentioned as

- $F_s > 1.3 \rightarrow$ Glass former
- $F_s > 0.4 \rightarrow$ Glass modifiers
- $0.4 < F_s > 1.3 \rightarrow$ Intermediates

5.7.1.5 Sun Theory

According to Sun, during the cooling process, the strong bonds prevent the structural orientation of the melt into crystalline geometry, hence favouring the formation of glass. The bond strength can be defined as the dissociation energy per number of cation–anion bonds in the coordination unit. The dissociation energy (E_d) for a molecule can be defined as the energy required to dissociate an oxide into its individual oxide as

$$A_xO_y \rightarrow xA + yO \quad (5.19)$$

$$\left[\text{Hence, Single bond strength (SBS)} = \frac{E_d}{\text{Coordination no.}}\ \text{kcal/bond}\right]$$

On the basis of SBS, the role of cations can be defined as

- SBS > 80 kcal/bond—Glass formers
- SBS < 60 kcal/bond—Glass modifiers
- 60 < SBS < 80 kcal/bond—Intermediates

The oxides with high SBS have high viscosity because they are difficult to be reformed into an ordered lattice upon cooling, and hence act as good glass formers.

5.7.1.6 Stanworth Theory

Stanworth used the concept of electronegativity of the cation to determine its role in glass network as follows:

(i) **Glass network formers**

- Possess partial ionic character near 50% concentration
- Usually belong to group I
- Possess high electronegativity
- Form the glass network

(ii) **Glass intermediates**

- More ionic character w.r.t. oxygen
- Usually belong to group II
- Cannot form glasses on their own, but can yield glass formation when combined with group I glass formers
- Lower electronegativity (between modifiers and formers)

(iii) **Glass modifiers**

- Highly ionic bonds w.r.t. oxygen
- Usually belong to group III
- Never form the network
- Very low electronegativity
- Modify the network structure by producing oxygen

5.7.1.7 Rawson Theory

Rawson considered temperature to be an important parameter in defining the nature of network oxides. He considered a low melting temperature with large bond strength for oxide to be a good glass former. Anyhow, the oxide with high melting temperature and similar bond strength is a comparatively less competitive network former.

5.7.2 Kinetic Theories of Glass Formation

The good glass formers are the ones which yield glass upon moderate cooling of melt, whereas the poor glass formers require a rapid cooling rate. Conclusively, the material is said to form glass when it is cooled rapidly such that the structural reorientation cannot take place due to insufficient time. Hence, the crystallisation process can be prevented. The crystallisation process involves nucleation of crystal sites and then growth of crystal on those sites to a detectable size. The crystallisation process can be homogeneous or heterogeneous. When the crystal is formed within the melt, the process is homogeneous, whereas when the crystal is formed due to some impurities, the crystallisation process is heterogeneous. There is equal probability of the formation of nuclei throughout the bulk melt for homogeneous nucleation. The material will form amorphous glass when no nuclei are present, hence preventing the crystallisation of the melt. To determine the concentration of nuclei, the sample can be reheated to a temperature where the nuclei grow to detectable sizes [This method is used for most instruments such as differential thermal analysis (DTA) and differential scanning calorimetry (DSC)]. The temperature at which the crystal grows to detectable size is known as the peak

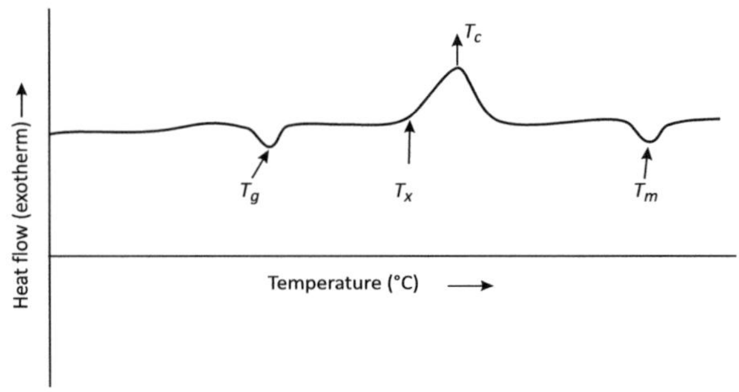

Fig. 5.21 Typical DSC thermogram for the glass

crystallisation temperature (T_c) (Fig. 5.21) and T_x represents onset crystallisation temperature.

By dividing the concentration of nuclei by total time of isothermal heat treatment at nucleation temperature, the nucleation rate (**I**) is obtained: $\mathbf{I} = \frac{\mathrm{d}_N}{\mathrm{d}_t}$. Usually, two barriers exist for the formation of a nucleus, the thermodynamic barrier and kinetic barrier. The thermodynamic barrier involves free energy change (ΔG), whereas the kinetic barrier involves the growth of an ordered phase from disordered lipid. The expression for the nucleation rate is

$$\mathbf{I} = A\ \exp[-(E^* + \Delta G)/kT] \tag{5.20}$$

where k is the Boltzmann constant, T is the absolute temperature (in K), A is the constant, E^* is the thermodynamic barrier, and ΔG is the kinetic barrier. The constant A is given as

$$A = 2n\mathbf{V}^{1/3}\left[\frac{kT}{h}\right]\left[\frac{\gamma}{kT}\right]^{1/2} \tag{5.21}$$

where

n is the formula unit for crystallising components

h is the Planck's constant

V is the volume/unit

γ is the interfacial free energy/area for crystal melt.

Using the approximations, A can be written as

$$A = n\left[\frac{kT}{h}\right] \tag{5.22}$$

The nucleus will grow only if it exhibits certain critical size; otherwise it will redissolve. When crystals are formed inside the melt, the surface free energy increases due to the formation of new interfaces (e.g., melt/crystal interface) within the melt, whereas volume for energy decreases because crystals are ordered than the melt. We can combine both the energies as

$$E = \underbrace{\frac{4}{3}\pi r^3 \Delta G}_{\text{Volume energy}} + \underbrace{4\pi r^2 \gamma}_{\text{Surface energy}} \tag{5.23}$$

When nucleus size increases, the energy E increases and nuclei become unstable. ΔG is negative below the melting point; hence, if the nucleus grows to a large size, then the first term will dominate and E will decrease (negative ΔG), which stabilises the nuclei. To determine the size of nuclei at which the nuclei become stable, we need to solve for 'γ' using the mathematical calculations:

$$\left.\frac{\mathrm{d}E}{\mathrm{d}\gamma}\right|_{\gamma=r} = 0$$

$$\Rightarrow \quad \Delta G = -\frac{2\gamma}{r^*}$$

$$\Rightarrow \quad \boxed{r^* = -\frac{2\gamma}{\Delta G}} \tag{5.24}$$

Using (5.24) and (5.23), we obtain

$$\Rightarrow \quad E = -\frac{16\pi\gamma^3}{3\Delta G^2} \tag{5.25}$$

The free energy change/mole, is given by $\Delta G^\circ = \mathbf{V}_\mathrm{m}\Delta G$, which should be substituted in (5.23) ($\mathbf{V}_\mathrm{m}$ = molar volume of crystal phase)

$$\Rightarrow \quad E^\circ = -\frac{16\pi\gamma^3 \mathbf{V}_\mathrm{m}^2}{3\Delta G^{\circ 2}} \tag{5.26}$$

Equation (5.26) gives the expression for the thermodynamic barrier for the nucleation. To discuss the kinetic barrier, we make use of diffusion coefficient D

$$\Rightarrow \quad D = \left(\frac{kT\lambda^2}{n}\right)\exp\left(-\frac{\Delta G}{kT}\right) \tag{5.27}$$

where λ and T are the atomic jump distance and temperature, respectively. According to Stokes's relation, $D = \frac{kT}{3\pi\lambda\eta}$, where η is viscosity. Hence, Eq. (5.27) becomes

$$\frac{h}{3\pi\eta\lambda^3} = \exp\left[-\frac{\Delta G}{kT}\right] \tag{5.28}$$

Using (5.28) in (5.20), we obtain

$$I = \left(\frac{nkT}{3\pi\lambda^3\eta}\right)\exp\left[-\frac{E^*}{kT}\right] \tag{5.29}$$

Equation (5.27) represents homogeneous nucleation, whereas the heterogeneous nucleation can be written as

$$\mathbf{I}^* = A^* \exp\left[-\frac{(E^* + \Delta G)}{kT}\right] \tag{5.30}$$

$$\left[A^* = \frac{nkT}{h} \text{ and } E^* \text{ depends on contact angle of nucleus with substrate}\right]$$

Figure 5.22 shows the shape of the nucleation rate versus temperature using the above equations.

When the melt is held at T_m, ΔG is small but still the nuclei size will be large and the probability of reaching large size is extremely low; hence, the melt remains free of nuclei. Upon decrease in T_m, the ΔG will increase and critical radius decreases (Eq. 5.24). Hence, the probability of formation of nuclei bigger than the critical size can be attained. If the melt has high viscosity, then there will be a high kinetic barrier and the nucleation rate subsequently decreases. Viscosity is temperature dependent; hence, when the temperature increases, the viscosity decreases, which leads to a decrease in the kinetic barrier. After the nucleation process, the crystal grows, and the growth rate for crystal ($\mathbf{V}'$) is given as

$$\mathbf{V}' = r_0 f \exp\left[-\frac{\Delta E}{kT}\right]\left[1 - \frac{\Delta G}{kT}\right] \tag{5.31}$$

where r_0 is the interatomic separation distance, f is the vibrational frequency, and ΔE and ΔG represent the thermodynamic and kinetic barriers, respectively. As shown in Fig. 5.22, the crystal growth follows the same pattern as that of the

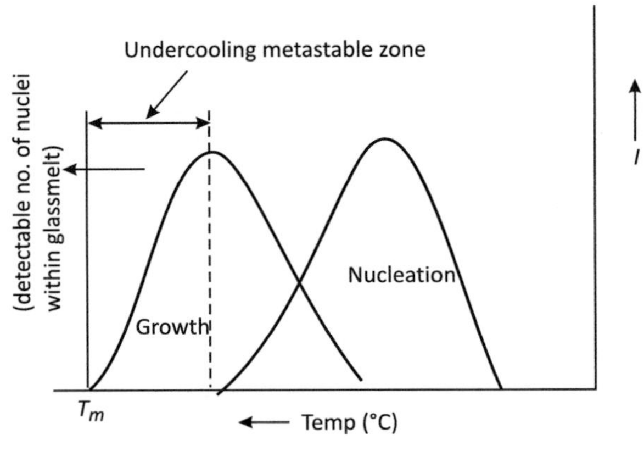

Fig. 5.22 Effect of temperature on the nucleation and growth rate of melt

nucleation rate. Anyhow, growth can take place below T_m, if the nucleus is available, even if the nucleus is formed during the heating process. As with the nucleation process, growth is also influenced by the viscosity, thermodynamic, and kinetic parameters.

5.8 Structures for Oxides Glasses

Experimental studies are being conducted to obtain the structure of glasses. The crystal structure can be determined accurately, whereas the glass structures cannot be determined precisely. Usually, the most common glasses are silicate, borate, aluminosilicate, phosphate and germinate. In this section, we discuss the structure of borate, silicate, and aluminosilicate glasses in detail which are within the scope of this book.

(a) **Silicate Glasses**

The basic building block for silicate glasses is a silicon–oxygen tetrahedral. These tetrahedrals are linked to each other to form a 3-D network (Fig. 5.23).

The Si–O bond angles are distributed over 120°–180° and centred at about 145°. The randomness in the SiO_2 structure is due to variation in Si–O bond angles and bond lengths. The shortest Si–O and O–O distance observed in the SiO_4 tetrahedron is 0.162 and 0.265 nm, respectively. Sometimes, stressed bonds and defects of oxygen vacancies are present in the SiO_4 tetrahedron. Typical oxygen defects include oxygen vacancies, pero-oxy defects and imperfections at the impurity site. When alkali/alkaline earth oxides are added to SiO_2, the oxygen to silicon ratio increases by more than 2. This leads to the formation of singly bonded oxygens which do not participate in the network as shown in Fig. 5.24. For univalent cations, one oxygen is required, whereas for divalent, two such oxygens are mandatory. Alkali oxides (<10%) usually containing glasses are difficult to melt due to their high viscosities. When the alkali oxide content increases, the CTE of glasses increases due to the formation of more NBO. The alkali ions usually occupy

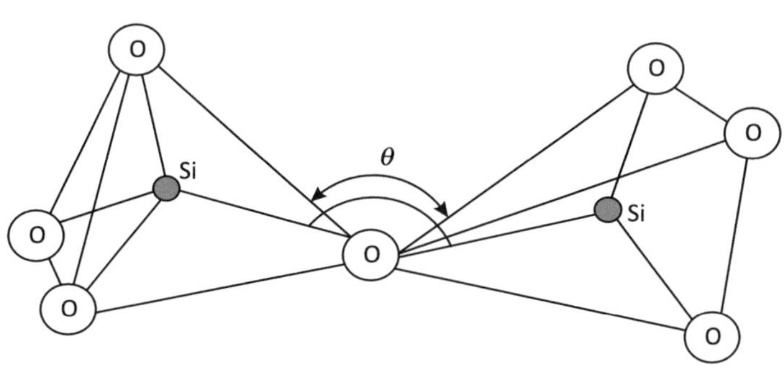

Fig. 5.23 SiO_4 tetrahedron and angle variation

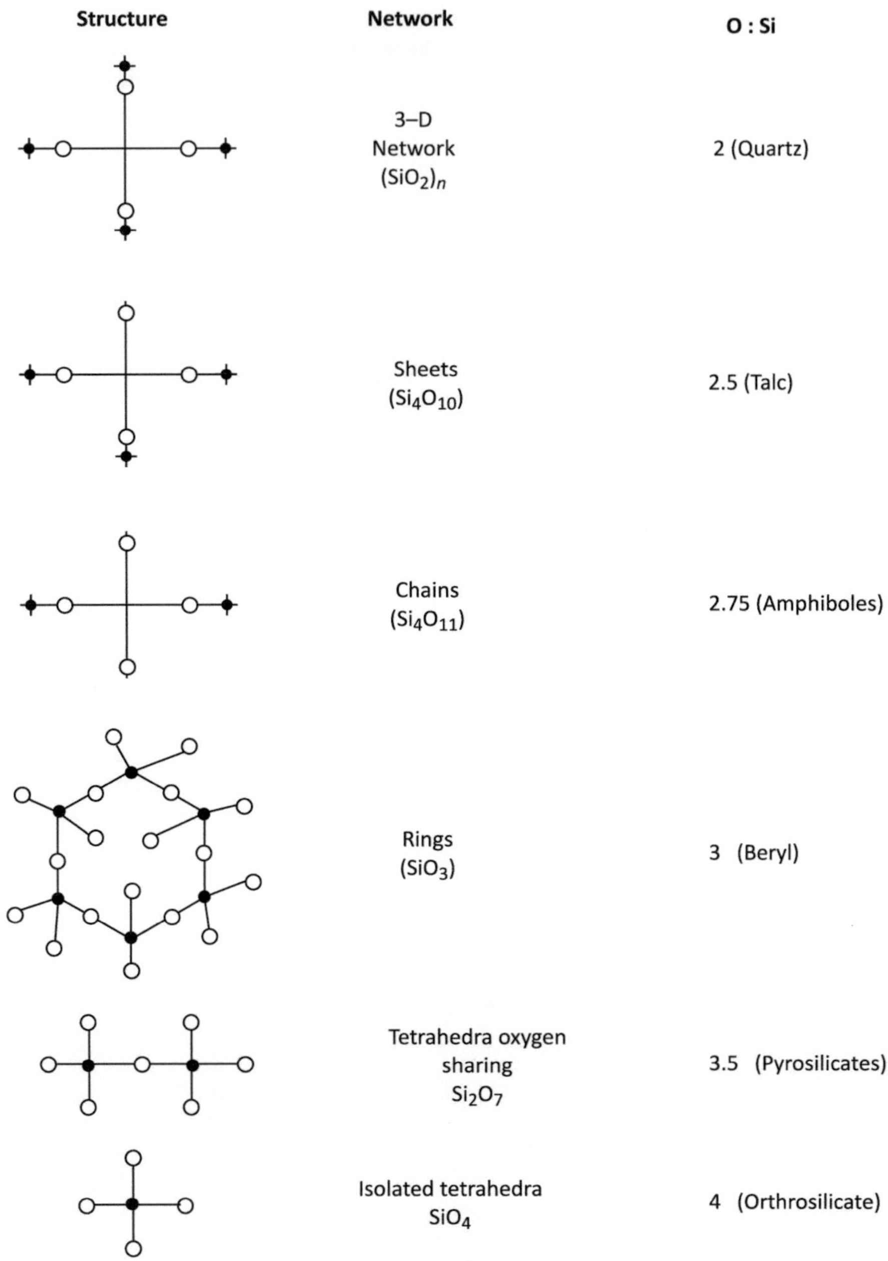

Fig. 5.24 Oxygen/silicon ratio in different silica structures

interstices of the network and NBO are associated with them. To express the concentration of bridging oxygen/tetrahedron, we use θ_n nomenclature. The tetrahedrons with 4, 3, 2, 1, and 0 linked oxygens are denoted by θ_4, θ_3, θ_2, θ_1, and θ_0;

hence, n would represent the bridging oxygens. The θ_n model makes an assumption that only two types of unit will exist in the composition. We obtain expression for the content of alkali oxide (A_2O), for which all the oxygens are non-bridging as follows.

Let us take the general composition to be $xA_2O + (100 - x)\ SiO_2$. Total oxygen content would be $[x + (100 - x)2]$. Every oxygen is shared by two tetrahedrons as shown in Fig. 5.23. For different oxygen sharings, we define θ_n as

- $\theta_0 \rightarrow$ 0 shared + 4 non-shared oxygens = 4 oxygen/tetrahedron
- $\theta_1 \rightarrow$ 1 shared + 3 non-shared oxygens = 3.5 oxygen/tetrahedron
- $\theta_2 \rightarrow$ 2 shared + 2 non-shared oxygens = 3 oxygen/tetrahedron
- $\theta_3 \rightarrow$ 3 shared + 1 non-shared oxygens = 2.5 oxygen/tetrahedron
- $\theta_4 \rightarrow$ 4 shared + 0 non-shared oxygens = 2 oxygen/tetrahedron

Usually, alkali oxide contributes in NBO. For the total oxygen to be shared between θ_4 and θ_3, we write

$$\begin{aligned} & \theta_4 + \theta_3 = 100 - x \\ \Rightarrow \quad & \theta_4 = (100 - x) - \theta_3 \end{aligned} \tag{5.32}$$

θ_4 shares 2 oxygens/tetrahedron and θ_3 shares 2.5 oxygen/tetrahedron, and hence we can write

$$\text{Total oxygen} = \theta_4 \times 2 + \theta_3 \times 2.5 = [(100 - x) - \theta_3] \times 2 + \theta_3 \times 2.5 \tag{5.33}$$

As already stated, total oxygen is $x(1) + (100 - x)2$; hence, using Eq. (5.33), we write

$$x + (100 - x)2 = (100 - x)2 - 2\theta_3 + 2.5\theta_3$$

$$\Rightarrow \quad \boxed{\theta_3 = 2x} \tag{5.34}$$

And

$$\boxed{\theta_4 = 100 - 3x} \tag{5.35}$$

For $\theta_4 = 0$, i.e., all the oxygens will be non-bridging

$$\Rightarrow \quad 100 - 3x = 0$$

$$\Rightarrow \quad x = 33.3\,\%$$

This implies that for concentration of 33.3%, the glass will contain no θ_4 unit. When alkaline atoms are also added to the glass composition, every A^{2+} alkaline earth metal will have two neighbouring NBO and every A^{+} alkali metal will have one neighbouring oxide. When A^{+} alkali metal ions are replaced by A^{2+} alkaline atoms, the net mobility of modifiers is reduced. For the alkali/alkaline aluminosilicate glasses, the alumina AlO_4 tetrahedra may act as a network former by replacing SiO_4 tetrahedra. Aluminium oxide is intermediate and does not form glass on its own but it forms glass when it replaces SiO_4. This can be demonstrated as follows.

According to the Pauling packing rule,

$$\frac{r(Al^{3+})}{r(O^{2-})} = \frac{0.53}{1.40} = 0.38 \quad [6 < CN < 4] \tag{5.36}$$

Al_2O_3 prefers octahedral coordination, and hence,

$$\frac{\text{Charge of } Al^{3+}}{\text{CN of } Al^{3+}} = \frac{3}{6} = \frac{1}{2}$$

$$\frac{\text{Charge of } O^{2-}}{\text{CN of } O^{2-}} = \frac{2}{4} = \frac{1}{2}$$

This volates the Zachariasen rule that an oxygen atom cannot be linked to more than two cations. For A^{2+}, two AlO_4^{-} units are required because the Al_2O_3 unit provides 1.5 oxygen/tetrahedron. Additional oxygen must be provided by A^{2+} or A^{+} to make the oxygen be 2 oxygen/tetrahedron.

(b) **Borate Glasses**

Boron occurs in tetrahedral (BO_4) and triangular (BO_3) coordination. To form a continuous structure for the vitreous boric oxide, the BO units should be connected at the corner. For pure B_2O_3, $T_g \approx 260°C$, hence, borate glasses have low T_g as compared to the silicate glasses ($T_g \approx 1100°C$). The B–O–B angles are variable and may twist to give the out-of-plane boroxyl group. Alkali atoms in addition to the B_2O_3-based glasses cause an increase of T_g and decrease in CTE. The alkali atom causes the change of boron coordination from triangular to tetrahedral (no NBO). Hence, no formation of NBO may lead to a decrease in CTE and increase in $T_g(BO_3 \leftrightarrow BO_4)$. When the addition of alkali oxide is done above the required concentration, the reversal of properties is observed: the CTE increases and T_g decreases. This particular behaviour of borate containing melts is known as the 'boron anomaly'. The boron anomaly can be explained as follows.

- The conversion of $BO_3 \leftrightarrow BO_4$ takes place only up to a certain critical concentration of alkali oxide.
- Above this critical concentration, the addition of alkali oxide causes the formation of NBO, which reverses the trend of properties.

The boron anomaly does not occur for the glasses containing $Al_2O_3{:}A_2O = 1{:}1$, i.e., for the alkali aluminoborate glasses where alkali and Al_2O_3 share the same ratio. If the alumina content is increased, then the boron anomaly should occur for higher alkali oxide content. Anyhow, experimental studies indicate that the boron anomaly occurs for the same concentration. The most commonly used glasses are borosilicate glasses. The alkali-containing borosilicate glasses are usually rich in the silica phase and alkali/boric oxide phases. Alkali borosilicate glasses are easy to manufacture. Alkaline earth oxide borosilicate glass systems are much exploited for their use as sealants.

5.9 Glasses as Sealants

Glass and glass-ceramic seals can prevent leakage by rigidly bonding to the cell components, avoiding mixing of gases, and are electrically insulating. Additionally, glass and glass-ceramic seals are easy to fabricate, and flexible in design, as well as cost competitive. By choosing a suitable compositional design of glass or glass-ceramic seal, a wide range of sealing material properties can be achieved. T_g, T_s, CTE, and devitrification resistance are the four important thermal properties while choosing a SOFC seal glass. A seal glass generally contains 40–70 mol% network formers (SiO_2/B_2O_3), 20–45 mol% network modifiers, and 5–10 mol% of intermediate oxides. Rare earth metal oxides and transition metal oxides are used as additives in the seal glass compositions to tailor the thermal properties. Some of these additives are listed in Table 5.2. In order to relieve thermal stress and self-heal the cracks, the T_g value of a seal glass should be slightly below the cell operating temperature. With increasing SiO_2 content, T_g and T_s generally increase due to the increased network connectivity caused by bridging oxygen. In contrast to this,

Table 5.2 Effect of additives on properties of seal glass

Oxide	Function
Al_2O_3	Allows control over viscosity through the rate of crystallisation
B_2O_3	Reduces T_g, T_s, and viscosity and improves wetting
BaO, SrO, CaO	Reduces T_g and T_s, and raises CTE in the glass-ceramic
MgO	Improves surface adherence, reduces T_g and T_s, and raises CTE in the glass-ceramic
La_2O_3, Nd_2O_3, Y_2O_3	Used as a viscosity modifier and long-term CTE stabiliser
CuO, CoO, NiO, MnO	Improves surface adherence
Na_2O, K_2O	Acts as effective flux but the alkali cations react vigrously with the fuel cell components such as cathodes; have undesirable TEC. Increases conductivity
TiO_2, ZrO_2, Cr_2O_3, Ni	Stimulates crystallisation
ZnO, PbO	Improves flux, reducing agent
P_2O_5	Decreases volatilisation, reduce, CTE and mechanical strength
Sb_2O_5	Oxidising agent

Table 5.3 Potential glass compositions for SOFC

Glasss-code	$Na_2O/K_2O/Li_2O$	MgO	CaO	BaO	SrO	B_2O_3	Al_2O_3	SiO_2	La_2O_3	Other dopants	mol% or wt%
14a	–	20	20	10	–	40	10	–	–		wt%
MAS		45				5	5	45			mol%
MAS10		40				5	10	45			mol%
MAST5		43				5	5	45		2 TiO_2	mol%
MAST10		38				5	10	45		2 TiO_2	mol%
MASZ5		43				5	5	45		2 ZrO_2	mol%
MASZ10		38				5	10	45		2 ZrO_2	mol%
MASN6		43.7				3.4	6.4	44.6		1, 9 Ni	mol%
MASC6		46.2				3.7	6.6	42.9		0, 6 Cr_2O_3	mol%
GS1		40				5	10	45		0 Y_2O_3	mol%
GS2		40				5	8	45		2 Y_2O_3	mol%
GS3		40				5	6	45		4 Y_2O_3	mol%
GS4		40				5	4	45		6 Y_2O_3	mol%
BAS				45		5	5	45			mol%
CAS			45			5	5	45			mol%
MA1	–	8.1	–	–	–	33.9	9.4	–	48.6		wt%
BM1	–	10.5	–	45.6	–	43	–	–	–		wt%
BMA1	–	6.0	–	6.7	–	32.2	8.9	–	46.2		wt%
BM2	–	9.7	–	50.1	–	40.2	–	–	–		wt%
BCAS4	–	–	9.2	55.6	–	–	5.5	25.3	–		wt%
BCAS5	–	–	8.8	57.4	–	–	5.4	22.1	–		wt%
7A	–	16.9	18.82	6.43	–	–	2.14	47.88	6.83		wt%
7(Sr)-2B	–	16.9	18.83	–	4.35	2.0	2.14	47.92	6.84		wt%

(continued)

Table 5.3 (continued)

Glasss-code	$Na_2O/K_2O/Li_2O$	MgO	CaO	BaO	SrO	B_2O_3	Al_2O_3	SiO_2	La_2O_3	Other dopants	mol% or wt%
7A-Cr	–	16.82	18.72	6.40	–	–	2.13	47.64	6.80	0.5 Cr_2O_3	wt%
10B	–	16.48	18.34	6.27	–	2	2.08	46.67	6.66	0.5 Cr_2O_3	wt%
10C	–	16.14	16.85	9.21	–	2	2.04	45.73	6.53	0.5 Cr_2O_3	wt%
G18	–	–	8.8	56.4	–	7.3	5.4	22.1	–		wt%
YSO-4	–	–	3.7	–	42.2	9.3	–	27.2	–	14.9 Y_2O_3 2.7 ZnO	wt%
Glass 27	–	–	15	–	26.6	1.9	4.1	35.2	–	15 ZnO	wt%
Mg 1.5–55	–	8.9	–	60.7	–	–	–	40.4	–		wt%
Mg 1.5–40.5B–10Pb	–	7.4	–	42	–	3.5	–	24.4	–		wt%
Mg 1.5–40.1B–8Zn	–	4.7	–	47.9	–	12.1	–	27.8	–	7.5 ZnO	wt%
Glass A		25.6				11.3	12.6			50.5 P_2O_5	mol%
Glass B			25.6			11.3	12.6			50.5 P_2O_5	mol%
Glass C		25.6				11.3	11.6			46.5 P_2O_5 5 Cr_2O_3	mol%
Glass D		25.6	5			11.3	10.6			42.5 P_2O_5 5 Cr_2O_3	mol%
SACNLZ	5.7		34.2				17.1	32.3	5		wt%
SACNMg	1.3	0.3	34.8	–	–	–	21.4	42.2	–	5.7 ZnO	wt%
SACNZn	1.2	0.27	31.6	–	–	–	19.4	38.4	–	9.1 ZnO	wt%
NAS	17.8	–	–	–	–	–	9.4	72.8	–		mol%
SACN	10-12		24-26				6-8	53-58			mol%
SACN1	18		18			10	9	45			mol%

(continued)

Table 5.3 (continued)

Glasss-code	$Na_2O/K_2O/Li_2O$	MgO	CaO	BaO	SrO	B_2O_3	Al_2O_3	SiO_2	La_2O_3	Other dopants	mol% or wt%
SACN2	23		18			10	9	40			mol%
KMBY	13.78 Na_2O + 1.67 K_2O	8.37	9.02			0.9	11.26	54.39		0.61 Y_2O_3	wt%
SCN-1	7.29 Na_2O + 10 K_2O	0.62	3.34	8.23		0.03	2.81	66.86		0.01 ZrO_2 0.22 Fe_2O_3 0.01 ZnO	wt%
JV1	13.67					8.52	9.42	68.39			wt%
JV2	11.73					8.53	9.42	70.32			wt%
LZSA	9.53 Li_2O						14.14	62.96		13.36 ZrO_2	wt%
YBA	–	–	–	60	–	19.26	7.21	3.53	–	10 Y_2O_3	wt%
SrLaB	–	–	–	–	15.2	20.5	9	3.5	51.8		wt%
MA12	–	8.4	–	–	–	8.8	9.7	22.7	50.4		wt%
VS1	–	–	–	–	33.9	15.2	–	26.2	–	24.7 Y_2O_3	wt%
VS2	–	–	–	–	29.8	13.4	–	23	33.8	–	wt%
VM1	–	16.6	–	–	–	19.2	–	33.1	–	31.1 Y_2O_3	wt%
VM3	–	20.1	–	–	–	23.1	16.9	39.9	–	–	wt%
BaBSi	–	–	–	50	–	19.9	–	24	2.2	1.9 Y_2O_3	wt%
A2	–	–	–	60.6	–	12.6	5	21.8	–		wt%
17	–	3.9	5.5	23.9	–	17.4	7.7	15.1	26.5		wt%
Glass 59	–	–	–	34.3	23.2	15.6	–	26.9	–		wt%
AO	–	–	–	61.8	–	14	–	24.2	–		wt%
CAS1	–	–	24.5	–	–	–	14	61.5	–	–	wt%
CAS2	–	–	37.5	–	–	–	20.3	42.2	–	–	wt%

(continued)

Table 5.3 (continued)

Glasss-code	$Na_2O/K_2O/Li_2O$	MgO	CaO	BaO	SrO	B_2O_3	Al_2O_3	SiO_2	La_2O_3	Other dopants	mol% or wt%
CAS1B	–	–	23.1	–	–	5.7	13.2	58	–	–	wt%
CAS2B	–	–	35.4	–	–	5.7	19.2	39.8	–	–	wt%
CAS2BT	–	–	34.74	–	–	5.81	5.81	38.13		2.36 TiO_2	wt%
CAS2B2V	–	–	34.54	–	–	6.13	17.74	39.18	–	$2.14V_2O_5$	wt%
CAS2BK	4.57 K_2O	–	35.52	–	–	5.34	16.90	37.82			wt%
CaAl			30			20	10	40			mol%
BaAl				30		20	10	40			mol%
ML		30				20		40	10		mol%
CL			30			20		40	10		mol%
SL					30	20		40	10		mol%
BL				30		20		40	10		mol%
14					24.56	40.29	6.92	8.11	20.13		mol%
42					26.80	46.04	17.69	5.72	3.75		mol%
43					41.76	27.01	16.27	11.23	3.74		mol%
45					35.31	18.50	14.26	28.29	3.65		mol%
47					36.61	10.88	19.25	33.20	0.06		mol%
K/Si					27.14	41.72	7.22	1.76	22.17		mol%
K3/Sr					30.89	35.18	7.51	4.98	22.17		mol%
G1			20	20		12		36		$Al_2O_3 + La_2O_3 = 12$	
G2			16	24		13.7		34.3		$Al_2O_3 + La_2O_3 = 12$	
G3			16	24		16		32		$Al_2O_3 + La_2O_3 = 12$	
G4			20	20		24		24		$Al_2O_3 + La_2O_3 = 12$	
G5			16	16		14		42		$Al_2O_3 + La_2O_3 = 12$	

(continued)

Table 5.3 (continued)

Glasss-code	$Na_2O/K_2O/Li_2O$	MgO	CaO	BaO	SrO	B_2O_3	Al_2O_3	SiO_2	La_2O_3	Other dopants	mol% or wt%
G6			16	16		28		28		$Al_2O_3 + La_2O_3 = 12$	
G7			16	28		15		30		$Al_2O_3 + La_2O_3 = 12$	
G8			24	24		10		30		$Al_2O_3 + La_2O_3 = 12$	
G9					36	13		41		$Al_2O_3 + La_2O_3 = 10$	mol%
G10					40	17		33		$Al_2O_3 + La_2O_3 = 10$	mol%
G11					40	25		25		$Al_2O_3 + La_2O_3 = 10$	mol%
G12					44	16		30		$Al_2O_3 + La_2O_3 = 10$	mol%
G13			16		16	15		43		$Al_2O_3 + La_2O_3 = 10$	mol%
G14			16		24	17		33		$Al_2O_3 + La_2O_3 = 10$	mol%
G15			16		28	15		31		$Al_2O_3 + La_2O_3 = 10$	mol%
G16			16		24	25		25		$Al_2O_3 + La_2O_3 = 10$	mol%
SALSZn06					27		6	34	4	29ZnO	mol%
SALSZn08					39		6	41	2	12ZnO	mol%
SALSZn010					20		12.5	40	7.5	20ZnO	mol%
SALSZn012					27		10	46	6	11ZnO	mol%
SALSTi07					40		4	35	11	$10TiO_2$	mol%
SALSTi09					38		8	41	5	$8TiO_2$	mol%
SALSTi11					30		10	45	5	$10TiO_2$	mol%
SALSTi13					29		11	48	4	$8TiO_2$	mol%
SALSZT07					38		5	38	2	14ZnO, $3TiO_2$	mol%
SALSZT09					20		10	40	5	20ZnO, $5TiO_2$	mol%
SALSZT10					29		6	45	4	10ZnO, $6TiO_2$	mol%
SALSZT12					26		9	47	5	6ZnO, $7TiO_2$	mol%

(continued)

Table 5.3 (continued)

Glasss-code	$Na_2O/K_2O/Li_2O$	MgO	CaO	BaO	SrO	B_2O_3	Al_2O_3	SiO_2	La_2O_3	Other dopants	mol% or wt%
SZS-1					51			40		9ZnO	wt%
SZS-4					51	10		30		9ZnO	wt%
SZS-6					49	8		30		9ZnO, $4V_2O_5$	wt%
SZS-7					51	5	3	30		9ZnO, $4V_2O_5$	wt%
SLABS-3					10	30	15	30	15		wt%
SLABS-4					15	25	15	30	15		wt%
SLABS-5					20	20	15	30	15		wt%
SLABS-6					25	15	15	30	15		wt%
SLABS-7					30	10	15	30	15		wt%
G1				46.5		12.5	8.8	18.2	14		wt%
G2				50.3		10.9	9.6	18.8	10.5		wt%
G3				49.		9.1	9.1	18.3	14.5		wt%
INM1				30		7.5		40		20ZnO, $2.5Mn_2O_3$	mol%
INM2				30		5.0		40		20ZnO, $5Mn_2O_3$	mol%
INM3				30		2.5		40		20ZnO, $7.5Mn_2O_3$	mol%
INM4				30		0		40		20ZnO, $10Mn_2O_3$	mol%
P00				12.62		16.67	7.59	63.12			mol%
P01				12.62		21.67	7.59	58.12			mol%
P02				17.62		16.67	7.59	58.12			mol%
P03				12.62		26.67	7.59	53.12			mol%
P04				17.62		21.67	7.59	53.12			mol%
P05				22.62		16.67	7.59	53.12			mol%
P06				12.62		31.67	7.59	48.12			mol%

(continued)

Table 5.3 (continued)

Glasss-code	$Na_2O/K_2O/Li_2O$	MgO	CaO	BaO	SrO	B_2O_3	Al_2O_3	SiO_2	La_2O_3	Other dopants	mol% or wt%
P07				17.62		26.67	7.59	48.12			mol%
P08				22.62		21.67	7.59	48.12			mol%
P09				27.62		16.67	7.59	48.12			mol%
L00				20		30	5	45			mol%
L01				20		40	5	35			mol%
L02				30		30	5	35			mol%
L03				20		50	5	25			mol%
L04				30		40	5	25			mol%
L05				40		30	5	25			mol%
L06				20		60	5	15			mol%
L07				30		50	5	15			mol%
L08				40		40	5	15			mol%
L09				50		30	5	15			mol%
BAS1				55			5	40			mol%
BAS2				45			5	50			mol%
BAS3				35			5	60			mol%
BAS4				25			5	70			mol%
BAS5				15			5	80			mol%
18			15	35		10	5	35			mol%
21			10	35		15	5	30			mol%
23			12.5	32.5		12.5	2.5	35	5		mol%
24			12.5	32.5		10	2.5	39	3.5		mol%

(continued)

Table 5.3 (continued)

Glasss-code	$Na_2O/K_2O/Li_2O$	MgO	CaO	BaO	SrO	B_2O_3	Al_2O_3	SiO_2	La_2O_3	Other dopants	mol% or wt%
BCAS1			15	35				44		Balance by B_2O_3, Ca_2O_3 and ZnO	wt%
BCAS2			7.8	50			4.8	28.1			wt%
BCAS3			8.8	52.9			5.1	25.3			wt%
A			8.6	41.4				34.7		Balance by Al_2O_3, ZnO, PbO, B_2O_3	wt%
D			5.7	49.1				27.9			wt%
BCAS			15	35		10	5	35			mol%
H			6.1	48.2				29.8		Balance by Al_2O_3, ZnO, PbO, B_2O_3	wt%
DT-1		14.87	25.86				9.40	49.87			wt%
DT-2		14.35	18.70			3.87	9.07	41.44		$12.55Y_2O_3$	wt%
DT-3		13.83	19.24	13.15		2.98	4.37	46.4			wt%
10		17.27	21.63			2	2.18	48.93	6.98	$0.5Cr_2O_3$, 1NiO	wt%
10A		17.19	21.52			2	2.17	48.67	6.95	$0.5Cr_2O_3$, 1NiO	wt%
10D		15.83	15.41	12.04		2	2.06	44.82	6.40	$0.5Cr_2O_3$, 1NiO	wt%
BASP0				30	20	10	5	30	5		mol%
BASP2				31.7	19.2	9.6	4.8	28.8	4.8	$0.9P_2O_5$	mol%
Mg 1.5–55		18		27				55			mol%
Zn 1.5–55				27				55		18ZnO	mol%
Zn 1.5–50				30				50		20ZnO	mol%
Mg 1.5–50–5B		18		27		5		50			mol%

(continued)

Table 5.3 (continued)

Glasss-code	$Na_2O/K_2O/Li_2O$	MgO	CaO	BaO	SrO	B_2O_3	Al_2O_3	SiO_2	La_2O_3	Other dopants	mol% or wt%
Mg 1.5–50–5B–8Zn		10		27		5		50		8ZnO	mol%
G36			26.48		26.48	1.92	1.96	41.16		$2TiO_2$	mol%
YSO-1			10		42.5	7.5		34		$6Y_2O_3$	mol%
YSO-11			5		42.5	7.5		34		$6Y_2O_3$, 5NiO	mol%
YSO-12					42.5	7.5		34		$6Y_2O_3$, 10NiO	mol%
YSO-13					37.5	7.5		34		$6Y_2O_3$, 15NiO	mol%
9		14.53	15.72		12.45	2.00	2.04	45.72	6.53	1NiO	wt%
9-Bi1		14.38	15.56		12.32	2.00	2.02	45.25	6.46	1NiO, Bi_2O_3	wt%
9-Bi3		14.08	15.24		12.07	2.00	1.98	44.31	6.32	1NiO, $3Bi_2O_3$	wt%
9Bi5		13.78	14.91		11.81	2.00	1.94	43.37	6.19	1NiO, $5Bi_2O_3$	wt%
AF8				8		20	6	60		3ZnO, $3PbO_2$	mol%
A01						10	10	50		$30Ca_2O_3$	mol%
A05						6.7	30	33.3		$30Ca_2O_3$	mol%
A09						10	30	50		$10Ca_2O_3$	mol%
MgY		30				20		40		10 Y_2O_3	mol%
CaY			30			20		40		10 Y_2O_3	mol%
SrY					30	20		40		10 Y_2O_3	mol%
BaY				30		20		40		10 Y_2O_3	mol%
L1				36.1		17.2	10.3	34.4	2.0		mol%
L2				35.0		16.7	10.0	33.3	5.0		mol%
Z1				36.1		17.2	10.3	34.4		2.0 ZrO_2	mol%
Z2				35.0		16.7	10.0	33.3		5.0 ZrO_2	mol%

(continued)

Table 5.3 (continued)

Glasss-code	$Na_2O/K_2O/Li_2O$	MgO	CaO	BaO	SrO	B_2O_3	Al_2O_3	SiO_2	La_2O_3	Other dopants	mol% or wt%
SZ1				36.1		21.5	10.3	30.1		2.0 ZrO_2	mol%
SZ2				35.0		20.8	10.0	29.2		5.0 ZrO_2	mol%
N1				36.1		17.2	10.3	34.4		2.0 NiO	mol%
N2				35.0		16.7	10.0	33.3		5.0 NiO	mol%

Note The last three entries on page 195, i.e., G1, G2, and G3 are studied by Ghosh and coworkers.

Table 5.4 Properties of glass sealants for SOFC

Glass code	T_g (°C)	T_c (°C)	T_s (°C)	Density (g/cm^3)	CTE (× 10^{-6}/°C)
18	–	630	685		10.8
SACNMg	780	–			8.1
SACNLZ	740	780, 900			9.7
SACNZn	700	950–1020	900	–	7.8–10
SACN	670	830, 940			9.4-9.8
GS1	670	865			
GS2	700	905			
GS3	704	945			
GS4	727	995			
Glass A	595				5.9
Glass B	602				7.9
Glass C	654				5.7
Glass D	637				5.9
BAS	628	827			
CAS	700	921			
MAS	717	921			
MAS10	720	962			
MAST5	712	913			
MAST10	710	954			
MASZ5	723	931			
MASZ10	721	973			
MASN6	735	939			
MASC6	738	946			
Mg 1.5–55	715	–	750	3.15	10.1
Mg1.5–40.5B–10Pb	–	–	512	5.44	–
Mg1.5–40.1B–8Zn	616	–	633	1.97	–
Glass A	470	–	560	2.5	10.4
Glass C	600	–	720	–	10.7
AO	622	763	907	–	10.83
A2	602	830	–	–	10.67
MBS	–	–	700	–	10
L2	655	862	710	–	11.1
10B	755	904	–	3.06	8.4
10C	756	906	–	3.1	8.71
17	620	–	–	–	8.8
14a	576	–	–	–	9.4
GC-9	668	820	864	745	–
BCAS4	609	755	672	3.89	11.9
BCAS5	608	720	660	3.93	12.4
7A	685	912	716	3.08	8.69

(continued)

Table 5.4 (continued)

Glass code	T_g (°C)	T_c (°C)	T_s (°C)	Density (g/cm^3)	CTE (× 10^{-6}/°C)
MA1	621	887	671	–	9.01
MA12	638	907	723	–	8.33
BMA1	620	785	656	–	9.66
BM1	530	804	616	–	10.06
BM2	582	845	593	–	11.1
7A–Cr	675	895	721	3.08	8.48
VS1	626	–	–	–	7.36
VS2	620	–	–	–	5.26
VM3	735	–	–	–	1.99
LO9	–	–	3.92	–	10.96
CAS1	782	–	817	–	–
CSA2	794	–	831	–	–
CAS1B	729	–	760	–	–
CAS2B	734	–	757	–	–
ML	700	832	726	3.32	–
CL	714	846	740	3.69	–
SL	722	858	748	3.69	–
BL	682	775	701	3.86	–
G1	–	706	678	–	10.13
G2	–	696	670	–	10.46
G3	–	672	655	–	10.55
G9	–	765	781	–	9.72
G10	–	752	751	–	10.59
G11	–	700	684	–	10.67
G14	–	740	655	–	10.12
G15	–	737	637	–	10.59
G16	–	661	656	–	10.87
SALSZn08	773	–	882	3.93	10.3
SALSZn12	788	–	894	3.86	8.8
SALSTi13	820	–	935	3.60	9.1
SALSZT10	769	–	879	3.82	9.7
SALSZT12	790	–	911	3.75	7.4
SALSZn06	733	–	822	4.19	9.5
SZS-1	722	890	745	3.68	10.8
SZS-4	643	820	676	3.71	10.5
SZS-6	640	830	664	3.66	11.5
SZS-7	671	885	707	3.78	11.2
G1	625.8	745.7	675.4	4.1	11
G2	621.6	735.4	669	4.0	10.8
G3	630.2	757.7	681	4.1	10.9

Note The last three entries on this page, i.e., G1, G2, and G3 are studied by Ghosh and coworkers.

(continued)

Table 5.4 (continued)

Glass code	T_g (°C)	T_c (°C)	T_s (°C)	Density (g/cm^3)	CTE (× 10^{-6}/°C)
INM1	676	791	–	–	–
CaAl	640	705	–	–	–
BaAl	664	766	–	–	–
BCAS1	665	785	712	3.33	9.9
BCAS2	623	762	679	3.82	11.2
BCAS3	617	758	661	3.86	11.6
BCAS4	609	755	672	3.89	11.9
BCAS5	608	720	660	3.93	12.4
BCAS	619		682	3.88	10.5
H	550	714	–	–	–
H + 10% ZrO_2	612	695	–	–	–
H + 30% ZrO_2	642	705	–	–	–
YSO-1	698	–	735	3.80	12.09
YSO-11	675	–	725	3.92	11.79
YSO-12	683	–	730	3.99	11.45
YSO-13	683	–	730	4.06	11.48
R	555	875	710	–	8.28
9-Bi1	535	871	695	–	8.38
9-Bi3	520	880	695		8.57
9-Bi5	526	877	705		8.33
L1	626	690	827		10.1
L2	655	710	862		11.1
Z1	623	684	825		9.7
Z2	655	730	886		9.3
SZ1	613	669	807		10.6
SZ2	628	691	835		9.9
N1	613	670	813		11.5
N2	610	659	763		12.8
MgY					
CaY	635	879			9.12
SrY	678	843			7.58
BaY	622	948			8.58

B_2O_3-containing glasses possess low T_g due to the presence of planar boroxyl structural groups and increased fraction of the NBO, which decreases network connectivity.

Devitrification leads to deterioration of the properties, and hence undesirable phenomena. Devitrification usually arises from the surface flaws, heterogeneity in glass network structure, small nuclei formation during glass making, and residual stress due to mechanical polishing. A glass is said to possess higher devitrification resistance, if higher temperature and longer time are required for it to devitrify. Borate and borosilicate glasses have lower devitrification resistance because of their lower glass network connectivity, whereas silicate glasses possess higher devitrication resistance. Glasses/glass-ceramics are potential sealants as their thermal properties, especially CTE, can be tailored in order to make them compatible with the other components of the fuel cell. Some potential compositions of glasses employed as sealants in SOFC are summarized in Table 5.3, and their selected properties are listed in Table 5.4.

5.10 Crystallisation of Glasses

Understanding the nucleation and crystallisation kinetics is necessary to gain insight into the amorphous glassy state. The SOFC operates at high temperature and during its operation, the glass gets converted into glass-ceramics. Glass-ceramics possess superior mechanical, thermal, and structural properties compared to the parent glass. The glass-ceramics usually have enhanced CTE attributed to the nucleation of different crystalline phases inside an amorphous matrix. Two types of crystallisation behaviour are observed in the glass-ceramics, i.e., slow and rapid. The slow crystallisation is stable, whereas the rapid crystallisation (which occurs progressively up to 1000 h) is unsuitable for long-term operation of SOFC. The fluidity of the glass decreases due to rapid and progressive crystallisation, and hence the SOFC performance deteriorates. If the CTE value of the crystalline phase differs considerably from that of the parent glass, then thermal mismatch can occur, which decreases the structural integrity of the SOFC stack. Hence, it is imperative to understand the crystallisation kinetics and nucleation behaviour to better understand the sealing properties and their chemical interactions when in contact with adjacent components of the cell. Table 5.5 lists some of the crystalline phases that precipitate in various glass systems.

Controlled crystallisation of glasses is beneficial for the glass-ceramics and it involves two-stage heat treatment: nucleation and crystallisation. In the nucleation stage, small nuclei are formed within the parent glass and after the formation of stable nuclei, crystallisation proceeds by growth of the new crystalline phase. Nucleation and crystallisation are very important parameters during glass preparation that should be completely understood to obtain desired microstructure and properties. For this purpose, various thermodynamic parameters such as the

Table 5.5 Various crystalline phases and their CTE

System	Crystalline phase	CTE ($\times 10^{-6}$/°C)
Ba–Si–O	Barium silicate ($BaSiO_3$)	9–13
	Barium orthosilicate ($BaSi_2O_5$)	14
Mg–Si–O	Enstatite ($MgSiO_3$)	9–12
	Clinoenstatite ($MgSiO_3$)	7.8–13.5
	Protoenstatite ($MgSiO_3$)	9.8
	Forsterite (Mg_2SiO_4)	9.4
Ca–Si–O	Wollastonite ($CaSiO_3$)	4–9
	Calcium orthosilicate (Ca_2SiO_4)	10–14
Ba–Ca–Si–O	Barium calcium orthosilicate ($Ba_3CaSi_2O_8$)	12–14
Mg–Al–Si–O	Cordierite ($Mg_2Al_4Si_5O_{18}$)	2
	Cristobalite (SiO_2)	2.32
Sr–Al–Si–O	Hexacelsian ($SrAl_2Si_2O_8$)	8–11
	Monocelsian ($SrAl_2Si_2O_8$)	3
	Orthocelsian ($SrAl_2Si_2O_8$)	5–8
Ba–Al–Si–O	Hexacelsian ($BaAl_2Si_2O_8$)	7–8
	Monocelsian ($BaAl_2Si_2O_8$)	2–3
	Orthocelsian ($BaAl_2Si_2O_8$)	5–7
Ba–Mg–B–Si–O (with Zn)	Barium silicate ($Ba_2Si_2O_4$)	–
	Barium magnesium silicate ($BaMgSiO_4$)	–
	Barium magnesium silicate (Ba_2MgSiO_3)	–
	Barium zinc silicate ($BaZn_2Si_2O_7$)	–
Ba–Ca–Al–B–Si–O (with chromia forming alloys)	$BaCrO_4$	$\alpha_a = 16.5$
		$\alpha_b = 33.8$
		$\alpha_c = 20.4$
Ba–Mg–Ca–B–Si–O (with electrolyte)	Barium zirconate ($BaZrO_3$)	7.9
	Calcium zirconate ($CaZrO_3$)	10.3
	Kurchatovate ($CaMgB_2O_5$)	10.4
Na–Ca–Al–B–Si–O	Gehlenite ($Ca_2Al_2SiO_7$)	–
	Nepheline ($AlNaSiO_4$)	–
	Corundum (Al_2O_3)	–

Lasocka parameter, activation energy of crystallisation, frequency factor, and Avrami parameter can be evaluated for different systems. Along with these parameters, crystallisation constants as a function of different temperatures as well as different heating rates (β) should be studied. T_g and its dependence on heating rates can be given by the following original empirical Lasocka equation,

$$T_g = A + B \ln \beta \tag{5.37}$$

Here, A is the glass transition temperature when the heating rate is 1 °C min^{-1}. B depends upon the composition of glass and gives the time response of various configurational changes occurring in the glass transition region. The parameter B depends on the chemical nature of the modifier. The activation energy is the threshold energy above which any transformation can take place in the activated complex. Generally, total activation energy is composed of two parts:

$$E_c = aE_N + bE_G \tag{5.38}$$

E_N is the activation energy of nucleation and E_G is the activation energy of growth. a and b are the parameters related to the Avrami parameter and $a + b = 1$. During the beginning process, nucleation dominates; i.e., $b = 0$. By the end of the crystallisation process, $a = 0$, only the growth process is going to dominate. The activation energy at the peak of crystallisation is obtained using the Augis and Benett relation:

$$\ln(T_c/\beta) = E_c/RT_c + \ln v \tag{5.39}$$

where v (frequency factor) is defined as the number of attempts which are made by the nuclei/second for overcoming the energy barrier, and β is the heating rate. v provides information regarding the available nucleation sites for crystal growth. The maximum value of v will indicate instability of the glass. The network modifier disrupts the glass network structure and occupies random positions between polyhedrals and maintains local charge neutrality. The properties of the glass seal depend upon the cation oxygen bond length and bond angles, CN of cations, degree of network connectivity, and local ordering in the network structure. The Avrami parameter, which is related to the morphology of crystal growth, can be calculated using the following equation,

$$[d \ln(\Delta y)/d(\Delta y)] = -nE_c/R \tag{5.40}$$

where Δy is the vertical displacement at temperature T of the DTA exotherm. A non-integer value of n indicates the possible occurrence of two crystallisation mechanisms simultaneously working during amorphous-to-crystalline transformation. The rate of crystallisation k can be evaluated using the relations:

$$k_y(T) = v \exp\left(-(E/RT)\left(T_x - T_g\right)/T_m\right) \tag{5.41}$$

and

$$k_b(T) = v \exp(-(E/RT)H_r) \tag{5.42}$$

where T_x is the crystallisation onset temperature and T_m is the melting temperature of glass and Hruby parameter $H_r = (T_x - T_g)/(T_m - T_c)$. Minimum values of k indicate a decrease in crystallisation, and hence an increase in glass-forming ability. The new k criterion of Chen takes $(T_c - T_f)$ into account, where T_f is the inflection point temperature and can be obtained from the peak temperature of differential DTA. It remains almost stable, and hence temperature as well as rate independent; i.e.,

$$k_f(T) = v\exp\left(-(E/RT)\left(T_c - T_f\right)/T_f\right) \tag{5.43}$$

This may be due to the fact that thermodynamic factors such as $(T_c - T_f)/T_f$ and $(T_x - T_g)/T_m$ are involved in the evaluation of $k_y(T)$ and $k_f(T)$ and parameter H_r is used to determine $k_b(T)$. Hence, $k_b(T)$ cannot be regarded as stability criteria in the present glasses. Hence, these parameters play a substantial role in determining the nature of glasses. In addition to this, the concept of fluctuation free volume, the transition of an amorphous substance from a liquid to solid state, is accompanied with a decrease in the fraction of fluctuation free volume, f_g,

$$f_g = (V_f/V) \tag{5.44}$$

where V_f is the fluctuation free volume and is the sum of the volumes V_h of N fluctuation holes ($V_f = V_h N$), which are formed due to local excitation caused by deviation of the interatomic distance between adjacent kinetic units from the mean distance r_0 (V is the volume of glass at T_g). The relation between the fraction of fluctuation free volume (f_g) and activation energy is

$$E_g/RT_g = 1/f_g \tag{5.45}$$

According to the Frenkel equation, the abrupt change in coefficient of thermal expansion α_f which is associated with an increase in number of microvoids (free volume) upon heating of glass can be determined merely by the fraction f_g:

$$\alpha_f T_g = f_g \ln\left(1/f_g\right) \tag{5.46}$$

Increasing f_g indicates a decreased number of bonds within this glass network. This ultimately leads to an increase in free volume for the rearrangement of kinetic units in the glass matrix. Consequently, the degree of connectivity of the glass-forming network decreases. In addition to this, higher covalency of the glass network leads to a low abrupt change in α_f. Materials with strong bonds show a small change in the coefficient of thermal expansion. Glass networks have the highest NBOs and hence loose network, which is usually responsible for its high α_f. One example is given, which demonstrates the use of these calculations as follows.

Example 1 Crystallisation kinetics of ML, CL, SL, and BL glass
The non-isothermal method is widely used to obtain various characteristic parameters from DTA and the glass sample is heated up at a fixed heating rate. The

Table 5.6 Characteristic temperatures of glasses obtained from DTA

Glass	Heating Rate β	T_g (°C)	T_x (°C)	T_c (°C)	T_m (°C)	T_f (°C)
ML	10	700	732	832	937	818
	20	728	746	843	949	824
	30	732	752	859	958	831
	40	746	758	866	967	839
CL	10	714	749	846	902	827
	20	722	753	858	913	832
	30	738	757	862	922	836
	40	752	761	869	944	842
SL	10	722	752	858	934	832
	20	736	759	867	941	839
	30	747	764	873	953	853
	40	762	770	882	960	864
BL	10	682	734	775	901	752
	20	718	741	782	927	773
	30	722	747	793	933	780
	40	731	759	813	941	802

characteristic transition temperatures of glass seals obtained from DTA at different heating rates are listed in Table 5.6. For obtaining T_f, DTA curves were differentiated, and hence derivative differential thermal analysis (DDTA) was obtained.

A graph is plotted between ln [β] and 1000/T_g, which gives a straight line and the data obtained were well fitted in straight lines. The slope and intercept of Eq. (5.37) give parameters B and A, respectively, which are listed in Table 5.7. The values of frequency are obtained from the intercept by the least squares fitting method. The maximum value of ν will indicate the instability of the glass. ML glass has the lowest ν followed by BL glass, hence exhibiting lowest nucleation sites. The value of n is calculated using a Piloyan plot between ln(Δy) and 1000/T and the non-integer value reveals two types of crystallisation mechanisms (Table 5.7).

Equations (5.41)–(5.43) yield the crystallisation constants as shown in Table 5.8. It is quite observable from Table 5.8 that $k_y(T)$ and $k_f(T)$ are quite independent of

Table 5.7 Constants A and B, frequency, Avrami parameter and activation energy for all the glasses

Glass	A (°C)	B	Frequency (s^{-1})	Avrami parameter (n)	Activation energy (kJ/mol)
ML	629.05	31.46	3.19×10^8	1.43	382
CL	647.65	26.72	1.54×10^{11}	1.68	273
SL	656.45	27.53	9.51×10^{11}	1.84	291
BL	606.09	34.59	7.53×10^{10}	1.72	308

Table 5.8 Crystallisation constants for all the glasses

	β	$k_y(T)$			$k_b(T)$			$k_f(T)$		
		$k_y(T_g)$	$k_y(T_c)$	$k_y(T_f)$	$k_b(T_g)$	$k_b(T_c)$	$k_b(T_f)$	$k_f(T_g)$	$k_f(T_c)$	$k_f(T_f)$
ML	10	3.186×10^8	3.187×10^8	3.189×10^8	3.144×10^8	3.150×10^8	3.149×10^8	3.188×10^8	3.188×10^8	3.188×10^8
	20	3.187×10^8	3.188×10^8	3.189×10^8	3.165×10^8	3.168×10^8	3.167×10^8	3.187×10^8	3.188×10^8	3.188×10^8
	30	3.187×10^8	3.188×10^8	3.189×10^8	3.161×10^8	3.164×10^8	3.164×10^8	3.186×10^8	3.187×10^8	3.187×10^8
	40	3.188×10^8	3.189×10^8	3.190×10^8	3.173×10^8	3.175×10^8	3.174×10^8	3.186×10^8	3.187×10^8	3.187×10^8
CL	10	1.538×10^{11}	1.539×10^{11}	1.539×10^{11}	1.508×10^{11}	1.511×10^{11}	1.511×10^{11}	1.539×10^{11}	1.539×10^{11}	1.539×10^{11}
	20	1.539×10^{11}	1.539×10^{11}	1.539×10^{11}	1.511×10^{11}	1.515×10^{11}	1.514×10^{11}	1.539×10^{11}	1.539×10^{11}	1.539×10^{11}
	30	1.539×10^{11}	1.539×10^{11}	1.539×10^{11}	1.524×10^{11}	1.526×10^{11}	1.526×10^{11}	1.539×10^{11}	1.539×10^{11}	1.539×10^{11}
	40	1.54×10^{11}	1.54×10^{11}	1.54×10^{11}	1.534×10^{11}	1.535×10^{11}	1.534×10^{11}	1.539×10^{11}	1.539×10^{11}	1.539×10^{11}
SL	10	9.501×10^{11}	9.502×10^{11}	9.502×10^{11}	9.375×10^{11}	9.390×10^{11}	9.388×10^{11}	9.502×10^{11}	9.503×10^{11}	9.503×10^{11}
	20	9.504×10^{11}	9.504×10^{11}	9.504×10^{11}	9.404×10^{11}	9.417×10^{11}	9.414×10^{11}	9.501×10^{11}	9.502×10^{11}	9.502×10^{11}
	30	9.505×10^{11}	9.509×10^{11}	9.506×10^{11}	9.439×10^{11}	9.446×10^{11}	9.445×10^{11}	9.505×10^{11}	9.505×10^{11}	9.505×10^{11}
	40	9.508×10^{11}	9.508×10^{11}	9.508×10^{11}	9.476×10^{11}	9.479×10^{11}	9.479×10^{11}	9.505×10^{11}	9.505×10^{11}	9.505×10^{11}
BL	10	7.517×10^{10}	7.518×10^{10}	7.518×10^{10}	7.412×10^{10}	7.422×10^{10}	7.420×10^{10}	7.524×10^{10}	7.524×10^{10}	7.524×10^{10}
	20	7.524×10^{10}	7.525×10^{10}	7.525×10^{10}	7.486×10^{10}	7.488×10^{10}	7.488×10^{10}	7.528×10^{10}	7.528×10^{10}	7.528×10^{10}
	30	7.524×10^{10}	7.525×10^{10}	7.525×10^{10}	7.480×10^{10}	7.484×10^{10}	7.483×10^{10}	7.527×10^{10}	7.527×10^{10}	7.527×10^{10}
	40	7.524×10^{10}	7.524×10^{10}	7.524×10^{10}	7.472×10^{10}	7.477×10^{10}	7.476×10^{10}	7.527×10^{10}	7.527×10^{10}	7.527×10^{10}

Table 5.9 Fluctuation free volume (f_g) and bulk thermal expansion coefficient (α_f) for all the glasses

Glass	Heating rate β (°C/min)	f_g	α_f
ML	10	2.12×10^{-2}	8.39×10^{-5}
	20	2.18×10^{-2}	8.33×10^{-5}
	30	2.19×10^{-2}	8.32×10^{-5}
	40	2.21×10^{-2}	8.29×10^{-5}
CL	10	3.01×10^{-2}	1.07×10^{-4}
	20	3.03×10^{-2}	1.06×10^{-4}
	30	3.08×10^{-2}	1.06×10^{-4}
	40	3.12×10^{-2}	1.06×10^{-4}
SL	10	2.84×10^{-2}	9.87×10^{-5}
	20	2.88×10^{-2}	9.77×10^{-5}
	30	2.91×10^{-2}	9.76×10^{-5}
	40	2.95×10^{-2}	9.75×10^{-5}
BL	10	2.57×10^{-2}	1.01×10^{-4}
	20	2.68×10^{-2}	1.01×10^{-4}
	30	2.69×10^{-2}	1.01×10^{-4}
	40	2.71×10^{-2}	1.01×10^{-4}

heating rates, whereas $k_b(T)$ shows inconsistent variation. This may be due to the fact that thermodynamic factors like $(T_c - T_f)/T_f$ and $(T_x - T_g)/T_m$ are involved in the evaluation of $k_y(T)$ and $k_f(T)$ and parameter H_r is used to determine $k_b(T)$. Hence, $k_b(T)$ cannot be regarded as stability criteria in the present glasses. CL and SL glasses posses higher crystallisation constants in comparison to ML and BL glasses, which is in accordance with their frequency factors (ν). Hence, these parameters play a substantial role in determining the nature of glasses.

Equations (5.45) and (5.46) yield the values of f_g and α_f, which are listed in Table 5.9. With increasing heating rate, f_g is increasing, which indicates that the number of bonds within this glass must have decreased. This ultimately leads to an increase in free volume for the rearrangement of kinetic units in the glass matrix. Consequently, the degree of connectivity of the glass-forming network decreases. In the present case, CL glass exhibits the highest fluctuation free volume. This is in agreement with the low activation energy of crystallisation required for this glass. f_g follows the trend f_g(CL) > f_g(SL) > f_g(BL) > f_g(ML). This signifies that a glass with higher activation energy possesses lower free volume. At the same time, ML and SL glasses have exhibited a small jump in α_f in comparison to CL and BL glasses. BL glass has exhibited the highest jump in α_f. The BL glass network has the highest NBOs and hence loose network, which may be responsible for its high α_f.

5.11 Glasses in Contact with Adjacent Components of SOFC

Desired thermal properties of SOFC seal glasses have been achieved in several glass systems but long-term stability remains an ongoing problem. Seal glass should be well bonded with the electrolyte, metallic interconnects, and electrodes without any pores, cracks, and detrimental reaction products at the interface. Thermomechanical properties such as interfacial strength play a key role in determining the performance, efficiency, and durability of the SOFCs. Therefore, interfacial reaction kinetics and thermomechanical properties of the interface should be discussed in light of the interfacial bonding strength for different glass systems. In order to obtain a pore and crack-free interface, the acceptable CTE difference between seal glass and the adjoining SOFC components should not differ by more than 1×10^{-6}/K. The stresses arise due to the CTE difference between the devitrified phase(s) and the amorphous glass phase, which lower flowability of the devitrified phases and hence lead to crack and pore formation in the glass at the interface. These pores and cracks within the seal and at the interface are major gas leakage paths.

Seal glasses generally wet ZrO_2 electrolyte well, whereas severe interaction is observed with the metallic interconnects. Seal glass can interact with yttria-stabilised zirconia and form $BaZrO_3$/$CaZrO_3$/$SrZrO_3$ and lead to high CTE chromate formation when in contact with interconnects. Hence, understanding interfacial reaction kinetics of seal glasses and different components of SOFC is very important to know about the various phases formed and their nature as well. In addition to this, additional work is required to look for the conditions which inhibit or avoid the formation of detrimental phases. Chapters 6, 7, and 8 are focusing on the chemical compatibility of glass seals with interconnect and electrolytes for different glass systems. The major concern is to gain insight into crystallisation kinetics, formation of various crystalline phases, effect of different phases on thermal expansion, and the mechanism of interfacial interaction/structural modification.

Bibliography

Chou YS, Stevenson JW, Chick LA (2002) J Power Sources 112:130
Chou YS, Stevenson JW, Gow RN (2007a) J Power Sources 168:426
Chou YS, Stevenson JW, Singh P (2007b) J Electrochem Soc 154:B644
Ediger MD, Angell CA, Nagel SR (1996) J Phys Chem 100:13200
Fergus JW (2005) J Power Sources 147:46
Mauro JC, Allan DC, Potuzak M (2009a) Nonequilibrium viscosity of glass. Phys Rev B 80:094204
Mauro JC, Yue Y, Ellison AJ, Gupta PK, Allan DC (2009b) Viscosity of glass-forming liquids. Phillips JC (Commun) Rutgers University, Summit, NJ, 9 Oct 2009
Weil KS (2006) JOM 58(8):37

Suggested Bibliography

Arora A, Goel A, Shaaban ER, Singh K, Pandey OP, Ferreira JMF (2008) Phys B 403:1738
Arora A, Kumar V, Singh K, Pandey OP (2011) Cera Int 37:2101
Augis A, Bennett JE (1978) J Therm Anal Calor 13:283
Bainova AB, Sanditov DS (2002) G Phys Chem 28(3):189
Bansal NP, Hyatt MJ, Drummond CH (1991) Cera Eng Sci Proc 12:1222
Bansal NP, Gamble EA (2005) J Power Sources 147:107
Batfalsky P, Haanappel VAC, Malzbender J, Menzler NH, Shemet V, Vinke IC, Stenbrech RW (2006) J Power Sources 155:128
Brochu M, Gauntt BD, Shah R, Miyake G, Loehman RE (2006) J Euro Cera Soc 26:3307
Calas G, Cormier L, Galoisy L, Jollivet P (2002) CR Chim. 5:831
Chang CYS, Wei WCJ, Hsue CH (2011a) J Non-cryst Solids 357:1414
Chang HT, Lin CK, Liu CK, Wu SH (2011b) J Power Sources 196:3583
Cheng Y, Xiao H, Guo W (2007) Mater Sci Eng A 464:210
Cheng K (1999) J Phys Chem B 103:8272
Chou Y-S, Stevenson JW, Chick LA (2003) J Am Ceram Soc 86(6):1003
Chou YS, Stevenson JW, Xia GG, Yang ZG (2010) J Power Sources 195:5666
Chou YS, Thomsen EC, Williams RT, Choi JP, Canfield NL, Bonnett JF, Stevenson JW, Shyam A, Curzio EL (2011) J Power Sources 196:2709
Dai Z, Pu J, Yan D, Chi B, Jian L (2011) Int J Hydrogen Energy 36:3131
Demirkesen E, Goller G (2003) Cera Int 29:463
Demirkesen E, Maytalman E (2001) Cera Int 27:99
Deubener J (2005) J Non-cryst Solids 351:1500
Donald JW, Metcalfe BL, Morris AEP (1992) J Mater Sci 27:2979
Dong DK, Bo ZL, Zhu JQ (1996) J Non-cryst Solids 201:47
Fergus JW, Lessing PA (2007) J Mater Sci 42:3465
Frenkel YI (1975) Nauka, Moscow
Galoisy L, Cormier L, Rossano S, Ramos A, Calas G, Gaskell P, Grand ML (2000) Miner Mag 64:409
Ghosh S, Sharma AD, Kundu P, Mahanty S, Basu RN (2008a) J Non-cryst Solids 354:4081
Ghosh S, Kundu P, Sharma AD, Basu RN, Maiti HS (2008b) J Euro Cera Soc 28:69
Ghosh S, Sharma AD, Mukhopadhyay AK, Kundu P, Basu RN (2010) Int J Hydrogen Energy 35:272
Goel A, Tulyaganov DU, Agathopoulos S, Ribeiro MJ, Basu RN, Ferreira JMF (2007a) J Euro Cera Soc 27:2325
Goel A, Tulyaganov DU, Kharton VV, Yaremchenko AA, Agathopoulos S, Ferreira JMF (2007b) J Am Cera Soc 90:2236
Goel A, Shaaban ER, Melo FCL, Ribeiro MJ, Ferreira JMF (2007c) J Non-cryst Solids 353:2391
Goel A, Tulyaganov DU, Kharton VV, Yaremchenko AA, Ferreira JMF (2008) Acta Mater 56:3065
Goel A, Tulyaganov DU, Kharton VV, Yaremchenko AA, Eriksson S, Ferreira JMF (2009) J Power Sources 189:1032
Goel A, Tulyaganov DU, Ferrari AM, Shaaban ER, Prange A, Bondioli F, Ferreira JMF (2010a) J Am Cera Soc 93:830
Goel A, Tulyaganov DU, Pascual MJ, Shaaban ER, Muñoz F, Lu Z, Ferreira JMF (2010b) J Non-cryst Solids 356:1070
Goel A, Pascual MJ, Ferreira JMF (2010c) Int J Hydrogen Energy 35:6911
Gross SM, Koppitz T, Remmel J, Bouche JB, Reisgen U (2006) Fuel Cell Bull 12
Gupta PK, Mauro JC (2009) J Chem Phys 130:094503
Gutzow I, Schmelzer J (1995) Springer, Berlin
Haanappel VAC, Shemet V, Gross SM, Koppitz T, Menzler NH, Zahid M, Quadakkers WJ (2005) J Power Sources 150:86

Hardy JS, Kim JY, Weil KS (2004) J Electrochem Soc 151(8):J43
Herma P, Han WT, Cooper AR (1988) J Non-Cryst Solids 102:88
Holand W, Beall G (2002) Glass ceramics technology. The ACS, Westerville
Howe JM (1993a) Int Mater Rev 38:233
Howe JM (1993b) Int Mater Rev 38:257
Hoyt KO, Gannon PE, White P, Tortop R, Ellingwood BJ, Khoshuei H (2012) Int J Hydrogen Energy 37:518
Huang S, Lu Q, Wang C (2011) J Alloys Compds 509:4348
Jiang SP, Christiansen L, Hughan B, Foger K (2001) J Mater Sci Lett 20:695
Jinhua P, Kening S, Naiqing Z, Xinbing C (2007) J Rare Earths 25:434
Kaur G, Kumar V, Pandey OP, Singh K (2012) J Electrochem Soc 159:B277–B284
Kerstan M, Muller M, Russel C (2011) Mater Res Bull 46:2456
Kingery WD, Bowen HK, Uhlmann DR (1976) Introduction to ceramics, 2nd edn. Wiley, New York
Kuhn B, Wessel E, Malzbender J, Steinbrech RW, Singheiser L (2010) Int J Hydrogen Energy 35:9158
Kumar V, Arora A, Pandey OP, Singh K (2008) Int J Hydrogen Energy 33:434
Kumar V, Sharma S, Pandey OP, Singh K (2010) Solid state Ionics 181:79
Kumar V, Rupali, Pandey OP, Singh K (2011a) Int J Hydrogen Energy 36:14971
Kumar V, Kaur G, Pandey OP, Singh K (2011b) Phys Chem Glasses 52:212
Lahl N, Singh K, Singheiser L, Hilpert K, Bahadur D (2000) J Mater Sci 35:3089
Lahl N, Bahadur D, Singh K, Singheiser L, Hilpert K (2004) J Electrochem Soc 149:A607
Lara C, Pascual MJ, Duran A (2004a) J Non-cryst Solids 348:149
Lara C, Pascal MJ, Prado MO, Duran A (2004b) Solid State Ionics 170:201
Larsen PH, James PF (1998) J Mater Sci 33:2499
Lasocka M (1979) J Therm Anal 16:197
Ley KL, Krumplet M, Kumar R, Meiser JH, Bloom IJ (1996) J Mater Res 11:1489
Liu WN, Sun X, Koeppel B, Khaleel M (2010) Int J Appl Cera Tech 7:22
Mahapatra MK, Lu K (2009) J Mater Sci 44:5369
Mahapatra MK, Lu K (2010a) Int J Appl Cera Tech 7:10
Mahapatra MK, Lu K (2010b) Int J Hydrogen Energy 35:7945
Mahapatra MK, Lu K (2011) J Power Sources 196:700
Mahapatra MK, Lu K, Reynolds WT (2008) J Power Sources 179:106
Malzbender J (2010) J Euro Cera Soc 30:3407
Malzbender J, Monch J, Steinbrech RW, Koppitz T, Gross SM, Remmel J (2007) J Mater Sci 46:6297
Meera BN, Ramakrishna J (1993) J Non-cryst Solids 159:1
Meinhardt KD, Kim DS, Chou YS, Weil KS (2008) J Power Sources 182:188
Menzler NH, Bram M, Buchkremer HP, Stoever D (2002) J Euro Cera Soc 23:445
Milhans J, Khaleel M, Sun X, Tehrani M, Al-Haik M, Garmestani H (2010) J Power Sources 195:3631
Milhans J, Li DS, Khaleel M, Sun X, Garmestani H (2011) J Power Sources 196:3846
Namwong P, Laorodphan N, Thiemsorn W, Jaimasith M, Wannakon A, Chairuangsri T (2010) Chiang Mai J Sci 37:231
Nielsen KA, Solvang M, Nielsen SBL, Dinesen AR, Beeaff D, Larsen PH (2007) J Euro Cera Soc 27:1817
Ogasawara K, Kameda H, Matsuzaki Y, Sakurai T, Uehara T, Toji A, Sakai N, Yamaji K, Horita T, Yokokawa H (2007) J Electrochem Soc 154:B657
Ojha PK, Ratha SK, Chongdar TK, Gokhale NM, Kulkarni AR (2011) J Power Sources 196:4594
Pascual MJ, Guillet A, Duran A (2007) J Power Sources 169:40
Peng L, Zhu Q (2009) J Power Sources 194:880
Pi SH, Lee SB, Song RH, Lee JW, Lim TH, Park SJ, Shin DR, Park CO (2011) Int J Hydrogen Energy 36:13735
Piloyan GO, Rybachikov JD, Novikova OS (1996) Nature 212:1229

Raghvan V (2008) Materials science and engineering, 5th edn
Sakuragi S, Funahashi Y, Suzuki T, Fujishiro Y, Awano M (2008) J Power Sources 185:1311
Scholze H (1991) Glass nature, structure and properties (Chap 2). Springer, New York
Shelby JE (2005) Introduction to glass science and technology, 2nd edn. The Royal Society of Chemistry, Cambridge
Singh RN (2006) J Mater Eng Perf 15:422
Singh RN (2007) Int J Appl Cera Tech 4:134
Singh K, Gupta N, Pandey OP (2007) J Mater Sci 42:6426
Smeacetto F, Salvo M, Ferraris M, Cho J, Boccaccini AR (2008a) J Euro Cera Soc 28:61
Smeacetto F, Salvo M, Ferraris M, Casalegno V, Asinari P, Chrysanthou A (2008b) J Euro Cera Soc 28:2521
Smeacetto F, Salvo M, D'He´rin Bytner FD, Leone P, Ferraris M (2010) J Euro Cera Soc 30:933
Sohn SB, Choi SY (2004) J Am Cera Soc 87:254
Sohn SB, Choi SY, Kim GH, Song HS, Kim GD (2002) J Non-cryst Solids 297:103
Steele BCH, Heinzel A (2001) Materials for fuel cell technologies. Nature 414:345
Stolen D, Monreal E, Miller W (1992) Fuel cell seminar abstracts. AZ, Tucsan, p 53
Sun T, Xiao H, Guo W, Hong X (2010) Cera Int 36:821
Touloukian YS (1967) Thermophysical properties of high temperature solid materials. The MacMillan Co., New York
Uhlmann DR (1980) J Non-cryst Solids 38:693
Vogel W (1992) Glass chemistry. Springer, Verlag
Volf MB (1984) Chemical approach to glass. In: Glass Science Technology (vol 7). Elsevier, Amsterdam
Wang R, Lu Z, Liu C, Zhu R, Huang X, Wei B, Ai N, Su W (2007) J Alloys Compd 423:189
Wang SF, Wang YR, Hsu YF, Chuang CC (2009) Int J Hydrogen Energy 34:8235
Wang SF, Lu CM, Wu YC, Yang YC, Chiu TW (2011) Int J Hydrogen Energy 36:3666
Weil KS, Hardy JS, Kim JY (2003a) In: Indacochea JE (ed) Joining of advanced and specialty materials V. Proceedings of Material Solutions 2002, ASM International, Materials Park, OH, pp 47–55
Weil KS, Deibler JE, Hardy JS, Kim DS, Xia GG, Chick LA, Coyle CA (2003b) J Mater Eng Perf 13:316
Weil KS, Kim JY, Hardy JS (2005) Electrochem. Solid State Lett. 8(2):A133
Widgeon SJ, Corral EL, Spilde MN, Loehman RE (2009) J Am Cera Soc 92:781
Yang Z, Stevenson JW, Meinhardt KD (2003) Solid State Ionics 160:213
Yang Z, Xia G, Well SK, Stevenson JW, Meinhardt KD (2004) J Mater Eng Perf 13:327
Ye Y, Yan D, Wang X, Pu J, Chi B, Jian L (2012) Int J Hydrogen Energy 37:1710
Yinnon H, Uhlmann DR (1983) J Non-crystal Solids 54(3):253
Zhang T, Fahrenholtz WG, Reis ST, Brow RK (2008) J Am Cera Soc 91:2564
Zhang T, Zhang H, Li G, Yung H (2010) J Power Sources 195:6795
Zhu JH, Geng SJ, Ballard DA (2007) Int J Hydrogen Energy 32:3682

Chapter 6
Interfacial Compatibility of Glasses and Interconnects

We have studied compressive, compliant, and rigid seals in the previous chapter. Compliant seals do not bond with the components of the solid oxide fuel cell (SOFC) leading to hydrogen embrittlement, whereas compressive seals require applied load. Rigid seals such as glass/glass ceramics, especially alkaline earth containing, are the preferred sealant materials because they can sustain in the stringent operating conditions of the SOFCs. Moreover, selecting proper initial constituents and processing conditions can tailor the glass properties. Glass gets converted into glass ceramics due to formation of crystalline phases under the SOFC operating conditions. The presence of an amorphous glass matrix in glass ceramics is beneficial for healing the cracks during the SOFC operation. In addition to this, the glasses are in contact with the adjoining components of the SOFC, hence the interdiffusion of ions can lead to the formation of different crystalline phases. The phases formed at the glass/component interface can be detrimental sometimes and hence cause deterioration of the SOFC stack. Due to these reasons, it becomes important to understand the crystallisation kinetics of glass seals when they are in contact with the other components of the SOFC in order to achieve long-term performance goals. This chapter gives insight of the interaction phenomena of glass seals with the interconnects to get a better idea of the interfacial interaction mechanism.

6.1 Different Interconnects for SOFC

SOFCs are among the most promising solutions and are considered the backbone of pollution-free power production and high energy efficiency attributed to their high operating temperature of 600–800°C. Anyhow, the high operating temperature raises the corrosion issues in the SOFC stack which affects the long-term stability in the cells. Interconnects are the main components which are responsible for successful operation of SOFC.

*All the glass compositions mentioned in this chapter are listed in Table 5.3.

G. Kaur, *Solid Oxide Fuel Cell Components*,
DOI 10.1007/978-3-319-25598-9_6

To obtain a desirable voltage output for practical mobile/stationary applications, single cell units are stacked in series or parallel or both using an interconnect. Hence, the interconnect separates the anode of one cell from the cathode of the other though maintaining the proper electrical conduction among cells. The interconnect also acts as a seal to prevent the mixing of harmful gases and hence avoids cathode reduction or anode oxidation due to leakage. Looking at these points, it is obvious that the interconnect fulfils a wide range of responsibilities and hence the material requirements for the SOFC interconnect are the most demanding among all cell components for the uninterrupted functioning of the SOFC stack.

As ceramic interconnects exhibit many disadvantages, the search for metallic interconnects geared up especially after the development of anode-supported planar SOFC. Metallic interconnects offer several orders of higher electronic conductivity than the $LaCrO_3$ due to the migration of valence electrons. The electrical conductivity of metallic interconnects is independent of oxygen partial pressure which stretches over a wide range for SOFCs. An oxide layer usually develops on the metal surface when exposed for a long time to the oxidising environment. The oxide layers such as spinels possess high electronic conductivity and hence the ohmic losses become almost insignificant. The oxide layer should be defect-free, dense, chemically stable, and adhere well to the substrate interface. Metallic interconnects not only collect and transfer the electrical charge built up in the PEN, but also provide the structural support system to the SOFC. Metallic interconnects should be economical, exhibit matching CTE with the components of SOFC have low area-specific resistance (ASR), and possess high chemical stability and low permeability with respect to hydrogen and oxygen and high resistance to wear as well as thermal shocks.

Cr-containing ferritic steels are widely used metallic interconnects due to their low cost and high oxidation resistance. Chromium vaporises at cell operating temperature and is deposited at triple phase boundaries leading to degradation of the SOFC stack performance and this phenomenon is called chromium poisoning. Iron from the interconnect side should not diffuse extensively into the glass side because formation of conductive Fe-containing phases may create a short circuit in the cell stack. Ferritic steels exhibit a body-centred cubic (bcc) structure and maintain their dimensions during the operation in a dual atmosphere, hence preventing the generation of stresses in the SOFC stack. Ferritic stainless steels are also highly resistant to oxidation in the gaseous atmospheres present in the cell. One of the most common interconnects is Crofer 22 APU, which is the foremost choice of researchers and fuel cell builders (Fig. 6.1). Figure 6.2 demonstrates the industrial process of manufacturing metallic interconnects. Tables 6.1 and 6.2 list the compositions of various metallic interconnects and CTE for ceramic interconnects, respectively. The CTE for all metallic interconnects lies in between 10 and 13×10^{-6}/K and hence no separate table is being given.

The thermal expansion mismatch between the metallic interconnect and adjoining SOFC components is a matter of deep concern especially during rapid heating and cooling in portable applications. These problems can be mitigated by lowering the working temperature of SOFC, hence efforts are also focused on lowering the operational temperature of SOFC. In addition to this, to reduce the

Fig. 6.1 Rolled Crofer 22 APU for current collection (Photo courtesy FIAXELL TECHNOLOGIES)

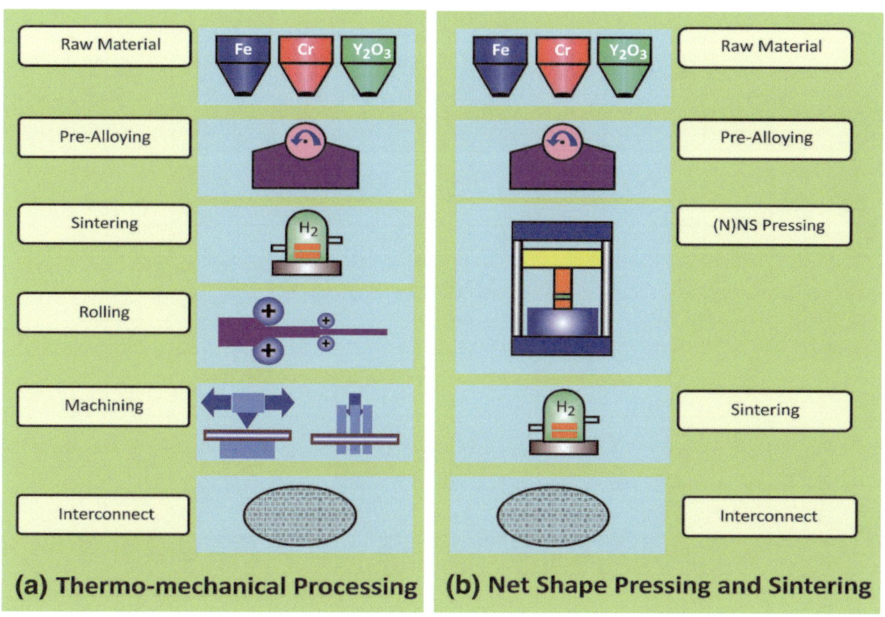

Fig. 6.2 Industrial processing technologies for the manufacture of metallic interconnects for SOFCs (Venskutonis and co-workers 2005)

chromium vaporisation on the cathode side and scale growth on the surfaces of chromia-forming alloys, coating with a perovskite or spinel ceramic layer should be done. Coating metallic interconnects with oxides of a complex structure has reduced the contact resistance and increased conductivity as well.

Table 6.1 Composition for different interconnects (mol%)

Interconnect	Cr	Fe	Mn	Ti	Si	Al	C	S	P	Ni	Others
Crofer 22APU	22.8	Bal	0.45	0.08			0.005	0.002	0.016		0.06 Re
AISI 441	17.6	Bal	0.33	0.18	0.47	0.045	0.01	0.001	0.024	0.20	0.46 Nb
446	25	Bal	1.50		1.00		0.20	0.03	0.04		
FeCrAlloy	22	Bal				5.0					0.1 Zr
AL29-4C	29	Bal	0.3		0.2		0.01	0.02	0.025		
JS3	23.3	Bal	0.4	0.05	<0.01	<0.01					0.1La
AISI446	25	Bal	1.5		1.0		0.20	0.04			
AISI1040	0.028	Bal	0.74		0.20		0.41		0.024		0.019 Mo
Nicrofer 6025HT	25	9.5	0.10		0.5		0.2			Bal	2.1 B
Ferritic T1	22.6	Bal	0.4		0.1	0.1				0.2	0.06 Zr/0.1 La
Hastelloy-XR	22	Bal	0.88		0.33					49.5	8.7 Mo
E-brite	26.5	Bal									1.0 Mo
TIMETAL-834			Bal		0.35	5.8	0.06				3.5 Zr/0.7 Ta
Nimonic AE435	19-22	1	0.7		0.8	0.15	0.12	0.1	0.015	Bal	0.07 Cu
ZMG232	22.1	Bal	0.51		0.40	0.19					0.22Zr
RA446	26.3	Bal	0.78	<0.05	0.16						
ITM14	26	Bal			<0.03	<0.03					
ITM	26	Bal			<0.03	<0.03					
Sanergy HT	21.7	Bal	0.3	1	0.12						0.24 Zr, 0.71 Nb, 0.96Mo
SUS430	16-18	Bal	<1		<1		<0.12	<0.03	<0.04		
Inconel600	14-16	Bal	1				0.15			72	0.5 Cu

Table 6.2 Ceramic interconnects with their CTE

Ceramic interconnect	CTE ($\times 10^{-6}$/K)
$LaCrO_3$	9.5
$La_{0.9}\ Sr_{0.1}CrO_3$	10.7
$La_{0.8}\ Sr_{0.2}CrO_3$	11.1
$LaCr_{0.9}\ Mg_{0.1}CrO_3$	9.5
$La_{0.65}Ca_{0.35}CrO_3$	10.8
$LaCr_{0.9}\ Co_{0.1}O_3$	13.1
$La_{0.8}Ca_{0.2}Cr_{0.9}\ Co_{0.1}O_3$	11.1
$YCrO_3$	7.8
$Y_{0.9}Ca_{0.1}CrO_3$	8.9
$Y_{0.8}Ca_{0.2}CrO_3$	9.6
$La_{0.7}Ca_{0.3}Cr_{0.5}\ Ti_{0.5}\ O_3$	10.1
$La_{0.7}Sr_{0.3}Cr_{0.8}\ Ti_{0.2}\ O_3$	10.7
$LaCr_{0.9}Ni_{0.1}O_3$	10.1
$LaCr_{0.8}Ni_{0.2}O_3$	11.2
$LaCr_{0.7}Ni_{0.3}O_3$	11.8
$La_{0.8}Ca_{0.2}Cr_{0.9}Co_{0.1}O_3$	11.1
$LaCr_{0.2}Co_{0.8}O_3$	20.9
$LaCr_{0.6}Co_{0.4}O_3$	18.7
$LaCr_{0.8}Co_{0.2}O_3$	14.6
$La_{0.7}Sr_{0.3}Cr_{0.8}Co_{0.2}O_3$	14.5
$La_{0.7}Sr_{0.3}Cr_{0.5}Co_{0.5}O_3$	19
$La_{0.85}Sr_{0.15}Cr_{0.98}Cu_{0.02}O_3$	9.91
$La_{0.87}Sr_{0.1}Cr_{0.95}Cu_{0.05}O_3$	9.81
$La_{0.9}Sr_{0.1}Cr_{0.93}Mg_{0.05}Cu_{0.02}O_3$	9.61
$La_{0.9}Sr_{0.1}Cr_{0.95}V_{0.05}O_3$	9.87
$La_{0.85}Sr_{0.15}Cr_{0.95}V_{0.05}O_3$	10
$La_{0.9}Sr_{0.1}Cr_{0.95}V_{0.05}Co_{0.03}O_3$	10.4
$La_{0.95}Sr_{0.05}Cr_{0.85}Mg_{0.1}V_{0.05}O_3$	9.22
$La_{0.9}Sr_{0.1}Cr_{0.9}Mg_{0.05}V_{0.05}O_3$	9.64
$La_{0.85}Sr_{0.15}Cr_{0.935}V_{0.05}Co_{0.015}O_3$	10.5
$La_{0.87}Sr_{0.1}Cr_{0.935}Cu_{0.05}Co_{0.015}O_3$	10.4

Though planar design of SOFC is more advanced than tubular design, one of the technological problems in the development of planar SOFCs is the sealing of the stack to prevent direct mixing of air and fuel gases (H_2, CO_2, CH_4) and avoid leakage of fuel into the air. The sealing material should be in intimate contact with the stack components with maximum chemical compatibility especially with the interconnect of SOFC as the seals must function for long time (5000–40,000 h) at high temperatures (between 600 and 900°C). Interaction of a seal glass with a metallic interconnect is much more severe than with other cell components. In the current chapter, the focus is on the chemical compatibility of glass/glass-ceramic seals with the interconnect because one area that currently needs more attention is the degradation due to chemical interactions at the interfaces among the glass sealants and other parts of SOFCs. This helps to gain insight into the mechanism of chemical interactions and their effect on crystallisation kinetics, formation of various crystalline phases, and their effect on thermal expansion and structural features. One can understand the unwanted effects of interaction between glass/interconnect, which leads to interdiffusion, reduced electrochemical activity of cell components, and degradation of SOFC performance. Many of the reaction products at the interface could be detrimental for SOFCs due to the structural modifications in various components. Various glass and glass ceramics have been studied as the candidate material for sealants. During SOFC operation, glass gets crystallised and transforms into glass ceramics. The formation of some detrimental phases after interaction among different components of SOFC leads to severe stack degradation and affects the efficiency of SOFC. In the following sections, we focus on the chemical compatibility of glass seals with the interconnect to understand the mechanism(s) involved for attaining the long-term performance goal.

6.2 Compound Formation at Glass/Interconnect Interface

The glass seal must exhibit strong bonding at the interface to keep the residual stresses as low as possible. CTE mismatch between the glass and components of SOFC lead to thermal stress, which is given by the relation:

$$\sigma = CE \int (\alpha_{g} - \alpha_{m})\mathrm{d}T \tag{6.1}$$

where E is the Young's modulus, C is the geometrical constant, and α_g and α_m are the CTE of the glass and material. Thermomechanical stresses arise from differences in the material properties and temperature gradient. CTE mismatch between the seal glass and in adjoining components can cause tensile thermal stress ($CTE_{comp} > CTE_{glass}$) or compressive thermal stress ($CTE_{glass} > CTE_{comp}$) at the interface. Cracks and pores may form at the interface due to tensile stress and delamination of the glass seal occurs due to compressive/shear stress. CTE increases due to asymmetric structure of borate structural groups that decrease network connectivity. Therefore, to minimise the interfacial residual stresses among the glass and components of SOFC, not only the thermal expansion characteristics of the metal and bulk glass-ceramic should be closely matched, but also interfacial reactions should be kept to a minimum such that they do not lead to the formation of detrimental reaction products within the interfacial reaction zone. Sometimes the reaction between diffusing ions and glass constituents is disadvantageous and hinders the functioning of SOFC. The following reaction is an undesirable reaction and leads to the formation of iron and/or chromium phosphide phase within the interfacial region,

$$9(Cr - Fe) + P_2O_5 \rightarrow 2(Cr - Fe)_2P + 5(Cr - Fe)O \tag{6.2}$$

Such kind of reactions may lead to a change in the microstructure near the interfacial region and hence, a large difference in the thermal expansion is obtained within the interfacial region and the rest of the bulk glass-ceramic leading to the development of localised residual stresses. These highly localised residual stresses can cause seal failure either by delaminating the seal or due to cracking and pore formation at the interface. The presence of water in the initial glass plays an important role in determining the seal/metal interface; i.e., certain metallic species including Cr, Y, Ti, and Ta may react with water to yield hydrogen gas as shown below (at 1300 K)

$$Cr + 3/2(H_2O) \rightarrow 1/2(Cr_2O_3) + 3/2(H_2) \quad (\Delta G = -129\ kJ/mole) \tag{6.3}$$

$$Ti + 2(H_2O) \rightarrow TiO_2 + 2(H_2) \quad (\Delta G = -355\ kJ/mole) \tag{6.4}$$

$$Y + 3/2(H_2O) \rightarrow 1/2(Y_2O_3) + 3/2(H_2) \quad (\Delta G = -501\ kJ/mole) \tag{6.5}$$

$$Ta + 5/2(H_2O) \rightarrow 1/2(Ta_2O_5) + 5/2(H_2) \quad (\Delta G = -298\ kJ/mole) \tag{6.6}$$

Such kind of reactions may lead to preferable gas leakage paths and hence affect the seal integrity. Severe interaction among the seal glass and the adjacent cell components increases the interfacial layer thickness, leading to the formation of pores and crack and at the interface and delamination of the glass.

Seal glass should wet and bond with the components uniformly to form good sealing without any pores or cracks at the interface. The interaction involves inter-diffusion and chemical reaction of the elements of the seal glass and the interfacing cell components. Hence, the formation of a porous layer and cracks at the interface is twofold: i.e., first the glass devitrification occurs causing a CTE mismatch among the glass and devitrified phase resulting in poor gas flow; second the glass components react with the volatile chromium species (CrO_3 and $CrO_2(OH)_2$) of the interconnect forming high CTE chromates (21.0–23.0 × 10^{-6}/K). Hence, glass/interconnect interaction depends primarily on the chosen composition, sealing conditions, and the exposed atmospheres. SOFCs can work under a variety of fuels such as hydrogen, syngas, and hydrocarbons. The seals have to work under stringent conditions of continuous oxidising and reducing atmospheres at high temperatures. Due to the presence of high water content at the anode, the glass devitrification may gear up changing the local microstructure along with vaporisation of some glass constituent ions. For example, B_2O_3 from a seal glass vaporises in the form of HBO_2 in non-alkali glasses and $B(OH)_3$ species in alkali oxide containing glasses more rapidly in the presence of water rather than in oxidising atmospheres. Gaseous borates and alkali/alkaline earth metaborates are the common by-products for the alkali oxide-containing borate glass and borosilicate glass, respectively, whereas alkaline earth oxide modifiers and intermediate oxides do not vaporise. Similarly, silica vaporises in the form of $Si(OH)_4$ species by reacting with water at 1000°C.

Hence, all these factors make a significant contribution to the lifetime behaviour of seal components. Therefore, a thorough understanding of the glass ceramic/metal interactions is essential to minimise interfacial reactions to avoid localised modifications of the glass–ceramic microstructure. Metallic component failure is rare because the plastic flow of the metal usually accommodates the stress, which is generated during operation at such high temperatures. Therefore, the composition–structure–property relations of seal glass should be well understood to design seal glass.

6.3 Chemical Compatibility of Glass/Interconnect

It is necessary to monitor the interfacial chemical reactions thoroughly to gain an insight of glass seal/interconnect interaction. Generally, diffusion occurs prior to chemical reaction. Sometimes chemical reactions are detrimental in nature leading to microstructural modifications, thereby hindering the functioning of SOFC. Chromium ion diffuses prominently as it follows a two-step process; first, at 800°C, chromium comes out of Crofer 22APU alloy and reacts with non-bridging oxygens (NBO) to form Cr_2O_3. During the second stage, the condensation of Cr_2O_3 may lead to formation of Cr_2O_3 films at the interface. The chromia layer may react with the glass seal to form undesirable chromate products at the interface. In addition to this, chromium diffusion into glass enhances devitrification of glass by breaking the glass network structure. In spite of the achievement of desired thermal properties for SOFC glass seals, long-term interfacial stability of a seal with the adjoining components still remains a persistent problem. Accordingly, the thermomechanical properties of the seal and interfacial reaction kinetics at the interface are discussed and understood for different glass systems.

6.3.1 Alkali Metal-Based Glasses/Interconnects Interface

Alkali metal glasses are successful in terms of thermomechanical performance, but possess severe corrosive action on the interconnects which puts a limitation on their applicability in a planar SOFC stack. The corrosion destabilises the protective chromia scale on the interconnect and hence affects the mechanical integrity of the stack. Nielsen and co-workers focused on the composite seals developed from the compliant glass (NAS—17.8 mol% sodium) dispersed with high CTE material such as MgO as MgO display slow crystallisation behaviour and desirable T_g values. Good adhesion between the NAS/MgO thin film composite glass composite with preoxidised Sandvik interconnect steel and (as received) Crofer 22APU is obtained at 950°C as shown in Fig. 6.3a, b.

The light grey and whitish deposits on the MgO particles in the seal composites are attributed to the chromium oxides (at % <15) accompanied by traces of iron

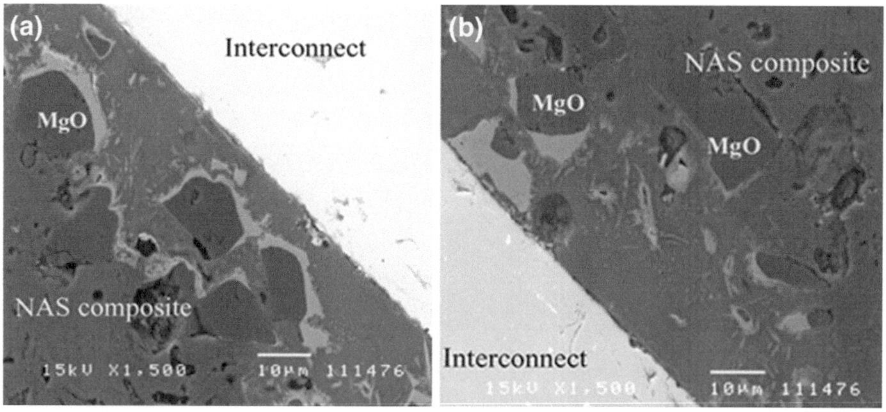

Fig. 6.3 Interface of **a** preoxidised surfaces of Sandvik steel and **b** Crofer 22APU (as received) with thin film NAS glass composite with 30 vol% MgO-filler after 170 h at 950°C (Nielsen and coworkers 2007)

oxides from the metal or oxides from the glass. Sodium chromate (Na_2CrO_4) is expected to be the predominant reaction product at the interface because sodium is in large concentration of about >17 mol%; anyhow no sodium and chromium are seen coexisting at the glass steel interface. The interface is gradually enriched in chromium and depleted in sodium when moving towards the metal surface. The thickness of the glass transition zone varies between 2 and 80 μm as a function of both glass composite composition and the sealing temperature. Hence, it can be assumed that the oxidation of metallic chromium leads to evaporation of sodium through open porosity before the glass softens and hence closes off the interconnect surface from the atmosphere. Chromium probably was transported via gaseous Cr (VI) to the MgO grains as

$$Na_2O_{(sol)} + O_{2(g)} + 2Cr_{(s)} \Leftrightarrow Cr_2O_{3(s)} + 2Na_{(l/g)} \tag{6.7}$$

One expects poor wetting and adhesion due to increased viscosity attributed to the evaporation of sodium from the glass. But, the NAS glass displayed good wettability which can be due to the dissolution of magnesium from the filler into sodium-depleted Na–Al–Si–O glass. However, excessive corrosion at the three-phase-boundary between glass, interconnect, and cathode as shown in Fig. 6.4a is probably due to the formation of sodium chromate as

$$Na_2O_{(sol)} + 1.5O_{2(g)} + Cr_{(s)} \Leftrightarrow Na_2CrO_{4(s/l)} \tag{6.8}$$

Sodium chromate is a low melting compound and readily decomposes at 950°C under reducing conditions. However, the contact cobaltite-based coating, which was applied to the metal surface before assembly, prevented the interface corrosion as shown in Fig. 6.4b.

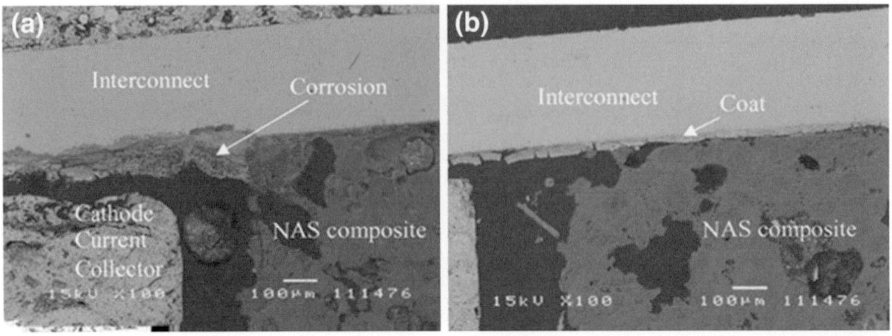

Fig. 6.4 Corrosion between cathode, NAS glass composite (45 vol% MgO-filler), and **a** uncoated Crofer 22APU interconnect steel, **b** coated Crofer 22APU interconnect steel after 3700 h at 750°C (Nielsen and coworkers 2007)

SACN1(18% Na_2O) and SACN2 (23% Na_2O) have higher sodium content than the NAS composition and possess CTE of 11.3×10^{-6} and 9.3×10^{-6}/K, respectively. The wettability tests of glasses as received and preoxidised AISI 430 steel in air/argon atmosphere was done and the aim of preoxidation is to form a transition oxide layer, which is adherent and compatible with the glass as well as metal substrate hence promoting chemical bonding between the metal and the glass seal. A 3–5 µm thick chromium–manganese oxide and chromium–iron oxide layer is formed as a result of peroxidation of the AISI 430 substrate. Due to low CTE of SACN1 glass, the interface between the glass matrix and the preoxidised AISI 430 is detached along with the presence of some cracks near the interface. In contrast to this, the interfacial adhesion of SACN2, with preoxidised AISI 430 reveals partial infiltration of the glass through the chromium–iron manganese oxides rich layer without any cracks or pores as evident from Fig. 6.5a. Therefore, SACN2 glass is chosen as a matrix for the composite seal. In spite of higher sodium content,

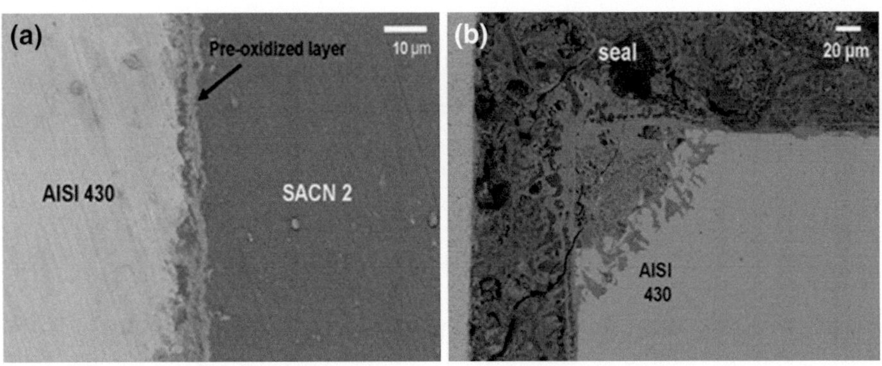

Fig. 6.5 SEM micrograph of **a** interface between preoxidised AISI 430 and SACN2 glass (1000°C, 30 min) and **b** three-phase boundary for AISI 430/seal sample, after exposure for 200 h at 800°C to H_2–3% H_2O atmosphere (Smeacetto and coworkers 2008a)

SACN2 seal displayed quite acceptable bonding characteristics. The enrichment of sodium in the preoxidation layer usually leads to the formation of Na_2CrO_4. The high vapour pressure of Na_2CrO_4 can result in its vapourisation along with the chromium depletion as well as decomposition and removal of the $(Mn,Cr)_3O_4$ spinel oxide. In addition, depletion of sodium from the glass seal at the interface with Crofer22 APU promotes cracks.

For obtaining the composite seal, 10–20 vol% alumina particles were added to the SACN2 glass at a temperature of 1020°C, for 30 min in an argon atmosphere. Furthermore, this process leads to the crystallisation of the SACN2 glass matrix precipitating out $Ca_2Al_2SiO_7$, $AlNa(SiO_4)$, and residual aluminium oxide as predominant phases. The CTE of the composite seal is calculated according to the rule of mixtures and comes out to be 10.84×10^{-6}/K.

Both the AISI 430/seal and AISI 430/composite seal joined samples were subjected to H_2−3% H_2O atmosphere exposure at 800°C for 200 h. A high degree of corrosion is observed with a high amount of chromium, iron, and manganese oxides diffusion into the seal around the joint region glassy seal with the metallic interconnect at the three-phase boundary (Fig. 6.5b). Hence, SACN2 is regarded as a new barium-free multi-component borosilicate glass, which successfully sealed AISI 430 to YSZ. The composite seal also depicted a good interface with preoxidised AISI 430 without any detrimental reaction at the interface (except at the three-phase boundary), though some residual porosity could be seen at the interface.

A very good adhesion between the SACN glass–ceramic (both barium and boron-free alkali glass with CTE 10.7×10^{-6}/K) and Crofer22 APU without any cracks near the interface could be observed. In SACN/Crofer22 APU system, no internal oxidation of the ferritic steel beneath the SACN glass–ceramic sealant was observed along with the outstanding bonding between the SACN and Crofer22 APU. Even after thermal ageing, the crystalline phases of the SACN glass did not change hence assuring the thermochemical stability of the SACN glass–ceramic seal. The heating rate also influences the glass–ceramic sintering behaviour. The Crofer 22 APU/SACN sample joined at the heating rate of 5°C/min revealed very low closed porosity in the seal region, assuring its high gas tightness and a hermetic structure. The SEM image of Crofer 22 APU/SACN interface is shown in Fig. 6.6a revealing a continuous and defect-free interface. The decrease in porosity could be seen when the heating rate that decreases (from 25 to 5°C/min) is attributed to effective removal of the gases entrapped in the sealant during the longer heat treatment. During the present studies, higher heating rates overcame concurrent crystallisation.

Exposure to air and humidified hydrogen at 800°C revealed no corrosion products at the interfaces between metallic interconnects (both Crofer22 APU and AISI 430) and glass–ceramic sealant. No chromium depletion or swelling of the metal with subsequent bulging at the surface of the Crofer22 APU close to the three-phase boundary was observed. Hence, the glass–ceramic sealant is effective in joining both Crofer22 APU and anode-supported-electrolyte.

Three new low-sodium, barium, and boron-free silica-based glass–ceramic sealants were developed doped with magnesium (SACNMg), lanthanum-zinc

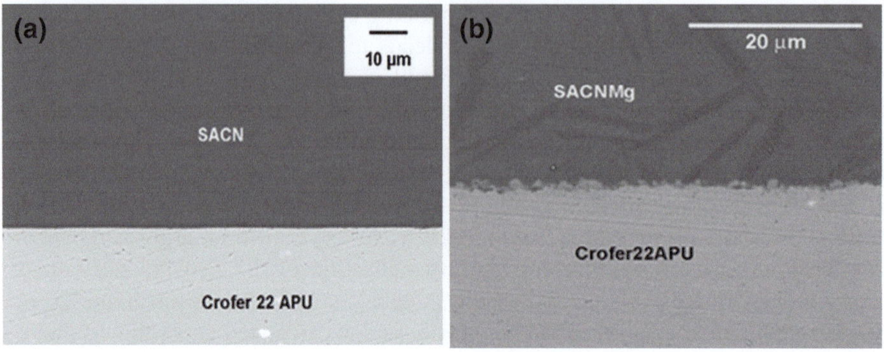

Fig. 6.6 Cross section of the interface between **a** Crofer 22 APU alloy and SACN glass–ceramic (heat treatment 5°C/min, 900°C, 30 min) (Smeacetto and coworkers 2008b), **b** Crofer22APU/SACNMg (Smeacetto and co-workers 2010)

(SACNLZ), and zinc (SACNZn) with CTE of 8.1 × 10^{-6}, 9.7 × 10^{-6}, and 7.8 × 10^{-6}/K, respectively. Lanthanum oxide and zinc oxide devitrified the parent glass (during the heat treatment for the joining process) into a glass–ceramic with a higher CTE as compared to SACNMg glass. Higher characteristic temperatures were obtained due to lower sodium content in these glasses. SACNMg glass exhibited excellent wettability on the Crofer22APU substrate with pore-free interface (Fig. 6.6b) in spite of the fact that SACNMg possesses low CTE. The SACNMg exhibits mostly amorphous state hence making it intrinsically brittle. The Crofer22APU/SACNLZ joined sample exhibited poor adhesion and large pores in the glass–ceramic region at the Crofer22APU/glass–ceramic interface.

Figure 6.7 reveals no new products at the Crofer22APU/SACNZn glass–ceramic interface indicating no reaction between the chromium–manganese oxide scale and

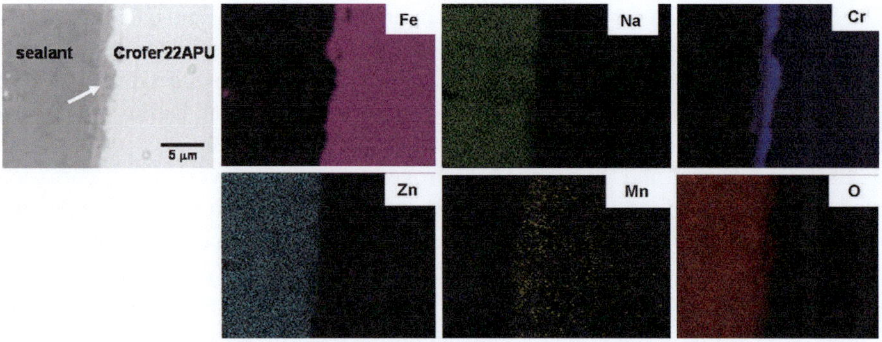

Fig. 6.7 SEM cross section of the Crofer22APU/SACNZn glass–ceramic sample exposed (at 800°C for 300 h) to H_2–3 vol% H_2O atmosphere along with elemental mapping (Smeacetto and co-workers 2010)

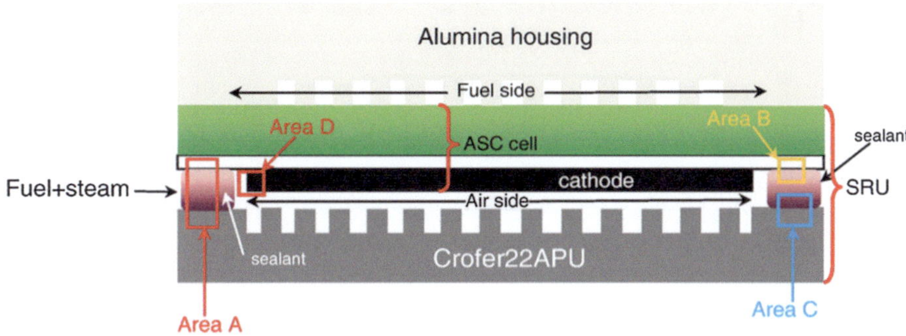

Fig. 6.8 Schematic illustration of Crofer22APU plate/SACN sealant/ASC joined SRU subjected to dual atmosphere thermal ageing test (Smeacetto and coworkers 2011)

the glass–ceramic components. This confirmed the absence of sodium chromate as there was no reaction of glass–ceramic elements with chromium. No migration of sodium away from the SACNZn glass–ceramic sealant was observed and hence good adhesion to the Crofer22APU substrate was observed.

The higher thermodynamic stability of the SACNZn glass–ceramic sealant is attributed to the presence of Na in one of the two crystalline phases. Even no undesired reactions or no corrosion products are formed at the Crofer22APU–SACNZn glass–ceramic sealant interface after exposure to the H_2−3 vol% H_2O atmosphere.

The joined single repeating unit (SRU) (Crofer22APU bipolar plate/SACN sealant/anode-supported cell) was obtained by placing the cell on the top of the slurry coated Crofer22APU with an alumina load of 7 kPa. Furthermore, this SRU was subjected to heat treatment in a tubular oven at 900°C for 30 min (Argon atmosphere) as shown in Fig. 6.8.

No detrimental reaction between metallic interconnect and glass sealant was observed, which could be due to the low amorphous residual phase limiting the alkali ion diffusion, presence of sodium in crystalline phase, and high stability of glass–ceramic as compared to the glass sealant. The thermodynamic stability and chemical durability of main crystalline phase nepheline ($NaAlSiO_4$), in humidified reducing atmosphere was evident in the present studies. Chou and co-workers also demonstrated that the electrical resistivity of the parent glass is not affected when the glass crystallises by the addition of K_2O (up to 5 mol%) in an alkali-earth containing glass and the fast transport kinetics gets slowed down due to crystallinity. The SEM cross-section of preoxidised Crofer22APU/glass-ceramic sealant interface revealed that sodium and chromium did not react.

Figure 6.9a shows the electrical behaviour for the SRU unit in an ageing experiment (460 h at 800°C), at open circuit voltage. The cell voltage is very close to the theoretical reversible voltage of 1080 mV (T = 820°C, pH_2 = 0.959; pH_2O = 0.041, pO_2 = 0.21 atm) at an operating temperature of 800°C. In addition to this, under standard experimental configurations, i.e., circular shaped cell without

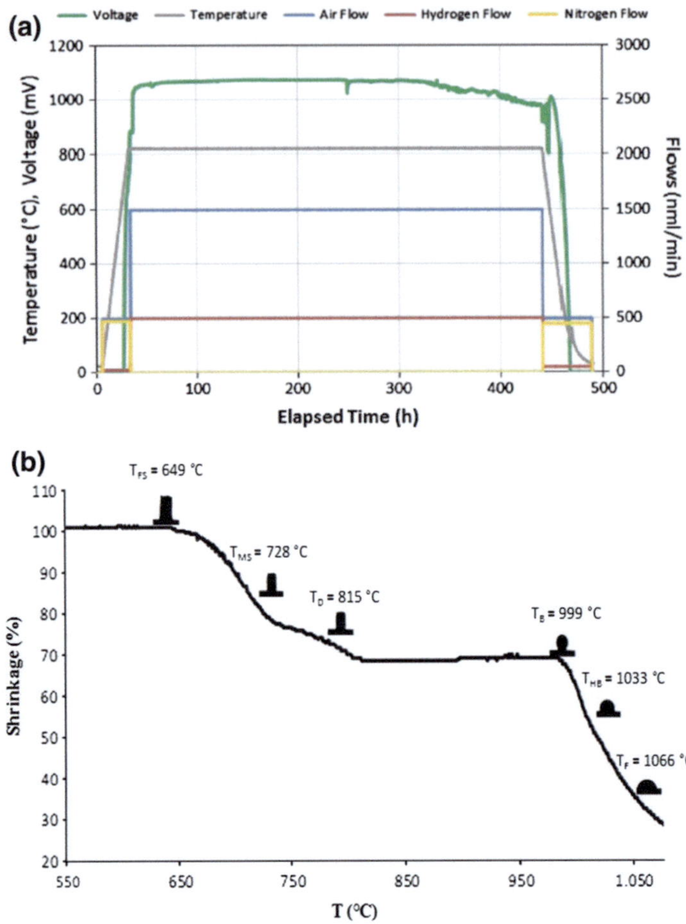

Fig. 6.9 **a** Electrical behaviour of the SRU with the glass–ceramic sealant in an ageing experiment (460 h at 800°C), at open circuit voltage (Smeacetto and coworkers 2011) and **b** HSM of KMBY glass (Smeacetto and coworkers 2014)

sealant, anodic and cathodic ceramic housing, the reversible voltage is obtained to be 1030–1035 mV indicating the effectiveness of the sealing process.

Another novel glass–ceramic KMBY (1.67 wt%) was developed and its compatibility with $Mn_{1.5}Co_{1.5}O_4$-coated Crofer22APU was found to be very good, even after 500 h of thermal cycling (RT-800) in air atmosphere, without any crack formation, unwanted interactions, or seal failure. The sintering behaviour of the KMBY glass was depicted that the maximum shrinkage was reached at around 850°C (heating rate: 5°C/min); and hence 850°C was chosen as the joining temperature, because it was expected that complete densification should be obtained at this point (Fig. 6.9b). KMBY sealant showed a dense and homogeneous microstructure, depicting its excellent sintering and densification behaviour. Diopside ($Ca_{0.89}$ $Mg_{1.11}$ Si_2O_6) is the main crystalline phase along with the residual amorphous matrix. The $Mn_{1.5}$ $Co_{1.5}$ O_4 coating effectively reduced adverse reactions at the Crofer22APU/$Mn_{1.5}$ $Co_{1.5}$ O_4/ KMBY interface.

Certain alkali galliosilicate compositions were also investigated with the aluminised stainless steel Al-441SS substrate. It was observed that with the T_g approaches 600°C with only 5 mol% of alkali. The composition with 5 mol% alkali oxide and <10 mol% of B_2O_3 ($GaBSi_2$) was chosen for the interfacial study with Al-441SS. The alumina coating was intact after 500 h treatment at 850°C, though diffusion of iron and chromium could be observed on the glass side (Fig. 6.10).

Another sealant SCN-1 (10 wt% K_2O and 7.29 wt% Na_2O) was used under mimicked SOFC conditions, and SS441 substrates were sprayed with a novel reactive air aluminising coating followed by drying and oxidising at 1000°C for 1 h in air. No high Si or alumina containing precipitates were found at the glass and aluminised SS441 interface, and the alumina coating remained intact without showing substantial dissolution into the glass matrix (Fig. 6.11). Hence, the alumina coating minimised Cr diffusion into the glass despite the fact that the coating was non-uniform and not fully dense. Some tiny white Fe-containing particles next to the alumina coating and the glass matrix revealed equiaxed light particles and darker rod-like precipitates. The dark rod-like precipitates were mostly of Si, Al, and K and the light particle contained primarily Si, Na, and Ca.

The high temperature leak rates during repeated thermal cycling were measured for the SCN-1 seal between 700 and 850°C at a differential pressure of 0.2–1.0 psi,

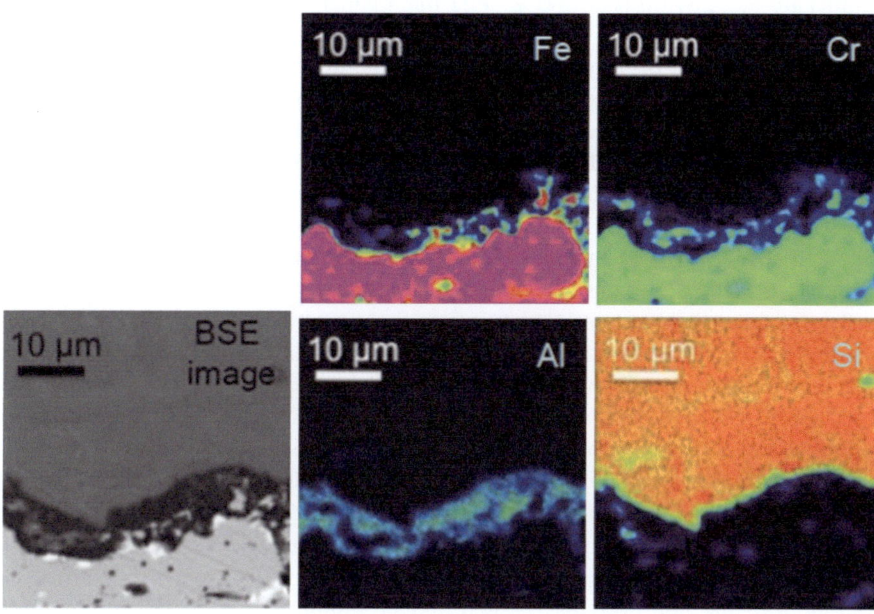

Fig. 6.10 Interface of GaBSi2 glass with Al-441SS steel after 500 h at 850°C (presentation of Scott Misture and James Shelby)

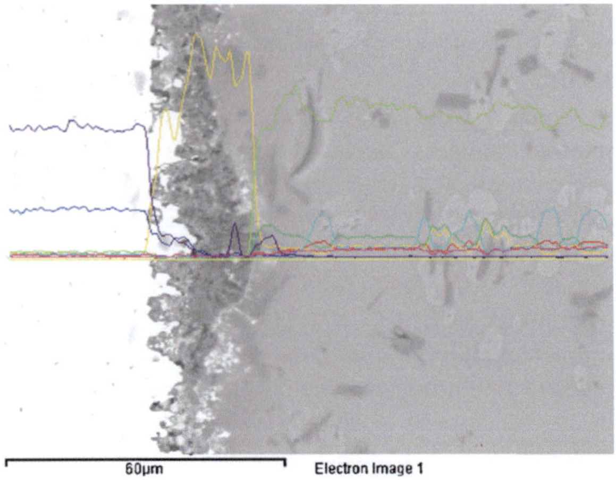

Fig. 6.11 SCN-1 glass/aluminised SS441 interface for sample tested at 800°C/1000 h followed by four thermal cycles in dual environment (Chou and coworkers 2011)

where Inconel600 and alumina were chosen as the primary fixture materials, with hybrid mica seals around the perimeter. The total leak rate during the leakage tests is given as the sum of the leak rate through the hybrid mica perimeter seal plus the leak rate through the glass seal. A hybrid mica seal is used around the perimeter because welding the Inconel600 fixture with SS441 substrate would induce severe thermal stresses inside the setup. In addition to this, the hybrid mica seal has shown fairly constant leak rates for 1000 thermal cycles along with long-term ageing stability up to ~28,400 h at 800°C. The mica paper (thickness of ~200–300 µm) used for testing SCN-1 seal was composed of discrete mica flakes overlapping each other and comprised of varying sizes. Hence, the different pore microstructure for every mica sheet leads to a range of leak rate measurements. All samples revealed good thermal cycle stability in air for 20 thermal cycles at different temperatures (between 700 and 850°C). The fairly constant leak rates at all temperatures were in correlation with the thermal cycle stability of hybrid micas. Anyhow, the leaks observed were much less than those for a typical glass seal failure as shown in Fig. 6.12. This indicates that the SCN-1 glass demonstrates good thermal cycle stability without fracture during cooling. The leak rates at various helium pressures (~3.4–6.8 kPa) display linear behaviour with respect to the pressure.

BiBaSi glass sealants with almost 5–10 mol% K_2O and 45–50 mol% Bi_2O_3 have been doped with different contents of Al_2O_3 in the range of 0–15 wt% to yield GA0, GA3, GA5, GA8, GA10, and GA15 glass compositions. BiBaSi glass sealants possess a low softening point that may help in releasing the stress due to CTE mismatching of the SOFC stack materials to produce different glasses. When a glass contains more than one alkaline metal ion or alkaline–earth metal ion, then they do not form the glass network but contribute to the NBO, hence making the glass more open. During the present investigation, Al_2O_3 acts as a glass network formed by reacting with the non-bridging oxygen and promoting the formation of bridging oxygen. Al_2O_3 cannot form a glass network until and unless non-bridging oxygen in the glass is not adequate. In addition to this, the electrical field of Al^{3+} ion may

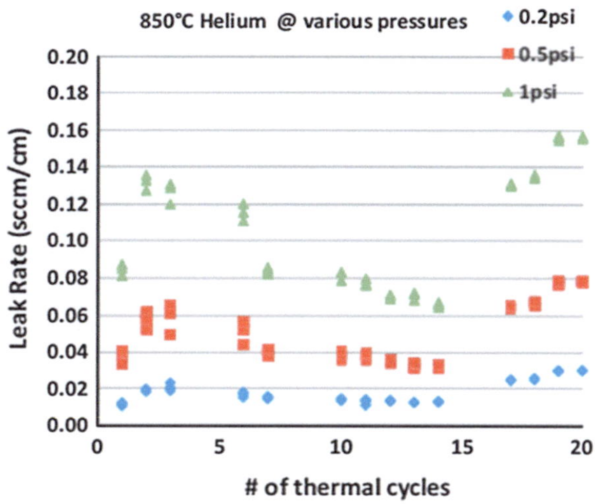

Fig. 6.12 Thermal cycle stability of SCN-1 glass seal at 850°C (seal between ceramic bilayers and aluminised SS441 substrate, and contained between two concentric YSZ rings) (Chou and coworkers 2011)

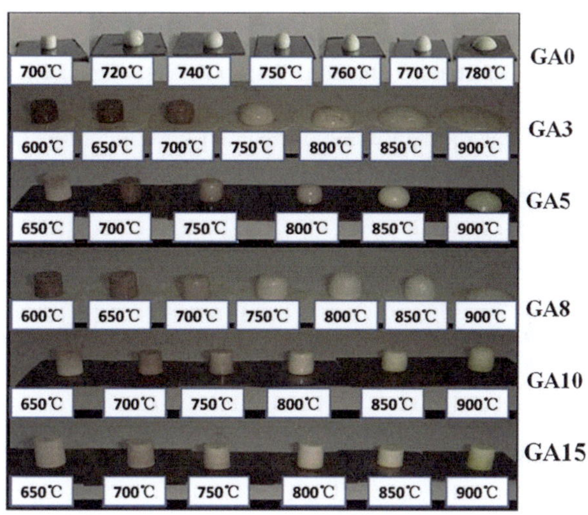

Fig. 6.13 Different glass sealants at different temperatures (Tong and coworkers 2012)

affect the glass structure, ultimately breaking it. Figure 6.13 shows the shapes of the cylindrical glass pellets at different temperatures indicating the shrinkage of glass pellets followed by the rounded corners and finally change of cylindrical shape to hemisphere. The images for GA0, GA3, GA5, and GA8 shows that the pellets completely collapse whereas GA10 and GA15 show the shrinking and the rounding at the corners without any further shape change below 900°C. The temperatures for rounded corner, hemisphere, and collapse are called rounded corner temperature (T_r), hemisphere temperature (T_h), and collapse temperature (T_c) and the corresponding viscosities are 6.3 ± 0.1, 4.1 ± 0.1, and 3.4 ± 0.1 dPa s, respectively.

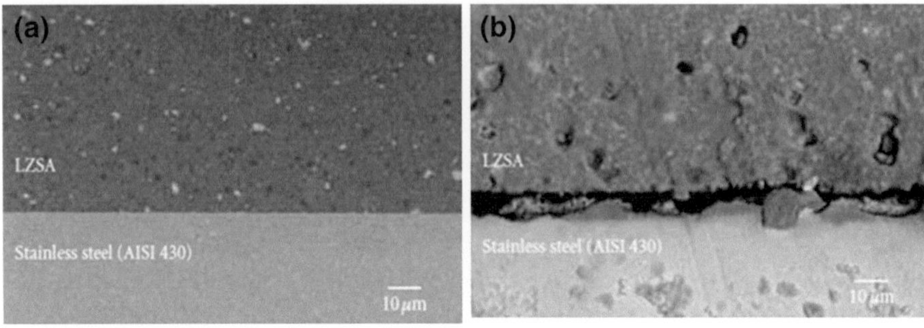

Fig. 6.14 Interface of LZSA with AISI 430 **a** 800°C and **b** 900°C (Nonnet and coworkers 2012)

LZSA glass–ceramic (LiO_2–ZrO_2–SiO_2–Al_2O_3) is a lithium-based glass sealant tested for interfacial compatibility with AISI 430. The four prominent phases formed in this glass are lithium metasilicate (Li_2SiO_3), zircon ($ZrSiO_4$), lithium aluminium silicate ($LiAlSi_3O_8$), and zirconium dioxide (ZrO_2). The $ZrSiO_4$ phase grew at higher temperatures when the ZrO_2 reacted with SiO_2 of the residual glass phase, leading to generation of pores due to devitrification. At 800 and 900°C, the LZSA laminates (obtained from slurries with 1285 mPas, 20% binder) were fired onto an AISI 430 substrate where the former represented maximum density/shrinkage temperature and the latter was operational temperature for the HT-SOFC. No delamination or discontinuity was observed at the LZSA-steel interface at 800°C, whereas the LZSA detached from the interface at 900°C attributed to the devitrification of LZSA at high temperature (Fig. 6.14).

6.3.2 Alkaline Earth Metal-Based Glasses/Interconnects Interface

6.3.2.1 Magnesium-Based Glasses/Interconnect Interface

Most research groups focus on the interaction studies between glasses and metallic interconnects, but Larsen and James chose calcium-doped lanthanum chromite, $La_{0.79}$ $Ca_{0.22}$ CrO_3 (LCC) ceramic interconnect for the interaction with magnesium phosphate glasses. The glass composition C contains 5 mol% Cr_2O_3 whereas MgO content is 25 mol% for both glass A and C. Addition of Cr_2O_3 decreased the CTE of glass C. For both glasses, the reaction is caused by vapour phase transport of P_2O_5 to the LCC surface. The interface for sample A reveals a 5–10 μm thick reaction zone along with grain extension from the LCC surface after heat treatment at 1000°C for 24 h as shown in Fig. 6.15a. The major reaction products on the LCC surface consist of Ca, P, and traces of La though the outer region of the glass surface is found to be depleted in P. Anyhow, no bonding of glass A has been observed with the LCC. Sample C also did not bond to the LCC and exhibited almost the same surface

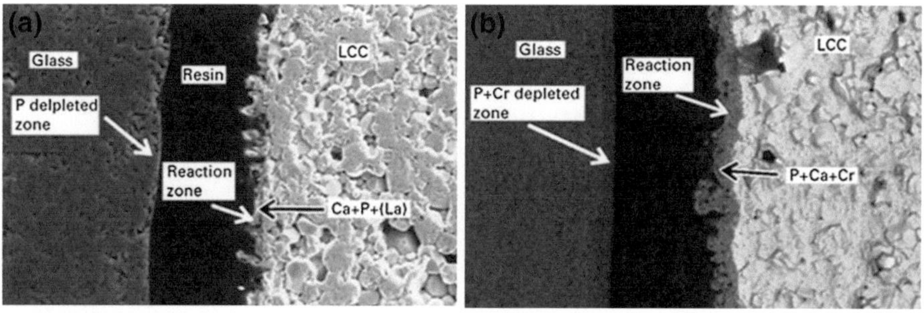

Fig. 6.15 SEM micrographs interface between **a** glass A and **b** glass C with LCC after heat treatment in air at 1000°C for 24 h (Larsen and James 1998)

morphology as that of glass A (Fig. 6.15b). The major reaction products are P, Ca, and traces of Cr. The outer 3 μm region on the glass surface is depleted in P and Cr. The reactions were more severe at 1200°C due to liquid–liquid interaction.

According to McDowell and Beall, Al_2O_3–SiO_2 glasses containing less than 5 mol% Al_2O_3 do not exhibit phase separation. In contrast to this, when Al_2O_3 content is increased to 10 mol% then the glass exhibits phase separation due to the structural role of Al^{3+} ion. Al^{3+} ion forms tetrahedral AlO_4 or octahedral AlO_6 groups when it is four or six coordinated with oxygen, respectively. Aluminium ion takes the role of network modifier in six coordinations and network former when tetrahedrally coordinated. This trend has been observed by the magnesium aluminosilicate (MAS)-based compositions reported by Lahl and co-workers. The DTA of glasses at the heating rate of 10°C/min for different particle size (<10 and >10 μm) is shown in Fig. 6.16. The temperature at which the rate of transformation is maximum represents the exothermal crystallisation peak. The crystallisation peak for the glasses with small particle size is at a higher temperature in

Fig. 6.16 DTA thermograms for glasses at 10°C/min (LG represents > 10 μm and SG < 10 μm particle size) (Lahl and coworkers 2000)

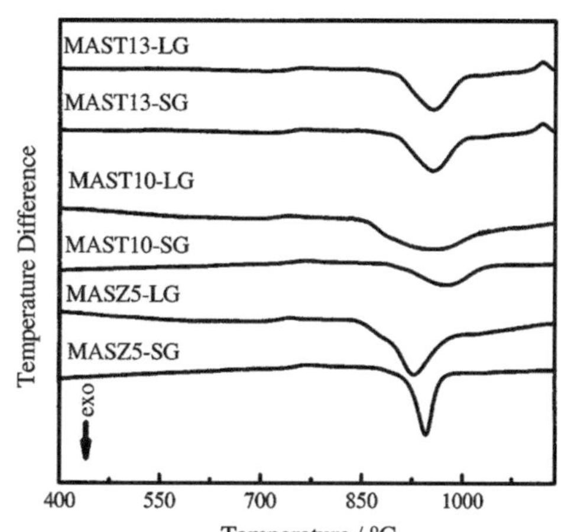

Table 6.3 Reaction products after mixing and annealing different sealants with ODS alloy at 1000°C for 100 h in air

Glass	Phases
MASC6	SiO_2, $MgSiO_3$, (Mg, Fe)(Cr, $Al)_2O_4$, Fe_2SiO_4
MASN6	(Mg, Fe)(Cr, $Al)_2O_4$, SiO_2, Fe_2SiO_4, $MgSiO_3$
MAS	(Mg, Fe)(Cr, $Al)_2O_4$, SiO_2, $MgSiO_3$, $ZrSiO_4$, (Mg, $Fe)_2$ SiO_4
MASZ10	(Mg, Fe)(Cr, $Al)_2O_4$, $ZrSiO_4$, (Mg, $Fe)_2$ SiO_4, $MgSiO_3$, $Mg_2Al_4Si_5O_{18}$

comparison to the glasses with large particle size. The exothermic peak may split up for the glasses with large particle size, which may correspond to the bulk nucleation as the dominant mechanism for these glasses. All the glass–ceramics show the formation of the $MgSiO_3$ phases and cordierite phase (Mg_2 Al_4Si_5 O_{18}) after heat treatment of 1000°C. The cordierite phase is suppressed when the glasses are heated below 1000°C. For the glass MAST5, the cordierite phase is well suppressed due to the presence of TiO_2 as nucleating agent.

Glass MASC6, MASN6, and MASZ10 contain Cr_2O_3, Ni, and ZrO_2 as nucleating agents. These glasses along with the base magnesium aluminosilicate glass yielded various phases after they were mixed with the ODS alloy and annealed at 1000°C for 100 h in air. The formed phases after chemical interaction are listed in Table 6.3. When zirconia is used as a nucleating agent, then the cordierite phase could be seen which can be attributed to the low T_g, T_c and activation energy for MASZ10. Cordierite phase is suppressed in MASC6 glass, which can be due to the affinity of chromium with magnesium leading to spinel phase formation. The interdiffusion layer between MASC6 and ODS alloy initially increases parabolically with time and then becomes almost constant after 20 h.

Neha and co-workers found that when Y_2O_3 was added at the cost of Al_2O_3, then only the formation of $MgSiO_3$ and Mg_2SiO_4 phases occurred after heat treatment. Sample GS1 also exhibited the presence of a cordierite phase, which is detrimental for SOFC. Furthermore, the volume fraction of the cordierite phase increases with an increase in the heat treatment time. In contrast to this, Y_2O_3 is found completely to suppress the formation of the cordierite phase along with enhancing the stability of the glasses.

Shabaan and coworkers also worked on the MAS-T system (21.21% MgO, 21.36% Al_2O_3, 53.32% SiO_2, and 4.11% TiO_2, in mol%) with TiO_2 as the nucleating agent. The formation of an indialite phase, which is a polymorph of high-temperature cordierite comprised of a pseudohexagonal lattice structure occurs after heat treatment at 950°C for 5 h and can be attributed to the impurities present in the glass. TiO_2 in glass composition affects the kinetics of phase separation because it operates as a surface active agent and also shifts the immiscibility boundary. Ti^{4+} ion can occupy the glass network forming units of the glass structure. The coordination number of titanium should not be greater than four in order to occupy glass network forming positions. Usually, due to the small size of titanium, i.e., 0.68 Å, the expected coordination number is six. As a general rule, the titanium coordination number becomes four at higher temperature or we can say that the coordination number decreases when the temperature increases. Hence, during the glass formation process, titanium may get separated from the network because of its tendency to attain a higher coordination number at lower temperatures. Hence, the titania rich phases may separate out during the initial cooling of the melt and during reheating of the glass these phases may recrystallise due to homogenous crystallisation. These phases are the basis of nucleation of hexagonal cordierite phase formation from the glass network at lower temperatures. For the present glass, two well-defined crystallisation peaks are obtained and the corresponding activation energies for crystallisation are 340 and 498 kJ/mol, respectively. When the glass sample is heat-treated at 850°C for 5 h, then the hexagonal cordierite ($Mg_2Al_4Si_5O_{18}$) phase precipitates out which gets converted to orthohmic cordierite at 1000°C. The $Mg_2Al_4Si_5O_{18}$ phase is attributed to the decomposition of the liquid phase present in the glass whereas the development of incipient remains and ordering transitions of Al and Si, via intermediate structural states leads to hexagonal-orthorhombic transitions. Conclusively, it is inferred that with the addition of 5 wt% TiO_2, a low value of activation energy of crystallisation for the α-cordierite phase is obtained. Hence, TiO_2 addition influences the nucleation of crystalline phases leading to a decrease of activation energy of crystallisation.

The Y_2O_3 and La_2O_3-doped magnesium borosilicate glasses designated as MgY and ML were tested for their compatibility with AISI 441 and Crofer 22 APU, respectively. For ML-coupled glass with Crofer 22 APU, no crystalline phase was obtained after 5 and 100 h heat treatment, respectively, whereas the $MgSiO_3$ phase could be seen after prolonged 750 h treatment at 850°C. For the $MgO–SiO_2$ glass system, the crystalline phase formation initiates with formation of silica-rich solid solution. Hence, the following reaction may follow,

$$3MgO + 2SiO_2 \rightarrow MgSiO_3 + Mg_2SiO_4 \tag{6.9}$$

In the above reaction, the substitution of Mg^{2+} at different sites of the unit cell leads to the formation of $MgSiO_3$ and Mg_2SiO_4 phases. When the heat treatment duration for the ML/Crofer 22 APU diffusion couple is increased, then the effective stability also increases. ML glass sealant exhibited a high residual porosity because of the amorphous nature of ML glass. Though no chromium deposition could be seen at the interface, the ML glass exhibited a bulging effect even after 5 h heat treatment at 350°C, which can be due to small CTE for ML glass.

Figure 6.17a shows the interface between MgY and AISI 441 steel after heat treatment of 500 h at 850°C in air atmosphere. A 4–5 μm thick layer is seen at the interface, which is devoid of pores and cracks. From the elemental mapping (Fig. 6.17b), it is evident that the interface has the highest concentration of chromium followed by yttrium. Anyhow, Cr has maximum concentration in a region of 10–12 μm inside the glass matrix from the interface. The diffusion of Fe is minimum diffusion towards the glass side. No magnesium accumulation is seen at the interface, hence the magnesium chromate formation is precluded.

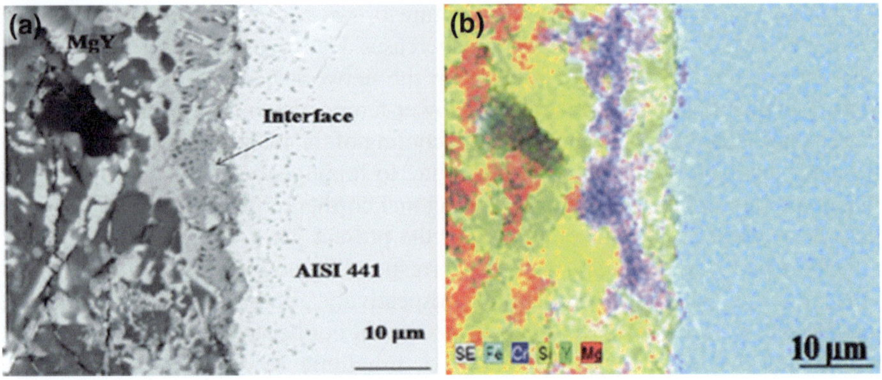

Fig. 6.17 **a** Microstructure and **b** elemental mapping at the interface between AISI 441 steel and MgY (Kumar and coworkers 2014)

6.3.2.2 Calcium-Based Glasses/Interconnect Interface

CaO is a viscosity and CTE modifier as well and hence the calcium-based glasses have shown excellent results. The calcium-based compositions B (25% CaO) and D (5% CaO) were evaluated by Larsen and James for their long-term stability when in contact with ceramic interconnect LCC. Sample D has a low concentration of calcium whereas lanthanum is found in significant amounts on the glass D/LCC interconnect. Almost 50 µm long narrow, Cr/Al-phosphate grains and small Al/Mg phosphate grains are observed. The reaction zone for sample B is more coherent and the interface of glass B/LCC comprises mainly P, Ca, and Cr. In contrast to this, the outer part of the zone near the glass has a high concentration of Al and P along with some calcium. The CaO content largely influences the bonding abilities of glasses with the interconnect. Even at 1000°C, both glass B and D exhibited good bonding behaviour as compared to the glasses containing MgO (Sect. 6.3.2.1, glass A and C) which can be due to the low melting point in the CaO system. The phases formed in the CaO-based glasses are expected to be ortho- and pyrophosphates. As the crystallisation proceeds, the vapour loss rate is limited in the interface region leading to an increase of P_2O_5 in the amorphous glass. Ultimately, this causes decreases in liquidus temperature of the residual glass and increased reactivity. For B and D glasses, the following reaction mechanism is proposed,

$$CaCrO_{4(s)} + (La, Ca)CrO_{3(s)} + glass \rightarrow Ca/Cr/La\text{-}Phosphates_{(l,s)} \quad (6.10)$$

At 1200°C, the reactions are more severe due to the reaction between the LCC liquid phase and the glass. It should be noted that for the samples heat-treated at 1200°C, the interface region is thicker in Cr_2O_3 concentration and the following reaction is proposed:

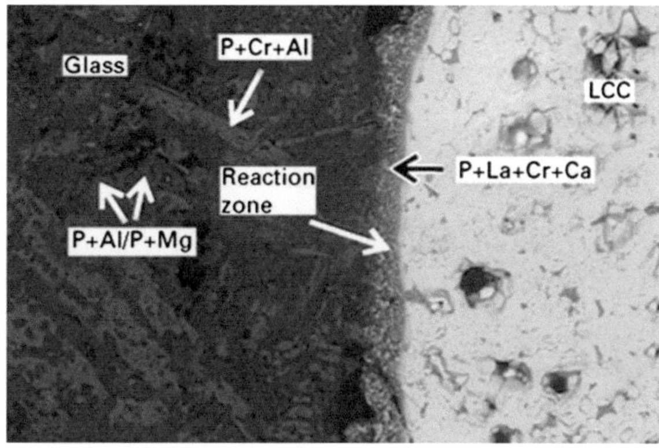

Fig. 6.18 SEM micrographs showing glass D/LCC interfaces heat-treated in air at 1000°C for 24 h (Larsen and James 1998)

$$CaCrO_{4(l)} + (La, Ca)CrO_{3(s)} + \text{glass} \rightarrow Cr_2O_{3(s)} + La_2O_3/LaO/Cr_2O_{3(s)} + O_{2(g)} + \text{glass} \tag{6.11}$$

Glass B is regarded as the suitable candidate for SOFC applications as it exhibits excellent bonding without severe reactions at the interface and can be attributed to comparatively high CTE (7.9×10^{-6}/K) among all the glasses (Fig. 6.18).

Nonnet and co-workers explored the interfacial stability among the $CaO–Al_2O_3–SiO_2–B_2O_3$ (CAS) system with Crofer 22 APU. TV1 and TV2 compositions were derived from Schott glass (64.48% SiO_2, 8.33% B_2O_3, 11.39% Na_2O, 9.2% Al_2O_3, 5.96% K_2O, 0.3% CaO, and 0.4% BaO in wt%). It was observed that after exposure to 800°C, Albeite phase ($Na(Si_3Al)O_8$) has been precipitated out. However, these glasses can effectively seal metals and ceramics at 800°C without cracking on cooling indicating structural stability of the seal during the thermal cycling. The CAS1 system glass–ceramic seals are effective for the sealing due to their chemical durability. CAS1 (24.5% CaO, 14% Al_2O_3, and 61.5% SiO_2, wt%) and CAS2 (37.5% CaO, 20.3% Al_2O_3, and 42.2% SiO_2, wt%) have the lowest melting points in the CAS system, i.e., 1170 and 1265°C, respectively. To decrease the T_g, B_2O_3 was added to the composition and CAS1B/CAS2B are obtained. The crystallisation kinetics is rapid for the CAS2B glasses because all the phases are formed in almost the first 4 h of the heat treatment whereas CAS1B remains almost amorphous. The main crystalline phases formed after 64 h heat treatment in CAS1B glass are $CaSiO_3$, $CaAl_2Si_2O_8$, and SiO_2, whereas for the CAS2B glass, the predominant phase is $Ca_2Al_2SiO_7$ in addition to the three-phases observed during crystallisation of CAS1B glass. These phases are consistent with those as observed from the phase diagram shown in Fig. 6.19a. The pressure tests on the glass ceramics revealed very good air tightness: 10^{-5} Pa m^3s^{-1} at 800°C for 300 h and 200 mbar. Both glass CAS1B and CAS2B reveal good bonding behaviour with the Crofer 22APU. For the CAS1B/Crofer 22APU interface, (Fig. 6.19b) a thin chromia-rich layer could be observed at the interface, though glass CAS2B glass had no such layer formation with Crofer 22APU. The EPMA revealed the presence of chromium for both

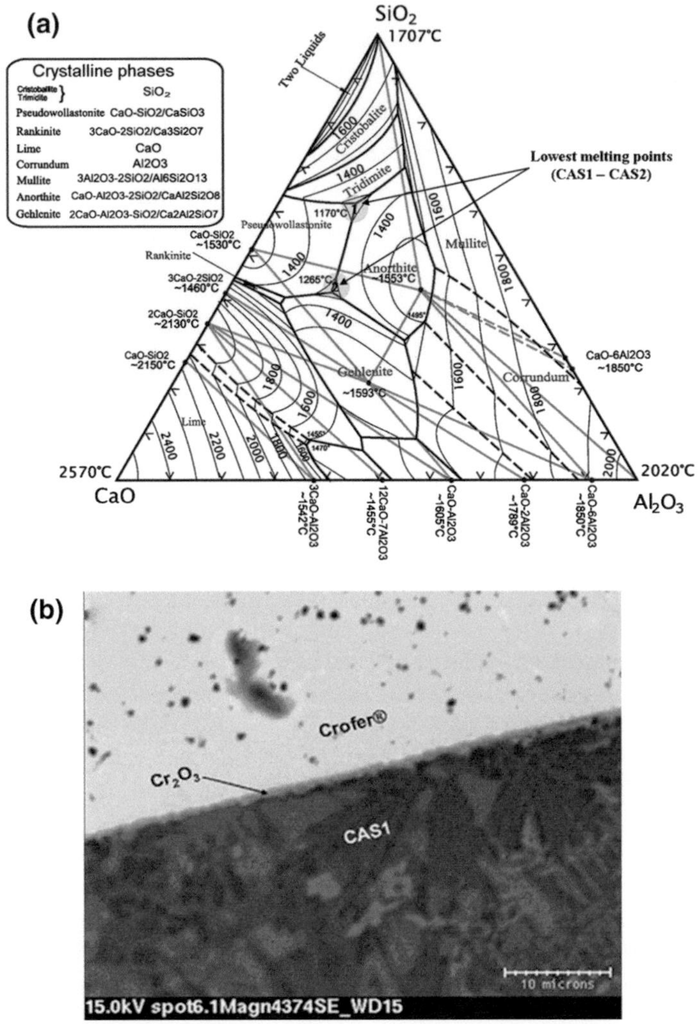

Fig. 6.19 **a** Pseudo ternary phase diagram for CaO–Al_2O_3–SiO_2 system (wt%) and **b** Cross-section of heat-treated CAS1B glass/metal interface (Khedim and coworkers 2012)

compositions. For CAS1B glass composition, this chromium-rich oxide passivation layer seems to be partially dissolved whereas for the CAS2B composition, it is completely dissolved. Usually, the melt alkalinity increases the chromium oxide solubility under reducing conditions at a particular temperature. For CAS2B glass, CaO content is higher than CAS1B glass and hence the alkalinity is higher for CAS2B glass (higher CaO/SiO_2 ratio). The glasses also display the desired range of CTE, i.e., $\sim 11 \times 10^{-6}$/K. Furthermore, optimisation of the CAS2B glasses has been done by adding K_2O, V_2O_5, and TiO_2 into their composition. It could be

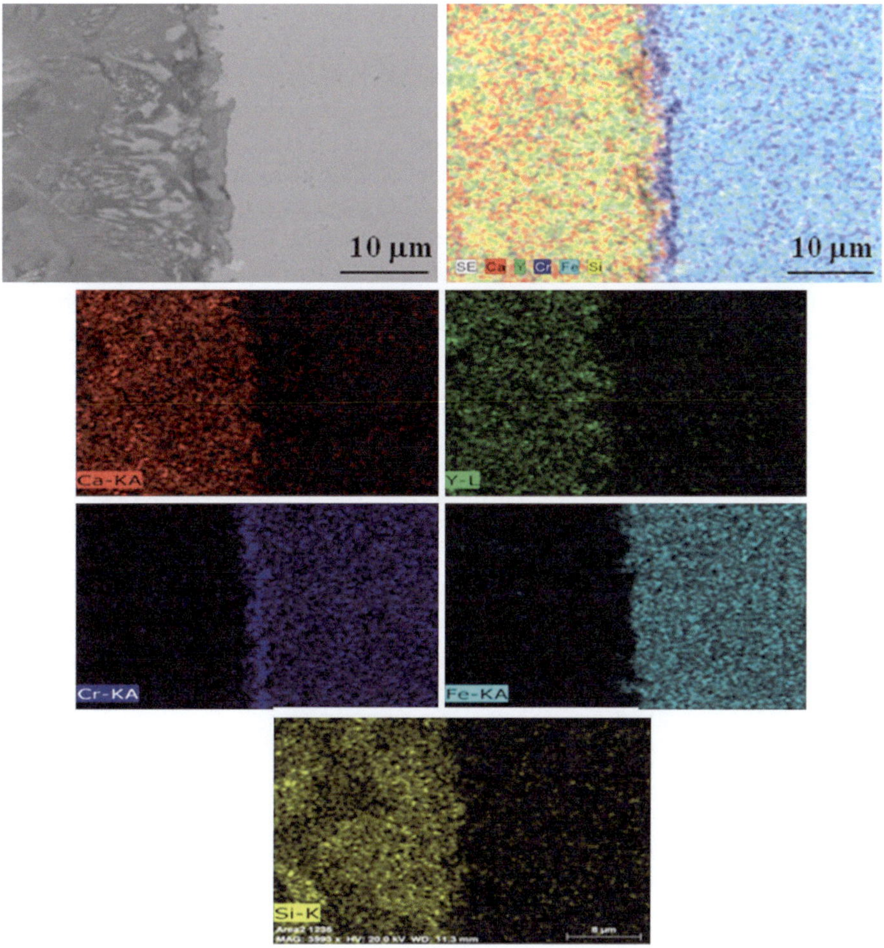

Fig. 6.20 Microstructure and elemental mapping of CaY/AISI441 steel after heat treatment at 800°C for 500 h (Kumar and coworkers 2014)

obtained that K_2O delayed the crystallisation but increased viscosity, whereas V_2O_5 addition yielded improved behaviour of the glass by enhancing crystallisation and reducing viscosity. In addition to this, TiO_2 could not yield any effect on the glass behaviour.

The CAS systems (45% CaO–45% SiO_2–5% Al_2O_3–5% B_2O_3, mol%) investigated by Lahl and co-workers has yielded $CaSiO_3$ and $CaAl_2Si_2O_8$ phases after 1000 h heat treatment at 1000°C. CAS sealants have revealed higher crystallisation as compared to the MAS sealants. In addition to this, the activation energy of MAS is higher than CAS, whereas CTE follows the reverse trend. The CAS system by Kumar and co-workers, has 30% CaO–40% SiO_2–20% B_2O_5–10%Al_2O_3

(CaAl) and 30% CaO–40% SiO_2–20% B_2O_5–10% Y_2O_3 (CaY) compositions (in mol%).

The CTE of the CaAl glass is 7.6×10^{-6}/K, whereas for glass ceramic (800°C for 100 h), it raises to 7.8×10^{-6}/K, CaY glass ceramic exhibits CTE of 8.21×10^{-6}/K after heat treatment at 800°C for 100 h. No crystallisation was observed even after prolonged heat treatment of CaY glass, which can be due to the capability of Y_2O_3 to bond into both the SiO_2 and B_2O_3 networks leading to a strained glass matrix. Hence, the competition among B_2O_3 and SiO_2 networks prevents glass crystallisation. For CaAl glass, crystallisation of Al_2SiO_5 is observed after 10 h of heat treatment at 800°C in spite of the fact that Al_2O_3 prevents crystallisation. The fraction of Al_2SiO_5 increases as the duration of heat treatment increases for the CaAl glass. The interface of CaY/AISI441 steel is shown in Fig. 6.20 revealing an interface of a 3–4 μm thick layer along with good adhesion. The chromium diffusion is the maximum at the interface, though it has not diffused inside the glass matrix. Moreover, no calcium accumulation is seen at the interface, precluding any detrimental $CaCrO_4$, formation. The chromate structure consists of zircon-type (I41/amd, No. 141) and scheelite-type (I41/a, No 88) structures with body-centred trigonal (Fig. 6.21a, b).

For both structures, the chromium ion is tetrahedrally coordinated by oxygen ions at different bond lengths and bond angles. For the scheelite structure, Ca and Cr atoms are located at $\left(0, {}^{1}\!/_{4}, {}^{5}\!/_{8}\right)$ and $\left(0, {}^{1}\!/_{4}, {}^{1}\!/_{8}\right)$ on the 4b and 4a Wyckoff sites, respectively. Oxygen atoms occupy 16f Wyckoff sites (x, y, z). For the zircon-type structure, Ca and Cr atoms are located at $\left(0, {}^{3}\!/_{4}, {}^{1}\!/_{8}\right)$ and $\left(0, {}^{1}\!/_{4}, {}^{3}\!/_{8}\right)$ on 4a and 4b sites, respectively. The Wyckoff site occupied by the oxygen atom is (0, y, z). The bulk modulus for the Scheelite phase (94 GPa) is higher compared to the zircon phase (81 GPa) and hence it becomes difficult to compress the zircon phase. It should be noted that the zircon phase is a direct band gap insulator and the sheelite phase is an indirect band gap insulator.

The XRD of coupled CaAl glass/AISI 441 steel (at 850°C for 500 n) is shown in Fig. 6.22 in a reducing and oxidising atmosphere. It is evident that the coupled CaAl/AISI 441 shows the formation of the $CaSiO_3$ phase in reducing the atmosphere and the calcium chromate ($Ca_5Cr_3O_{12}$) phase also precipitates out in the oxidising atmosphere along with the calcium silicate phase. Hence, the oxidising atmosphere promotes chromate formation. The small peaks in the diffractograms of all the coupled glasses are attributed to the microcrystalline phases present in them. The oxidising atmosphere is seen to promote the crystallisation kinetics though no alumina-rich phase could be obtained. Figure 6.23a, b gives the microstructure of CaAl/AISI 441 under oxidising and reducing conditions, respectively (heat-treated at 800°C for 100 h). The CTE of CaAl glass ceramic is 7.8×10^{-6}/K for 100–650°C range. No delamination could be seen at the CaAl/AISI 441 steel in both oxidising as well as reducing atmosphere. Anyhow, a chromia-rich zone could be seen at the interface (EDS not shown here) attributed to the $Ca_6Cr_3O_{12}$ phase formation for the oxidising conditions. No chromate phase could be detected in the coupled

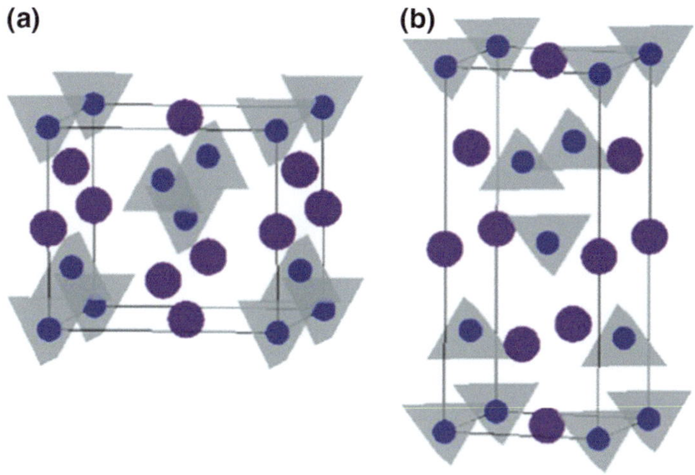

Fig. 6.21 The crystal structure of **a** zircon-type and **b** scheelite-type $CaCrO_4$ *Purple* is Ca, *blue* one is Cr, and the oxygen atom is at the vertices without drawing (Li and coworkers 2006)

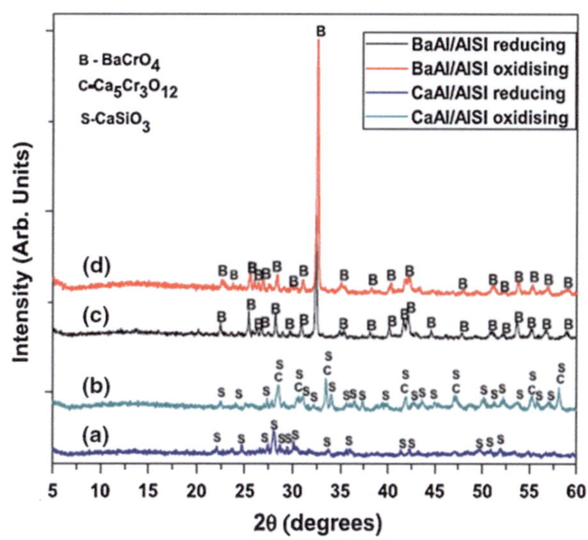

Fig. 6.22 XRD for coupled CaAl/AISI 441 and BaAl/AISI441 glass after heat treatment of 500 h at 850°C under reducing and oxidising atmosphere (Kumar and coworkers 2015)

CaAl/AISI 441 for reducing conditions, but the small chromia-rich zone could be seen at the interface. Small porosity could be seen at the glass side (under reducing conditions), though the interface is smooth and intact which can be explained as follows. The glass crystal stage involves neck formation between powder particles, followed by the formation of open porosity with continuous solid phase (intermediate stage) and then the third final stage of removal of closed pores embedded in a dense solid. Hence, the third stage can be regarded as the pore determining stage. The pores in CaAl glass (reducing conditions) could be attributed to the defects

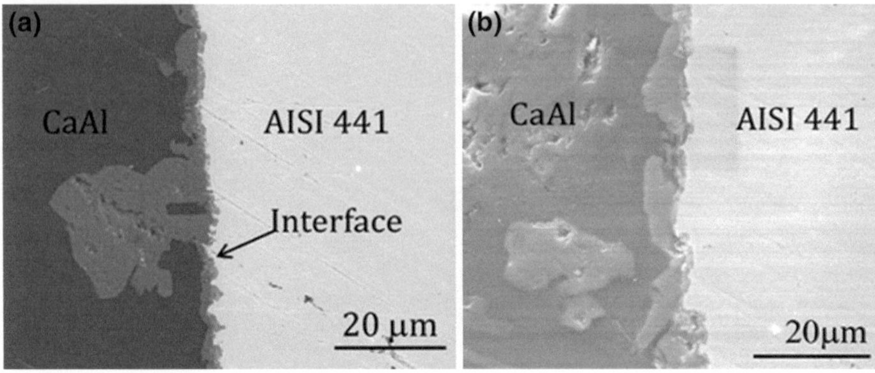

Fig. 6.23 CaAl/AISI441 steel interface after 500 h heat treatment at 850°C under **a** oxidising and **b** reducing conditions (Kumar and coworkers 2015)

introduced due to the partial removal of closed pores during the sintering process. The pore formation may also take place when the alloy elements such as Cr interact with the dissolved water species of the glass.

The CL glass (30% CaO—40% SiO—20% B_2O_3—10% Ca_2O_3) investigated by Kaur and co-workers shows a sharp increase in CTE when the duration of heat treatment is increased from 5 to 750 h. As CTE is attributed to the asymmetric potential well in the system, materials with strong bonds indicate a small change in their CTE. For coupled CL/Crofer 22 APU glass, the lanthanum silicate ($La_2Si_2O_7$) phase precipitated out after 5 h heat treatment at 850°C. Upon the prolonged heat treatment, the phase fraction of the phase increased. The mechanism of $La_2Si_2O_7$ phase formation can be given as

$$La_2O_3 + 2SiO_2 \rightarrow La_2Si_2O_7 \quad (6.12)$$

The CTE of the lanthanum-based calcium boroxilicate glass is higher than the corresponding yttrium/aluminium-doped glasses which can be attributed to the larger ionic radii of La^{3+} ion. The interface of CL/Crofer 22 APU after 5 h heat treatment at 850°C revealed observable delamination though no chromate accumulation could be seen at the interface. Extensive cracking in the glass region could also be seen within the CL glass. After 750 h heat treatment at 850°C, an intact interface devoid of cracks could be obtained. No substantial change in the interface could be seen due to the absence of detrimental reaction products. The crystallised fraction for CL glass is higher due to its small activation energy.

A glass with very high B_2O_3 content, i.e., 33% CaO—67% B_2O_3 exhibited self-healing crystalline phases. CaB_4O_7 and CaB_2O_4 are the major crystalline phases formed after 1 week heat treatment at 850°C. Boron is volatile in nature; hence the volatility of boron may lead to boron-deficient zones and consequently lead to CaB_2O_4 species. B_2O_3 reduces T_g and T_c of the glass but also exhibits volatility. Hence, their content in glass is usually restricted up to 10 mol%. For instance, the reaction between B_2O_3 and water vapour led to $B_3H_3O_6$ species ultimately causing weight loss in the glass. The self-healing characteristics of 33% CaO—67% B_2O_3 glass are evident as the micro-indentation on their surface disappears after increasing the heat treatment temperature from 700 to 800°C. The self-healing properties of these glasses can be related to the viscoelastic flow of crystalline phases at temperatures close to the melting point temperatures. Hence, the boron-containing species exhibit excellent self-healing property at operational temperatures of SOFC.

6.3.2.3 Strontium-Based Glasses/Interconnect Interface

SrO–La_2O_3–Al_2O_3–B_2O_3–SiO_2-based glasses were some of the initial glasses to be investigated for the SOFC applications. The compositions designed by Ley and co-workers contained boron as the primary glass modifier. Alumina usually retards the formation of strontium borate phases, whereas lanthanum is a viscosity modifier and strontium increases the CTE of the glass composition. The sealants investigated by Ley and co-workers yielded CTE in the range of 8–13 × 10^{-6}/K and the glass transition temperature lies in the range of 500–800°C. Glass 42 (26.8% SiO_2, 3.75% La_2O_3, 17.69% Al_2O_3, 46.04% B_2O_3, 5.72% SiO_2, mol%) have T_g of 560°C which is minimum among the series because its B_2O_3: SiO_2 ratio is higher compared to other glasses. In contrast to this, glass 43 (41.76% SrO, 3.74% La_2O_3, 16.27% Al_2O_3, 27.01% B_2O_3, 11.23% SiO_2, mol%) and glass 45 (35.31% SrO, 3.65% La_2O_3, 14.26 Al_2O_3, 18.50% B_2O_3, 28.29% SiO_2, mol%) have T_g in the range of 740–780°C. Therefore, glass 43 and 45 are useful when operated at high temperatures whereas glass 43 is suitable for operation at the working temperature of SOFC. Two compositions K1Si (27.14% SrO, 22.17% La_2O_3, 7.22% Al_2O_3, 41.72% B_2O_3, 1.76% SiO_2 mol%) and K31Sr (30.89% SrO, 22.17% La_2O_3, 7.51% Al_2O_3, 35.18% B_2O_3, 4.98% SiO_2 mol%) yielded CTE of 12.6 × 10^{-6} and 13.1 × 10^{-6}/K, respectively, which are almost same as the metallic interconnects. An oxygen concentration cell using composition 14 (24.56% SrO, 20.13% La_2O_3, 6.92% Al_2O_3, 40.29% B_2O_3, 8.11% SiO_2, mol%) with CTE 11.5 × 10^{-6}/K was tested in the operating range of 800–1000°C. The change in EMF, E was used to calculate the pO_2 change on the fuel side and oxygen permeation constant, K as

$$p_1 = p_2 \exp\left[-\frac{4FE}{RT}\right] \tag{6.13}$$

$$q = \frac{KAt(p_1 - p_2)}{d} \tag{6.14}$$

where q is the flux of oxygen molecules, A is the area through which formation occurs, p_1 and p_2 are the internal and external oxygen pressures (in Pa), d is the permeability distance, and F is the Faraday constant. While making the permeation measurements, the fuel gas flow was interrupted to observe the cell EMF as a function of time. The permeation coefficients at 800, 900, and 1000°C are 2×10^{-27}, 5×10^{-24}, and 4×10^{-21} cm^3/Pa s, respectively. These values are acceptable for glass 14 to be used as the SOFC sealant.

Lu and co-workers studied $SrO–La_2O_3–Al_2O_3–B_2O_3–SiO_2$ glass compositions where $SrO + La_2O_3 + Al_2O_3$ constituted 40 mol% and $B_2O_3 + SiO_2$ constituted 60 mol% with the B_2O_3: SiO_2 ratio between 0 and 1.4. Five compositions were studied, i.e., SABS–0, SABS–5, SABS–10, SABS—15, and SABS–35, where the number 0–35 represents boron content. When boron content increases from 0 to 35 mol%, then the T_g and T_d decreases from 775 to 635°C and 815 to 670°C, respectively. Usually B_2O_3 glasses exhibit low T_g/T_d. For SABS–0 glass, the silica θ^4, θ^3, θ^2, θ^1, and θ^0 units are present and the glass network is rigid as well as stable. Upon the introduction of B_2O_3 in the glass network, the trigonal $[BO_3]$ polyhedra causes the $[SiO_4]$ glass network to rearrange resulting in a less connected 3D structure. With the increasing $B_2O_3 : SiO_2$ ratio, the three and four coordination boron units increase. The network connectivity also affects the devitrification resistance of the glass. SABS–O has highest devitrification resistance as no phases are formed even after 200 h treatment, whereas for SABS–5, glass $La_2Si_2O_7$, $SrSiO_4$, and α–$SrSiO_3$ phase precipitated out after 100 h treatment at 850°C. Hence, due to decreased glass connectivity the $[SiO_4]$ unit interacts with the La_2O_3 and SrO units of glass. For other glasses, $La_2Si_2O_7$ (SABS–10), $La_2Si_2O_7/AlBO_3/Sr_2B_2Al_2O_8$ (SABS–15), and $SrB_2O_4/LaBO_3$(SABS–35) phases are identified after 50 h heat treatment at 850°C. For the lanthanum glasses, La^{3+} ions either link with La–O bonds forming La–O–La units or with silica tetrahedral forming Si–O–La units. These structural units are energetically favourable and act as the nucleation sites. For the strontium boroaluminate glasses, the interaction of four coordinated aluminium and three coordinated boron leads to formation of boroaluminate phases. The (SABS–0) seal exhibited desirable thermophysical properties and thermal stability; hence it was further studied for the interfacial interaction with Crofer 22 APU for 700–850°C and thermal treatment time of 0–100 h. The T_g for SABS-0 seal is 775 ± 2.75°C and the CTE is $10.8 \pm 0.5 \times 10^{-6}$/K which is almost comparable to the CTE of Crofer 22 APU. Figure 6.24a shows the SEM image of Crofer 22APU/SABS–0 sample heat-treated at 850°C for 100 h. The elongated phases were present very near to the interface region for the sample heat treatment temperature. The region of elongated phases does not extend below 10 μm. Small black spots could be seen on the interconnect due to the chromium evaporation from the interconnect.

Figure 6.24b gives the elemental distribution for the Crofer 22 APU/SABS-0 sample heat-treated at 700, 750, 800, and 850°C for 100 h. The microstructure (Fig. 6.24a) gives the denitrification of glass and line profile that indicates the occurrence of diffusion. All the glass elements have diffused up to 2–3 μm on the Crofer 22APU alloy side. Chromium has diffused almost ~7 and ~5 μm for SABS–0 at 800/850 and 700/750°C, respectively. It could be seen that manganese and titanium are maximum at the interface (Mn and Ti are minor constituents for the Crofer 22 APU alloy). Manganese and titanium exhibit higher affinity for oxygen than the chromium, hence, they form an oxide layer/spinal layer on Crofer

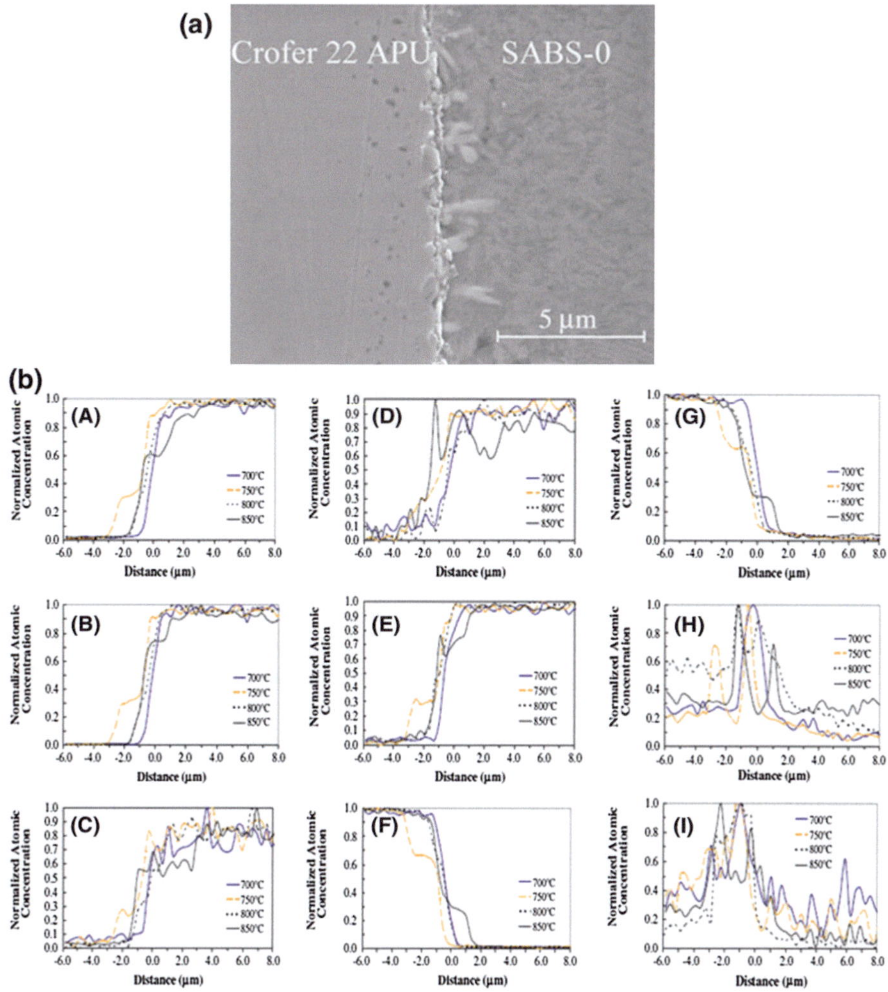

Fig. 6.24 **a** Sem of sabs-0/crofer22apu interface heat-treated at 850°C for 100 h. **b** Elemental analysis for different elements (Mahapatra and Lu 2009)

22APU. In addition to this, titanium improves the electrical conductivity of the alloy and manganese suppresses the chromium evaporation and diffusion.

The XRD of the thermally treated sample (Crofer 22APU/SABS-0) indicates the presence of $La_2Cr_2O_9$ (major) and $Sr_7Al_{12}O_{25}$ (minor) phases. This can be illustrated as follows. Al^{3+} is a glass former and Sr^{2+} compensates the charge for the nearest localised $[AlO_4]^-$ tetrahedral units. These localised AlO_4^- units cause the devitrification of SABS–0 glass leading to formation of the $Sr_7Al_{12}O_5$ crystalline phase. In contrast to this, La_2O_3-based glasses form energetically favoured small isolated clusters in the form of La–O–La units which act as nucleation sites as well due to the acidic nature of chromium (acidobasicity value of 3.50) and basic nature of lanthanum (acidobasicity value of 1.95); the $La_2Cr_2O_9$ formation is favoured as proposed by the following reaction

$$La_2O_3 + 2Cr + 6O_{glass} \rightarrow La_2Cr_2O_9 \tag{6.15}$$

In the case of the presence of residual H_2O in the glass, $CrO_2(OH)_2$ is the dominant and thermodynamically favourable vapour species. Hence, the chromate formation may take place as

$$La_2O_3 + 2CrO_2(OH)_{2(g)} \rightarrow La_2Cr_2O_9 + 2H_2O_{(g)} \tag{6.16}$$

Ojha and co-workers also devised five compositions for SLABS glass with varying SrO (10–30 wt%) and B_2O_3(30–10 wt%) oxides. The network former B_2O_3 is partially substituted by network modifier SrO and SiO_2 is kept constant. Figure 6.25a shows the glass CTE and dilatometer for SLABS-3 [15% La_2O_3, 15% Al_2O_3, 30% SiO_2, 10% SrO, and 30% B_2O_3, wt%], SLAB-5 [15% La_2O_3, 15% Al_2O_3, 30% SiO_2, 20% SrO, and 20% B_2O_3, wt%] and SLABS-7 [15% La_2O_3, 15% Al_2O_3, 30% SiO_2, 30% SrO, and 10% B_2O_3, wt%], respectively. The devitrification temperature of SLABS glass with the SrO content is shown in Fig. 6.25b. With the increase in SrO content, the density CTE, T_g and T_d of glasses increases. In DTA, two devitrification peaks are observed for each glass: the first devitrification between 913 and 821°C and second devitrification between 1092 and 1021°C (Fig. 6.25b). Hence, the devitrification temperatures decrease with increasing SrO_2 and decreasing B_2O_3/SiO_2 ratio. Furthermore, this indicates the network breakage due to increasing modifier content, therefore favouring devitrification and lowering devitrification temperature.

The SL composition (30% SrO—40% SiO_{2}—20% B_2O_3—10% La_2O_3) developed by Kaur and co-workers has CTE in the range of (89.7–9.81) × 10^{-6}/K for heat

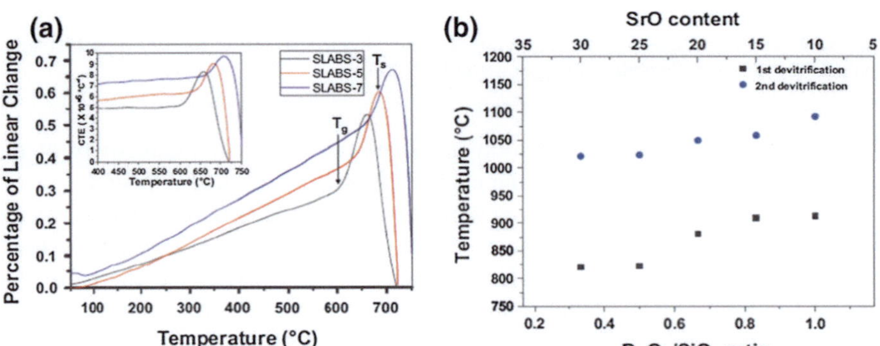

Fig. 6.25 **a** Dilatometer of different glasses with CTE in the *inset*. **b** Devitrification temperature of SLABS glass (Ojha and coworkers 2011)

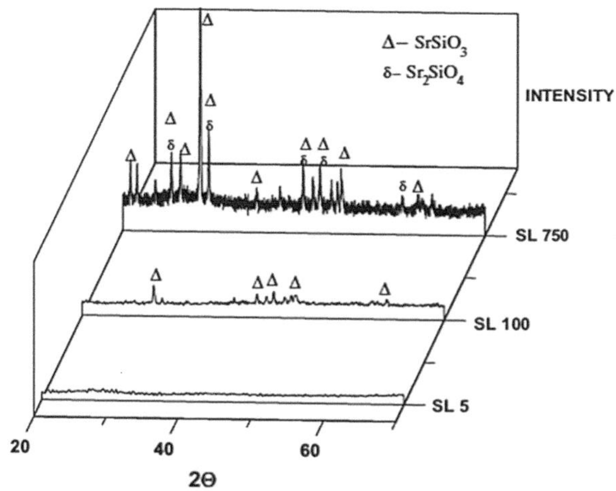

Fig. 6.26 XRD diffractogram of coupled SL glass/Crofer 22APU (Kaur and coworkers 2012b)

treatment duration between 5 and 750 h at 850°C. The coupled SL glass/Crofer 22APU shows the formation of the $SrSiO_3$ and Sr_2SiO_4 phase after 750 h heat treatment at 850°C (Fig. 6.26). Initially, the $SrSiO_3$ phase is formed after 100 h heat treatment, which is converted to the thermodynamically stable Sr_2SiO_4 phase. The phase fraction for the $SrSiO_3$ phase also increases upon heat treatment. The high nucleation rate and low activation energy of SL glass are primarily responsible for higher devitrification. The phases are sometimes desirable to increase the CTE and viscosity of the glass. The mechanism of the phase formation in SL glass could be given as

$$SrO + SiO_2 \rightarrow SrSiO_3 \tag{6.17}$$

$$SrO + 2SrSiO_3 \rightarrow SrSiO_3 + Sr_2SiO_4 \tag{6.18}$$

The microstructure and dot mapping of the diffusion couple for SL glass/Crofer 22APU (heat-treated at 850°C for 750 h), is shown in Fig. 6.27a, b. Seal glass is adhered to the interconnect and exhibits good bonding characteristics. However, the pit formation could be seen on the glass side attributed to the volatility of entrapped gases. The formation of unwanted reaction products such as $SrCrO_4$ at the interface assures the stability of glass seal (Fig. 6.27b). An interesting thing should be noted, i.e., with the increase in heat treatment duration from 5 to 750 h, the extent of chromium diffusion decreased. This phenomenon could be explained as follows.

(a) Initially, due to a large concentration gradient between the glass seal and Crofer 22 APU, the chromium diffusion towards either side of the interface is fast. In addition to this, the ionic radius of chromium is smaller than the iron and other cations, hence promoting the fast diffusion kinetics of chromium.
(b) The functional dependence of diffusion coefficient (D) on time (t) is represented by the parabolic law of oxidation; i.e.,

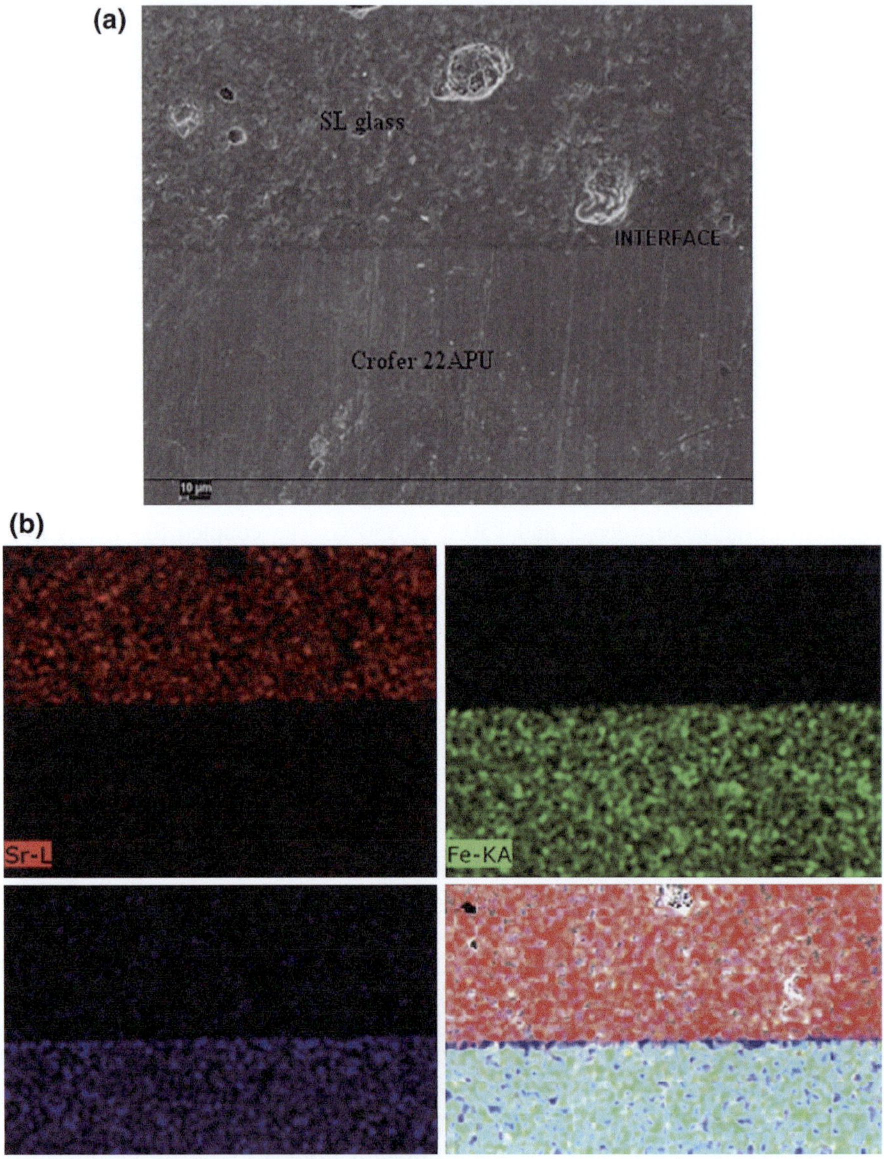

Fig. 6.27 **a** Crofer22APU/SL glass couple. **b** Elemental analysis after heat treatment at 850°C for 750 h (Kaur and coworkers 2012b)

$$D = \frac{x^2}{t} \tag{6.19}$$

where x represents interface thickness. As is clear from Eq. (6.19), the diffusion decreases with increase in time which goes in accordance with the experimental results. It should be noted that cations such as lanthanum and iron also migrate towards the interface up to a certain extent during prolonged heat treatment time. The heavy metal ions possess slow diffusion kinetics and they tend to remain in the glass matrix itself.

Hao and co-workers demonstrated that SrO can improve the wetting ability of glass by decreasing the contact angle between glass and Crofer 22APU substrates for SSA series of glass. SSA series consist of four glasses: G9 (36% SrO, 41% SiO_2, 13% B_2O_3, 10% (Al_2O_3 + La_2O_3), mol%), G10 (40% SrO, 33% SiO_2, 17% B_2O_3, 10% (Al_2O_3 + La_2O_3), mol%), G11 (40% SrO, 25% SiO_2, 25% B_2O_3, 10% (Al_2O_3 + La_2O_3), mol%), and G12 (44% SrO, 30% SiO_2, 16% B_2O_3, 10% (Al_2O_3 + La_2O_3), mol%), respectively. With the decrease of ratio Si/B from 3 to 1, the T_c of glass decreases from 765 to 700°C. G9, G10, and G11 show CTE of 9.72×10^{-6}, 10.56×10^{-6}, and 10.67×10^{-6}/K, whereas the T_s of these glasses decreases. Figure 6.28 shows the shape change of a cube glass specimen on Crofer 22 APU substrate indicating the wetting characteristics from room temperature to melting temperature at a heating rate of 20°C/min. The sintering temperature, melting temperature, and flowing temperature are maximum for G9 samples with values 850, 1100, and 1240°C, respectively. For glass G9, the optimum sealing temperature is between 930 and 980°C, whereas for G10 and G11 glass, the optimum sealing range is 930–980 and 800–840°C, respectively. The true melting temperature for G11 is ~840°C. At 920°C, the contact angle for G11 increases up to 68°, whereas at ~840°C, the contact angle increases up to 122°. Therefore, the wetting capability of G11 is expected to be more.

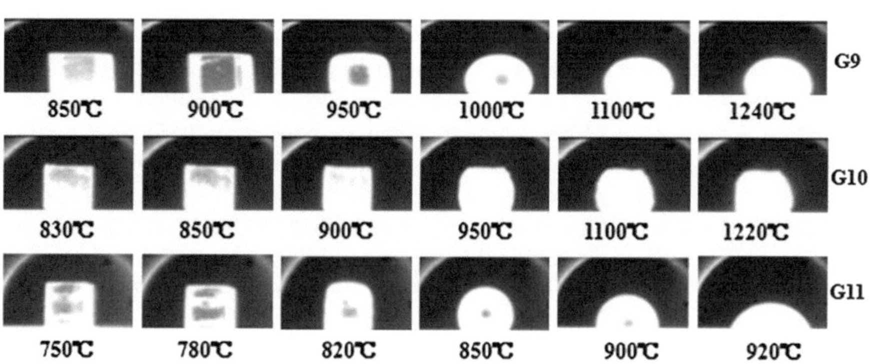

Fig. 6.28 Shape changes of the glass specimens with increasing temperature at the heating rate of 20°C/min (Hao and coworkers 2012)

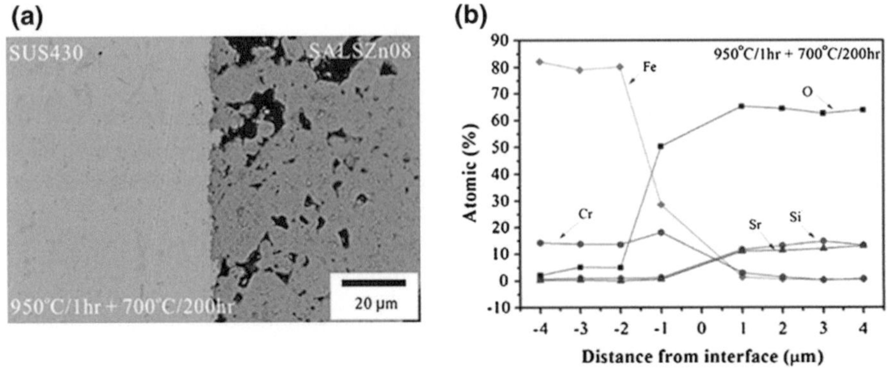

Fig. 6.29 **a** SEM micrographs. **b** Elemental analysis across the interface between SALSZn08 and SUS430 (Wang and coworkers 2012)

Boron-free SrO–La_2O_3–SiO_2–Al_2O_3 with zinc and titania doping glasses have been studied by Wang and co-workers. Upon adding NiO to the glass composition, the T_g decreases and denitrification increases, crippling the thermal stabilisation of glasses. Anyhow, ZnO and TiO modify sealing properties by affecting the crystallisation behaviour of glasses. The B_2O_3 in glass may raise the tendency of denitrification due to microhomogeneity in different structural units. The B_2O_3 vaporisation may occur in the form of HBO_2 and vaporisation may occur in the form of HBO_2 and B $(OH)_2$ species in a wet reducing atmosphere. Hence, some researchers try to avoid using B_2O_3 as glass constituents. The interface of Zn-based composition SALSZn08 (41% SiO_2, 6% Al_2O_3, 2% La_2O_3, 39% SrO, and 12% ZnO, mol%) with SUS 430 plate after heating at 950°C for 1 h is shown in Fig. 6.29a, b along with chemical analysis. The glass showed poor bonding characteristics with SUS430 and exhibited no densification. The sintering was inhibited because a significant amount of crystallites cropped up and only a limited amount of glass was left for the interaction. The adhered glass could be easily peeled off, but in spite of that, the bilayer structure could exhibit mechanical integrity attributed to its CTE of 10.3×10^{-6}/K close to CTE of SUS430 (12.5×10^{-6}/K). The EDS showed no significant diffusion of elements across the interface. The SALSZn10 glass (40% SiO_2, 12.5% Al_2O_3, 7.5% La_2O_3, 20% SrO, 20% ZnO, mol%) exhibited a delaminated interface and huge porosity when in contact with SUS430. When titanium was used instead of zinc, then good bonding characteristics were observed of glass/steel. For instance, SALTSTi11 (45% SiO, 10% Al_2O_3, 5% La_2O_3, 30% SrO, 10% TiO_2, mol%) exhibited strong bonding with the SUS430. When both zinc and titania were used, a decrease in CTE was observed for SALSZT09 composition (40% SiO_2, 10% Al_2O_3, 5% La_2O_3, 20% SrO, 20% ZnO, and 5% TiO_2, mol%). SALSZT09 exhibited CTE of 6.6×10^{-6}/K, which can be attributed to the formation of the $SrAl_2Si_2O_8$ phase in this glass. Severe cracking and flaking off of the glass SALSZT09 was observed on the SUS430 interface. Anyhow, for the SALSZT10 glass (45% SiO_2, 6% Al_2O_3, 4% La_2O_3, 29% SrO, 10% ZnO, and 6%

TiO_2, mol%) and SUS 430 interface, a good adhesion and chemical stability could be obtained.

Tiwari and co-workers reported lanthanum free strontium borosilicate zinc glasses with V_2O_5, Al_2O_3, and Cr_2O_3 as potential additions. All the glasses yielded $Sr_2ZnSi_2O_7$ phase formation and CTE in the range (9–11) × 10^{-6}/K. Raman studies revealed the presence of θ^1 and θ^2 silicate structural units in the glass network. SZS–6 composition (49% SrO, 9% ZnO, 30% SiO_2, 8% B_2O_3, 4% V_2O_5, wt%) was chosen for the interaction with Crofer 22 APU at 950°C for 1 h. The CTE of SZS–6 glass is maximum, i.e., 11.5 × 10^{-6}/K among all other glass compositions. The softening temperature and T_g for SZS–6 are 664 and 640°C, respectively, and also lower than other compositions. The microstructure and elemental analysis of SZS–6/Crofer 22APU (950°/1 h) is shown in Fig. 6.30a, b. The interface is smooth depicting good wettability of glass on the steel surface. The line scan indicates that Cr and Fe have diffused from the metal side to the glass side and Si from glass has diffused into the metal side which is attributed to good bonding behaviour. The interfacial region is almost ~4 μm thick. SZS–6 seals were leak tested with Crofer 22APU at elevated temperatures up to 950°C in steps of 100°C for 30 min (at each temperature). The scales can withstand a vacuum of 10–6 Torr in the whole temperature range indicating high seal tightedness.

The SZS–1 composition (51% SrO, 9% ZnO, 40% SiO_2 wt%) has been studied using non-isothermal techniques by heating glasses at different temperatures and different durations. The heat treatment schedules of the SZS–1 glass ceramics are given in Table 6.4.

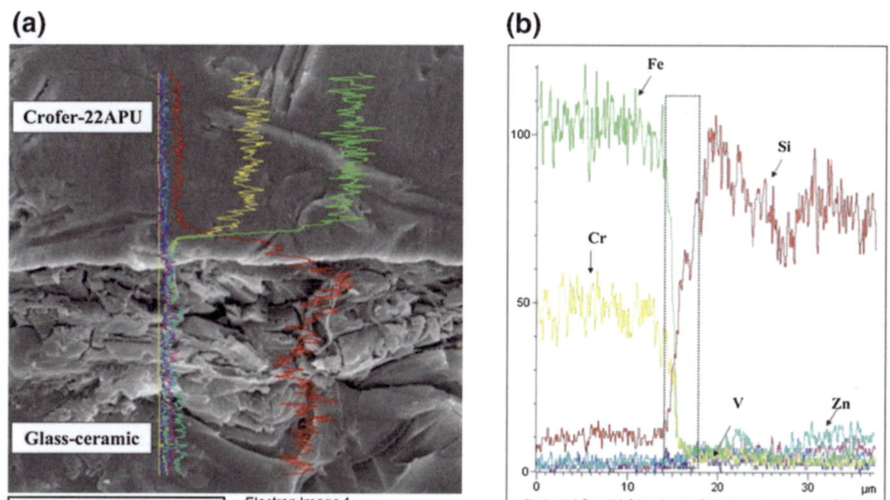

Fig. 6.30 **a** SEM micrographs. **b** EDAX line scans of the SZS-6 glass–ceramic to Crofer 22 APU interface after sealing at 900°C (Tiwari et al. 2012)

Table 6.4 Heat treatment schedules of SZS-1 glass ceramic

Sample name	Dwell time (h)	Temperature (°C)	Phases formed
GC-1	1	750	No phase
GC-2	5	750	No phase
GC-3	1	820	No phase
GC-4	2	820	$ZnSiO_3$, $Sr_3Si_3O_9$, $Sr_2ZnSi_2O_7$, SiO_2
GC-5	2	850	$Sr_2Si_3O_9$, $ZnSiO_3$ (minor)
			$Sr_2ZnSi_2O_7$, SiO_2 (minor)
GC-6	2	925	$Sr_3Si_3O_9$, $Sr_2ZnSi_2O_7$

For the GC–6 Sample, only $Sr_3Si_3O_9$ and $Sr_2ZnSi_2O_7$ phases are obtained with complete absence of SiO_2 and $ZnSiO_3$ phases. These phases must have been formed by simultaneous crystallisation; i.e., initially SiO_2 and $ZnSiO_3$ metastable phases are formed and at later stages more stable $SrSiO_3/Sr_3Si_3O_9$ and Sr_2ZnSiO_7 phases precipitate out according to the following reactions

$$SiO_{2(hexagonal)} + SrO_{(glass)} \rightarrow SrSiO_{(monoclinic)}/Sr_3Si_3O_{9(monoclinic)} \tag{6.20}$$

$$2SrO_{(glass)} + SiO_{2(glass)} + ZnSiO_{3(orthorhombic)} \rightarrow Sr_2ZnSi_2O_{7(tetragonal)} \tag{6.21}$$

The SZS–1 glass shows a spinoidal decomposition mechanism prior to crystallisation during glass heating, which results in a highly interconnected microstructure of glass.

6.3.2.4 Barium-Based Glass/Interconnect Interface

The glasses which can withstand high temperatures usually possess high T_s and T_g, and have low CTE. For instance, the glasses in $MgO/SiO_2/Al_2O_3$, $Y_2O_3/Al_2O_3/SiO_2$ system usually have low CTE in the range of $(4–5) \times 10^{-6}$/K and $T_g \approx 800°C$. Glasses in the system $BaO–Al_2O_3–SiO_2$ are advantageous as seals, and mostly binary phases are precipitated in these systems whose CTE can be adjusted by adding ZnO, La_2O_3, and CaO as additives. Kerstan and co-workers provided a detailed study on the thermal expansion behaviour using dilatometry and high-temperature X-ray diffraction. Rietveld was performed for the barium silicate phases and the various parameters obtained are listed in Table 6.5.

All the barium silicate phases exhibit CTE > 10×10^{-6}/K, especially orthorhombic barium silicate. $BaSiO_3$ and Ba_2SiO_4 have high CTE (mean) of 15.4×10^{-6}/K and also large CTE $(12.7–19.1) \times 10^{-6}$/K in all crystallographic directions. BAS (barium aluminosilicate) have been evaluated by many research groups as the potential SOFC sealants due to their high CTE. Puig and co-workers developed a BAS series using the sol–gel technique to reduce the formation temperature of glasses up to 100–200°C (Fig. 6.31). For BAS glasses system,

Table 6.5 Cell parameters for the barium silicate phases

Phases	*a* (Å)	*b* (Å)	*c* (Å)	*β* (°)	*V* (Å °)
$Ba_5Si_8O_{21}$	32.6805	4.699	13.9147	98.146	2115.57
Monoclinic	32.660	4.693	13.9068	98.199	2110.2
C2/C	32.6750	4.695	13.894	98.10	2110.2
$Ba_2Si_3O_8$	12.4864	4.6888	13.9482	93.5645	815.046
Monoclinic	12.478	4.6875	13.9400	93.5371	813.85
$P2_1/a$	12.477	4.685	13.944	93.54	813.54
Ba_2SiO_4	5.81098	10.22012	7.50828	90	445.908
Orthrombic	5.798	10.1989	7.4895	90	442.88
P_{mcn}	5.805	10.200	7.499	90	444.02
$BaSiO_3$	4.5835	5.61638	12.4418	90	320.286
Orthrombic	4.5703	5.6093	12.4167	90	318.32
$P2_12_12_1$	4.580	5.611	12.431	90	319.46
$BaSi_2O_5$	4.63612	7.69258	13.53075	90	482.556
Orthrombic	4.6301	7.6807	13.5103	90	480.46
P_{cmn}	4.629	7.688	13.523	90	481.25

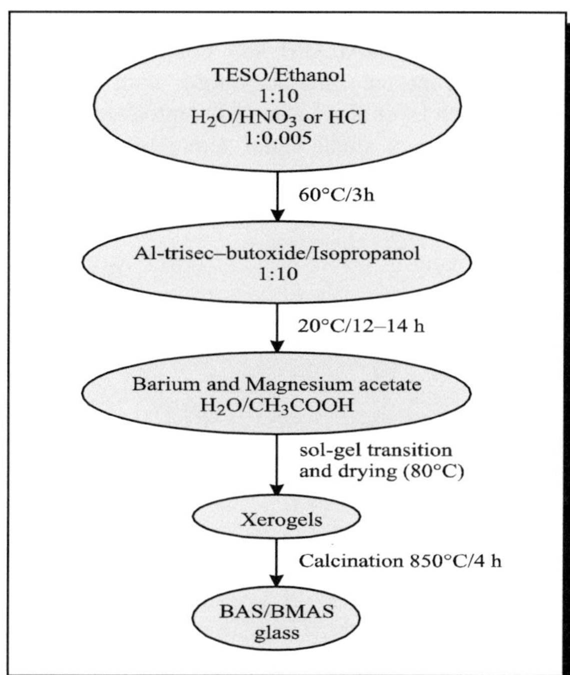

Fig. 6.31 The route for glass preparation

(5% Al_2O_3, (55 − x)% BaO, Bal. SiO_2%, mol%, x = 0, 10, 20, 30, 40). Ba_2SiO_4 appeared as a major phase along with $Ba_6Si_{10}O_{36}$ as a minor phase and large residual amorphous matrix. The hexacelsian phase is completely suppressed indicating the homogeneity of BAS glass. The CTE of glasses increased with increasing barium content. Glass BAS2 (45% BaO), BAS3 (35% BaO), and BAS4 (25% BaO) are exhibiting acceptable properties as sealant materials in spite of their 10–20 vol% porosity. The CTE of BAS4 is around 12.3 × 10^{-6}/K, which is mainly attributed to the high content of $BaSiO_3$ crystals for BAS4 glass ($CTE_{BaSiO_4} \approx 12.5 \times 10^{-6}/K$).

Kumar and co-workers investigated boron containing BAS composition designated as BaAl (30% BaO, 40% SiO_2, 20% B_2O_3, 10% Al_2O_3, mol%) along with its interaction with AISI 441 steel under reducing and oxidising atmospheres. As clear from Fig. 6.22, the BaAl/AISI 441 coupled glass heat-treated at 850°C for 500 h in both oxidising and reducing atmosphere indicated $BaCrO_4$ phase formation. $BaCrO_4$ phase is a detrimental orthorhombic structure with linear CTE in the range of (16–31) × 10^{-6}/K in the temperature range of 20–820°C. Figure 6.32a, b shows the interface microstructure and elemental mapping for the BaAl glass/AISI 441 couple heat-treated at 850°C for 500 h in an oxidising atmosphere. Due to the high CTE of $BaCrO_4$ phase and low CTE of BaAl glass/glass–ceramic (6.98–7.67 × 10^{-6}/K) delamination could be observed at the interface along with a chromate-rich zone at the interface. Figure 6.32b indicates high barium and chromium concentration at the interface supporting the $BaCrO_4$ formation as given in XRD. Even for BaAl/AISI 441 couple heat-treated under reducing conditions, discontinuity at the interface along with pits and pores could be seen. The chromate-rich layer also forms at the interface under reducing conditions though the layer thickness is small under reducing conditions as compared to that of the oxidising conditions. It should be noted from the XRD pattern of BaAl/AISI 441

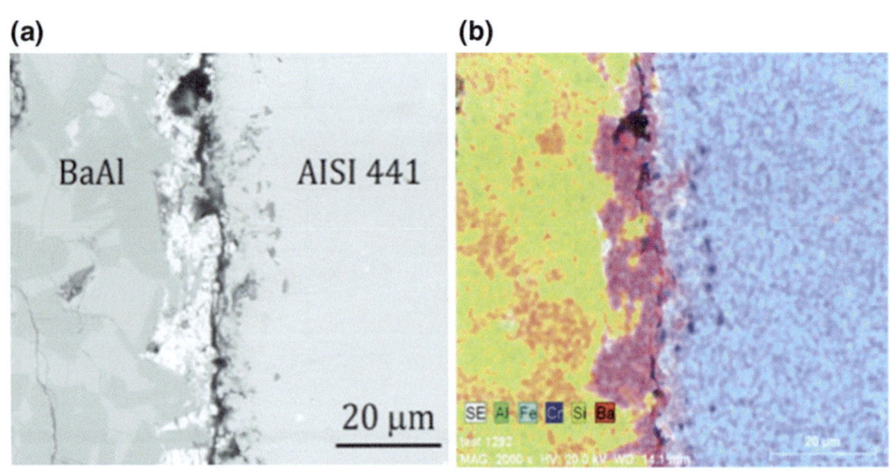

Fig. 6.32 **a** Microstructure. **b** Elemental mapping of BaAl/AISI 441 diffusion couples after heat treatment at 850°C for 500 h under oxidising atmosphere (Kumar and coworkers 2015)

couple glass (Fig. 6.22) that the oxidising atmosphere promotes chromate formation. Hence, the microstructural studies are in correlation with the diffractogram analysis.

Wang and co-workers formulated P series and L series of glass using BaO–Al_2O_3–SiO_2–B_2O_3 base compositions. For P and L series Al_2O_3 was fixed at 7.59 and 5 mol%, respectively. P series exhibit silica in the range of 63.12–48.12 mol%, Ba_2O_3 in the range of 16.67–31.6 mol% and BaO in the range 12.62–27.62 mol% and BaO in 20–50 mol% range. The densities for P series vary from 2.54 to 3.24 g/cm^3, whereas for L series it varies between 2.74 and 3.92 g/cm^3. For the same BaO content, silica-rich glasses have higher densities than the B_2O_3-rich glasses. Interestingly, the CTE of glasses exhibit wide range, i.e., from 4.92 to 10.98 × 10^{-6}/K from P00 (63.12% SiO_2, 7.59% Al_2O_3, 16.67% B_2O_3, and 12.62% BaO, mol%) glass to L09 glass (15% SiO_2, 5% Al_2O_3, 30% B_2O_3, 50% BaO, mol%). The leakage test for L06 glass (15% SiO_2, 5% Al_2O_3, 60% B_2O_3, 20% BaO, mol%), L07 glass (15% SiO_2, 5% Al_2O_3, 50% B_2O_3, 20% BaO, mol%), L08 glass (15% SiO_2, 5% Al_2O_3, 40% B_2O_3, 20% BaO, mol%) and L09 glass sealants is shown in Fig. 6.33 measured at 800°C after heating at 1000°C for 1 h. The detectable limit of gas flow is <0.5 (sccm/cm) (kg/cm^2)/min. For L06–20% MgO, L08 30 vol% MgO and L09–40 vol%. For MgO glass composites, the leakage rates are less than the detectable limit. Ceramic fillers such as MgO have been used to enhance the CTE. The use of MgO also enhanced the thermal stability and structural uniformity with time for the glasses.

It has been explained in this section that the BaO–Al_2O_3–SiO_2–B_2O_3 glass system may exhibit a range of CTE as Kumar and co-workers obtained low CTE of almost 7.68 × 10^{-6}/K for this system. Ba–B–Si–Al–O (26.67% SiO_2, 20.95%

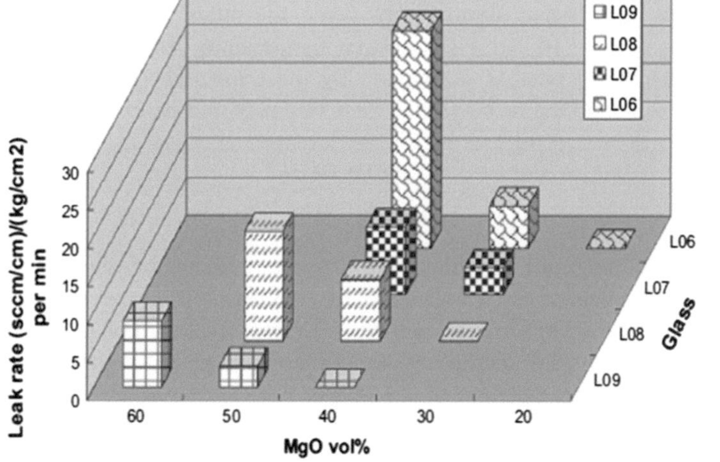

Fig. 6.33 Leakage rate of the glass seals at 800°C after heating at 1000°C for 1 h (Wang and coworkers 2009)

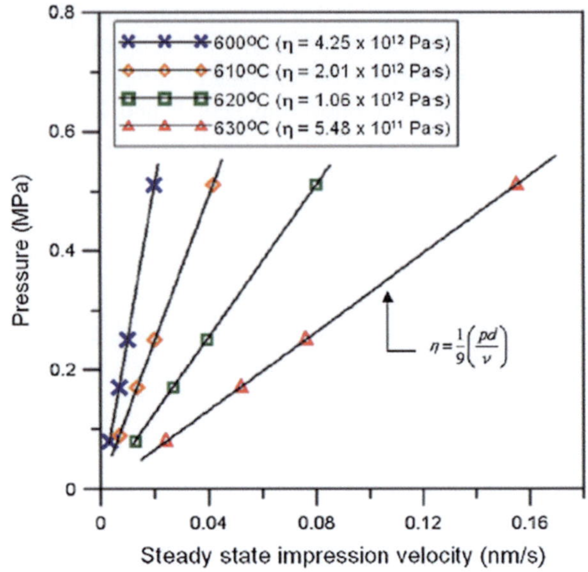

Fig. 6.34 Viscosity as determined by the pressure steady-state impression velocity (Chang and coworkers 2011)

B_2O_3, 47.62% BaO, 4.76% Al_2O_3, mol%) glass has been investigated with average particle size of 4 μm. The T_s and T_g for Ba–B–Si–Al–O is 605 and 571°C, respectively. Two phases precipitated out as $BaSiO_3$ and hexacelsian $BaAl_2Si_2O_8$ at temperatures 695 and 735°C, respectively. The indentation creep test revealed viscosity change from 4.25×10^{12} to 5.48×10^{11} Pa s upon temperature increase of 600–630°C. The viscosity is almost stress independent indicating Newtonian viscous flow for the Ba–B–Si–Al–O glass (Fig. 6.34). Figure 6.34 indicates that the viscosity did not change significantly when the strain rate was changed from 3.3×10^{-9} to 7.7×10^{-8} s^{-1} at constant temperature. The viscous flow can be considered as a rate process dominated by the transition state of energy. When external stress is applied to a system, then viscosity can be explained as

$$\eta = \frac{\tau \exp(\Delta E/kT)}{2v \sinh(\tau V_0/2kT)} \tag{6.22}$$

When ΔE is the height of the energy barrier in the absence of stress, V_0 is the flow volume, τ is the shear stress, k is Boltzmann constant, T is temperature, and v is the number of times the barrier is attempted (frequency factor). For large stresses, the viscosity decreases with increasing temperature whereas for small stresses, the viscosity is almost independent of stresses.

Acceptable CTE are obtained when the BAS compositions are doped with La_2O_3, i.e., in the range of (10.13–10.90) × 10^{-6}/K for glass heat-treated at 800°C for 100 h. The T_g and T_s values lie between 624.6–625.8 and 669–681°C for these glasses. The density, T_g, T_s, and CTE displayed almost independent behaviour w.r.t. the BaO concentration. The Gl composition (46.5% BaO, 14% La_2O_3, 8.8% Al_2O_3,

Fig. 6.35 Interface of the Crofer22APU/G1 glass (Ghosh and coworkers 2008)

12.5% B_2O_3, and 18.2% SiO_2, wt%) possessed maximum CTE of 11 × 10^{-6}/K within room temperature to glass transition range. The viscosity value for Gl is obtained to be $10^{7.96}$ dPa s which is within the sealing range of 10^6–10^9 dPa s for SOFC applications. The glasses exhibit hexacelsian $BaAl_2Si_2O_8$ phase formation at heat treatment temperature of 800°C. As the heat treatment duration is prolonged (up to 1000 h), the metastable hexacelsian phase starts transforming to the thermodynamically monoclinic celsian phase. The monoclinic celsian is detrimental as compared to the hexacelsian phase due to its low CTE of ≈2.29 × 10^{-6}/K for 30–1000°C. Hence, the rapid and progressive transformation from hexacelsian to celsian is undesirable. Due to the suitable thermophysical properties of Gl glass, it was further investigated with Crofer 22APU as shown in Fig. 6.35. The interface of Gl/Crofer 22APU revealed a well adhered interface with minimum chemical reactivity. The interface is defect-free and devoid of cracks, hence indicating the structural stability.

Another lanthanum containing composition BL (30% BaO, 40% SiO_2, 20% B_2O_5, 10% La_2O_3, mol%) has revealed a very porous glass microstructure upon its interaction with Crofer 22APU at 850°C for 750 h (Fig. 6.36a, b). The BL yielded high CTE ≈ 10.71 × 10^{-6} after heat treatment of 750 h at 850°C. No phase formation could be observed for BL/Crofer 22APU coupled glass after 100 h heat treatment. After 750 h heat treatment at 850°C, sanbomite ($BaSi_2O_5$) and $BaSiO_3$ phase could be obtained indicating controlled crystallisation for BL glass. The following reaction mechanism may take place,

$$2BaO + 3SiO_2 \rightarrow BaSiO_3 + BaSi_2O_5 \quad (6.23)$$

From Fig. 6.36a, it is evident that the interface is clear and devoid of reaction products. The microstructure inside the glass region is different from that of the joint region. Closed pits are observed which could be due to the vapour species via the interaction of chromium with dissolved water of the glass. Besides larger pits, a population of small pits could be seen surrounding them. Anyhow, the closed pits do not lead to enhancement of the leak rate. Elemental analysis (Fig. 6.36b)

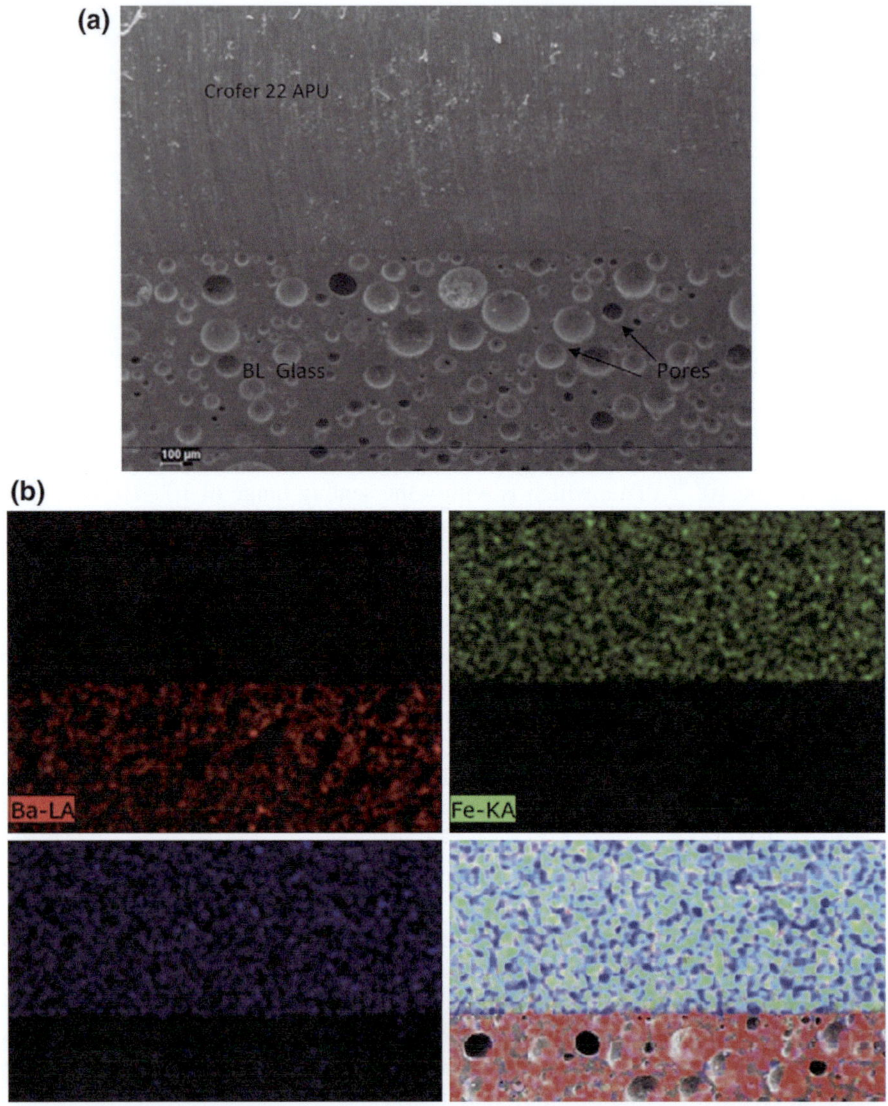

Fig. 6.36 **a** Microstructure. **b** Elemental analysis of Crofer 22APU/BL couple at 850°C for 750 h (Kaur and coworkers 2012b)

indicates no elemental segregation and hence precluding the $BaCrO_4$ formation at the interface. The biggest pit size observed in BL glass have $d = 0.23$ µm and area 41,247 µm^2.

Lanthanum-free and zinc-doped BAS glass (BZAS–30% BaO, 20% ZnO, 40% SiO_2, 7.5% B_2O_3, and 2.5% Al_2O_3, mol%) exhibited low CTE of 7.8×10^{-6}/K. After heat treatment at 850°C, barium zinc silicate ($Ba_2Zn_3Si_3O_{11}$), barium aluminium silicate ($BaAl_2Si_2O_8$), and barium zinc borate ($BaZn_2(BO_3)_2$) appeared. Two polymorphs were observed for barium aluminium silicate: hexacelsian (HC) and orthocelsian (OC). Orthocelsian was observed at 900°C whereas HC was observed at 850°C. The activation energy of the glass transition was 280 kJ/mol. The crystallisation peaks were overlapped by two exotherms with respective activation energy of crystallisation (E_c) to be 245 and 193 kJ/mol. Manganese doped barium zinc borosilicate glasses were also investigated due to the variable oxidation states of Mn_2O_3. The use of such compositions containing manganese is rarely reported. Arora and co-workers studied a manganese containing composition, i.e., 40% SiO_2^2–30% BaO–20% ZnO–x% B_2O_3–$(10 - x)Mn_2O_3$, mol% with $x = 7.5$, 5, 2.5, 0 and samples named as INMI ($x = 7.5$), INM2 ($x = 5$), INM3 ($x = 2.5$), and INM4 ($x = 0$), respectively. INM4 glass exhibited relatively less crystallisation whereas all three glasses exhibited Al_2SiO_5, $MnSiO_3$, $BaZnSiO_4$, $BaMn_2O_8$, $MnSiO(SiO_4)$, and $Ba(Al_2Si_2O_8)$ phase formation. The addition of Mn_2O_3 leads to enhanced T_c and T_g up to 7.5 mol% addition. In contrast to this, for 10% Mn_2O_3, the glass exhibited phase separation and decreased T_g. This indicates the suppressed phase separation for the glass containing both B_2O_3 and Mn_2O_3. The competition between B^{3+} and Mn^{3+} cation leads to an increase in T_g and suppressed phase separation. Moreover, with increasing modifier content in the glass, the nucleation sites occur at different places in the glass matrix. Due to different sites, phases in the glass matrix possess different kinetics which induces phase separation. INM2 exhibits CTE $\approx 8.3 \times 10^{-6}$/K whereas the viscosity value of glasses lies $\approx 10^{4.2}$ dPa s which is much lower than required for the sealing applications. Another rare composition (51% BaO–9.0% ZnO–$(40 - x)$/SiO_2–$x Nd_2O_3$ with $x = 1$, 2 wt%) is also studied for sealing applications. The density values are obtained to be 3.91 and 3.86 gm/cm^3 whereas the CTE increases from (10.39–15.06) $\times 10^{-6}$/K for $x = 1$, 2 wt%, respectively. The microhardness for BZS–1 and BZS–2 are 4.25 and 4.53 GPa, respectively. Anyhow, due to poor flow properties of these glasses, they cannot be regarded as suitable sealing materials.

Bibliography

Chang CYS, Wei WCJ, Hsueh CH (2011) J Non-Cryst Solids 357:1414

Chou YS, Thomsen EC, Williams RT, Choi JP, Canfield NL, Bonnett JF, Stevenson JW (2011) J Power Sources 196:2709 (A. Shyam, E. L. Curzio)

Ghosh S, Kundu P, Sharma AD, Basu RN, Maiti HS (2008) J Eur Ceram Soc 28:69

Hao J, Zan Q, Ai D, Maa J, Deng C, Xu J (2012) J Power Sources 214:75

Kaur G, Pandey OP, Singh K (2012b) Int J Hydrogen Energy 37(8):6862

Khedim H, Nonnet H, Méar FO (2012) J Power Sources 216:227

Kumar V, Kaur G, Lu K, Pandey OP, Singh K (2014) Int J Appl Glass Sci 5(4):329

Kumar V, Kaur G, Lu K, Pickrell G (2015) Int J Hydrogen Energy 40:1195

Lahl N, Singh K, Singheiser L, Hilpert K, Bahadur D (2000) J Mater Sci 35:3089

Larsen PH, James PF (1998) J Mater Sci 33:2499

Li L, Yu W, Long Y, Jin C (2006) Solid State Commun 137:358

Mahapatra MK, Lu K (2009) J Mater Sci 44:5569

Nielsen KA, Solvang M, Nielsen SBL, Dinesen AR, Beeaff D, Larsen PH (2007) J Euro Cera Soc 27:1817
Nonnet H, Khedim H, Mear F (2010) J Aust Ceram Soc 48(2):205
Ojha PK, Ratha SK, Chongdar TK, Gokhale NM, Kulkarni AR (2011) J Power Sources 196:4594
Smeacetto F, Salvo M, Ferraris M, Casalegno V, Asinari P (2008a) J Euro Cera Soc 28:611
Smeacetto F, Salvo M, Ferraris M, Cho J, Boccaccini AR (2008b) J Euro Cera Soc 28:61
Smeacetto F, Salvo M, Bytner FDDH, Leone P, Ferraris M (2010) J Euro Cera Soc 30:933
Smeacetto F, Salvo M, Leone P, Santarelli M, Ferraris M (2011) Mater Lett 65:1048
Smeacetto F, Miranda AD, Chrysanthou A, Bernardo E, Secco M, Bindi M, Salvo M, Sabato AG, Ferraris M (2014) J Am Ceram Soc 97(12):3835
Tiwari B, Dixit A, Pillai CGS, Gadkari SC, Kothiyal GP (2012) J Am Ceram Soc 95(4):1290
Tong JJ, Han MF, Singhal SC, Gong Y (2012) J Non-Cryst Solids 358:1038
Venskutonis A, Glatz W, Kunschert G (2005) In: P/M processing of ODS Cr- and FeCr-based alloys for solid oxide fuel cell applications. PLANSEE AG, Reutte, Austria, RM, vol 56, pp 534
Wang SF, Wang YR, Hsu YF, Chuang CC (2009) Int J Hydrogen Energy 34:8235
Wang SF, Hsu YF, Lu HC, Lo SC, Cheng CS (2012) Int J Hydrogen Energy 37:5901

Suggested Bibliography

Arora A, Goel A, Shaaban ER, Singh K, Pandey OP, Ferreira JMF (2008) Physica B 403:1738
Arora A, Kumar V, Singh K, Pandey OP (2011) Cera Int 37:2101
Donald IW (2007) Glass-to-metal seals. Society of glass technology, Sheffield (Amazon edn, 2009).
Donald JW, Mallinson PM, Metcalfe BL, Gerrard LA, Fernie JA (2011) J Mater Sci 46:1975
Fergus JW (2005) J Power Sources 147:46
Fergus JW, Lessing PA (2007) J Mater Sci 42:3465
Kaur G, Pandey OP, Singh K (2012a) Int J Hydrogen Energy 37(4):3883
Kaur G, Kumar V, Pandey OP, Singh K (2012c) J Electrochem Soc 159:B277
Kerstan M, Russel C (2011) J Power Sources 196:7578
Kerstan M, Muller M, Russel C (2011) Mater Res Bull 46:2456
Kumar V, Arora A, Pandey OP, Singh K (2008) Int J Hydrogen Energy 33:434
Kumar V, Sharma S, Pandey OP, Singh K (2010) Solid State Ionics 181:79
Kumar V, Pandey OP, Singh K (2011a) Int J Hydrogen Energy 36:14971
Kumar V, Kaur G, Pandey OP, Singh K (2011b) Phys Chem Glasses 52:212
Lahl N, Bahadur D, Singh K, Singheiser L, Hilpert K (2004) J Electrochem Soc 149:A607
Ley KL, Krumplet M, Kumar R, Meiser JH, Bloom IJ (1996) J Mater Res 11:1489
Mahapatra MK, Lu K (2010a) J Power Sources 195:7129
Mahapatra MK, Lu K (2010b) Int J Appl Cera Tech 7:10
Mahapatra MK, Lu K, Bodnar RJ (2009) Appl Phys A 95:493
Mahapatra MK, Lu K, Liu X, Wu J (2010) Int J Hydrogen Energy 35:7945
Puig J, Ansart F, Lenormand P, Antoine L, Dailly J (2011) J Non-Cryst Solids 357:3490
Shelby JE (2005) Introduction to glass science and technology, 2nd edn. The Royal Society of Chemistry, Cambridge
Smeacetto F, Salvo M, Ferraris M, Casalegno V, Asinari P, Chrysanthou A (2008c) J Euro Cera Soc 28:2521
Tietz F (1995) Ionics 5:129
Uhlmann DR (1980) J Non-Cryst Solids 38:693
Zhang T, Zhang H, Li G, Yung H (2010) Reduction of chromate formation at the interface of solid oxide fuel cells by different additives. J Power Sources 195:6795
Zhang T, Tang D, Yang H (2011) J Power Sources 196:1321

Chapter 7
Mixed Alkaline/Composite Glasses and Coated Interconnects

As seen in Chap. 6, the interaction of seal glass with metallic interconnect is much more severe due to chromate formations. The current chapter is a continuation of Chap. 6, which focuses on the chemical compatibility of glass/glass–ceramic seals with the mixed alkaline earth glasses. In addition, the spinel-coated interconnects and their interaction with glasses have been discussed.

7.1 Mixed Alkaline Earth Glasses and Their Interaction with Interconnects

7.1.1 Magnesium–Calcium Glasses/Interconnect Interface

It has been elucidated that composites exhibit good wetting characteristics with the interconnect material. Smeacetto and co-workers used alumina to increase seal viscosity, CTE, and thermal check resistance along with enhanced mechanical properties. Kaur and co-workers proposed a new novel approach of making composites without using any unreactive particle in the glass matrix; i.e., the glasses have been mixed in fixed weight ratio and then ball-milled for 5 h. The idea behind this novel approach was to increase the sealing capability, improve viscosity and reduce chromate formation without dispersing any reactive filler material. At high temperatures, the segregation of filler materials may lead to increased stress and hence loss of structural integrity. The ML, CL, SL, and BL (mentioned in Sect. 6.3.2 and Table 5.3) are chosen for the composite formation and the compositions are listed in Table 7.1.

Composite GG1 is obtained by mixing ML and CL in a 1:1 weight ratio and then ball mill for 5 h. The obtained particle size and density of GG1 are 1.52 μm and 3.42 g/cm^3, respectively. The viscosity for GG1 is $10^{8.05}$ dPas and the CTE is

*All the glass compositions mentioned in this chapter are listed in Table 5.3

G. Kaur, *Solid Oxide Fuel Cell Components*,
DOI 10.1007/978-3-319-25598-9_7

Table 7.1 Composition of glass composites along with density, CTE, and characteristic temperatures

Sample name	Composition (1:1)	Density (g/cc)	CTE (RT-softening) × 10^{-6}/K	CTE after 1000 h (RT-softening) × 10^{-6}/K	T_g (DIL in °C)	T_g (DTA in °C)	T_c (DTA in °C)	Viscosity η (dPas)
GG1	ML + CL	3.42	8.31	9.36	712	694	828	8.05
GG2	CL + SL	3.68	8.43	9.92	701	686	843	7.61
GG3	SL + BL	4.01	8.21	10.89	775	741	867	9.05
GG4	BL + ML	3.82	8.50	9.32	727	701	834	8.71
GG5	BL + CL	3.91	8.52	7.05	785	759	848	10.59
GG6	ML + SL	3.54	8.74	10.01	716	693	861	8.12
GG7	ML + CL + SL + BL	3.85	8.19	9.77	774	704	856	8.51

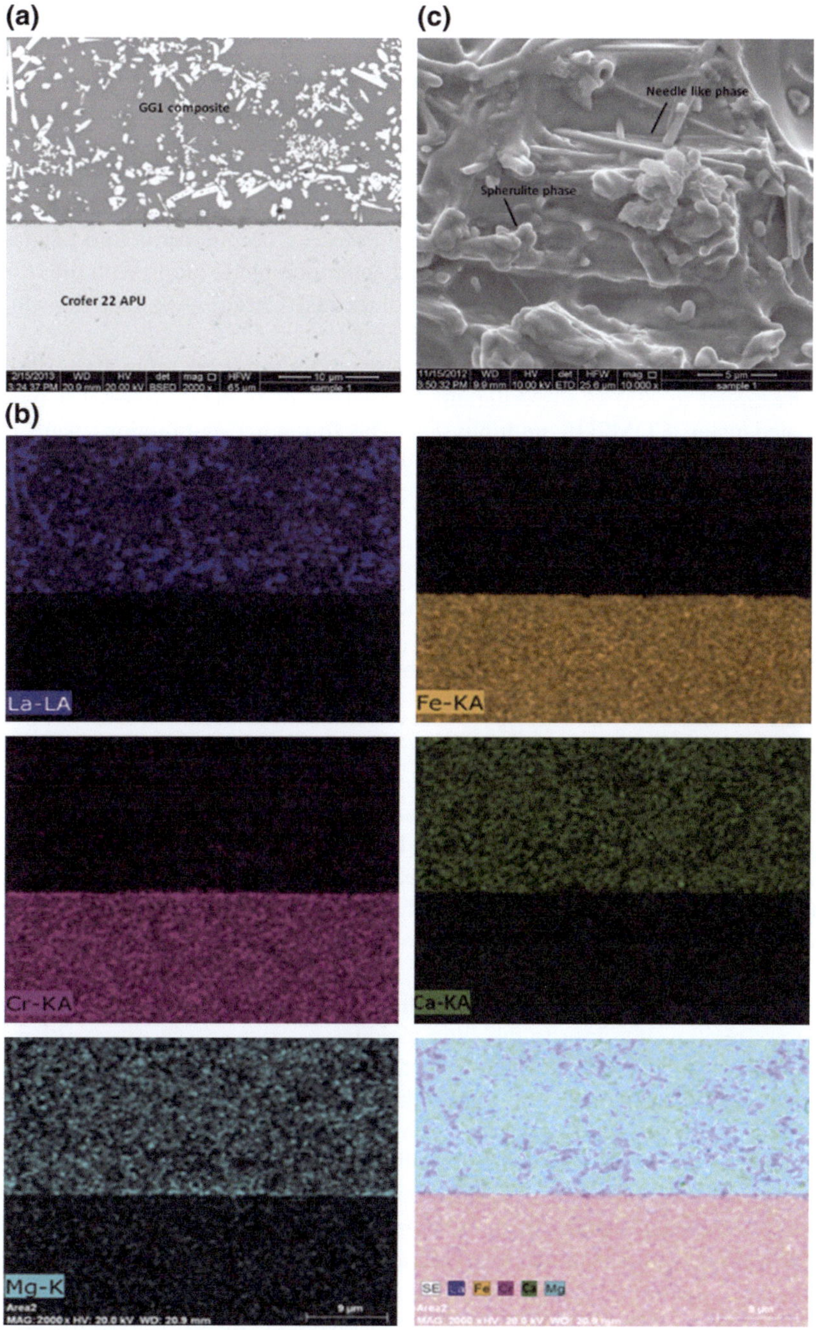

Fig. 7.1 **a** Microstructure, **b** dot mapping and **c** phase formation in GG1/Crofer 22APU couple after heat treatment at 850°C for 1000 h (Kaur and co-workers 2013a)

8.31×10^{-6}/K, which indicates its acceptability as a sealant. Figure 7.1a, b indicates the microstructure and elemental analysis for the GG1/Crofer 22 APU interface after heat treatment at 850°C for 1000 h. Good adhesion could be seen though some closed pits could be seen towards the glass side attributed to the internal residual stress within the amorphous glass. No diffusion of lanthanum and iron could be seen on either side of the interface, though Mg^{2+} and Ca^{2+} diffused out of the glass towards the interconnect side due to their small ionic radii. No reaction products or chromate formation could be seen at the interface. The microstructure of glass in Fig. 7.1c shows the needle shape phase and spherulite phase along with the residual amorphous glassy matrix. XRD has also indicated $CaMgSi_2O_6$ and $Ca_{9.33}Si_6O_{26}$ phase formation along with amorphous matrix.

Furthermore, the thermal stress generated within modeled seals after 10,000 h heat treatment at 800°C for different thicknesses of 0.25, 0.5, and 1.5 mm are simulated using COMSOL multiphysics. The following assumptions were made during the finite element simulation using COMSOL.

- No exposure to external forces.
- The temperature range is RT–800°C.
- Static linear elastic theory.
- The geometry is a two-layered structure exhibiting homogeneous isotropic material properties.
- The simulation has been done to obtain the thermal stress for 10,000 h of operation of SOFC.

Figure 7.2a–c gives the stress distribution for a GG1 sample surface for varying thickness. For $t = 0.25$ mm, the stress varies within 2.64×10^{-5} to 5.58×10^{-3} N/m^2 whereas for $t = 0.5$ mm, the stress limits are 1.87×10^{-6} to 4.55×10^{-3} N/m^2. For $t = 1.5$ mm, the stress is within 2.27×10^{-6}–2.74×10^{-3} N/m^2. Because Crofer 22APU is stiffer, it is stress free. After heat treatment for 1000 h, the CTE of GG1 glass–ceramic is 9.36×10^{-6}/K. The difference in CTE ($\Delta\alpha$) for GG1 and Crofer 22APU is 1.94×10^{-6}/K. Anyhow for $t = 0.25$ mm, the stress is almost 10 times more than that of $t = 0.5$ mm and $t = 1.5$ mm. For $t = 0.25$ mm, the high-stress region is concentrated on the surface of the glass whereas for $t = 0.5$ and 1.5 mm, the stress regions are close to the edge regions.

The effect of TiO_2 on the MgO–CaO–SiO_2–P_2O_5–F glass system is also studied by Park and co-workers. Interestingly, these glasses are also bioactive in nature. The DTA of all glasses at a heating rate of 10°C/min is shown in Fig. 7.3. The thermographs show two well-defined exotherms: the first crystallisation peak lies in the temperature range of 860–930°C corresponding to apatite crystallisation and the second crystallisation peak lies in the temperature range of 950–1050°C attributed to wollastonite crystallisation. Both bulk and surface crystallisation of glasses have taken place. Apatite formation is a 3D bulk process whereas wollastonite is a 2D surface process. When the TiO_2 content is increased up to 4 wt% for Glass 3 (4.42% MgO, 42.91% CaO, 32.64% SiO_2, 15.55% P_2O_5, 0.48% CaF_2, and 4% TiO_2, mol%), the activation energy decreases to a minimum of 320 and 408 kJ/mol indicating TiO_2 as an effective nucleating agent. The tetrahedral [TiO_4] enhances

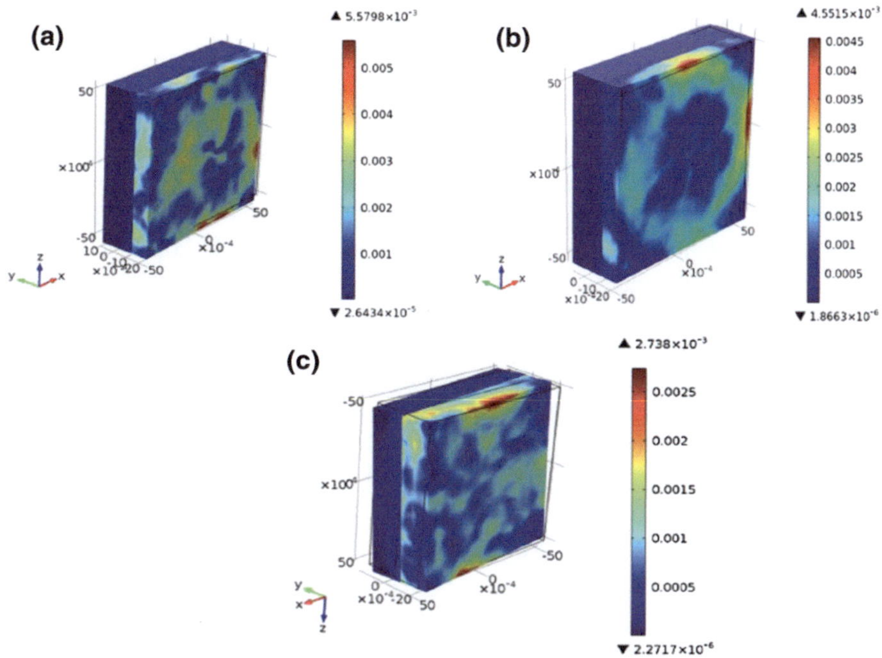

Fig. 7.2 The stress contours for GG1/Crofer 22APU couple for different thickness of **a** $t = 0.25$, **b** $t = 0.5$, and **c** $t = 1.5$ mm (simulated up to 10,000 h, step size 1) (Kaur and co-workers 2013b)

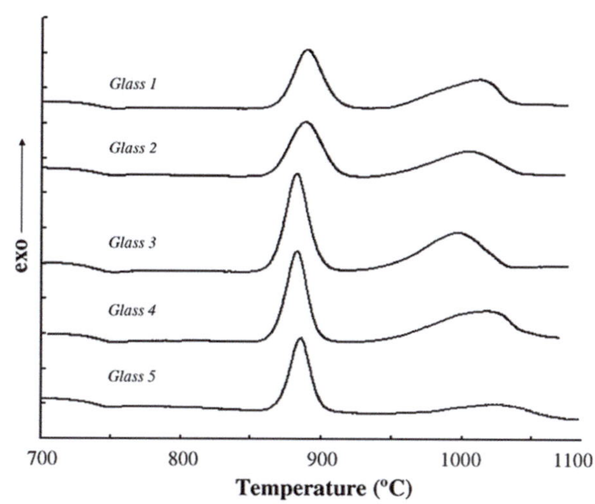

Fig. 7.3 DTA *curves* for the glasses at the heating rate of 10°C/min (Park and Ozturk 2008)

glass formation ability by networking with [SiO_4] groups and hence no NBO are produced. However, octahedral [TiO_6] exist in sixfold coordination and cause loosening of the structure. Therefore, one can expect an initial decrease and then increase in the thermal parameters such as T_g and T_c with increased TiO_2 content.

7.1.2 Calcium–Strontium Glasses/Interconnect Interface

A new glass formulation G36 (26.48% SrO, 26.48% CaO, 1.92% B_2O_3, 1.96% Al_2O_3, 41.16% SiO_2, and 2% TiO_2, mol%) was prepared by Zhang and co-workers and further reacted with Cr_2O_3 powder to stimulate the interfacial reactions between glass and the chromium containing the interconnect. The G#36 glass exhibits a CTE of 12.5×10^{-6}/K. The G#36 is bonded to the 430SS and Fig. 7.4 gives the optical image of the G#36/430SS sample after heat treatment in air at 800°C at different time durations. At the interface, a yellow reaction product could be seen growing in from the edge of the glass sample. This indicates the chromate formation at the interface. The XRD results also affirmed the presence of the $SrCrO_4$ phase along with the $CaSiO_3$ and Sr_2SiO_4 phases. It should be noted that the glass G#36 used for interaction is made of glass G36 mixed with 10 wt% Cr_2O_3. When SrO containing glasses reacts with both Cr_2O_3 powder and with 430SS substrate under air atmosphere for 800°C, then the $SrCrO_4$ formation takes place as shown below:

$$SrO_{(g)} + \frac{1}{2}Cr_2O_3 + \frac{3}{4}O_2 \rightarrow SrCrO_{4(s)} \tag{7.1}$$

For the G#36 samples, the formation of Cr^{6+} in the samples was analysed using UV-visible spectroscopy as shown in Fig. 7.5. Two distinct peaks at 270 and 370 nm represent the Cr^{6+} ions. The percentage of Cr^{6+} increases when heat

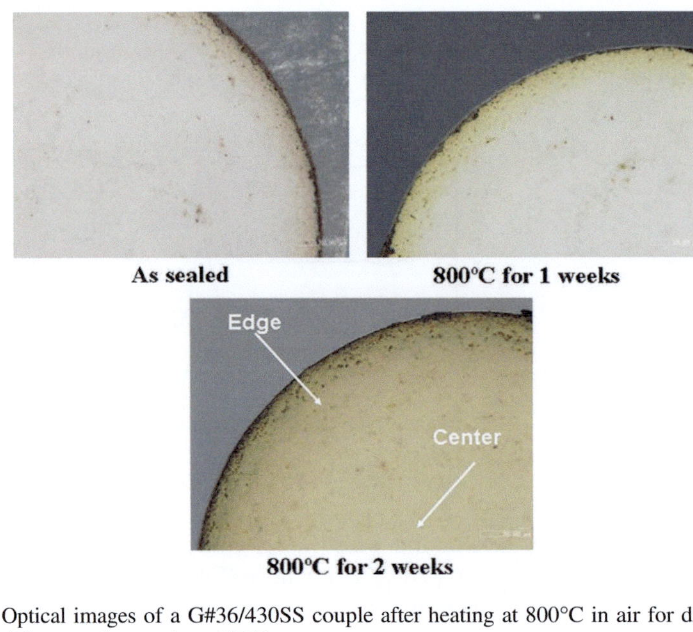

Fig. 7.4 Optical images of a G#36/430SS couple after heating at 800°C in air for different time durations (Zhang and coworkers 2012)

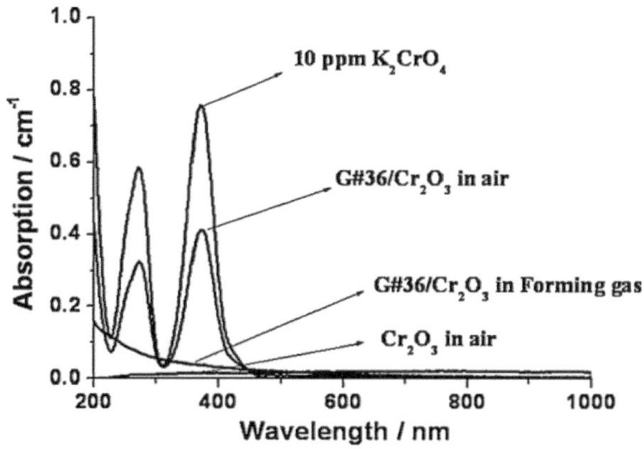

Fig. 7.5 The absorption spectra of a solution of K_2CrO_4 (10 ppm), and of mixtures of G#36 and 10 wt% Cr_2O_3 held in air and forming gas (10% H_2/90% N_2) along with Cr_2O_3 solution held in air at 950°C for 24 h (Zhang and coworkers 2012)

treatment temperature increases from 800–950°C. After the prolonged heat treatment at 950°C, the maximum percentage of 70% for Cr^{6+} was observed. Another reaction mechanism for $SrCrO_4$ formation involves CrO and Cr vapour interaction as

$$Cr_2O_{3(s)} + \frac{3}{2}O_{2(g)} \rightarrow 2CrO_{3(g)} \tag{7.2}$$

$$CrO_{3(g)} + SrO_{(g)} \rightarrow SrCrO_{4(s)} \tag{7.3}$$

Equation (7.3) is the thermodynamically favoured reaction, which occurs in the presence of an oxygen atmosphere. The seals upon exposure to oxygen have yielded maximum corrosive action at seal edges instead of the seal centre. Anyhow, the Cr dissolves into the seal centre to form a <10 μm Cr-rich region.

The effect of transition metal oxide NiO on the thermal and sealing properties of the Sr–Ca–B system was extensively investigated by Chou and co-workers. Two techniques were employed to synthesise the sample: introducing NiO in the glass melt and making glass/Nio composite. The NiO content did not affect the glass properties much but in the composite approach, the CTE followed an increasing trend with increased NiO concentration. In contrast to this, the sintered glass powder exhibited a decrease in CTE from the parent glass. For glass YSO-1 (42.5% SrO, 10% CaO, 6% Y_2O_3, 7.5% B_2O_3 and 34% SiO_2, mol%) and YSO-11 (42.5% SrO, 5% CaO, 5% NiO, 6% Y_2O_3, 7.5% B_2O_3 and 34% SiO_2, mol%). Ca_2SiO_4 and Ca_3SiO_5 phases precipitated out. For YSO-12 (42.5% SrO, 10% NiO, 6% Y_2O_3, 7.5% B_2O_3 and 34% SiO_2, mol%) and YSO-13 (37.5% SrO, 15% NiO, 6% Y_2O_3, 7.5% B_2O_3 and 34% SiO_2, mol%) glasses, no calcium phases are seen because these glasses are calcium-free glasses. For the YSO-12 and YSO-13 sample, the main phases which precipitated out are $SrSiO_3$, NiO, $SrNiO_3$, and YBO_3. No Ni_2SiO_4 could be found in the glasses, which is the only stable phase for the NiO–SiO_2 system. The thermodynamics for different phases is given as

$$2CaO + SiO_2 \rightarrow Ca_2SiO_4 \quad \Delta G = -137\,\text{kJ/mol} \tag{7.4}$$

$$3CaO + SiO_2 \rightarrow Ca_3SiO_5 \quad \Delta G = -170\,\text{kJ/mol} \tag{7.5}$$

$$SrO + SiO_2 \rightarrow SrSiO_3 \quad \Delta G = -132\,\text{kJ/mol} \tag{7.6}$$

$$SrO + SiO_2 \rightarrow SrSiO_3 \quad \Delta G = -132\,\text{kJ/mol} \tag{7.7}$$

It is evident from the above reactions that $SrSiO_3$, Ca_2SiO_4, and Ca_3SiO_5 are thermodynamically more favourable than the nickel olivine (Ni_2SiO_4) phase at a particular temperature. Based on the above studies YSO-1 was chosen for making composite glass; i.e., 5, 10, and 15 mol% NiO are mixed directly with the glass yielding composite YSO-14, YSO-15, and YSO-16, respectively. YSO-15 exhibited the closest CTE of 12.01×10^{-6}/K to that of the SOFC components; hence, this composite was chosen for the sealing and leakage studies. The YSO-15 sample sealed at 1050°C/2 h gave leakage of 2.9×10^{-4} sccm cm^{-1} for the sealed samples whereas after 10 thermal cycles, the leakage rate is 1×10^{-4} sccm cm^{-1}. The leakage rates of the order of 10 sccm cm^{-1} are considered as high and usually occur due to the brittle fracture. Hence the sample treated at 1050°C/2 h is the hermetic seal due to the leak rate equivalent to or smaller than the background leakage. Figure 7.6a, b gives the microstructure of Crofer 22APU/YSO-15 as a sealed interface at 950 and 1050°C for 2 h. At 950°C, the chromium layer is thinner as compared to the sample treated at 1050°C. There appeared a depletion zone of about 10–15 μm of NiO particles near the glass–metal interface; NiO may have reacted with Crofer 22 APU elements via the mechanisms:

$$NiO + Mn_2O_2 \rightarrow NiMn_2O_4 \quad \Delta G = 180\,\text{kJ/mol}(800\,°C) \tag{7.8}$$

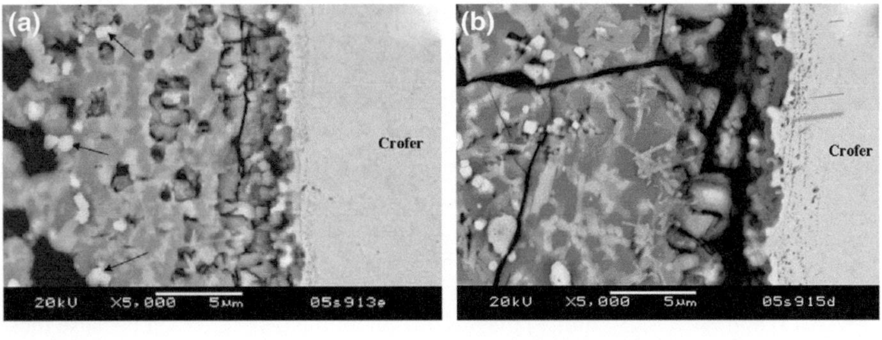

Fig. 7.6 Interfacial microstructure of composite glass (YSO-15) sealed at **a** 950°C/2 h and **b** 1050°C/2 h with Crofer 22APU metal (Chou and coworkers 2007a, b)

$$NiO + Cr_2O_3 \rightarrow NiCr_2O_4 \quad \Delta G = -8\,\mathrm{kJ/mol}(800^\circ\mathrm{C}) \tag{7.9}$$

$$NiO + Fe_2O_3 \rightarrow NiFe_2O_4 \quad \Delta G = -24\,\mathrm{kJ/mol}(800^\circ\mathrm{C}) \tag{7.10}$$

Reactions (7.9) and (7.10) are thermodynamically favourable; hence, the NiO particles may have migrated towards the interface to react with Cr and Fe. The fully reduced samples also showed no $SrCrO_4$ formation.

Another glass (YSO77) was developed at the Pacific Northwest National Laboratory (PNNL) by modifying the parent YSO-1 sealing glass. YSO-77 (Sr–Ca–Y–B–Si with 6% Ba) was used to seal the cell to the window frame and hybrid mica with glass interlayers. To evaluate the long-term stability of materials, a three-cell short stack was assembled and tested for 6000 h at 800°C (Fig. 7.7a). The cell materials were AISI 441 hybrid mica seal, (Mn, Co) spinel coatings, and sealing glass whereas the cell was an anode supported with LSM-based cathode. Figure 7.7a indicates four interconnect plates, hybrid mica seals, and three window frame plates in a three-cell short stack assembly. Inconel600 super alloy was used to make heat exchanges (one for incoming air and other for the fuel; the gases are pumped in cross-flow pattern). A compressive load of 250 kPa was applied onto the stack. To avoid any substantial leakage, both outgoing air and fuel lines were connected to the water bubbler, which provides indication of gas flow. The stack was disassembled for post-mortem analysis (after the test). The cell electrochemical performance indicated a degradation rate of 1.4 kh^{-1} for top cell and no glass seal failure could be seen. The bottom and middle cells showed rapid degradation, which was further supported by the formation of $SrCrO_4$ in the seal apparent from post-mortem analysis. The seal failure could have been due to inadequate dimensional control and incomplete alumina coating. The YSO77 glass exhibited good thermal stability and matching CTE as that of its parent glass YSO-1. Figure 7.7b is the microstructure of YSO77-aluminised AISI441 interface indicating a 5–10 μm thick interface layer. No large precipitates of detrimental phases could be seen on the interface region or on either side of interface, though corrosion of alumina layer was evident. A barium celsian phase ($BaAl_2Si_2O_8$) could be expected; anyhow, barium is only 6% for this glass which precludes the possibility of occurrence of a stable Ba celsian phase. We can expect a Sr celsian phase ($Sr \cdot Al_2 \cdot Si_2 \cdot O_8$) as per the following reaction kinetics.

$$SrO + 2SiO_2 + Al_2O_3 \rightarrow SrAl_2Si_2O_8 \quad \Delta G = -19\,\mathrm{kJ/mol} \tag{7.11}$$

$$SrSiO_3 + SiO_2 + Al_2O_3 \rightarrow SrAl_2Si_2O_8 \quad \Delta G = -59\,\mathrm{kJ/mol} \tag{7.12}$$

In spite of the thermodynamically favourable Sr-celsian phase formation, it is a kinetically complex process. Hence, no such phase formation is evident for the YSO77/AISI 441 microstructure. The steady-state volatility data indicated lifetime weight loss (40,000 h) of the sealing glass <0.1 wt% indicating the potential viability of YSO77 as a sealant material.

The lanthanum-based Sr–Ca–Si–B–La–O composite GG2 yielded good compatibility and no chromate formation with Crofer 22APU after 1000 h heat

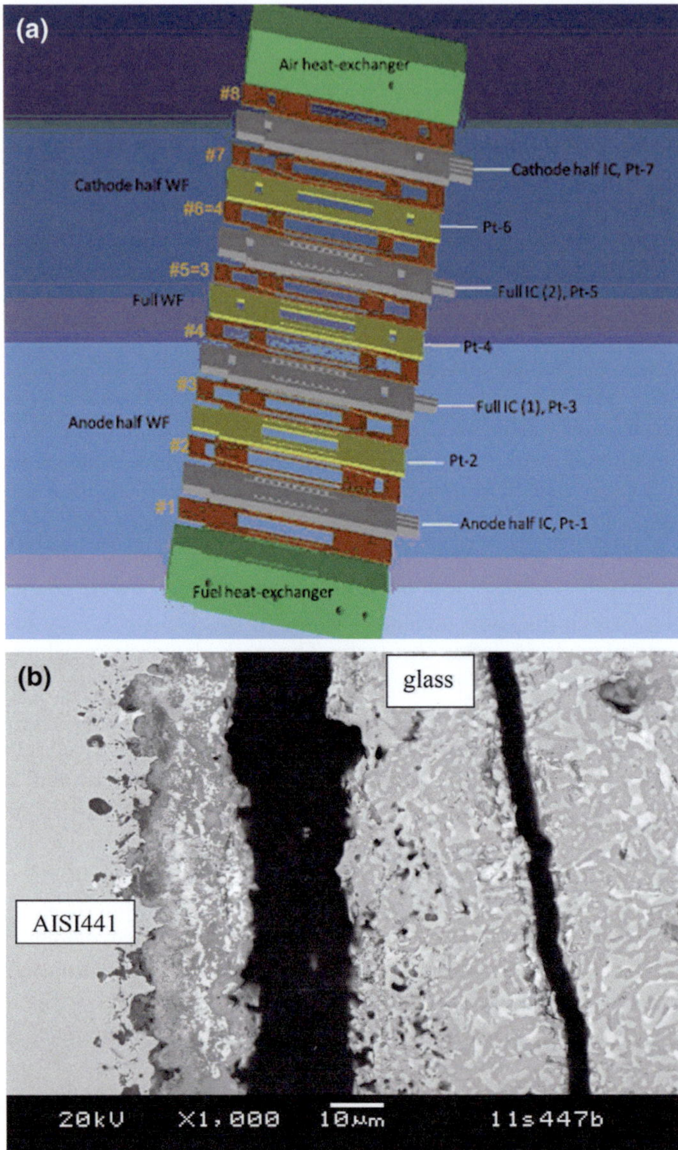

Fig. 7.7 **a** A three-cell short stack, *yellow* window frame plates (WF); *grey* interconnect (IC) plates; *brown* mica seals; *green* heat exchanger; *Pt* platinum wire (Chou and coworkers 2014a). **b** Microstructure of the YSO77 glass/aluminised AISI441 interface near air side (Chou and coworkers 2007b)

treatment at 850°C (Fig. 7.8a, b). GG2 composites are made by mixing SL (30% SrO, 40% SiO_2, 20% B_2O_3 20% La_2O_3, mol%) and CL (30% Cao, 40% SiO_2, 20% B_2O_3, 10% La_2O_3, mol%) in 1:1 weight ratio and then ball milling for 5 h. The CTE

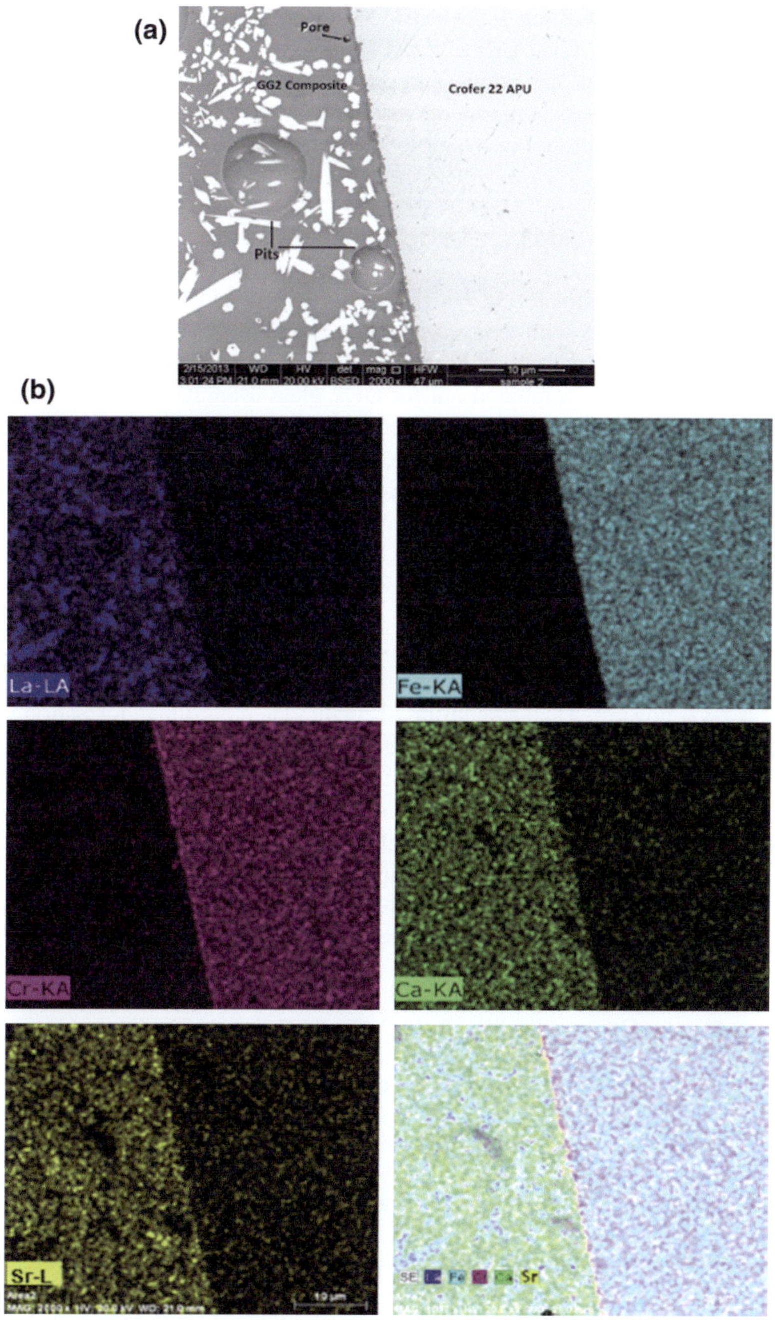

Fig. 7.8 **a** Microstructure. **b** Elemental analysis of GG2/Crofer22APU after heat treatment at 850°C for 1000 h (Kaur and co-workers 2013a)

and particle size of GG2 are 9.36×10^{-6}/K (after heat treatment at 800°C) and 3.22 μm, respectively. GG2/Crofer 22APU is devoid of any cracks, though the interior of GG2 has some closed pits and pores due to the thermal stresses generated within the glass. The glass components may react with the alloy component leading to gaseous phase formation. For example, the hydrogen gas phase may form via the following reaction.

$$xM_{(alloy)} + yH_2O \rightarrow M_xO_y + yH_2 \tag{7.13}$$

where $\Delta G = -222$ kJ/mol for M = Cr, and $\Delta G = -55$ kJ/mol for M = Fe

Hence, the reaction with chromium is energetically more favourable. Some phase formation can take place during prolonged heat treatment of glass, e.g., $Ca_{0.8}Sr_{0.2}\ SiO_4$ phase precipitated out for GG2 glass after 1000 h heat treatment at 850°C. These phases may force the pores to migrate into sealing glass. The GG2 has also high amorphous content, hence, it may contribute towards internal brittleness leading to pit formation. The dot-mapping (elemental analysis, Fig. 7.8b) indicates no Fe diffusion, though Cr^{2+} ions have diffused out towards the glass side without any segregation at the interface. La^{3+}, Sr^{2+}, and Ca^{2+} have also diffused out of the glass matrix and in spite of diffusion, no detrimental reaction or phase could be seen at the interface.

As previously described in this section, Zhang and co-workers studied the mechanism of chromate formation at the glass/steel interface by reacting glass with 20 wt% Cr_2O_3 powder. The same group also elucidated chromate formation studies by incorporating different additives in the glass including ZnO, MnO_2, and La_2O_3. Every glass was melted and then mixed with 10 wt% Cr_2O_3 and heated at 900°C for different time durations. Using UV–visible spectroscopy, it was found that the percentage of Cr^{6+} in the mixture of glass + 10 wt% Cr_2O_3 decreased in the presence of additives implying that additives suppress chromate formation. The order of chromate phase reduction is on the order $MnO_2 > ZnO > La_2O_3$. The thermal–chemical model proposed for reduced chromate formation is shown in Fig. 7.9 and the possible reactions are given as follows (at 900°C).

$$SrO + 0.5Cr_2O_3 + 0.75O_2 \rightarrow SrCrO_4 \quad \Delta G = -150\,kJ/mol \tag{7.14}$$

$$ZnO + Cr_2O_3 \rightarrow ZnCr_2O_4 \quad \Delta G = -57\,kJ/mol \tag{7.15}$$

$$MnO_2 + Cr_2O_3 \rightarrow MnCr_2O_4 + 0.5O_2 \quad \Delta G = -2\,kJ/mol \tag{7.16}$$

$$0.5La_2O_3 + 0.5Cr_2O_3 \rightarrow LaCrO_3 \quad \Delta G = -79\,kJ/mol \tag{7.17}$$

The reduction of $SrCrO_4$ can be accompanied by other phase formations such as $LaCrO_3$, $ZnCr_2O_4$, and $MnCr_2O_4$. Another lanthanum-containing system SCSA (Sr–Ca–Si–B–Al–La) studied by Hao and co-workers yielded CTE between 9.77

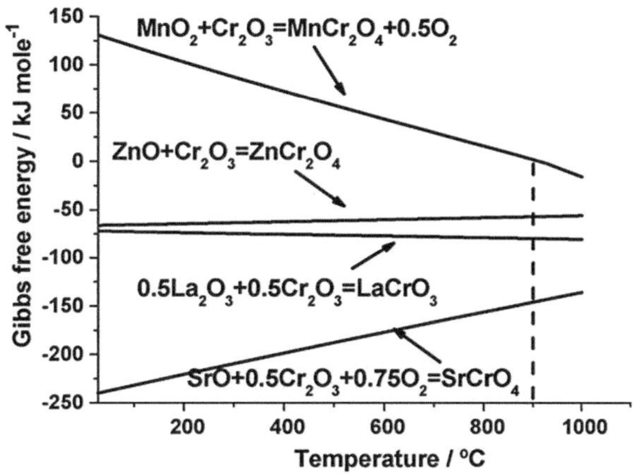

Fig. 7.9 Possible reactions for chromate formation and their Gibbs free energy as a function of temperature in air (Zhang and coworkers 2010)

and 10.87×10^{-6}/K. The welting characteristics of four glasses, i.e., G13 (16% SrO, 16% CaO, 43% SiO_2, 15% B_2O_3, 10% (Al_2O_3 + La_2O_3), mol%), G14 (24% SrO, 16% CaO, 33% SiO_2, 17% B_2O_3, 10% (Al_2O_3 + La_2O_3) mol%), G15 (28% SrO, 16% CaO, 31% SiO_2, 15% B_2O_3, 10% (Al_2O_3 + La_2O_3), mol%) and G16 (24% SrO, 16% CaO, 25% SiO_2, 25% B_2O_3, 10% (Al_2O_3 + La_2O_3), mol%) are given in Fig. 7.10. The G16 sample exhibits minimum sintering and melting temperature of 730 and 840°C, respectively. Sample G16 exhibits a sealing range of 760–810°C whereas for the G15 sample, it is 920–960°C. Sample G13 and G14 have an optimum sealing range of 880–930 and 900–950°C, respectively. The Si/B ratio decreases from G13 to G16, hence the decrease in characteristic temperatures could

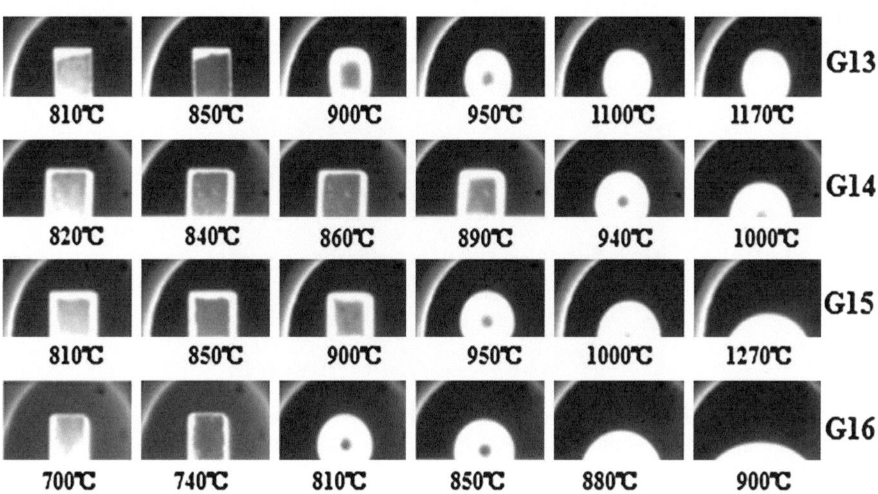

Fig. 7.10 Shape changes of the glasses with increasing temperature for G13–G16 at a heating rate of 20°C/min (Hao and coworkers 2012)

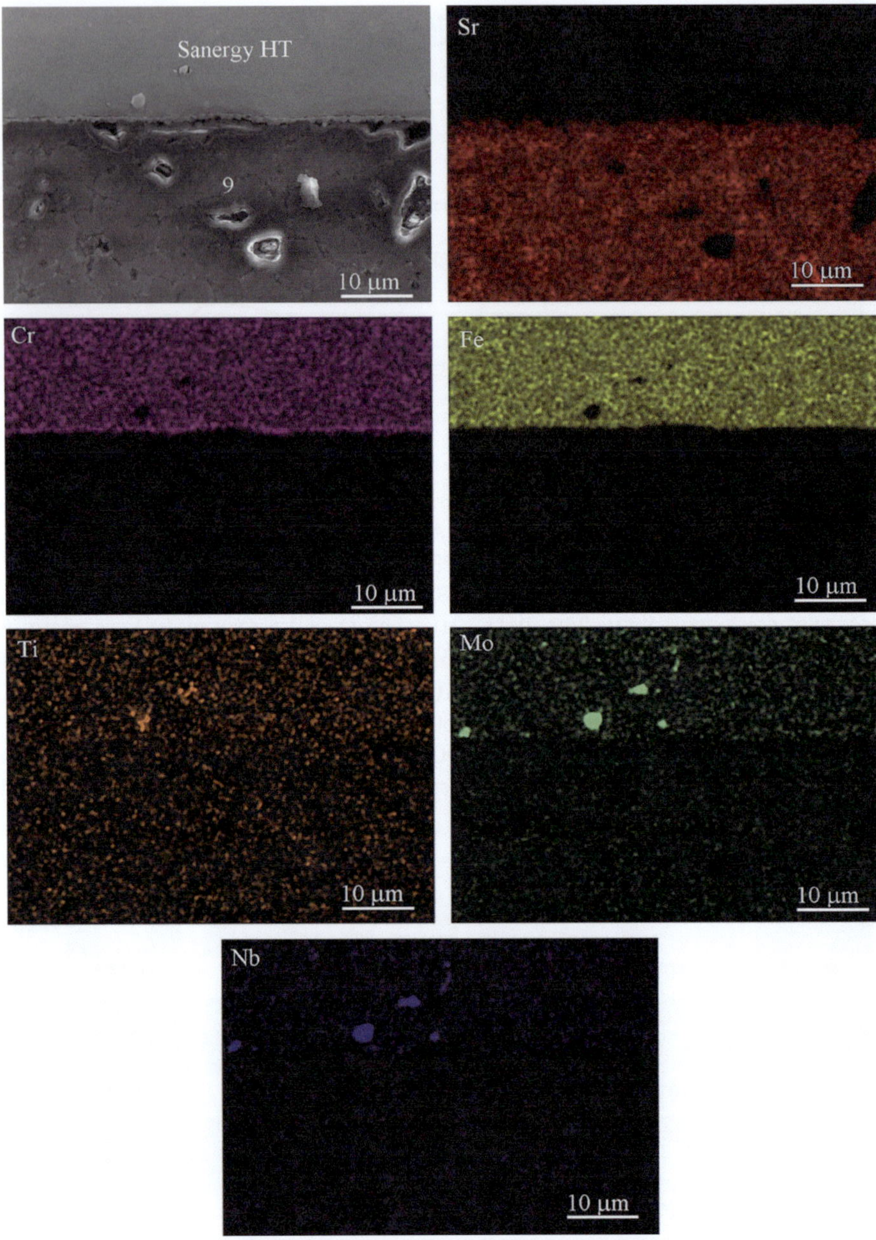

Fig. 7.11 SEM image and element mapping at the interface between GC 9/Sanergy HT after heat treatment at 850°C for 1 h followed by 800°C for 200 h (Goel and coworkers 2010a)

be observed. In addition to this, no visible flowing temperatures could be seen for the samples. G16 exhibit very good wetting abilities; 800°C is almost the turning point of the contact angle. As could be seen in Sect. 6.3.2.3, the glass G11 from the SSA system yielded good wetting properties, which indicate that strontium enhances wetting of the interface.

Goel and co-workers have worked extensively on the diopside ($CaMgSi_2O_6$) system. They formulated an SrO-based aluminosilicate composition in the primary field of diopside via the substitution scheme of 0.2 Ca^{2+} + 0.1 $Mg^{2+} \leftrightarrow$ 0.3 Sr^{2+} and 0.1 (Ca^{2+} + Si^{4+}) $\leftrightarrow$ 0.1 (La^{3+} + Al^{3+}) resulting in $Sr_{0.3}Ca_{0.7}Mg_{0.9}Al_{0.1}La_{0.1}Si_{1.9}O_6$ theoretical composition. A Bi_2O_3 addition was also made to the glass compositions due to its low melting temperature. The CTE of all compositions lies between 8.28 and 8.57×10^{-6}/K and increasing the Bi_2O_3 content increases the density of glasses. Bi_2O_3-free glass composition 9 exhibited diposide and augite ($Ca(Mg_{0.70}Al_{0.30})(Si_{0.70}Al_{0.30})O_6$) phase formation. For the Bi_2O_3 containing compositions (9–Bi1, 9–Bi3, 9–Bi5), with 1, 3, and 5% Bi_2O_3, respectively, Sr containing diopside ($Ca_{0.8}Sr_{0.2}MgSi_2O_6$) along with a minor augite phase could be observed. The SEM microstructure of Glass 9(22.47% MgO, 17.48% CaO, 7.49% SrO, 47.44% SiO_2, 25% Al_2O_3, 1.25% La_2O_3, 1.79% B_2O_3, 0.83% NiO, mol%) with Sanergy HT after heat treatment at 850°C for 1 h followed by heat treatment at 800°C for 200 h in air is shown in Fig. 7.11. No detrimental reaction products could be observed at the interface. Some voids could be seen in the glass region attributed to removal of some amorphous or crystalline materials from the glass samples during mechanical grinding/polishing of samples. Elemental analysis indicates a very thin chromia layer at the interface, which can be due to the spinel structure. During Glass 9 interaction with Crofer 22 APU, a titania rich zone could also be seen at the interface. Usually, titanium oxide layer formation is thermodynamically more favourable than the spinel layer formation.

7.1.3 Strontium–Barium Glasses/Interconnect Interface

The GG3 composite (Table 7.1) studied by Kaur and co-workers revealed the chromia-rich layer at the interface with Crofer 22APU after 1000 h heat treatment at 850°C (Fig. 7.12a, b). The GG3 composite exhibited CTE of 10.89×10^{-6}/K, which is in the acceptable range. Anyhow, the $BaCrO_4$ formation as revealed from elemental mapping (Fig. 7.12b), is responsible for the discontinuous interface because $BaCrO_4$ is a high CTE phase. The following reaction kinetics may have been followed for $BaCrO_4$ formation:

$$2Cr_2O_3 + 4BaO + 3O_2 \rightarrow 4BaCrO_4 \quad (7.18)$$

The formation of $SrCrO_4$ is completely precluded in the present composite indicating competition between Ba^{2+} and Sr^{2+} to form the chromate. When BaO is replaced with SrO, it is expected that it will mitigate the $BaCrO_4$ formation.

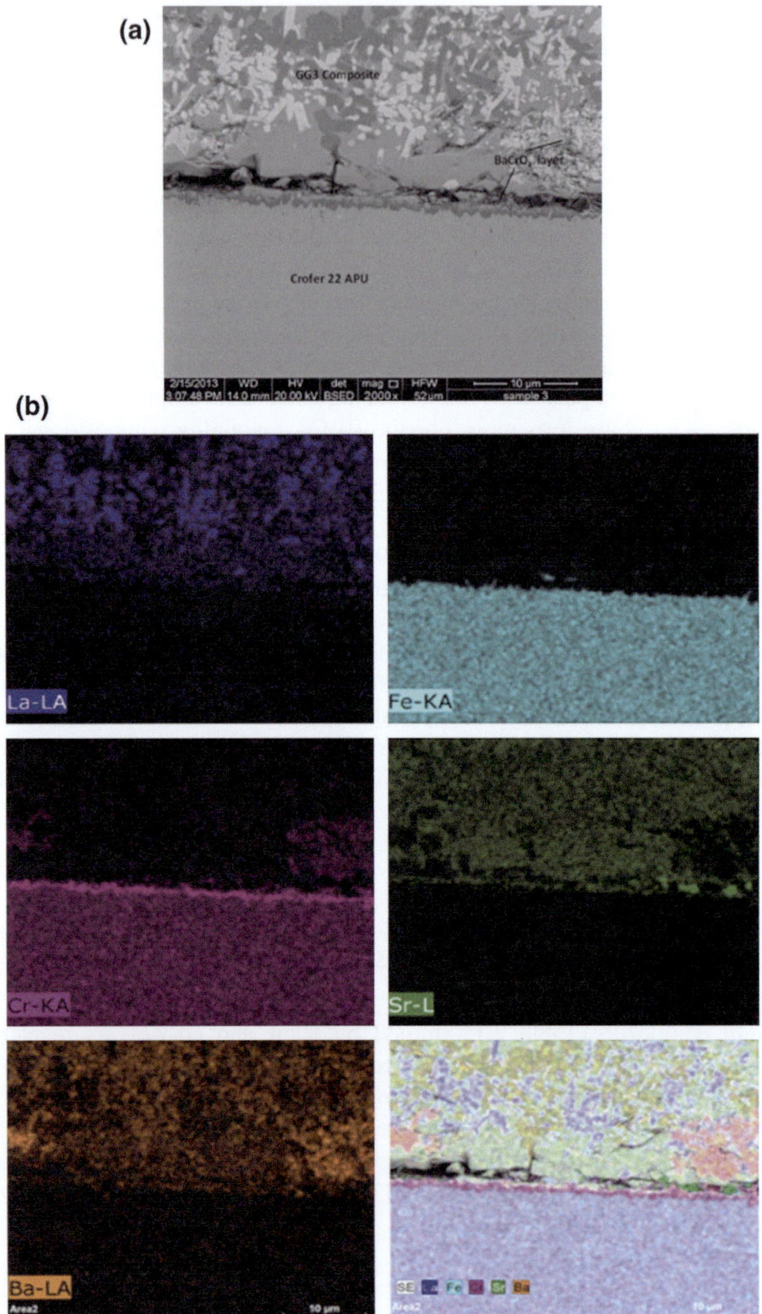

Fig. 7.12 **a** Microstructure. **b** Elemental analysis of GG3/Crofer22APU interface after 1000 h heat treatment at 850°C (Kaur and co-workers 2013a)

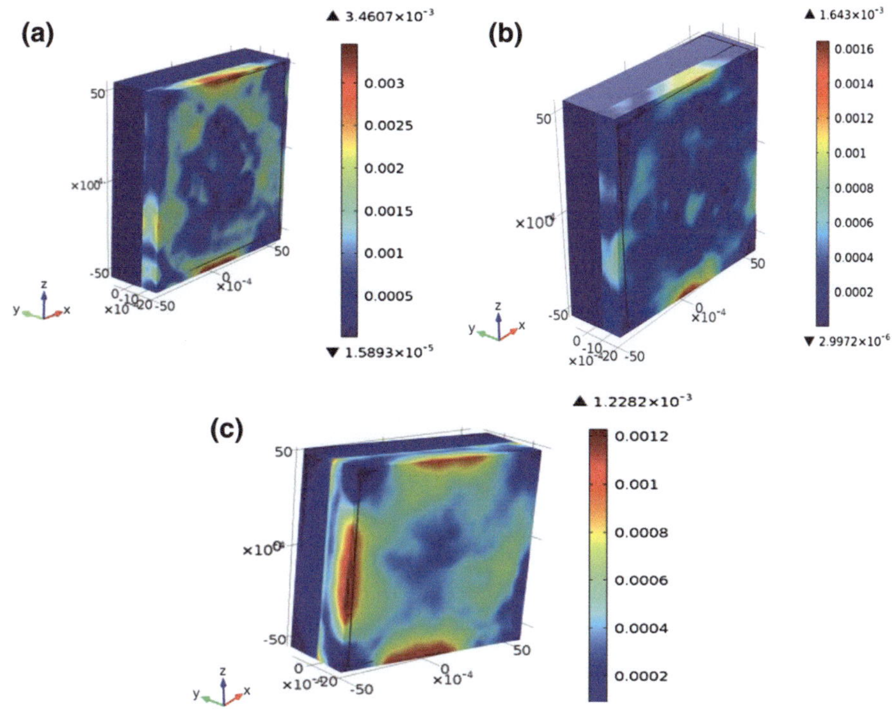

Fig. 7.13 The stress contours and their magnitude for the GG3/Crofer 22APU diffusion couple for thickness **a** $t = 0.25$, **b** $t = 0.5$, and **c** $t = 1.5$ mm (simulated up to 10,000 h, step size 1) (Kaur and co-workers 2013b)

Lanthanum has a diffused out-of-glass matrix though no iron has diffused out of the Crofer 22APU. Figure 7.13a–c indicate that the stress range is small for GG3 composite/Crofer 22APU couple due to its small mismatch with Crofer 22APU alloy. The major stress regions are concentrated on the edges when the thickness increases from 0.25 to 0.15 mm, though the magnitude of stress decreases. The XRD of the GG3 composite indicates $BaSrSi3O_8$ and $BaSiO_3$ phase formation. Usually, barium silicate phases are high CTE phases and hence help in reducing the thermal stresses. With the increase in diffusion couple thickness from $t = 0.25$ to 1.5 mm, the stress decreases from 3.46×10^{-3} to 1.22×10^{-3} N/m^2. For the GG3 couple, the extent of corrosion due to chromate formation is not extensive.

Sharma and co-workers have also studied the Ba–Sr–Si–B–La–Al–P glass system. P_2O_5 usually exhibits insufficient durability but the addition of silica can increase the durability. P_2O_5 is useful as a nucleating agent and also scavenges the modifier cations and increases network polymerisation. All the glasses such as BASP0 (30% SiO_2, 30% BaO, 20% SrO, 10% B_2O_3, 5% La_2O_3, 5% Al_2O_3, mol%), BASP1 (28.8% SiO_2, 31.7% BaO, 19.2% SrO, 9.6%B_2O_3, 4.8% La_2O_3, 4.8% Al_2O_3, 0.9% P_2O_5, mol%) and BASP2 (27.7% SiO_2, 33.3% BaO, 18.5% SrO, 9.2% B_2O_3, 4.6% La_2O_3, 4.6% Al_2O_3, 1.8% P_2O_5, mol%) exhibited CTE in the range of

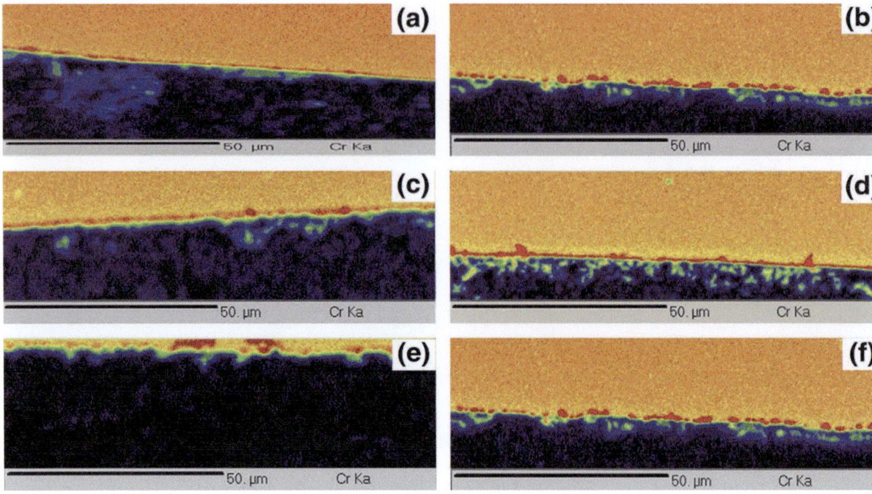

Fig. 7.14 Cr mapping at the BASP glass Crofer-22 APU interface after heat treatment at 800°C: **a** BASP0 (for 100 h); **b** BASP0 (for 500 h); **c** BASP1 (for 100 h); **d** BASP1 (for 500 h); **e** BASP2 (for 100 h); **f** BASP2 (for 500 h) (Sharma and coworkers 2012)

(11.4–13) × 10^{-6}/K whereas the T_g decreased from 650 to 584°C, respectively. The XRD of all glasses exhibit Sr_2SiO_4, Ba_2SiO_4, and $BaAl_2Si_2O_8$ phase formation. All the glasses show good bonding behaviour with the Crofer 22APU without any cracks or bubbles at the interface. BASP0 exhibits coarse microstructure whereas BASP1 and BASP2 have homogeneous microstructure Fig. 7.14a–f indicates the chromium mapping for samples after heat treatment at 800°C for 100 and 500 h, respectively. The chromium diffusion is limited, not extensive, and did not delaminate the interface. From XRD analysis, it is clear that P_2O_5 did not participate in network formation and remained isolated orthophosphate species. With the addition of P_2O_5, decrease in flow and sealing temperatures has been observed indicating the improvement of sealing properties.

7.1.4 Barium Magnesium Glasses/Interconnect Interface

Lara and co-workers have extensively studied the barium–magnesium system of glasses. The glass-forming region of the MgO–BaO–SiO_2 system is shown in Fig. 7.15a (the solid region inside the phase diagram). For determining the glass-forming region, different experimental conditions have been taken into consideration. Almost 1 g of materials were melted into platinum crucibles (or 80% Pt + 20% Rh) between temperatures of 1400 and 1700°C. Upon cooling of the melt to room temperature, the results were obtained by naked eye. For the larger batches of almost 775 g with SiO_2 < 45 mol%, the melt partially crystallised under the same

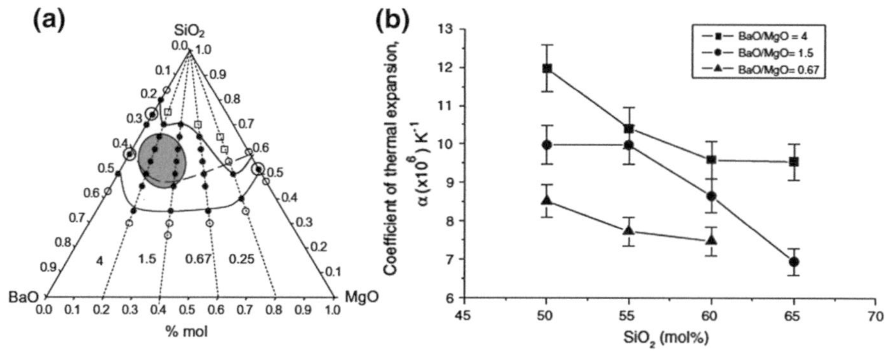

Fig. 7.15 **a** Phase diagram for MgO–BaO–SiO_2 system (dots indicate glassy system). **b** Variation of CTE with BaO/MgO ratio (Lara and coworkers 2004a)

experimental conditions. For the MgO–BaO–SiO_2 system, the BaO/MgO ratio affects CTE of the glasses as shown in Fig. 7.15b. For instance, for the BaO/MgO = 0.67, the CTE is minimum and lies between 8.5 × 10^{-6} and 7.5 × 10^{-6}/K, whereas for BaO/MgO = 4, the CTE varies between 12 × 10^{-6} and 9.5 × 10^{-6}/K. At the same time, the CTE also decreases with increasing silica content. The T_g and T_d also increase with increasing SiO_2 content. BaO is a network modifier which usually increases the CTE of silicate glasses by 10% and decreases viscosity. The network of smaller cations such as Mg^{2+} also increases in the presence of other ions in its vicinity. For the glasses with BaO/MgO = 4, the tendency of crystallisation versus sintering increases upon decreasing SiO_2 content. The composition with BaO/MgO = 1.5 (55% SiO_2, 27% BaO, 18% MgO, mol%) is designated as Mg1.5–55 and is regarded to be the best suitable candidate for SOFC sealing based on DTA and HSM analysis. Mg 1.5–55 was further modified using B_2O_3, PbO, and ZnO to enhance its properties as suitable sealing materials with acceptable compatibility with the SOFC components.

The PbO-based composition Mg 1.5–40–5B–10 Pb (with 40% SiO_2, 5% B_2O_3, 10% PbO, mol%) exhibited highest density and CTE of 5.44 g/cm^3 and 12.3 × 10^{-6}/K, respectively along with minimum dilatometric softening temperature T_s of 512°C. With increasing B_2O_3 content, the density increased except for the glass with 15% B_2O_3 whose density is smaller than the glass with 10% B_2O_3 probably due to change in coordination number. Mg 1.5–55, Mg 1.5–50–5B (with 5% B_2O_3, mol%), and Mg 1.5–50–5B–8Zn (with 5% B_2O_3 and 8% ZnO, mol%) exhibit $BaCrO_4$ formation, when in contact with the Crofer 22APU, hence no bonding could be observed for these glasses. Anyhow, the Mg 1.5–40–15B–8Zn (with 15% B_2O_3, 8% Zn, mol%) presented perfect adherence to the Crofer 22 APU (Fig. 7.16) indicating a homogeneous interphase after 100 h heat treatment at 850°C. Some diffusion of Fe and Cr was observed onto the glass side up to 50 μm of the joining interphase (Fig. 7.16).

A unidimensional crystallisation growth mechanism was observed for the glasses. The XRD of Mg 1.5–40–5B–8Zn glasses revealed the formation of barium zinc silicate ($BaZn_2Si_2O_7$) and barium silicate ($BaSiO_3$, $BaSi_2O_5$) phases. A very thin

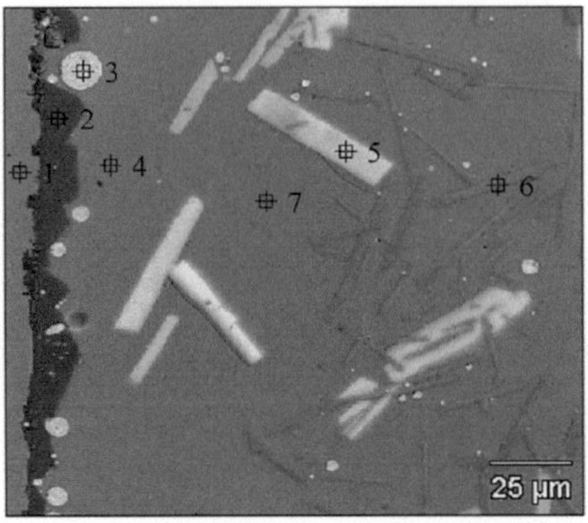

% Weight

	SiO_2	BaO	MgO	ZnO	Fe	Cr	Mn
1	0.26				84.16	12.35	2.77
2	35.58	36.28	0.43	4.64	3.02	19.54	0.52
3		0.69		97.77	1.33	0.20	
4	24.79	63.46	1.49	6.72	1.80	1.74	
5	34.21	65.79					
6	25.17	50.33	4.35	20.14			
7	25.28	65.28	1.46	6.74	0.29	0.79	0.15

Fig. 7.16 Microstructure and EDS analysis for the Mg 1.5–4–15B–8Zn glass/Crofer 22APU interface after 100 h heat treatment at 850°C (Pascual and coworkers 2007)

chromia and manganese-rich interlayer could be observed for the Mg 1.5–40–15B glass with Crofer 22APU after 100 h heat treatment at 850°C. The crystals correspond to the formation of barium magnesium silicate, which is evident from the XRD analysis. The glass Mg 1.5–40–15B results in a partially crystallised material with relatively large glassy phase volume unlike glass Mg 1.5–55 which gets fully crystallised after 100 h heat treatment at 850°C. No $BaCrO_4$ could be seen at the interface, which is attributed to the slower reaction kinetics of this glass. Though barium silicates have precipitated out in all the glasses, magnesium silicate and barium magnesium silicate are also present in glasses without zinc. Conclusively, the B_2O_3 addition has resulted in the decrease of viscosity and delay in crystallisation leading to better wetting and bonding of glasses on the steel interface inhibiting $BaCrO_4$ formation. The oxygen permeation fluxes through Mg 1.5–55 glass–ceramics under air/H_2–H_2O gradients are also measured. The oxygen permeation flux density (*j*) through a membrane for steady-state conditions is given by:

$$j = \frac{V}{RT_V S} \frac{p(H_2)_{in} - p(H_2)_{out}}{2} \tag{7.19}$$

where V is the gas flow rate, S is the membrane surface area, T_V is flow metre temperature $p(H_2)_{in,}$ and $p(H_2)_{out}$ are the inlet and outlet hydrogen partial pressures of the measuring cell. The hydrogen pressure is given by

$$p(H_2)\frac{Cp_{\text{total}}}{1+Kp(O_2)^{1/2}} \tag{7.20}$$

where C is the volume fraction of hydrogen-containing species such as H_2/H_2O (determined from initial compositions), p_{total} is the total pressure, K is the equilibrium constant of H_2 oxidation reaction, and $p(O_2)$ is the oxygen partial pressure (determined by the oxygen sensor). Figure 7.17a reveals the permeation fluxes through the Mg 1.5–5.5 sample with thickness 0.65 mm under air/H_2–H_2O–N_2 gradient. With time, the permeability undergoes gradual increase; i.e., after 200 h, the permeation fluxes get 15–40% higher than the respective initial values. The SEM analysis of Mg 1.5–5.5 (Fig. 7.17b) indicates that the pore formation may occur on the surface exposed to the H_2 due to the silica volatilisation from the surface causing substrate reconstruction of the surface layer. When the porosity of surface layers increases, the oxygen fluxes also increase. The pore formation can also cause degradation of conductivity under reducing conditions (predominant ionic conduction). Hence, to optimise conductivity and permeability and avoid microstructural changes, the operating temperature must be reduced and kept under check sample. Mg1.5–55 is found to be an insulator with electrical resistances higher than 10^4 Ωm. For an ideal glass ion conductor, the log σ (conductivity) varies linearly with temperature (for $T < T_g$). During phase transition, the viscosity and transport properties are affected which influence $-(d \log \sigma/d(1/T))$ versus temperature. During the glass transition range of the sample, usually a broad hump in the conductivity is observed. In glasses, the vacancy motion mechanism may also be around by the SiO network, which can be explained as follows.

The [SiO_4] tetrahedral does not exclusively form six-member rings but also forms considerable six-numbered rings arranged in an alternate manner with the

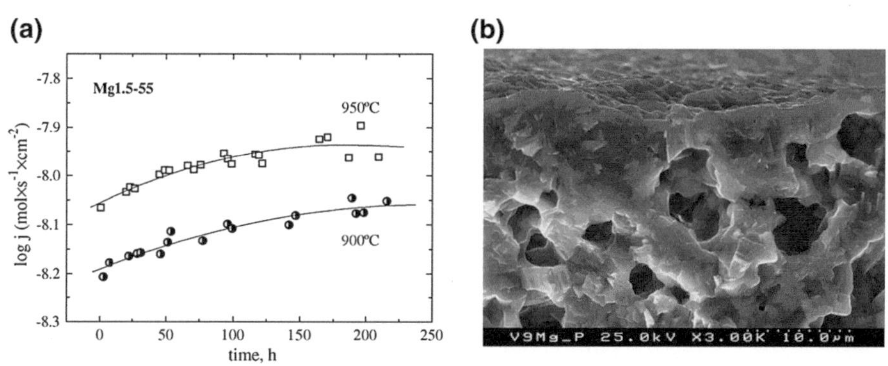

Fig. 7.17 **a** Oxygen permeation fluxes through Mg 1.5–55 glass–ceramic membrane (thickness of 1.0 mm) under oxygen partial pressure of 21 kPa. **b** Microstructure of exposed glass (Pascual and coworkers 2006)

barium layers. Tetrahedral chains and isolated rings may have formed silica-like networks promoting vacancy mechanism in the glass.

The BMAS (barium magnesium aluminosilicate) glasses are synthesised using the sol–gel route by Puig and co-workers. The BaO/MgO ratio was kept at 3.5, 2.5, and 0.3. All the prepared BMAS glass–ceramics were so porous that CTE could not be performed on them, which also affects their integrity for SOFC applications. The XRD of BMAS1 (10% BaO, 35% MgO, 5% Al_2O_3, and 50% SiO_2, mol%), BMAS2 (25% BaO, 10% MgO, 5% Al_2O_3, and 60% SiO_2, mol%), and BMAS3 (35% BaO, 10% MgO, 5% Al_2O_3, and 50% SiO_2, mol%) are shown in Fig. 7.18. The majority of glasses were amorphous and different crystalline phases precipitated out of the glass. MgO plays a key role as a BaO substitute to promote the crystallisation of barium silicates.

Apart from the aluminosilicate system, the lanthanum aluminate borosilicate system was also investigated as La_2O_3 is well known for viscosity modification and improvement of the mechanical properties of glasses. Table 7.2 lists the composition of glasses (mol%) along with their T_g and CTE.

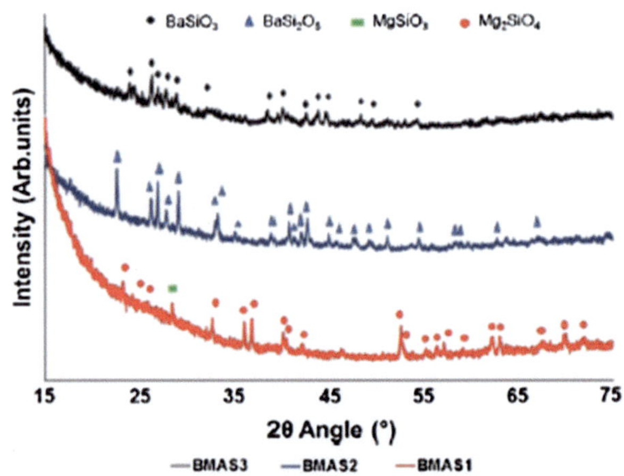

Fig. 7.18 XRD of the BMAS glasses (Puig and coworkers 2011)

Table 7.2 Glass compositions for the lanthanum-based aluminoborosilicate composition

Sample	MgO	BaO	Al_2O_3	La_2O_3	SiO_2/B_2O_3	T_g (°C)	CTE ($\times 10^{-6}$/K)
BMA1	17	5	10	15	Borate	620	9.66
BMA2	12	10	10	15	Borate	610	9.95
BM1	22	25	–	–	Borate	570	10.06
BM11	22	25	–	–	0.3	581	10.64
BM12	22	25	–	–	3	623	11.08
BM13	22	25	–	–	12.3	605	10.26
BM2	22	30	–	–	Borate	582	11.20

When the BaO content was increased up to 30 mol%, the T_g and CTE values decreased due to the conversion of trigonal BO_3^- into a rigid tetrahedral BO_4^- structure within the glass network. For BMA1 and BMA2 glasses, the major phase precipitation is due to $LaBO_3$ and BaB_2O_4 phases along with $MgAl_2O_4$, $AlBO_3$, $BaAl_2O_7$, and MgO as the minor phases. Usually for borate glasses, the BaB_2O_4 precipitates out, forming large needle-shaped phases having preferred surface orientations. However, for the glasses with higher BaO content (up to 30% BM2) barium magnesium borate ($Ba_2Mg(B_3O_6)_2$), barium borate ($Ba_4B_2O_7$), and magnesium borate ($Mg_2B_2O_5$) precipitate out. Figure 7.19a indicates the SEM microstructure of a cast BM2 glass indicating needle-shaped dendrite growth preferred towards the surface. The growth of such crystals within the amorphous glass matrix may promote extensive cracking through the stress generation. Upon incorporation of SiO_2 in lie of B_2O_3, barium silicate phases such as $Ba_3B_6Si_2O_{16}$, $Ba_2Si_2O_4$, $Ba_3Si_3O_9$, $BaMg_2Si_2O_7$, and $Mg_2Si_2O_4$ precipitated out in BM1, BM11, BM12, and BM13 glasses after heat treatment of 100 h at 800°C. All the glasses have CTE in the (9–11) $\times$ 10^{-6}/K range which is acceptable for sealing purposes. The joint region for BM12/Crofer 22APU is shown in Fig. 7.19b. Porosity could be observed in the glass side. BM12 is considered to be a suitable sealing material though the BM12 glass is more or less defect-free. Some reaction zone could be seen at the interface, which is rich in barium, magnesium, chromium, oxygen, and a little iron. The chemical interaction between glass and Crofer 22APU is limited up to 2–3 μm region only. BM12 glass also exhibits the highest nano-hardness and Young's modulus. In addition to this, the lowest leak rate of 4.3 $\times$ 10^{-6} Pa m^2/s has been obtained for the BM12 glass.

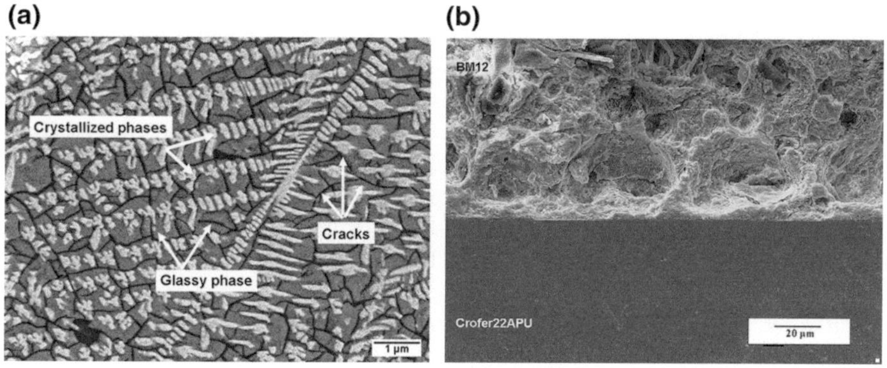

Fig. 7.19 SEM micrograph of **a** As prepared BM2 glass and **b** cross-section of the BM12/Crofer22APU-sealed specimen (Ghosh and coworkers 2010)

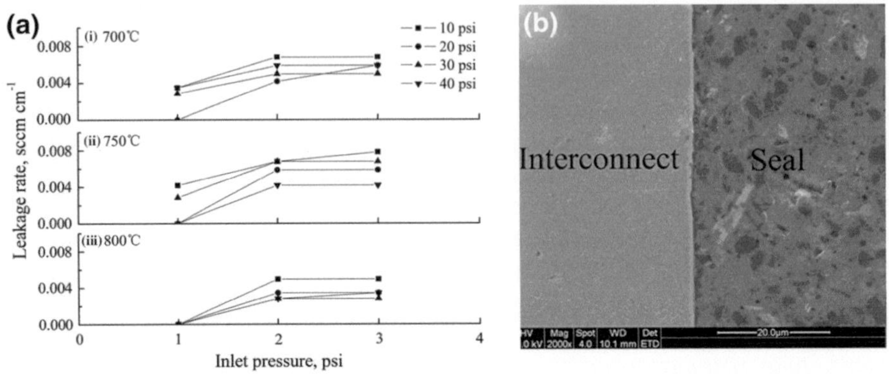

Fig. 7.20 **a** Leakage rates of HGS-1 seals with different inlet pressures and temperature and **b** SEM micrographs of cross-sections of seal/interconnect after 750°C for 5 h (Ye and coworkers 2012)

Many barium–magnesium composites also yielded promising results. The composition HGS1 (8.3% B_2O_3, 12.1% BaO, 21.8% MgO, 12.1% ZnO, and 45.7% SiO_2, mol%) was used to make a composite with 30% Al_2O_3. The leak test for HGS1 glass at an inlet pressure of 1, 2, and 3 psi under compressive stress from 10–40 psi at 700, 750, and 800°C, respectively, are shown in Fig. 7.20a. All the leakage rates are lower than 0.01 sccm/cm and the leakage rate increased when the inlet pressure increased from 1 to 2 psi. Under a compressive stress of 10 psi, the leakage rates were higher than those observed at 20 psi. This indicates that a compressive stress of 20 psi on the seal has already provided enough stress such that a good interface could be obtained between the seal and the interconnect. The seal may get broken under high compressive stress, thus leading to new leakage paths. Figure 7.20b reveals the interface between HGS1/interconnect after dwelling at 750°C for 5 h under compressive stress of 20 psi. No cracks or delamination could be seen on the interface. The thermal stress of glasses could be dispensed due to the flexible nature of glasses.

The lanthanum-containing GG4 composite (Table 7.1) exhibited a CTE of 9.32×10^{-6}/K with T_g of 692°C. Two phases precipitated out, i.e., $Ba_2MgSi_2O_7$ and $La_{9.31}(Si_{1.04}O_4)_6O_2$ of the glass matrix. The GG4/Crofer 22APU glass interface revealed no pores/cracks, thus revealing adhesion (Fig. 7.21a). The do-mapping of GG4/Crofer 22APU (Fig. 7.21b) reveals the migration of Mg^{2+} ions and chromium ions towards the interface. Magnesium has higher diffusion than the Ba^{2+} ions due to the radii of Mg^{2+}; hence, Mg^{2+} follows rapid reaction kinetics. Formation of $MgCrO_4$ at the interface may be possible, though no barium chromate could be seen at the interface.

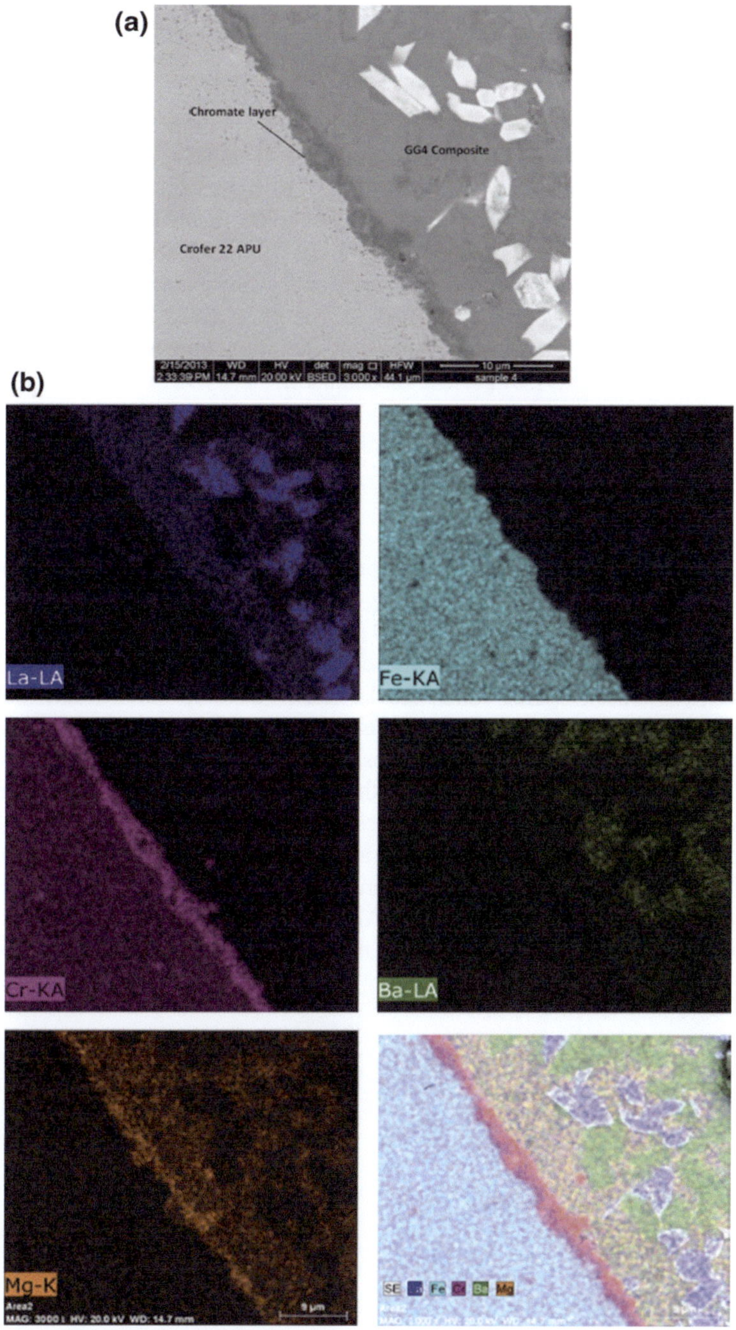

Fig. 7.21 **a** Microstructure. **b** Elemental analysis of GG4 glass/Crofer 22APU after heat treatment at 850°C for 1000 h (Kaur and coworkers 2013a)

7.1.5 Barium–Calcium Glasses/Interconnect Interface

The barium–calcium glasses have been widely investigated, especially G18, which was designed at PNNL, Richland, WA and possess a composition of 56.1% BaO, 7.19% CaO, 5.39% Al_2O_3, 6.66% B_2O_3, and 21.45% SiO_2 (in mol%). Yang and co-workers worked extensively on G18 glasses to figure out their interaction with the glass components. The BCAS system G18 glass reacted extensively with the Ni-base superalloy (Nicrofer 6025) and ferrite stainless steel 446 as shown in Fig. 7.22a, b. The composition of interconnect alloy and atmosphere widely determines the extent of corrosion action for the glass seals. Near the edge region, oxygen is more easily accessible; hence, upon interacting with chromia-forming alloys, they form $BaCrO_4$ as per the mechanism:

$$2Cr_2O_{3(s)} + 4BaO_{(s)} + 3O_2 \rightarrow 4BaCrO_{4(s)} \tag{7.21}$$

$$CrO_2(OH)_{2(g)} + BaO_{(s)} \rightarrow BaCrO_{4(s)} + H_2O_{(g)} \tag{7.22}$$

Equation (7.21) refers to the high-temperature reaction occurring in the presence of oxygen near the edge area yielding whitish barium chromate precipitates. Eq. (7.22) yields the yellowish $BaCrO_4$ on the surface involving the reaction between the glass component BaO and chromium oxyhydride. In addition to the chromate formation, pores and cracks also develop along the interface. Thermodynamically, the gas phase such as hydrogen can be formed when the metal component of the alloy reacts with the residual water of sealing glass (Eq. 7.22). It could be seen that the reaction of Cr with H_2O is thermodynamically favoured as compared to that with Fe or Ni. The pores are formed in the glass when hydrogen gets trapped inside the glass matrix. Yang and co-workers found that upon cooling of the diffusion couple, the steel matrix separated from the glass–ceramic, which could be due to larger CTE mismatch among AISI 446 and $BaCrO_4$. Chromia-rich solutions could be obtained in the interior of the glass region, where the air access was

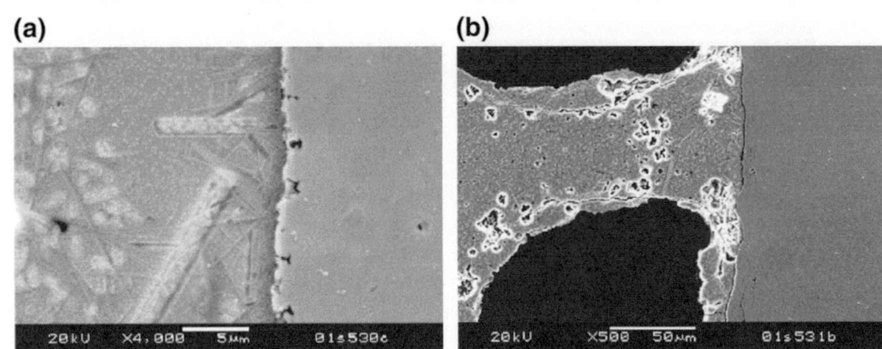

Fig. 7.22 Interfacial regions between G18 sealing glass and **a** Nicrofer 6025, **b** FeCr alloy after 750°C for 4 h in air (Yang and coworkers 2003)

blocked. The scale growth of Cr_2O_3 and (Cr, Mn)$_3O_4$ spinel layers on the alloy surface depends on their composition: i.e., 446 has scale growth rate of 1.332×10^{-13} g^2/cm^4 s whereas Nicrofer 6025 exhibits a rate of 0.022×10^{-13} g^2/cm^4 s at 800°C in ambient air. This infers that presence of NiO may have hindered $BaCrO_4$ formation. In contrast to this no new compounds or $BaCrO_4$ could be seen at the glass. G18/FeCr alloy interface. At 800°C in air, the FeCralloy developed Al_2O_3 scale, though Fe_2O_3 and Cr_2O_3 were also detectable up to some extent. The G18 exhibited well-bonded behaviour with the FeCr alloy. Large pore formation could be due to the following reactions (alloy elements with dissolved water in glass):

$$\mathrm{M} + \frac{3}{2}\mathrm{H_2O} \rightarrow \frac{1}{2}\mathrm{M_2O_3} + \frac{3}{2}\mathrm{H_2} \tag{7.23}$$

$$\begin{bmatrix} \mathrm{M} = \mathrm{Y} & \Delta G = -591\,\mathrm{kJ/mol} \\ \quad = \mathrm{Al} & \Delta G = -436\,\mathrm{kJ/mol} \end{bmatrix}$$

ΔG for reaction between Cr and H_2O is −222 kJ/mol suggesting that reaction (7.23) is highly favourable with Al and Y. Figure 7.22b also indicates large pore formation when G18 interacts with FeCr alloy. Evaporation of alkali metals may also lead to the pore formation

$$\mathrm{M} + \frac{3}{2}\mathrm{Na_2O} \rightarrow \frac{1}{2}\mathrm{M_2O_3} + 3\mathrm{Na}\uparrow \tag{7.24}$$

$$\begin{bmatrix} \mathrm{M} = \mathrm{Fe}, & \Delta G = 176\,\mathrm{kJ/mol} \\ \mathrm{M} = \mathrm{Cr}, & \Delta G = 18\,\mathrm{kJ/mol} \\ \mathrm{M} = \mathrm{Y}, & \Delta G = -350\,\mathrm{kJ/mol} \\ \mathrm{M} = \mathrm{Al}, & \Delta G = -323\,\mathrm{kJ/mol} \end{bmatrix}$$

$$\mathrm{M} + \mathrm{Na_2O} \rightarrow \mathrm{MO} + 2\mathrm{Na}\uparrow \tag{7.25}$$

$$\begin{bmatrix} \mathrm{M} = \mathrm{Fe}, & \Delta G = 104\,\mathrm{kJ/mol} \\ \mathrm{M} = \mathrm{Ni}, & \Delta G = 155\,\mathrm{kJ/mol} \end{bmatrix}$$

The reaction with M = Y and Al (the major elements of FeCr alloy) are highly favourable whereas those with Fe, Cr, and Ni are unfavourable. Hence, it could be concluded that the extent of glass reaction with alloys depends upon the nature of the alloys used and exposure conditions as well.

Another BCS system (41.4% BaO, 8.6% CaO, 34.7% SiO_4, mol%) was also studied for the symmetric shear testing at an operation temperature of 800°C. The viscoelastic model used to obtain the shear modulus considered total displacement to be the sum of viscous and elastic contribution as

$$\delta l_{\mathrm{tot}} = \delta l_{\mathrm{vis}} + \delta l_{\mathrm{elast}} \tag{7.26}$$

where

$$\delta l_{\text{vis}} = \frac{h\tau}{\eta}\delta t \tag{7.27}$$

$$\delta l_{\text{elast}} = \frac{h\tau}{G} \tag{7.28}$$

G is the shear modulus, τ is the shear stress, and η is the viscosity. For the glass–ceramics, the viscous flow is hindered due to interlocking of crystalline phases and the viscous displacement gets modified by τ_i contribution.

$$\Rightarrow \quad \delta l_{\text{vis}} = \frac{h(\tau - \tau_i)}{\eta}\delta \tag{7.29}$$

To determine shear modulus G, the following equation is used

$$G = \frac{h}{v_{\text{total}}}\dot{\tau} \tag{7.30}$$

where

$$v_{\text{total}} = \frac{\delta l_{\text{tot}}}{\delta t} = \text{constant} \tag{7.31}$$

As-joined sample exhibited critical shear stress of 0.3 MPa whereas the annealed samples exhibited 4 MPa, which enhances the mechanical stability of the glass. With increasing crystallisation, the viscous shear becomes more difficult. Glass A and B were evaluated for their suitability with different alloy combinations (Table 7.3) under SOFC stack conditions. Glass sealant A (49.1% BaO, 5.7% CaO, 27.9% SiO_2, 3.3Al_2O_3, wt%) and B (41.4% BaO, 0.6% CaO, 34.7% SiO_2, 2.1Al_2O_3, wt%) were applied by dispenser on the circumference of the sheet. Both glasses contained transition metal additives as balance. The specific resistance of sandwich glass samples is measured in H_2–air atmosphere under zero applied voltage as shown in Fig. 7.23. It could be seen that the specific resistance reduced significantly for T-1 from 10 kΩ m to <1 Ω m, after 50 h of exposure whereas with that for T-2, the specific resistance was always >10 kΩ m during the entire exposure time.

Table 7.3 Alloys used and their composition

Alloy	Cr	Fe	Mn	Ti	Si	Al	Ni	Remaining
T-1	22.6	Bal	0.4	0.06	0.1	0.1	0.2	La-0.1
T-2	23.3	Bal	0.4	0.05	–	–	–	La-0.1
T-3	22.1	Bal	0.5	–	0.4	0.2	0.3	La-0.1
T-4	17.3	Bal	0.3	0.01	0.9	1.0	0.2	–
T-5	22.1	Bal	0.4	0.05	0.1	–	–	La-0.1
T-6	22.0	Bal	0.4	0.05	–	0.1	–	La-0.1
T-7	22.7	Bal	0.4	0.04	–	–	–	La-0.05

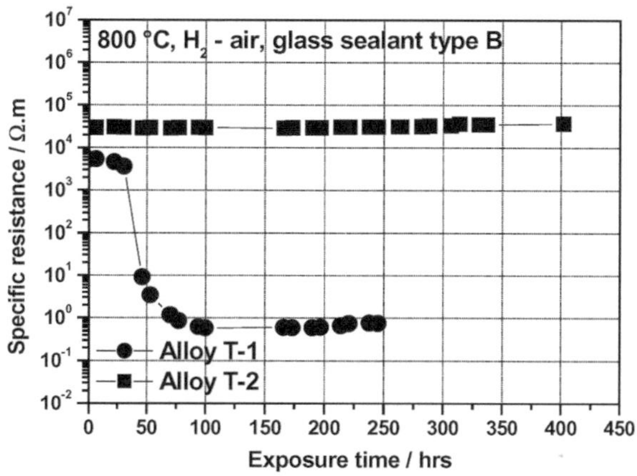

Fig. 7.23 Specific resistance of Alloy T-1 and T-2 in a dual atmosphere at 800°C (no externally applied voltage) (Haanappel and coworkers 2005)

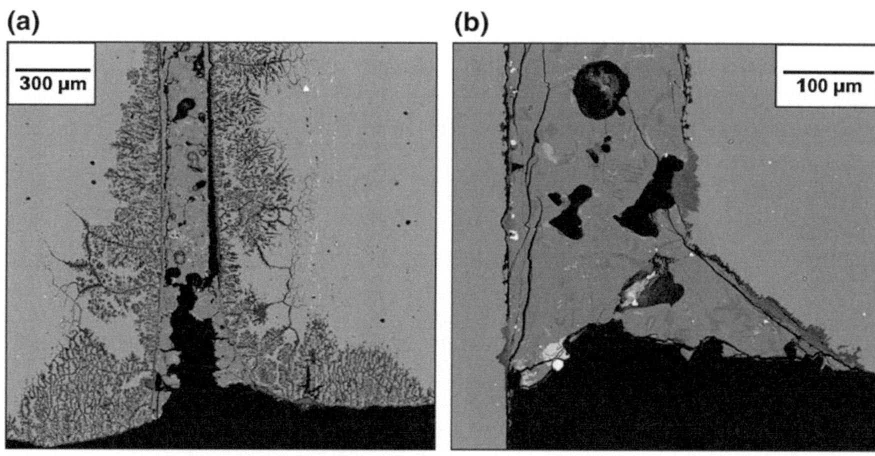

Fig. 7.24 Microstructure of the cross-section of alloy T-1 **a** and alloy T-2 **b** with glass sealant type B after 400 h of exposure in hydrogen at 800°C (Haanappel and coworkers 2005)

The cross-section of glass B with alloy T-1 and T-2 (Fig. 7.24a, b) revealed severe internal oxidation of the alloy for T-1 whereas satisfactory adhesion could be obtained for glass A after 100 h exposure to oxidising conditions at 800°C. The electrical and chemical behaviour of samples showed significant difference under reducing and oxidising conditions. During the conductivity measurement of all the samples, no degradation could be observed after 400 h exposure indicating no highly conductive oxide formation. The cross-sectional analysis revealed formation of a yellowish reaction product along the edges. Ferritic steels T-1, T-3, and T-5 exhibited decrease of specific resistance indicating the presence of electric shunt between metallic sheets. T-2 and T-6 exhibit high specific resistance even after 400 h exposure at 800°C under dual atmospheres. For alloy T-1 with glass B, internal oxidation could be seen. Reduced distance between the metallic sheets could be observed, i.e., initially T-1 sheets were 200 μm apart only. After long hour operation, the sheets may come very close leading to the short-circulating of the stack.

Four glass compositions named G18, G21, G23, and G24 were synthesised for their application as a sealing material. G18 glass exhibited superior properties among all the glasses with CTE of 11.8×10^{-6}/K and T_g of 630°C. In the case of G18 glass, a good balance between glass softening and the rate of crystallisation could be obtained. Under a modest load, G18 wets the substrate surface and fills the gap. Though the primary phase in G18 is barium silicate, a small decline in the CTE of G18 was observed due to the transformation of the hexacelsian phase into the thermodynamically stable celsian phase. G-18 yielded a contact angle of >90° at 750°C but it decreased below 90°C when the temperature increased to 1000°C. Hence, for effective sealing at 750°C, the joints should be compressively loaded. The devitrification of glass decreases its flexural strength though the bulk strength of G18 glass in aged condition is acceptable for transient and steady-state conditions. Another variant of glass A and B are used to investigate the performance of two-plane SOFC stack of F-design (FZJ) (Fig. 7.25).

The various stacks tested with the steel and glass–ceramic type are listed in Table 7.4. It was found that severe degradation occurred for stack B and C at 800°C with an electrical load of 0.3 A/cm^2 after 1000 hours of operation. Stack A revealed severe performance degradation after 100–200 h of operation at 800°C with an electrical load of 0.3 A/cm^2. Strong internal oxidation could be observed at the three phase boundary of steel, glass, and fuel gas. Rapid internal chromia formation

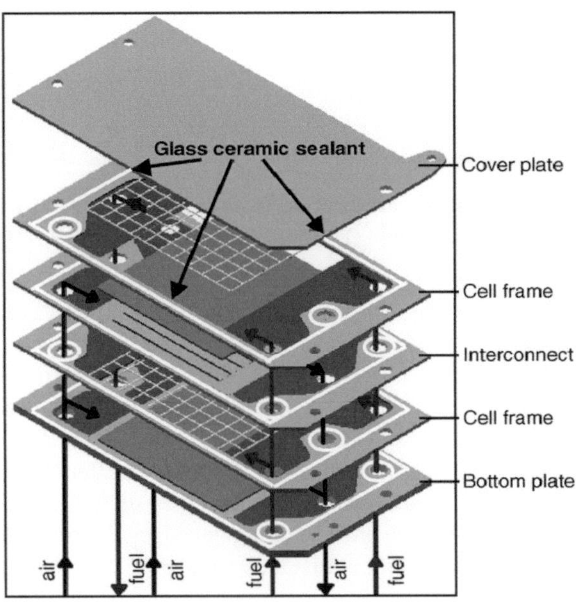

Fig. 7.25 Schematic view of a two-plane SOFC stack of F-design (FZJ) (Batfalsky and coworkers 2006)

Table 7.4 Different steel and glass–ceramic for stacks

Stack type	Steel type	Glass–ceramic (Al_2O_3, PbO, B_2O_3, N_2O_3, and ZnO additives in all of them)
A	Crofer 22APU	34.7 SiO_2–8.6 CaO–41.4 BaO
B	JS-3	27.9 SiO_2−5.7 CaO–49.1 BaO
C	JS-3	34.7 SiO_2–8.6 CaO–41.4 BaO
D	Crofer 22APU + YSZ	34.7 SiO_2–8.6 CaO–41.4 BaO

occurred, which followed intergranular grain boundaries along with dendritic growth towards the centre of the grains. Ferrite steel also showed volume increase along with the bulging manifested as the surface uplifts of several hundred microns. Nodules rich in iron and oxygen dominated the internal side of ferritic steel. The open-circuit voltage measured at the end of stack lifetime is obtained in no correspondence to the theoretical values in spite of progressive degradation in the stack. When PbO is used as an additive in glasses, it was found that increase in silica content of ferritic steel causes a higher rate of corrosion attack. Lowering the stack temperature possibly lowered the degradation rate triggered by corrosion. The reduction of PbO into Pb can be favoured in the presence of hydrogen-3 vol.% H_2O atmosphere. The Pb which reaches the steel surface contributes to severe internal oxidation of chromium resembling liquid metal corrosion. Chromium is continuously removed from the steel due to chromia formation along grain boundaries and chromia-rich dendrites inside the grains. When the critical concentration of 16% is reached, the steel cannot form the chromia layer further and then rapid formation of iron-oxide nodules with extremely small electrical resistance especially Fe_3O_4/FeO will lead to bridge formation between metals causing short-circuiting. The excessive cracking and degradation could be suppressed by YSZ coating separating glass–ceramic from the steel. The glasses without PbO and steels with low Si content diminish external and internal oxidation of the steel.

Another BCAS composition 35% BaO—15% CaO—5% Al_2O_3—10% B_2O_3—35% SiO_2 (mol%) yielded density of 3.88 g/cm^2 and T_g of 619°C: The CTE for this BCAS composition is 10.5 × 10^{-6}/K with Young's modulus and microhardness value of 72 GPa and 5.96 GPa, respectively. The $BaSiO_3$ phase precipitated out of this glass followed by hexacelsian $BaAl_2Si_2O_8$, minor celsian $BaAl_2Si_2O_8$ and $(Ba_xCa_y)SiO_4$ phases. Surprisingly, crystallisation activation energy is obtained to be 259 kJ/mol, which is much lower than those of barium aluminosilicate and strontium aluminiumosilicate compositions. The large difference in activation energy is attributed to the different first crystalline phase formed in different glasses: for the BCAS system, $BaSiO_3$ precipitates out, whereas for the BAS and MAS systems, $BaAl_2Si_2O_8$ and $MgSiO_3$ come out, respectively. The value of the Avrami parameter n come out to be 2.6, which corresponds to 2D growth mechanism in these glasses. For other BCAS systems investigated, Ghosh and co-workers revealed that high barium content of glass BCAS 4(55.6% BaO, 9.2% CaO, 5.5% Al_2O_3, 25.3% SiO_2, wt% and B_2O_3/La_2O_3/ZnO as additives) could lead to high

CTE and low T_g of glass, which makes it compatible with the Crofer 22APU alloy. The precipitation of the barium silicate phase along with the minor hexagonal barium aluminosilicate yield a glass matrix with compatible CTE with that of Crofer 22APU. BCAS 4 also revealed satisfactory flow and shrinkage behaviour though low barium-containing glass revealed better electrical insulation at elevated temperature. The GG5 composite prepared by Kaur and co-workers. (Table 7.1), yielded CTE of 7.05 × 10^{-6}/K and $CaSiO_3$, $Ba_{1.3}Ca_{0.7}SiO_4$ phase came after 1000 h heat treatment at 850°C. Due to the very low CTE and high viscosity of GG5 ($10^{10.59}$ dPas), it delaminated from the Crofer 22APU surface and hence no sealing could be observed.

The strengthening of G18 glass is done by reinforcing it with boron nitride nanotubes (BNNT) (4 wt%) using hot pressing. Figure 7.26a shows the SEM of as-synthesised BNNT yielding diameter ranging from tons to hundreds of nanometers and possessing bamboo-like features. X-ray analysis of G18 glass-BNNT composite yielded its amorphous nature. Figure 7.26b gives the TEM of G18-BNNT composite indicating the orientations of several BNNT in all possible directions. Both the toughness and flexural strength of the reinforced G18 glass increased by 35 and 90%, respectively, as that of non-reinforced G18 glass. Anyhow the density, Vickers microhardness, and elastic modulus of reinforced composite was lower than that of pristine glass.

Chemical composition GC-9 (0–40% BaO, 0–15% B_2O_3, 0–10% Al_2O_3, 0–40% SiO_2, 0–15% CaO, 0–15% La_2O_3, and 0–5% ZrO_2, mol%) was developed to check its compatibility with Crofer 22H. The joint strength for Crofer 22H/GC-9/Crofer 22H with two sides of glass–ceramic sealants to greater than that of single-sided ones. At 800°C, an increase of 195% in tensile strength for double-sided specimens is observed as compared single-sided specimens. Decrease in tensile joint strength

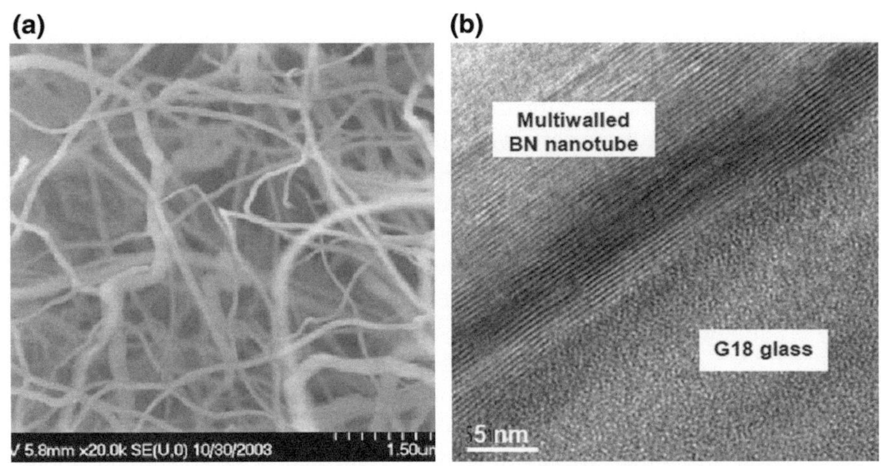

Fig. 7.26 **a** SEM micrographs of BNNT. **b** TEM of G18 glass-BNNT composite (Choi and coworkers 2007)

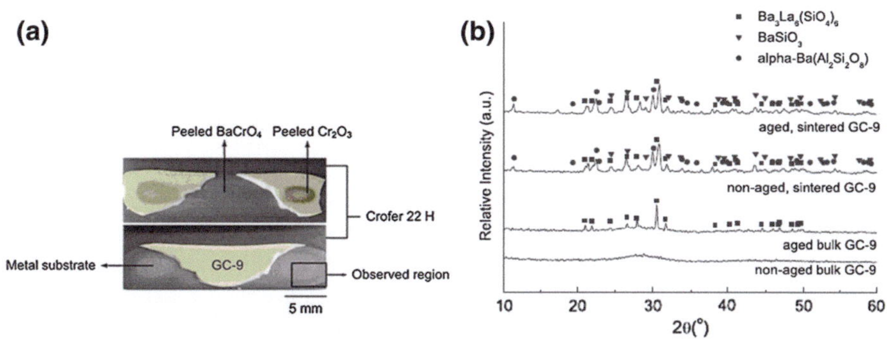

Fig. 7.27 **a** Failure patterns of single-sided shear specimens tested at room temperature. **b** XRD of aged/non-aged bulk GC-9 glasses (Lin and coworkers 2012)

by an extent of 45% is observed after pre-oxidation treatment of Crofer 22H for 20 h. $BaCrO_4$ formation occurs at the GC9/Crofer 22 H interface and is the weakest layer to resist a shear loading for joint specimens. The joint strength degrades by 17% due to continuous $BaCrO_4$ layer formation. Figure 7.27a shows that the Crofer 22H substrate is still well adhered to the GC-9 glass–ceramic layer and counterpart regions with metallic lustre of the metal substrate. The yellow region in the lower fracture surface respects the leftover of the GC-9 delaminated layer (identified as the $BaCrO_4$ layer). Figure 7.27b gives the XRD of aged/non-aged bulk and sintered glass. It is observed that non-aged, as-cast GC-9 is totally amorphous. For aged bulk and aged/non-aged sintered different phases are formed. Aged bulk GC-9 exhibited $Ba_3La_3(SiO_4)_6$ phase formation. For both sintered glasses, $BaSiO_3$ and α-Ba $(Al_2Si_2O_8)$ also precipitated out. The aging time of 100 h did not generate additional crystalline phases in sintered GC-9 glass. Sintered glass has yielded higher crystallisation attributed to the difference in the crystallisation process. For sintered GC-9 glass flexural strength improvement with increasing temperature at temperature $<T_g$ was observed due to the crack-healing effect from residual glass in the matrix. Anyhow GC-9 exhibits a large number of pores which leads to its lower Young's modulus than the bulk glass.

The BCSA system investigated by Hao and co-workers yielded CTE within $10.13–11.28 \times 10^{-6}$/K and T_s decrease from 678 to 645°C. The sintering temperature lies between 740 and 850°C for the BCSA system whereas T_m is between 850 and 1100°C. A strontium containing BCAS system known as G-73 glass (Ba–Ca–Sr–Al–Si–B) has been studied for long-term interaction with the aluminated SS441 steel. It was observed that with the increase in heat treatment duration, the Al concentration on surface decreased. This depletion could be modeled by Al diffusion into the steel matrix and migration of Al into the glass matrix. $BaAl_2Si_2O_8$ crystals might have formed near the glass/SS441 interface. No significant oxide growth could be seen on the air side of alumised SS441 even after 1000 h at 800°C.

7.1.6 Barium–Calcium–Magnesium–Strontium Glasses/Interconnect Interface

Goel and co-workers have extensively investigated the barium–calcium–magnesium and diopside system ($CaMgSi_2O_6$). The D–Ts system (diopside–Tschermak's) solid solutions exhibit general formula of $Ca[Mg_{1-x}Al_x](Si_{1-x/2}Al_{x/2})_2O_6$, where the parenthesis and square brackets denote tetrahedral and octahedral sub-lattices, respectively. In addition to this, the effect of BaO on the clinopyroxene-based GCs in the $CaO–MgO–Al_2O_3–SiO_2–B_2O_3$ system have been investigated by preparing four compositions, i.e., BaCMAS1 ($Ca_{0.95}Ba_{0.05}Mg_{0.8}Al_{0.4}Si_{1.6}O_6$), BaCMAS2 ($Ca_{0.9}Ba_{0.1}Mg_{0.8}Al_{0.4}Si_{1.8}O_6$), BaCMAS3 ($Ca_{0.8}Ba_{0.2}Mg_{0.8}Al_{0.4}Si_{1.8}O_6$) and BaCMAS4 ($Ca_{0.8}Ba_{0.2}Mg_{0.8}Al_{0.2}B_{0.2}Si_{1.8}O_6$), where BaCMAS4 is derived from BCMAS 3 by 50% substitution of Al^{3+} by B^{3+}. The glass density increases with the BaO content whereas T_g, T_s, and CTE are almost constant from boron-free glass. The boron-containing glass BCMAS4 possesses lowest T_g, T_s, and CTE. The well-sintered glass–ceramics exhibited orthocelsian and hexacelsian along with $D–T_s$ solid solutions after sintering at 850–900°C for 1 h. BaCMAS3 exhibited higher conductivity than BaCMAS1 and BaCMAS2, though the glass–ceramics possessed insulating properties. This could be attributed to the large size of Ba^{2+} cations compared with Ca^{2+} which decreases the energy barrier for ions and increases specific free volume. The effect of La_2O_3 and Cr_2O_3 on the diopside-based glasses was also investigated. The Cr_2O_3 has an impact on the physiochemical and electrical properties whereas La_2O_3 attend the thermal characteristics of the glass. Two compositions were derived which are named 7A and 7B with substitution as

$$\begin{gathered} 7A(Ca_{0.8}Ba_{0.1}MgAl_{0.1}La_{0.1}Si_{1.9}O_6) \\ \downarrow \\ 0.2Cao + 0.15SiO_2 \leftrightarrow 0.1BaO + 0.05La_2O_3 + 0.05Al_2O_3 \end{gathered} \tag{7.32}$$

$$\begin{gathered} 7B(Ca_{0.8}Ba_{0.1}Mg_{0.9}Al_{0.2}La_{0.2}Si_{1.8}O_6) \\ \downarrow \\ 0.2Cao + 0.1MgO + 0.2SiO_2 \leftrightarrow 0.1BaO + 0.1Al_2O_3 + 0.1La_2O_3 \end{gathered} \tag{7.33}$$

For obtaining glasses 7A-Cr and 7B-Cr, 0.5 wt% chromium was added to both 7A and 7B glass. To improve the adhesion behaviour of glasses, NiO (1 wt%) was added to all the compositions. With the increase in La_2O_3 content, increase in density could be seen for the glasses. The CTE of 7A, 7A-Cr, 7B, and 7B-Cr glasses is higher than the BaCMAS series of glasses which could be due to the decrease in connectivity of the glass structure with increasing La_2O_3 content. 7A glass also exhibits higher activation energy, which is due to higher field strength of La^{3+} compared to Ca^{2+}. With further increases in La_2O_3 content for glasses 7B, decrease in activation energy is observed and can be attributed to the clustering of La^{3+} ions leading to network-modifying effects. The clustering effect of La_2O_3 has been reported by Lu and co-workers, which was the cause of precipitation of lanthanum-containing phases. Cr_2O_3 played the role of nucleating agent, thus leading to decrease of activation energy for 7A-Cr and 7B-Cr. Cr_2O_3 also promotes fast spinel precipitation, which becomes the nucleator for epitaxial pyroxene growth. The sintered glass–ceramics exhibit augite phase formation after sintering at 900–1000°C for 1 h. No monocelsian phase was observed though monomineral could be observed for 7A and 7A-Cr glass–ceramics. The glasses could be seen for 7B-Cr/Crofer 22APU, couple after heat treatment at 850°C for 1 h, followed by 800°C for 10 h in air. Barium–calcium aluminosilicate glasses (Sect. 7.1.5) exhibit $BaCrO_4$ formation with chromia-forming alloys, which causes depletion of barium from the glass matrix resulting in delamination of glass–ceramic from the alloy surface due to large CTE mismatch. 7A, 7A-Cr, 7B, and 7B and 7B-Cr exhibited considerable decrease in Cr diffusion towards the glass side even after 300 h which demonstrated good scaling capability with negligible CTE mismatch. The glass–ceramics exhibited good insulating properties as evident from Fig. 7.28a. 7A and 7B GC possess low conductivity as 1.1–1.3 × 10^{-7} S/cm (780°) and 2.2–2.5 × 10^{-7} S/cm (800°C), which further decreases with lanthanum addition. Almost 1.6–2.0 times higher conductivity is observed after 0.5 wt% Cr_2O_3 addition to the glass matrix. The lanthanum and chromium-free analogues (BaCMAS series) show higher conductivity than the 7-series (Fig. 7.28b).

To further extend the investigation, boron and strontium containing glass with 7A/7B as the parent composition were studied for crystallisation kinetics and interaction mechanism with high temperature alloys. 7-2B, 7-5B, 7-10B, and 7-15B represented the glasses with 2, 5, 10, and 15 wt% of B_2O_3 whereas 7(Sr)–2B, 7 (Sr)–5B represented glasses with 2 and 5 wt% strontium (BaO ↔ SrO). With the increase in B_2O_3 concentration, T_g values also decreased. The same effect is

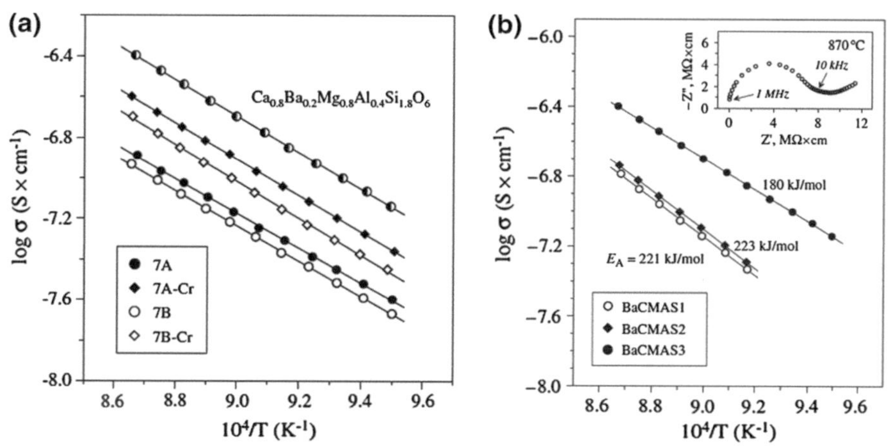

Fig. 7.28 Conductivity for **a** lanthanum-containing (Goel and coworkers 2008) and **b** lanthanum-free glass–ceramics (Goel and coworkers 2007a)

observed when BaO is replaced by SrO for 7(Sr)–2B and 7(Sr)–5B glasses. No detrimental reaction products could be observed at the interface of GC sealant/Sanergy HT interconnect even after 300 exposures in humidified atmosphere at 800°C. Cr- and Mn-rich oxide layers could be seen at the interface representing thermodynamically stable chromium–manganese spinel. The leak tests for 7A and 7–2B are performed as 7B and 7(Sr)–2B exhibited very low CTE after long-term heat treatments. The leak rate data are calculated using the equation:

$$\theta = V\frac{q_n}{l_s}\Delta p \tag{7.34}$$

where $V = 22.4$ L/mol, l_s is the sealing length, q_n is the rate of oxygen leakage, and Δp is the pressure difference. The OCV shown versus time in Fig. 7.29a is obtained under a different pumping current whereas the pO_2 in the chamber is calculated

$$pO_2 = p_o \exp\left(\frac{4FE}{RT}\right) \tag{7.35}$$

Figure 7.29b gives the partial pressure variation versus time under different pumping currents. Under high-pumping current conditions an almost linear relation could be obtained for pO_2 versus time. The leak rate for 7–2B glass lies in the range of (2.6–3.0) × 10^{-15} mol/s, whereas for 7 h glass it lies in between

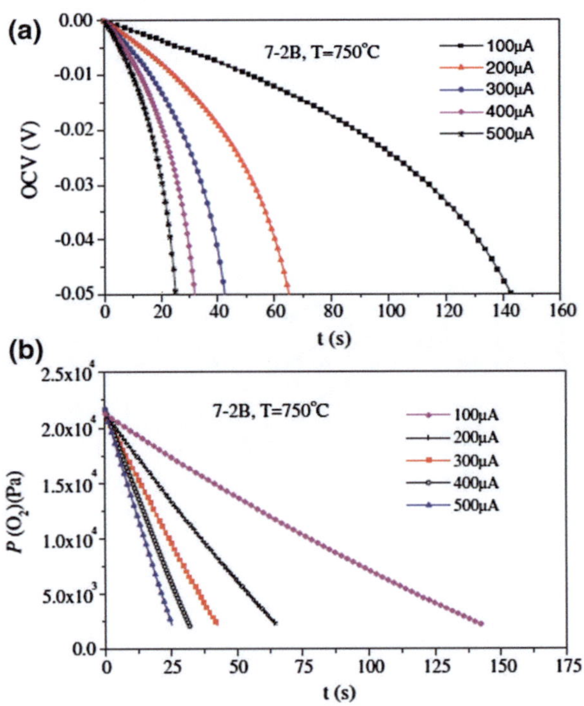

Fig. 7.29 **a** Open-circuit voltage. **b** Oxygen partial pressure versus time for composition 7-2B at 750°C (Goel and coworkers 2010b)

(8.8–10.4) × 10^{-15} mol/s. The compositions exhibit an acceptable leak rate, thus making them suitable for sealing.

Barium-, strontium-, calcium-, and magnesium-based lanthanum glass composite were investigated by Kaur and co-workers. The GG7 composition (ML:CL:SL: BL = 1:1:1:1 weight ratio, where M, C, S, and B represent MgO, CaO, SrO and BaO, respectively, Table 7.1) has viscosity and CTE of $10^{8.51}$ dPas and 8.19 × 10^{-6}/K, respectively. Though Ba_2SiO_4 and $Ca_{0.75}Sr_{0.2}Mg_{1.05}Si_2O_6$ phases precipitated out of the GG7 composite after 1000 h heat treatment 850°C, the Ba_2SiO_4 did not increase the CTE of the composite much. No lanthanum-rich phase could be seen for GG7 glass composite which exhibits very good bonding characteristics as evident from Fig. 7.30a. No delamination or detrimental reaction could be seen at the interface as supported by dot-mapping (Fig. 7.30b). Though calcium and magnesium diffusion could be seen on the other side of the interface, no chromate formation at the interface could be seen. Due to bigger size of strontium, barium, and lanthanum, they have shown sluggish diffusion kinetics. Figure 7.30c indicates a needle-like crystalline phase (probably Ba_2SiO_4) embedded inside the amorphous matrix indicating controlled crystallisation beneficial for the sealing applications. Figure 7.31a–c represents the stress contours for GG7/Crofer 22 APU couple for different thickness for the sealing GG7 components keeping Crofer thickness the same. It is evident that high-stress regions are concentrated on the edges rather than the surface of the GG7 composite. The maximum stress is 4.06 × 10^{-3}, 2.89 × 10^{-3}, and 2.05 × 10^{-3} N/m^2 for GG7 thickness of 0.25, 0.5 and 1.5 mm, respectively. Seal thickness decreases the stress, which is also shown by Weil and co-workers. The simulated heat flux for GG7/Crofer 22APU sample is shown in Fig. 7.32, which indicates that the contours are isothermal. The heat flux is simulated to see whether any temperature gradient is present about the interface or within the layers for the diffusion couple.

Goel and co-workers also modified the parent glass composition $Ca_{0.9}MgAl_{0.1}La_{0.1}Si_{1.9}O_6$ to form a series of five glasses designated 10, 10A, 10B, 10C, and 10D. Glass 10 was obtained by adding 2 wt% B_2O_3 addition to the parent glass. Glass 10 A was derived by adding 0.5 wt% Cr_2O_3 to composition 10, 10B, 10C, and 10D also contain 0.5 wt% Cr_2O_3 and are a partial substitution of CaO by BaO in composition 10. Augite phase $(Ca(Mg_{0.85}Al_{0.15})/(Si_{1.70}Al_{0.30})O_6)$ precipitated out of all GCs sintered at 850°C for 1 h. 10C exhibits the highest CTE of 9.5 × 10^{-6}/K, whereas for 10D the CTE is 8.8 × 10^{-6}/K. BaO addition increased CTE for 10C glass and its further addition decreased CTE of 10D. For glasses 10B and 10C, the hexacelsian phase also appeared along with the augite phase. For glass 10D, $LaBO_3$ and solid solution isostructural to Cr_5O_{12} also crystallised as minor phases in addition to the augite and hexacelsian phase. Good bonding of all GCs with the metallic interconnect Crofer 22APU is observed. No iron-, manganese-, or chromium-rich phases could be seen at the GC10/Crofer 22APU joint region. All diffusion couples except glass 10D/Crofer 22APU yield a Ti-rich zone near the interface. The titania-rich layer can control the migration of chromium to the cathode and hence work positively. In addition to this, the titania-rich phases are

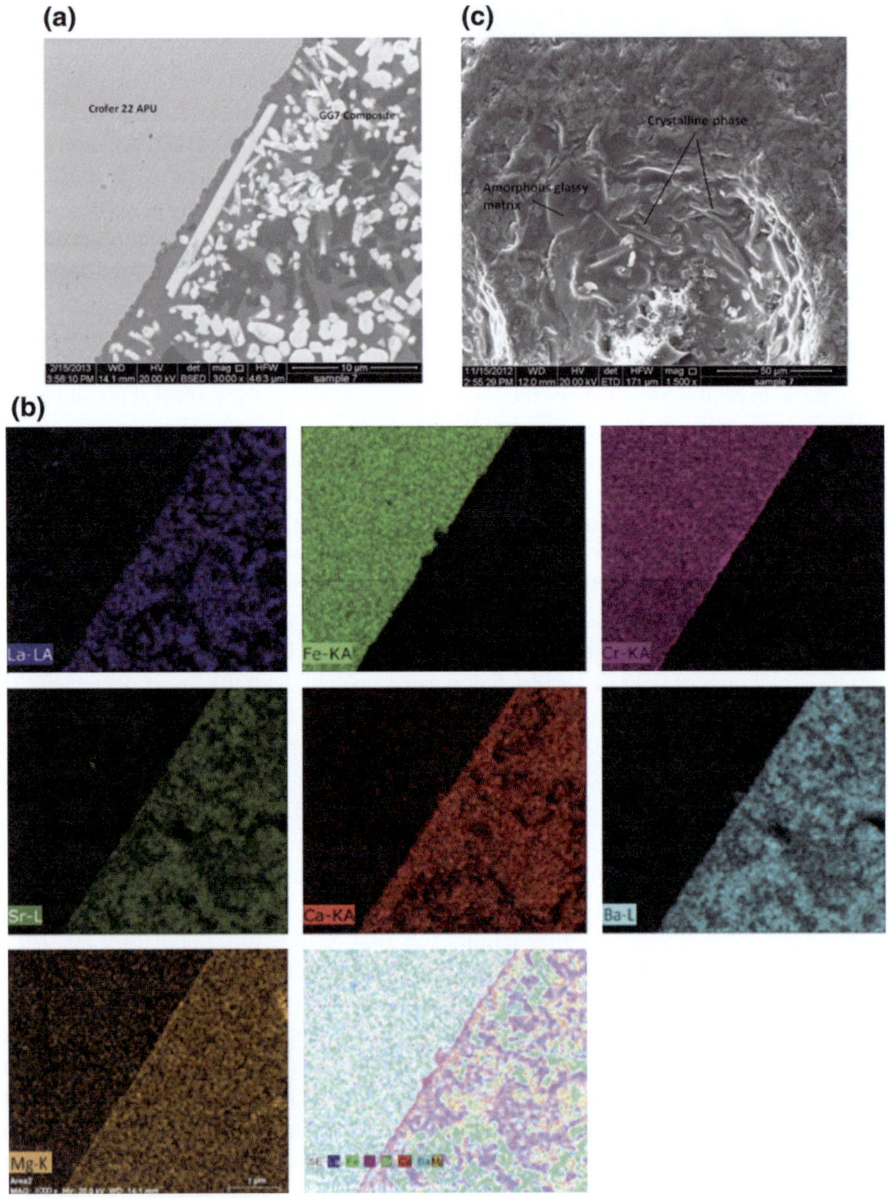

Fig. 7.30 **a** Interface of GG7/Crofer 22APU, **b** elemental mapping of GG7/Crofer 22APU and **c** microstructure of GG7 glass after heat treatment at 850°C for 1000 h (Kaur and co-workers 2013a)

thermodynamically more favourable from a thermodynamic point of view as compared to the formation of Mn/Cr-rich species. For 10B and 10C diffusion couple with Crofer 22APU, substantial segregation of Mn also takes place along

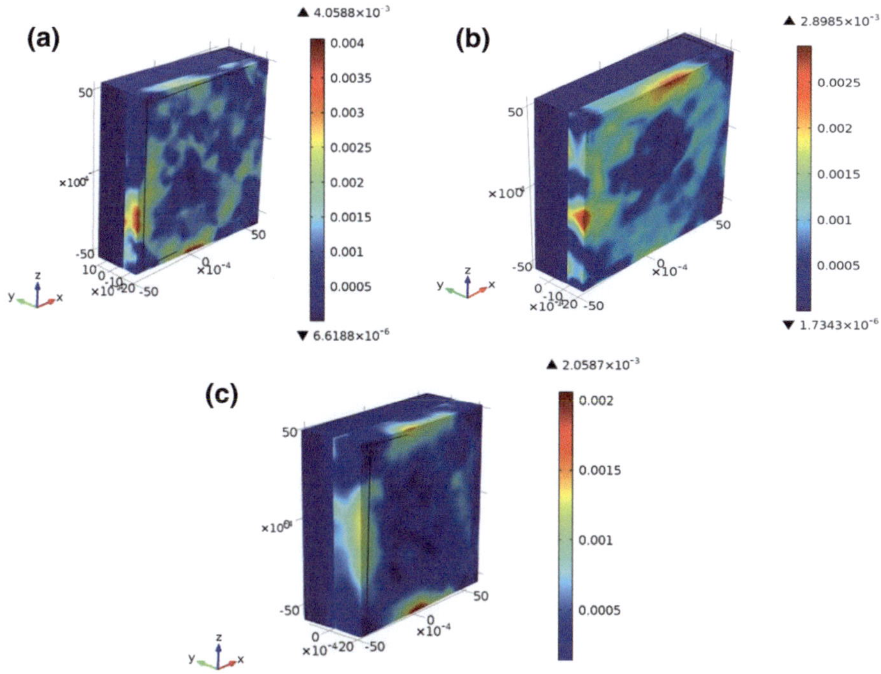

Fig. 7.31 The stress contours and their magnitude for the GG7/Crofer 22APU diffusion couple for thickness **a** *t* 0.25, **b** *t* 0.5, and **c** *t* 1.5 mm (simulated upto 10,000 h, step size 1) (Kaur and coworkers 2013b)

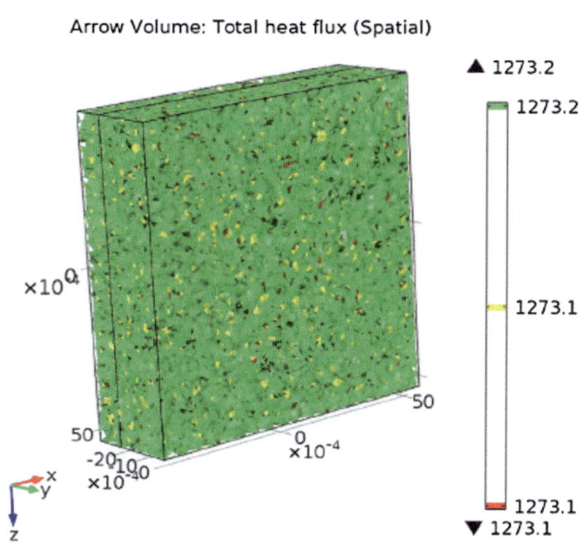

Fig. 7.32 Isothermal contours for the GG7/Crofer 22APU couple (Kaur and coworkers 2013b)

with Ti. Hence, possible Mn-rich phase formation or Mn–Ti rich phases could take place as per the mechanism.

$$MnO + TiO_2 \rightarrow MnTiO_3 \quad \Delta G = 298\,\text{kJ/mol} \tag{7.36}$$

$$MnO_2 + TiO_2 \rightarrow Mn_2TiO_4 \quad \Delta G = 276\,\text{kJ/mol} \tag{7.37}$$

MnO, MnO_2, Mn_2O_3, and Mn_3O_4 exhibit Gibbs free energy of formation to be 217, 208, 221, and 224 kJ/mol, respectively. Hence, the possible formation of $MnTiO_3$/$MnTiO_4$ is thermodynamically more favourable to occur as compared to Mn–O phases. Additionally, no chromium-rich zones could be observed at the interface, whereas Ba accumulation is present for the 10B/10C diffusion couple with Crofer 22APU. This also indicates the possible formation of $BaTiO_3$ phase which is a relatively stable phase and does not have any detrimental effect on the SOFC performance. B_2O_3 addition causes an increase in conductivity; anyhow all the glasses exhibit excellent insulating properties as shown in Fig. 7.33. For parent glass composition 10, the conductivity is low as 4×10^{-7} S/cm (850°C), which ascertains absence of any short-circuiting. Activation energies increased and conductivity decreased upon barium addition to the glasses.

Three other novel glass compositions based on the diopside were investigated: DT1 ($CaMg_{0.8}Al_{0.4}Si_{1.8}O_6$), DT2 ($Ca_{0.75}Y_{0.25}Mg_{0.8}Al_{0.4}Si_{1.55}B_{0.25}O_6$), and DT3 ($Ca_{0.8}Ba_{0.2}Mg_{0.8}Al_{0.2}B_{0.2}Si_{1.8}O_6$). The following substitutions were done to obtain these novel compositions.

For DT-1

$$0.2\left(Si^{4+} + Mg^{2+}\right) \leftrightarrow 0.4\,Al^{3+} \tag{7.38}$$

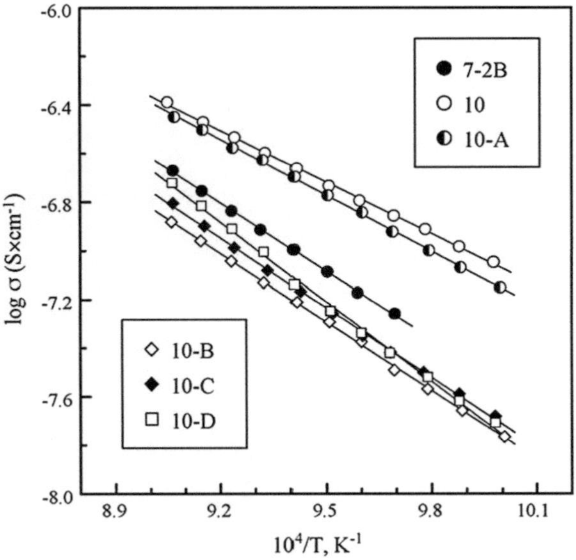

Fig. 7.33 Conductivity of glass ceramics in air (Goel and coworkers 2010c)

For DT-2

$$0.25\left(Ca^{2+}+Si^{4+}\right)\leftrightarrow 0.25\left(Y^{3+}+B^{3+}\right) \tag{7.39}$$

and

$$0.2\left(Si^{4+}+Mg^{2+}\right)\leftrightarrow 0.4\,Al^{3+} \tag{7.40}$$

For DT-3

$$0.2\,Ba^{2+}\leftrightarrow 0.2\,Ca^{2+} \tag{7.41}$$

and

$$0.2\left(Si^{4+}+Mg^{2+}\right)\leftrightarrow 0.2\left(B^{3+}+Al^{3+}\right) \tag{7.42}$$

The CTE for DT-1, DT-2, and DT-3 is 9.57×10^{-6}/K, 8.01×10^{-6}/K, and 9.62×10^{-6}/K, respectively. Lara and co-workers studied barium, calcium, and

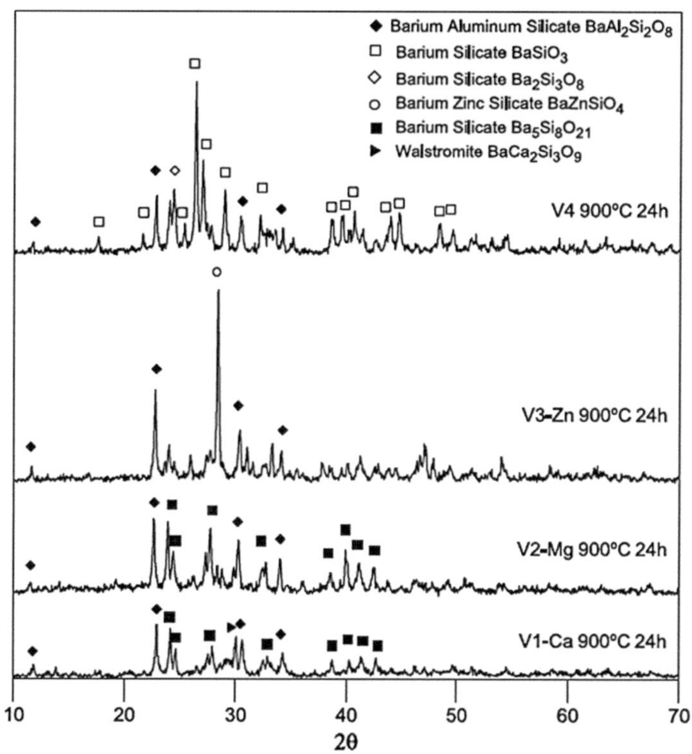

Fig. 7.34 XRD of glasses after 24 h heat treatment at 900°C (Lara and coworkers 2004b)

magnesium-based $30BaO–15RO–50SiO_2–5Al_2O_3$[R = Ca, Mg, Zn] composition. The parent glass composition is V4(50% SiO_2, 45% BaO, 5% Al_2O_3, in mol%), whereas for V1–Ca, V2–Mg, V3–Zn, 5% BaO is replaced by calcium, magnesium, and zinc, respectively. Figure 7.34 indicates the XRD of all glass heat treated at 900°C for 24 h. For parent glass, maximum devitrification could be observed with $BaSiO_3$, $Ba_2Si_3O_8$, and $BaAl_2Si_2O_8$ phase formation. For calcium-doped glass, $Ba_5Si_8O_{21}$, $BaCa_2Si_3O_9$, and $BaAl_2Si_2O_8$ phases are denitrified. For Zn-containing V3–Zn glass, $BaZnSiO_4$ is also detected. Ca- and Mg-containing glasses exhibit undesirable thermal expansion whereas zinc-based glass exhibits good CTE as required for sealing applications.

7.2 Coatings on Interconnect

As mentioned earlier, the interconnect is a critical component, which requires the most stringent requirements be fulfilled. For an ideal situation, the ohmic loss due to the interconnect is small so that the power density of the stack does not show much drop. The oxygen partial pressure on the cathode side varies from 10^{-4} to 10^{-7} atm, whereas on the anode side it varies from 10^{-18} to 10^{-8} atm. The interconnect should not contain volatile species in order to avoid their possible reacting with H_2S, CO_2 etc. in fuel. The interdiffusion/reaction process between the interconnect and its adjacent components should be minimum to avoid the ohmic and polarisation loss along with minimising corrosion. The interconnect should exhibit excellent sulphidation, oxidation, and carburisation resistances. Usually, these processes are not a problem for the perovskite-type ceramic oxides due to negligible small lattice diffusion coefficients. For the metallic interconnects, the growth of thermodynamically favourable metal oxide scales might enhance these issues. But the growth of these spinel layers retards chromium vaporisation thereby increasing cell performance. The surface of the interconnect can be modified via a coating of conductive oxides on the cathode side because the coating acts as a protection layer minimising the interfacial contact area-specific resistance between the cathode and interconnect. The alloy oxidation kinetics can be reduced and low conductivity Cr-based oxide growth can also be hindered by using protective coatings on the interconnect surface. Common coatings include ferrite-, chromite-, and strontium-doped lanthanum manganites. Usually, the perovskite coatings may degrade cell performances by decreasing the thermomechanical stability due to the extensive scale growth by inward diffusion of oxygen. Spinel layers such as $(Mn, Co)_3O_4$ are promising barriers to chromium migration and overall increase the surface stability of ferritic stainless steels.

Yang and co-workers have optimised the performance of $(Mn, Co)_3O_4$ spinels for coating applications. Figure 7.35a shows the XRD for $Mn_{1+x}Co_{2-x}O_4$ spinels with $x = -0.5, 0, 0.5, 1$, and 1.5 at room temperature. $x = 1, 1.5$ spinel exhibits a tetragonal structure whereas $x = 0$ and -0.5 demonstrated a cubic spinel structure. For $x = 0.5$, the spinel was a dual-phase material exhibiting both cubic and tetragonal phases. This indicates that for the $x = 0.3–0.9$ range, the cubic $\leftrightarrow$ tetragonal phase transformation

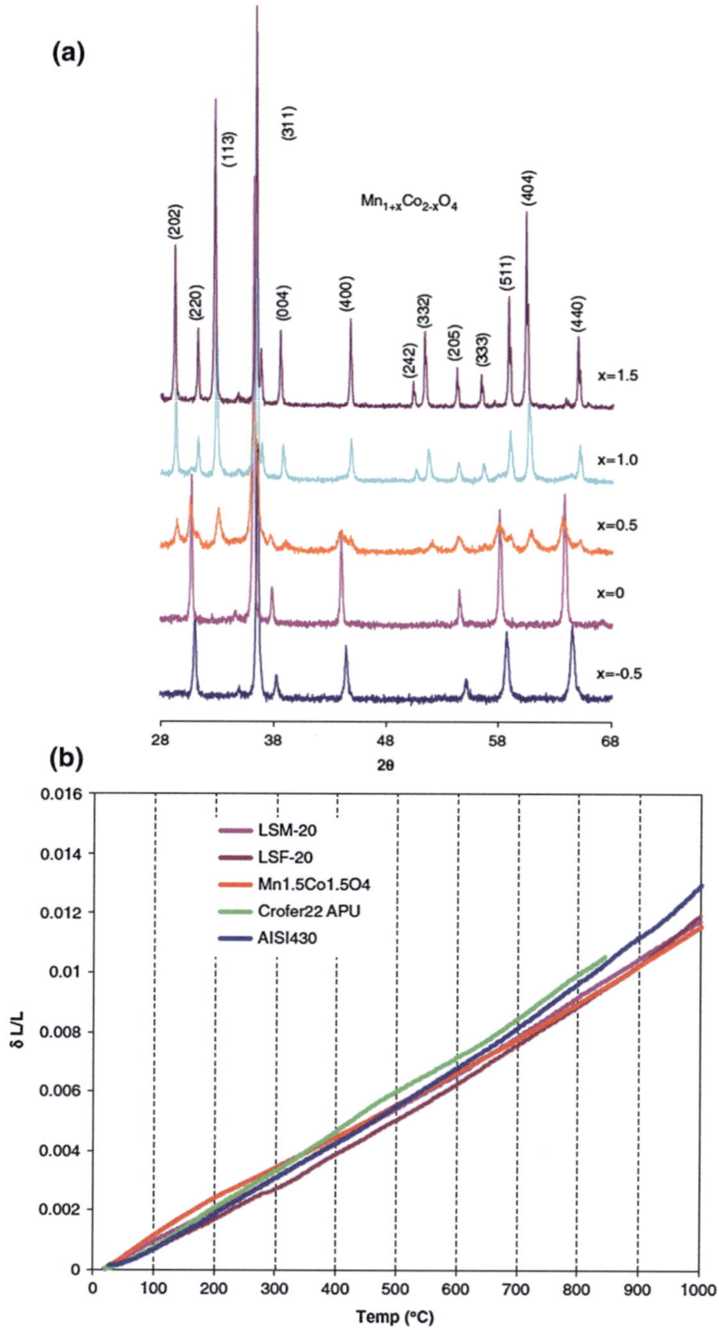

Fig. 7.35 **a** XRD of $Mn_{1+x}Co_{2-x}O_4$ for $x = -0.5$, 0, 0.5, 1.0, and 1.5 and **b** CTE of different interconnects (Yang and coworkers 2007)

takes place. These phase transformations negligibly affect the CTE as is clear from Fig. 7.35b. This spinel layer exhibits matching CTE with the AISI 430 and Crofer 22APU as well as with perovskite cathodes such as $La_{0.8}Sr_{0.2}FeO_3$ and $La_{0.8}Sr_{0.2}MnO_3$. $Mn_{1.5}Co_{1.5}O_4$ spinel demonstrates excellent thermal stability, good electrical conductivity, and CTE match with ferritic steel interconnects.

At high temperatures, ferritic stainless steels develop chromia-based thermally grown oxide which is more conduction than alumina forming alloys. Steels such as 441 HP develop a $(Mn, Cr, Fe)_3O_4$ spinel layer which exhibits chromium volatility and hence is not applicable for SOFC. The oxide layers provide a barrier to hydrogen permeation through the steel in order to keep the gas flow compartmentalised. Preoxidised 441 HP steel with magnetron-sputtered CoMn coatings was studied. It was observed that thinner surface oxide layers developed on the pre-oxidised coatings. It was also obtained that the power density of $(Mn, Co)_3O_4$-coated SUS430 interconnect after 300 h at 850°C is 275 mW/cm^2, whereas the uncoated SUS430 interconnect yields 175 mW/cm^2. Lu and coworkers studied the compatibility of SABS-0/$(Mn, Co)_3O_4$-coated Crofer 22APU and AISI441. During the interface formation oxidation of alloys, interdiffusion of elements, and chemical reaction along with glass denitrification occurred. Better interfacial compatibility was exhibited by coated Crofer 22APU/SABS-0 than the coated AISI441/SABS-0 couple. Figure 7.36 shows the elemental mapping for the spinel-coated Crofer 22APU/SABS-00 couple for argon, air, and H_2/H_2O atmosphere. The chromium diffusion cannot be eliminated completely, but the extent decreases from air to argon to H_2/H_2O atmosphere indicating that atmosphere plays a key role in the interfacial kinetics. Titanium is enriched beneath the chromium-enriched layer due to internal oxidation whereas Mn and Co are enriched at the interface (thickness highest for air treated samples and lowest for the H_2/H_2O treated samples). Oxygen enrichment occurs as a granular phase at the interface. The iron oxidation accelerates $(Mn, Co)_3O_4$ coating deterioration for the AISI441/SABS-0 couples. Minor elements have accumulated at the interface of coated AISI441/SABS-0 couples indicating the formation of more stable oxides than the Crofer 22APU alloy. For the AISI 441/SABS-0 couple, the glass network structure is disrupted enhancing the glass denitrification extent. SABS-0 has shown compatibility with Crofer 22APU alloy at 800°C regardless of thermal treatment atmosphere but inferior compatibility could be seen with the coated Crofer 22APU. The long-term interfacial compatibility of coated and bare Crofer 22APU/SABS-0 in air and high water containing H_2 atmosphere at 800°C for 1000 h is shown in Fig. 7.37. It can be seen that the uncoated Crofer 22APU/SABS-0 yield severe interaction and hence poor stability at the interface. In H_2/H_2O atmosphere, the interfacial stability is better due to limited alloy oxidation of both coated/uncoated Crofer 22APU. The SBAS-0 glass is denitrified with strontium/lanthanum-rich phases. Anyhow the thermal treatment atmospheres do not affect the denitrification extent.

The compatibility of alkali-based KMBY glass was investigated by Smeacetto and co-workers for $Mn_{1.5}Co_{1.5}O_{44}$-coated Crofer 22APU. The $Mn_{1.5}Co_{1.5}O_{44}$-coated Crofer 22APU/sealant/anode supported electrolyte was subjected to thermal treatment of 500 h from RT to 800°C. Figure 7.38 gives the dot-mapping of KMBY

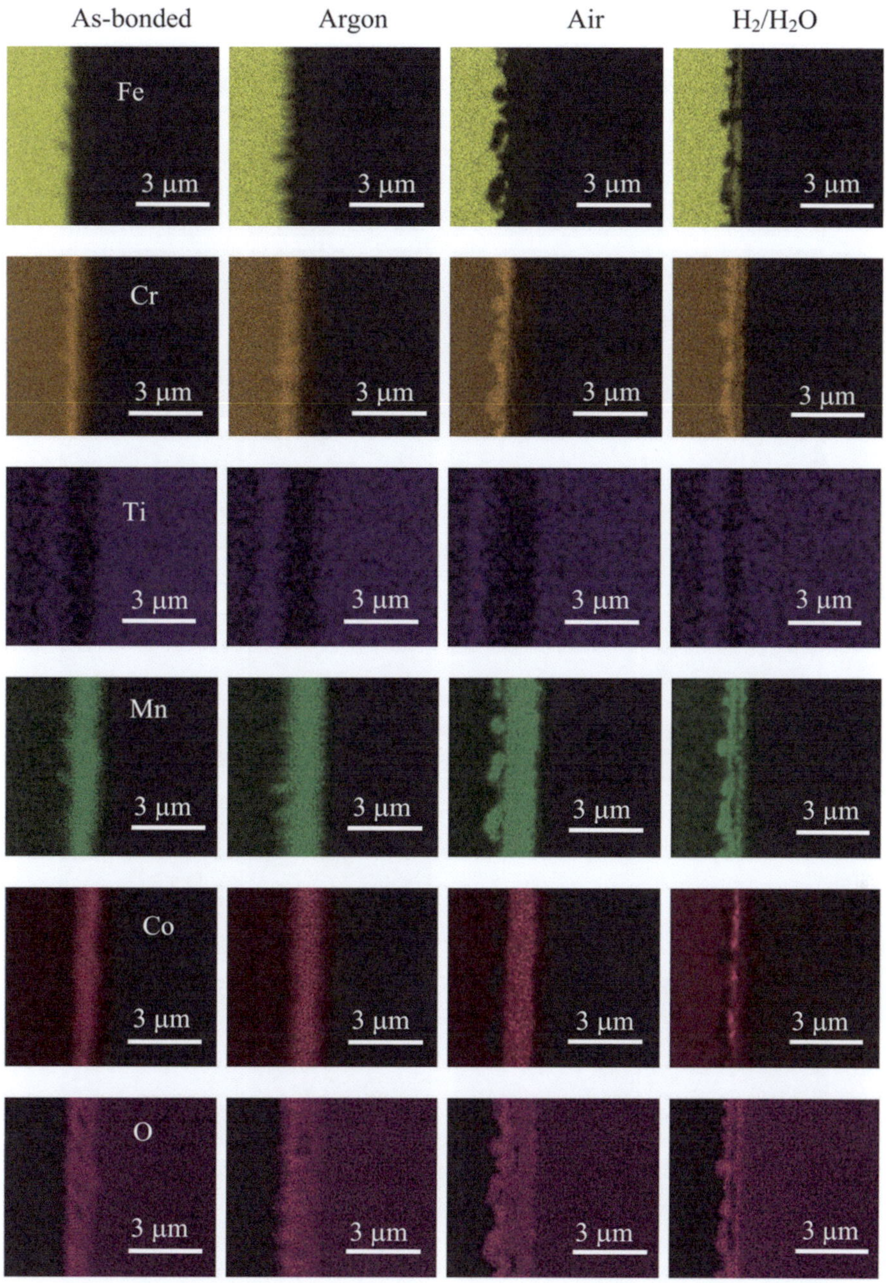

Fig. 7.36 Elemental mapping for the MC-coated Crofer 22 APU/SABS-0 couples (Mahapatra and coworkers 2010)

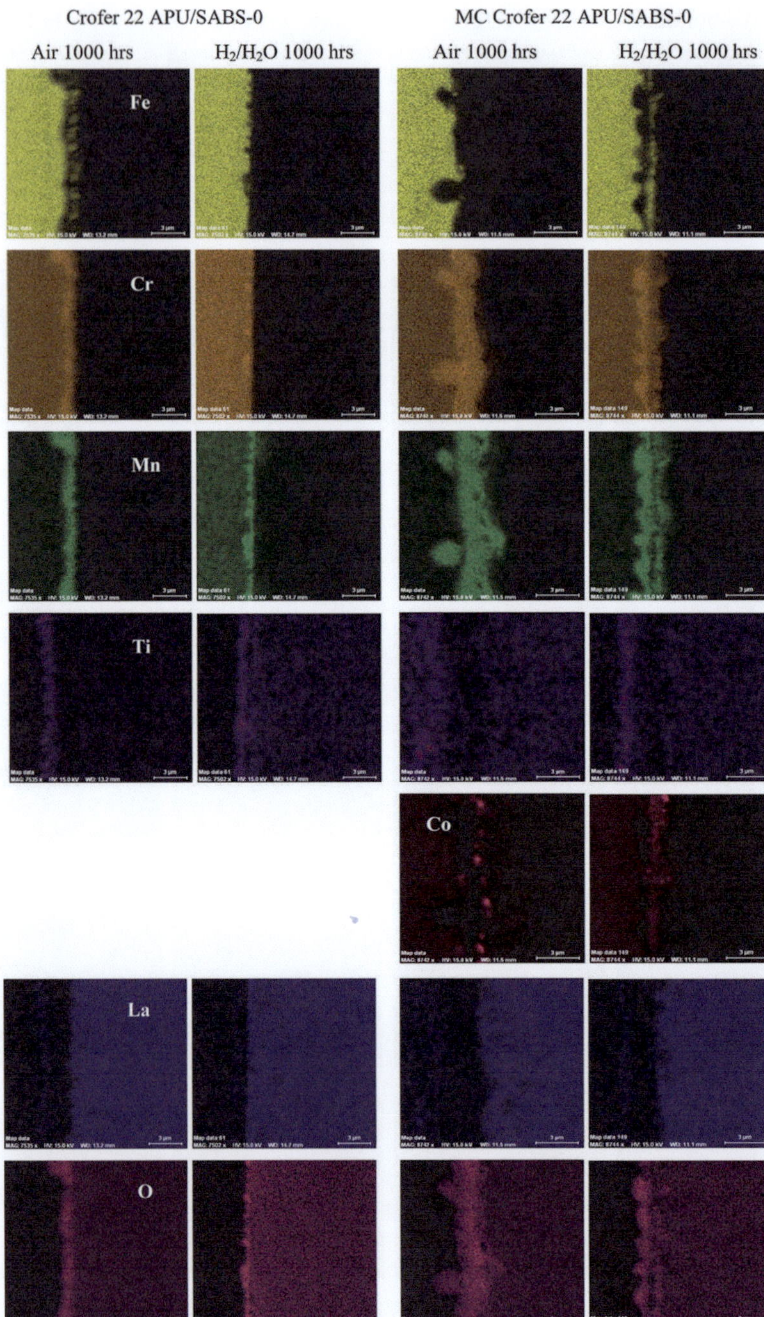

Fig. 7.37 Elemental mapping for the Crofer 22 APU/SABS-0 and MC Crofer 22 APU/SABS-0 couples thermally treated at 800°C for 1000 h in air and H_2/H_2O atmospheres (Mahapatra and Lu 2011)

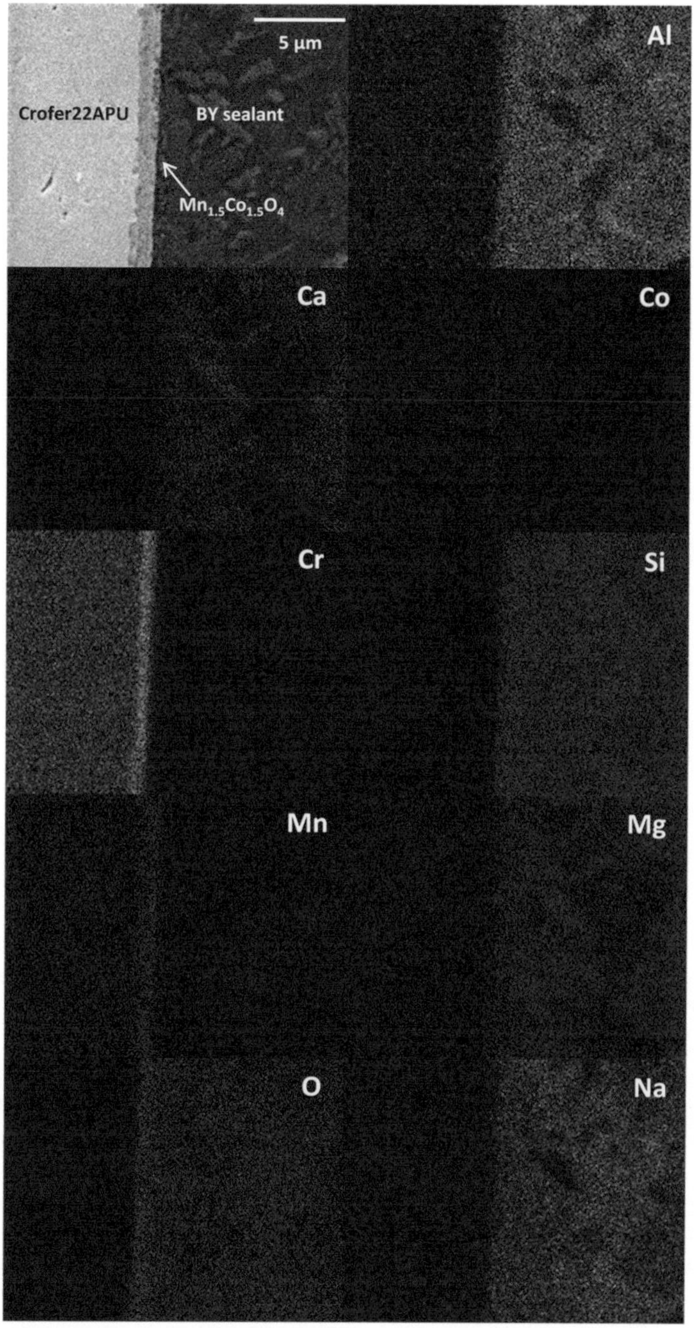

Fig. 7.38 Interface of $Mn_{1.5}Co_{1.5}O_4$-coated Crofer22APU/glass–ceramic after 500 h of thermal cycling (Smeacetto and coworkers 2014)

glass sealant/coated Crofer 22APU interface after 500 h. The sealant exhibit homogeneous and dense microstructure and diopside is the main phase formed in glass. The sealant and interconnect exhibit a well-bonded interface without evidence of coating spallation. $Mn_{1.5}Co_{1.5}O_4$ is very effective in preventing chromium diffusion into the glass–ceramic sealant. Chromium is found beneath the $Mn_{1.5}Co_{1.5}O_4$ layer, which is usually the behaviour expected by it. In addition to this, the Mn and Fe also did not diffuse inside the glass and hence, the $Mn_{1.5}Co_{1.5}O_4$ layer is used in avoiding a possible reaction between the alkali and chromium (No Na_2CrO_4 or K_2CrO_4 could be seen at the interface).

BaO–CaO–MgO-based glass such as Glass 14a (10% Al_2O_3, 10% BaO, 40% B_2O_3, 20% CaO, 20% MgO, mol%) has also shown promising results with $Mn_{1.5}Co_{1.5}O_4$-coated 441SS. Mn, Co, and Fe diffused slightly into the glasses and Fe from 441SS formed a partially substituted Fe–$Mn_{1.5}Co_{1.55}O_4$ spinel structure. Glass bonded tightly to the coating and no Cr diffusion into the glasses was observed. Anyhow, Cr and Ti accumulated at the interface and migrated only a few µm inside the coating. Presence of water vapour accelerates the oxidation of steels such as 430SS and 441SS because water vapour increases the corrosiveness of the environment and increases the critical chromium content required to form a Cr-rich scale. Upon exposure of steels to wet air, chromia can be lost by the reaction:

$$Cr_2O_{3(s)} + 2H_2O_{(g)} + \frac{3}{2}O_2(g) \rightarrow 2CrO_2(OH)_{2(g)} \quad (7.43)$$

and Cr can also vaporise from $MnCr_2O_4$ by the reaction:

$$2MnCr_2O_{4(s)} + 3H_2O_{(g)} + \frac{5}{2}O_{2(g)} \rightarrow Mn_2CrO_{4(s)} + 3\,CrO_2(OH)_{2(g)} \quad (7.44)$$

When the Cr supply from the alloy to the oxide scale is insufficient, then iron diffuses outward to maintain a high concentration of chromia and hence Fe-rich nodules are formed. Usually, low-CTE Fe–Co–Ni alloys display poor oxidation resistance at temperatures higher than 600°C. The oxidation resistance of some alloys have been improved by adding chromium, for example, Thermo-span with 5.5 wt% Cr was developed by Carpenter Technology Corporation and HRA929C with 2 wt% Cr was developed by Hitachi Metals Pvt. Ltd. Inco Alloys Internationals also developed EXP 4005 with 5.3 wt% Al and three phase with 6 wt% Al to improve their oxidation resistance. Figure 7.39a–d shows the microstructure of these four Fe–Co–Ni alloys after oxidation at 800°C for three weeks. For all of the alloys, a relatively thick, Cr-free, and electrically conductive Co_3O_4/$CoFe_2O_4$ surface layer is formed after the thermal heat treatment. The spinel layer provided the electron conduction path as well as acted as surface seal to block the Cr evaporation. Thermo-span and HRA929C are low Cr alloys and then ASR is lower than the Crofer 22APU. The alloys depict poor oxidation resistance yielding continuous mass gain with increase in oxidation time, In contrast to this, the low-Al alloys EXP4005 and three phase display acceptable ASR and oxidation resistance. For three phase and EXP 2005,

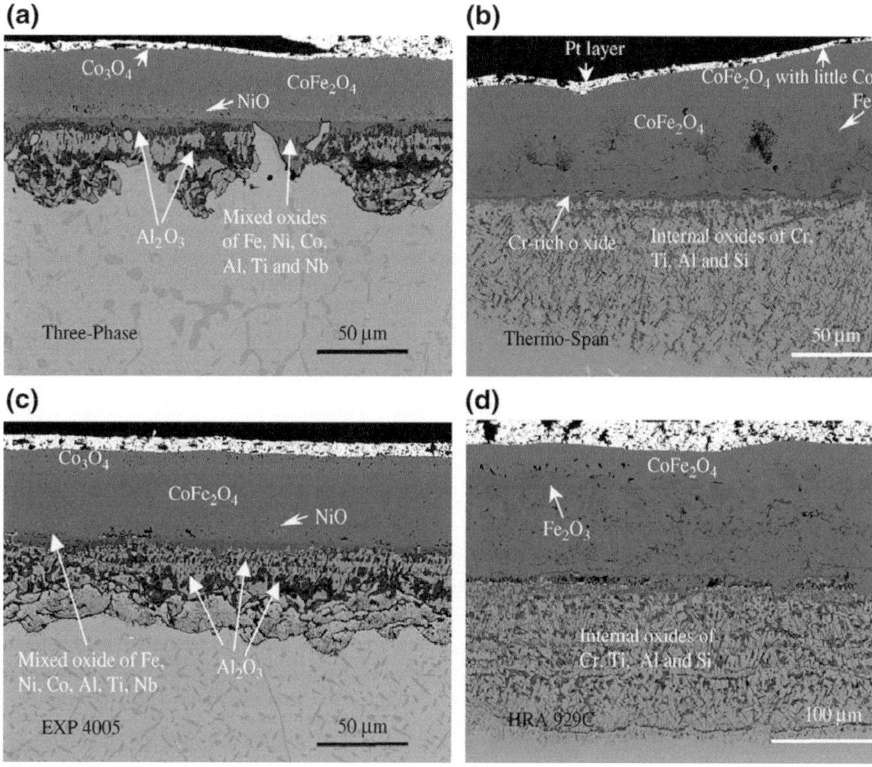

Fig. 7.39 Cross-sections of **a** three phase; **b** Thermospan; **c** EXP 4005; **d** HRA 929C after oxidation for three weeks at 800°C in air (Zhu and coworkers 2007)

some discrete NiO precipitates and a semicontinuous Al_2O_3 layer could be seen. The Al_2O_3-rich layer is the key factor responsible for low oxidation rate of three phase and EXP4005 alloys. The top oxide layer of Co_3O_4 is 10–12 μm and 2–4 μm for three phase and EXP2005, respectively. Thermospan and HAR 929C exhibited a higher degree of oxide scale growth with internal oxide formation of Cr, Ti, Al, and Si.

YSO77 glass developed by Chou and co-workers was also investigated for stack testing containing (Mn, Co) spinel coating, AISI steel and alumina coating. Ce–(Mn, Co) spinel coating is applied in the central section facing the cathode and aluminisation is applied on the AISI section requiring no electrical conduction. The spinel coating showed densification along with some chromium diffusion after 600 h duration treatment at 800°C. Si and Ti segregation could be observed at the interface, but no continuous layer was formed. No spallation of the spinel coating or underneath Cr-oxide scale could be seen. The AISI441 perimeter contained a protective alumina layer coating, which appeared more dense at the air side compared to the fuel side. A small quantity of Cr could be seen in the alumina coating

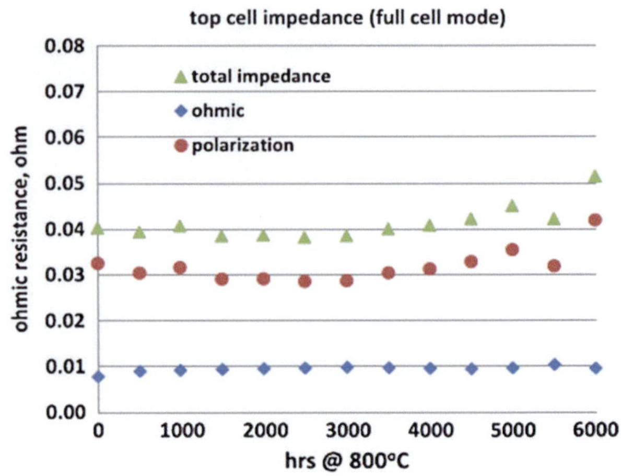

Fig. 7.40 Impedance of the top cell during the stability test at 800°C for 6000 h (Chou and coworkers 2014c)

keeping the alumina coating intact and strongly bonded to the metal substrate. Ni particle worsening could not be seen and hence the loss of connectivity between Ni grains and reduced triple-phase boundaries could be minimised. The impedance measurement of the top cell after 6000 h is shown in Fig. 7.40.

The polarisation part contributes mainly to the impedance part. The ohmic resistances increased a few milliohms within the first 500 h and then remained almost constant. The conductivity of Cr_2O_3 is almost 10^{-3}–10^{-4} S/cm at 800°C and by the end of the stack degradation test, the Cr scale is almost 5 μm. For an interconnect plate of area 25 cm^2 and oxide scale ~5 μm thick, an increase in ohmic resistance up to 10–100 mΩ could be expected. The cathode side of (Mn, Co) spinel-coated interconnect plate might contribute to the ohmic resistance.

Apart from the spinel coatings, Cr_2O_3 coatings are also preferred due to their chemically inert nature, good optical characteristics, and high mechanical strength. Cr_2O_3 exhibits hardness of 29.5 GPa, hence is well suited for the wear applications. Figure 7.41a shows the TEM microgrpah of coating, Cr interlayer, and the steel substrate. The Cr interlayer exhibits columnar grains which increase with the interlayer and the steel substrate. Both the coating and interlayer are dense in spite of some defects present in the substrate as well as the steel/Cr interlayer. Some thermal stresses may arise due to mismatch of coating and substrate CTE. The magnitude of thermal stress under the assumption of stress-free coating at depositions temperature is given as

$$\sigma_{\text{thermal}} = \frac{\Delta\alpha\,\Delta TE}{1 - \mathbf{f}} \tag{7.45}$$

where **f** is the coating Poisson's ratio, ΔT is the difference between deposition temperature and room temperature, $\Delta\alpha$ is the CTE difference of coating and substrate, and E is the elastic modulus of coating. Usually, for Cr_2O_3, the CTE range is $(5.4–7.5) \times 10^{-6}$/K with fof 0.25 and $E \approx 230$ GPa. Hence, almost 337 MPa

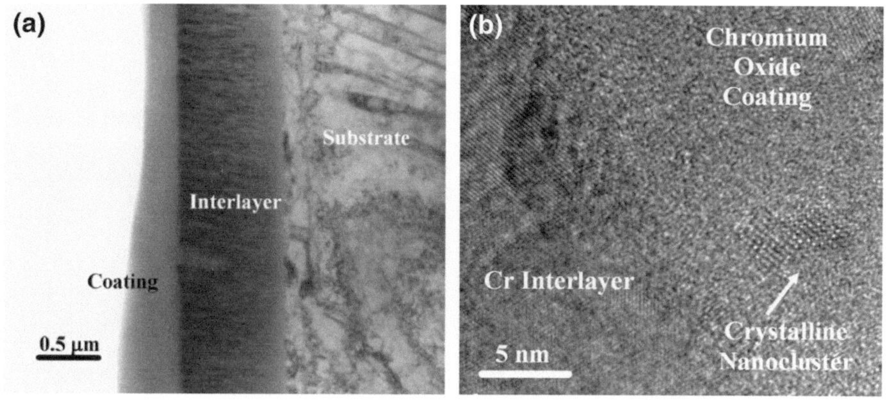

Fig. 7.41 **a** TEM micrograph of the coating cross-section and **b** HRTEM micrographs of the interfaces between the Cr interlayer and the coating (Pang and coworkers 2007)

compressive residual stress is expected for the chromium coating, which is quite high. To reduce the thermal stress in the coating, a pure Cr interlayer is deposited prior to the Cr_2O_3 layer. The chromium interlayer enhances the growth of thick coatings without delamination (Fig. 7.41b). Nanocrystalline clusters could be seen in between the amorphous layer within the Cr/substrate and Cr/coating. These amorphous layers could block coating columnar grains growth and relieve growth stress in the coating without any delamination. The Cr_2O_3 nanocrystals are high-stress regions within the amorphous layer of coatings.

Ni and Cobalt deposited Crofer 22APU interconnects are also widely investigated for SOFC applications. Figure 7.42 shows that the Co coating adheres well to the substrate and appears dense at the outer surface. Porosity could be seen at the innermost part of the layer attributed to the cobalt atom diffusion during oxidation. Fe diffusion is also observed which could enhance the formation of pores. The SEM of Co coating indicates a multilayer structure with ~9 μm thick Co_3O_4 top layer and a Cr_2O_3 layer. An almost 3 μm thick Co-rich layer could be seen at the interface. However, the coating covered the whole interconnect and adhered well to the substrate. These layers are chromium free and hence provide high hermeticity at the interface. NiO possesses high electrical conductivity and is considered to be a potential coating to hiderite Cr_2O_3 evaporation from the stainless steel.

Figure 7.43 shows the oxidation kinetics in air at 800°C within the cathode environment of SOFC. The Ni-coated steel yields a large mass gain due to rapid oxidation kinetics during initial 100 h of heat treatment. This indicates a protective layer formation between the substrate and surface oxides formed during the initial oxidation stage. XRD confirmed the formation of Cr_2O_3 and $(Mn, Cr)_3O_4$ on the bare steel. The coated steel developed NiO, $(Ni, Fe, Cr)_3O_4$, and Cr_2O_3 phases on its surface. The surface morphologies of bare steel and Ni-coated steel are shown in Fig. 7.44a, b. No spallations could be observed in the bare steel whereas severe spallation areas are evident on the coated steel. The noteworthy point is the

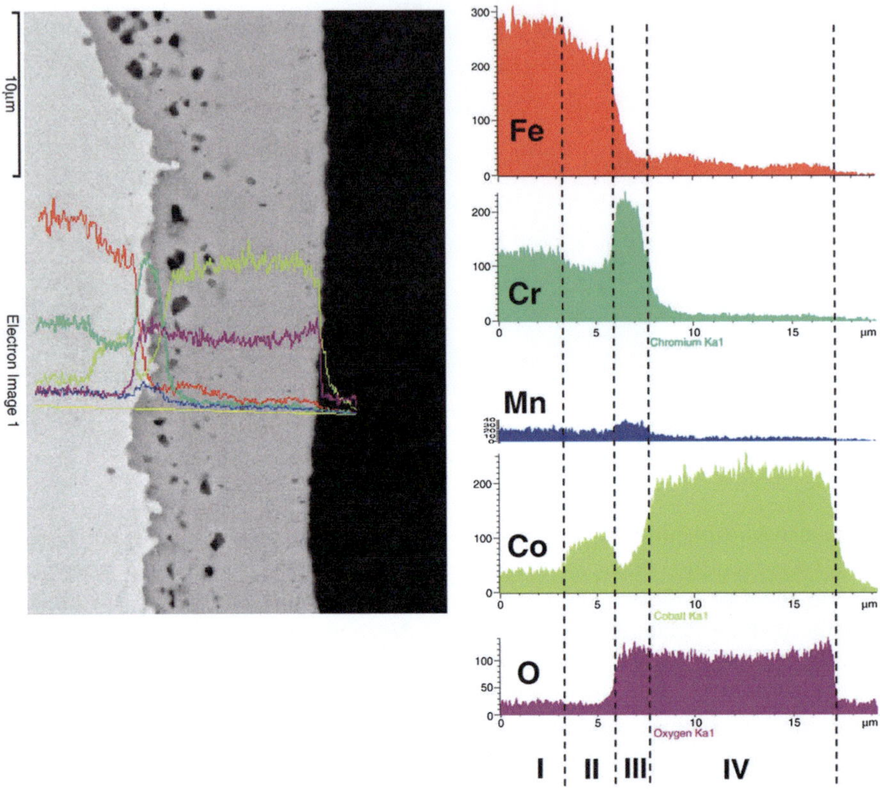

Fig. 7.42 EDX line scan of the oxidised Co layer on Crofer 22APU substrate (Fu and coworkers 2011)

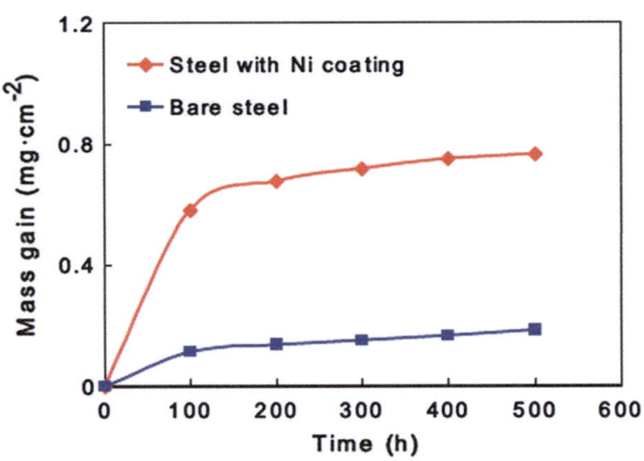

Fig. 7.43 Oxidation kinetics of steel with nickel coating after heat treatment at 800°C (Geng and coworkers 2012)

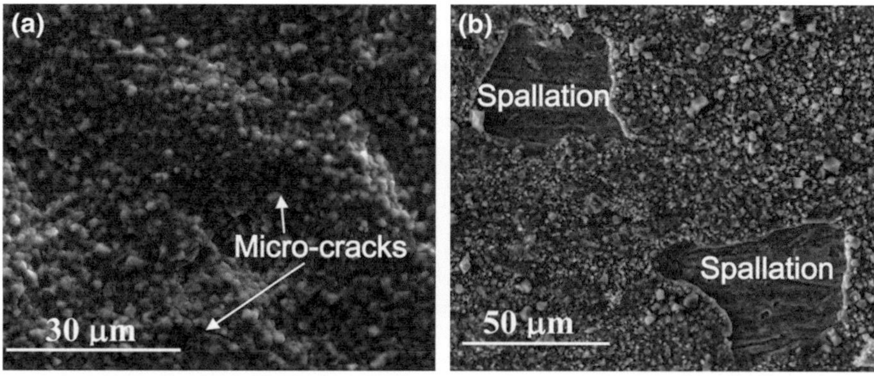

Fig. 7.44 SEM surface morphologies of oxide scale formed on steel with Ni coating **a** and bare steel **b** after oxidation for 500 h at 800°C in air (Geng and coworkers 2012)

difference in oxide layer morphology for bare and coated steel. A double layer oxide structure formed on the Ni-coated steel. Diffusion of Fe and Cr from steel substrate accrued. The Ni coating not only lowered the surface oxide scale ASR but also suppressed the evaporation of the inner Cr_2O_3 layer formed from the steel substrate.

Bibliography

Batfalsky P, Haanappel VAC, Malzbender J, Menzler NH, Shemet V, Vinke IC, Stenbrech RW (2006) J Power Sources 155:128

Choi SR, Bansal NP, Garg A (2007) Mater Sci Eng A 460–461:509

Chou YS, Stevenson JW, Gow RN (2007a) J Power Sources 168:426

Chou YS, Stevenson JW, Gow RN (2007b) J Power Sources 170:395

Chou YS, Stevenson JW, Choi JP (2014a) J Power Sources 255:1

Chou YS, Stevenson JW, Choi JP (2014c) J Power Sources 257:444

Fu QX, Sebold D, Tietz F, Buchkremer HP (2011) Solid State Ionics 192:376

Geng S, Wang Q, Wang W, Zhu S, Wang F (2012) Int J Hydrogen Energy 37:916

Ghosh S, Sharma AD, Mukhopadhyay AK, Kundu P, Basu RN (2010) Int J Hydrogen Energy 35:272

Goel A, Tulyaganov DU, Kharton VV, Yaremchenko AA, Agathopoulos S, Ferreira JMF (2007a) J Am Ceram Soc 90:2236

Goel A, Tulyaganov DU, Kharton VV, Yaremchenko AA, Ferreira JMF (2008) Acta Mater 56:3065

Goel A, Pascual MJ, Ferreira JMF (2010a) Int J Hydrogen Energy 35:6911

Goel A, Tulyaganov DU, Pascual MJ, Shaaban ER, Muñoz F, Lu Z, Ferreira JMF (2010b) J Non-Cryst Solids 356:1070

Goel A, Tulyaganov DU, Kharton VV, Yaremchenko AA, Ferreira JMF (2010c) J Power Sources 195:522

Haanappel VAC, Shemet V, Vinke IC, Gross SM, Koppitz TH, Menzler NH, Zahid M, Quadakkers WJ (2005) J Mater Sci 40:1583

Hao J, Zan Q, Ai D, Maa J, Deng C, Xu J (2012) J Power Sources 214:75

Kaur G, Singh K, Pandey OP, Homa D, Scott B, Pickrell G (2013a) J Power Sources 240:458. http://dx.doi.org/10.1016/j.jpowsour.2013.03.142
Kaur G, Homa D, Singh K, Pandey OP, Scott B, Pickrell G (2013b) J Power Sources 242:305. http://dx.doi.org/10.1016/j.jpowsour.2013.05.094.
Lara C, Pascual MJ, Duran A (2004a) J Non-Cryst Solids 348:149
Lara C, Pascal MJ, Prado MO, Duran A (2004b) Solid State Ionics 170:201
Lin CK, Chen JY, Tian JW, Chiang LK, Wu SH (2012) J Power Sources 205:307
Mahapatra MK, Lu K (2011) J Power Sources 196:700
Mahapatra MK, Lu K, Li X, Wu J (2010) Int J Hydrogen Energy 35:7945
Pang X, Gao K, Yang H, Qiao L,Wang Y, Volinsky AA (2007) Adv Eng Mater 9(7):594
Park J, Ozturk A (2008) Thermochim Acta 470:60
Pascual MJ, Kharton VV, Tsipis E, Yaremchenko AA, Lara C, Duran A, Frade JR (2006) 26(15):3315
Pascual MJ, Guillet A, Duran A (2007) J Power Sources 169:40
Puig J, Ansart F, Lenormand P, Antoine L, Dailly J (2011) J Non-Cryst Solids 357:3490
Sharma K, Kothiyal GP, Montagne L, Mear FO, Revel B (2012) Int J Hydrogen Energy 37(15):11360
Smeacetto F, Miranda AD, Chrysanthou A, Bernardo E, Secco M, Bindi M, Salvo M, Sabato AG, Ferraris M (2014) J Am Ceram Soc 97(12):3835
Yang Z, Stevenson JW, Meinhardt KD (2003) Solid State Ionics 160:213
Yang Z, Xia GG, Li XH, Stevenson JW (2007) Int J Hydrogen Energy 32:3648
Ye Y, Yan D, Wang X, Pu J, Chi B, Jian L (2012) 37(2):1710
Zhang T, Zhang H, Li G, Yung H (2010) J Power Sources 195:6795
Zhang T, Brow RK, Fahrenholtz WG, Reis ST (2012) J Power Sources 205:301
Zhu JH, Geng SJ, Ballard DA (2007) Int J Hydrogen Energy 32:3682

Suggested Bibliography

Bansal NP, Gamble EA (2005) J Power Sources 147:107
Chou YS, Stevenson JW, Choi JP (2014b) J Power Sources 250:166
Ghosh S, Sharma AD, Kundu P, Mahanty S, Basu RN (2008) J Non-Cryst Solids 354:4081
Goel A, Tulyaganov DU, Agathopoulos S, Ribeiro MJ, Basu RN, Ferreira JMF (2007b) J Eur Ceram Soc 27:2325
Hoyt KO, Gannon PE, White P, Tortop R, Ellingwood BJ, Khoshuei H (2012) Int J Hydrogen Energy 37:518
Malzbender J (2010) J Eur Ceram Soc 30:3407
Malzbender J, Monch J, Steinbrech RW, Koppitz T, Gross SM, Remmel J (2007) J Mater Sci 46:6297
Widgeon SJ, Corral EL, Spilde MN, Loehman RE (2009) J Am Ceram Soc 92(4):781
Zhang HH, Zeng CL (2014) J Power Sources 252:122

Chapter 8
Interaction of Glass Seals/Electrodes and Electrolytes

SOFC contains solid electrolyte, which reduces any material corrosion problems along with the increased portable applications. A good/ideal SOFC electrolyte has high ionic conduction, no electronic conduction, good mechanical properties, and stability in reducing as well as oxidising atmospheres. The compatibility electrolytes with the adjoining components of SOFC play a key role in determining the cell efficiency. Hence, research is being pursued for fabricating the electrolytes which can be used for low temperature–intermediate temperature applications. High temperature SOFC enhances the fuel flexibility whereas low temperature SOFC relaxes stringent material requirements. This chapter focuses around the compatibility of different electrolytes with the sealants glasses and electrolytes.

8.1 Different Types of Electrolytes

Various types of electrolytes have been investigated depending upon the structure type, i.e., fluorite type, perovskite, brownmillerites, aurivillius, LAMOX, and apatites (discussed in Chap. 3). Yttria stabilised zirconia (YSZ) is the most commonly used material, where yttria is added to enhance the concentration of oxygen vacancies and thus ionic conductivity along with stabilising the cubic fluorite phase. At 300°C, the bulk ionic conductivity for Y_2O_3–ZrO_2 solution is maximum at 3 mol% Y_2O_3 (the conductivity increases dramatically). This is attributed to the fact that the phase boundary between the cubic and tetragonal phases occurs at 3 mol% concentration of Y_2O_3, and the cubic phase is stabilised. Although YSZ gives high conductivity of 0.1 S/cm at 1000°C, the high temperature raises certain material issues such as maintaining chemical as well as mechanical stability. ScSZ (Scandia stabilised zirconia) offers higher conductivity than the YSZ at intermediate temperatures. Below 500°C, the conductivity of ScSZ is lower than YSZ because the activation energy for conduction in ScSZ increases as temperature decreases. For higher scandia contents

*All the glass compositions mentioned in this chapter are listed in table 5.3.

G. Kaur, *Solid Oxide Fuel Cell Components*,
DOI 10.1007/978-3-319-25598-9_8

of almost 10–12 mol%, the decrease in conductivity occurs at lower temperatures because the higher conductivity cubic phase changes to the lower conductivity rhombohedral phase. For avoiding the phase change, ScSZ can be co-doped with yttersium/bismuth etc., along with keeping the scandium content up to 8 mol%. YSZ and ScSZ possess almost similar mechanical properties although their strength can further be improved by adding niobates and alumina. Partially stabilised zirconia also shows enhanced mechanical properties due to a toughening mechanism attributed to the phase transformation from tetragonal to monoclinic phase. The mechanical and electrochemical properties of electrolytes can also be increased by adding alumina and calcium/magnesium to their matrix. For example, alumina addition increases mechanical strength to YSZ, whereas calcium addition leads to the reduction in activation energy of conduction. Even alumina addition also influences the conductivity to a larger extent depending upon its doping concentration.

Gadolinium and samarium-based electrolytes are regarded as potential candidates for the intermediate temperature SOFC due to their high ionic conductivity and higher compatibility with the perovskite cathode materials. As compared to YSZ, ceria exhibits lower polarisation resistance and high conductivity at lower temperatures. Anyhow, due to the *n*-type electronic conduction in doped ceria electrolytes (under reducing conditions), the cell output is reduced leading to lower efficiency. Gadolinium and samarium exhibit lower size mismatch with the ceria, leading to higher conductivity. The values of conductivity for gadolinium-doped ceria (GDC) and samarium-doped ceria (SDC) at 750°C are 6.7×10^{-2} S/cm and 6.1×10^{-2} S/cm, respectively. The doping of CeO_2 is not only restricted to Gd and Sm, but yttrium, lanthanum, ytterbium, and neodymium have been also tried as dopants. Along with the nature of the dopant, the oxygen partial pressure, processing conditions, and sinterability also decide the conductivity and other electrical properties of the electrolyte. The systems such as $Ce_{0.8}Sm_{0.2-x}Y_xO_{1.9}$, $Ce_{1-x-y}Gd_xPrO_{2-z}$, $Ce_{1-a}Gd_{a-y}Sm_yO_{2-0.5a}$, $Ce_{0.85}Gd_{0.1}Mg_{0.05}O_{1.9}$, $Ce_{1-x-y}Sm_xLa_yO_2$, and $Ce_{0.8}Gd_{0.2-x}Y_x$ have yielded promising results. Anyhow, the conductivities of doped ceria other than gadolinium and samarium are lower. Co-doping increases the conductivity: i.e., samarium replacing gadolinium and praseodymium replacing cerium yield increased conductivity of ceria. Apart from this, the effect of Gd doping concentration on the GDC conductivity is shown in Fig. 8.1.

For enhancing the stability of the doped ceria electrolyte, the reduced conductivity could be observed (Fig. 8.1). The densification and sinterability of doped ceria electrolytes also play a major role in their electrochemical properties. A 1% CaO addition to the $Ce_{0.9}Gd_{0.1}O_{1.95}$ promotes its densification and lowers the sintering temperature due to segregation of cobalt at grain boundaries. Additionally, dopants such as Fe_3O_4 scavenge on SiO_2 impurities (commonly ceria electrolytes and inhibits its performance) leading to improved conductivity at grain boundaries.

Doped perovskite structures such as $La_{1-x}Sr_xGa_{1-y}Mg_yO_3$ (LSGM) are promising candidates for low temperature oxide ion conductivity. LSGM possess superior conductivity as compared to the YSZ/ScSZ, whereas the conductivity of GDC is in range of LSGM. At low oxygen partial pressures, LSGM offers higher conductivity as it does not contain easily reducible Ce^{4+} ion. Ishihara and co-workers showed that

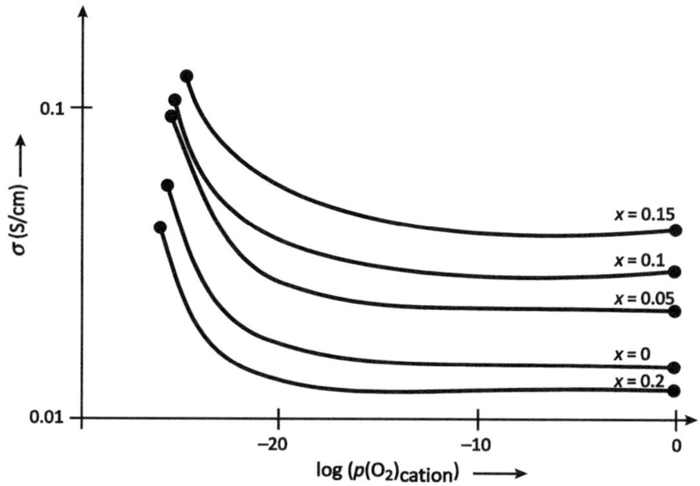

Fig. 8.1 Total conductivity of GDC w.r.t. oxygen partial pressure for different doping concentrations of Gd (At 873 K)

$La_{0.9}Sr_{0.1}Ga_{0.8}Mg_{0.2}O_{3-x}$ exhibit higher oxide ion conductivity than the ScSZ along with almost negligible electronic conductivity for the oxygen partial pressure range of (1×10^{-20}) atm. Good enough showed that $La_{0.9}Sr_{0.1}Ga_{0.8}Mg_{0.2}O_{2.85}$ exhibit $\sigma > 1 \times 10^{-2}$ S/cm (600°C) with the transport number of unity. For another composition $La_{0.9}Sr_{0.2}Ga_{0.83}Mg_{0.17}O_{0.2815}$, the conductivity is 0.17 and 0.08 S/cm at 800 and 900°C, respectively. The chemical stability of LSGM under reducing condition needs to be investigated because $LaSrGaO_4$ and $La(OH)_3$ are some unknown phases which precipitate out due to gallium oxide evaporation. Co-doping of LSGM with Co, Fe, and Ni, etc., increases its conductivity. Figure 8.2 demonstrate the conductivity of YSZ, GDC, and LSGM.

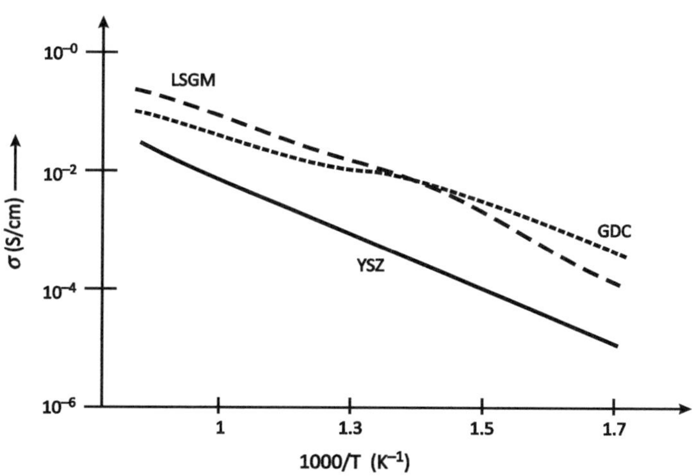

Fig. 8.2 Conductivity of LSGM, GDC, and YSZ

In addition to LSGM, YSZ, and GDC, another category of electrolyte named the LAMOX family ($La_2Mo_2O_9$) has also been reported. As does Bi_2O_3, LAMOX also shows phase transition from non-conductive monocline α to the highly conductive cubic β phase. In order to suppress the phase transitions and enhance the stability in reducing atmospheres, substitutions on the lanthanum and molybdenum site have been investigated. For example, neodymium substitution on the La site, monoclinic α-phase is stabilised over the entire compositional range. In contrast to this, for yttrium and gadolinium substitutions, only a small concentration would stabilise the cubic β form at room temperature and suppresses the phase transition. Molybdenum is substituted by the tungsten (W) for improving phase stability at lower oxygen partial pressure. $La_2Mo_{2-y}W_yO_9$ shows high CTE and chemical reactivity with the electrode materials.

The parent compound is $La_2Mo_2O_9$ and its high temperature β-phase polymorph exhibits isostructure with β-$SnWO_4$. The ionic transport mechanism takes place by a lone pair substitution mechanism according to which the oxygen vacancies could be stabilised by lone electron pairs of cations. For $La_2Mo_2O_9$, the La^{3+} cations occupying Sn^{2+} sites have no lone pair electrons, whereas for β-$SnWO_4$, the outer electrons Sn^{2+} project towards a vacant oxygen site. Therefore, more anion sites are occupied by oxygen ions as half of the cation site occupied by lone pairs in β-$SnWO_4$ are not occupied by the $La_2Mo_2O_9$ lone pair cations. The $\beta \leftrightarrow \alpha$ phase transition usually occurs at 853 K, which can be suppressed by minor doping, for example, $La_2Mo_{1.95}B_{0.05}O_{9-\delta}$, $La_{1.7}Bi_{0.3}Mo_2O_{9-\delta}$, and $La_2Mo_{1.7}W_{0.3}O_{9-\delta}$ exhibit quite high conductivity as shown in Fig. 8.3.

It should be noted that below 1000–1100 K, the $La_2Mo_2O_9$-based materials exhibit electronic conductivity (n-type) which is comparable to the p-type conduction in LSGM electrolytes. With reducing oxygen partial pressure and increasing temperature, the electronic contribution of LAMOX increases.

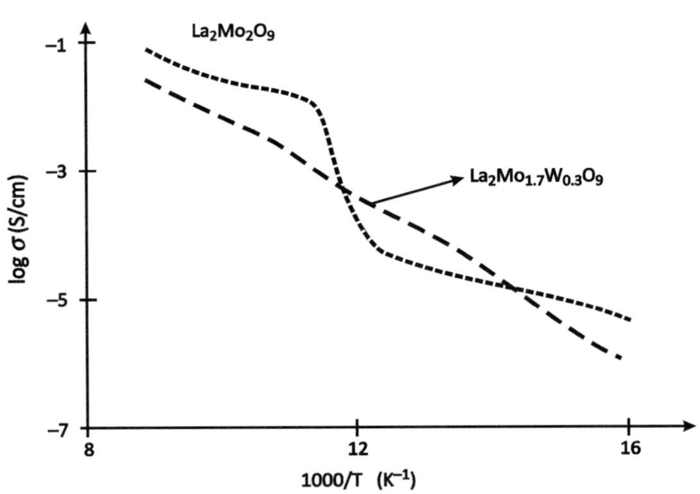

Fig. 8.3 Conductivity of LAMOX family of electrolytes

Bismuth oxide electrolytes have high ionic conductivity, i.e., greater than 15 S/cm at 800°C (β-Bi_2O_3), but it decomposes for low oxygen partial pressures. The major drawback of Bi_2O_3 as an electrolyte is its low melting point of 804°C and the stability temperature up to ≈750°C. At low temperatures, the oxygen vacancies are ordered in Bi_2O_3 which gets disordered when heated above 730°C. The β-Bi_2O_3 phase can be stabilised by doping and lowers the conductivity. The rare earth-doped Bi_2O_3 yields high conductivity but the phase transformation to vacancy ordered rhombohedral phase usually results. For instance, the $(Bi_2O_3)_{0.8}$ $(Er_2O_3)_{0.2}$ yield conductivity of 3.7×10^{-1} Ω cm^{-1} and 2.3×10^{-2} Ω cm^{-1} at 700 and 800°C, respectively, along with the phase transformation. 25–40 mol% doping of Y_2O_3 in Bi_2O_3 suppresses the phase transformation though the conductivity values are lower than the erbium-doped Bi_2O_3. The conductivity of $(Bi_2O_3)_{0.75}$ $(Y_2O_3)_{0.25}$ is 1.6×10^{-1} Ω cm^{-1} and 1.2×10^{-2} Ω cm^{-1} at 700 and 500°C, respectively. Bismuth-based electrolytes get reduced as bismuth metal under anode conditions and using bilayer anodes (anode/electrolyte layer) can reduce this problem.

Rare earth apatites have also been investigated as oxide ion conducting electrolytes. The apatites have general formula M_{10-x} $(YO_4)_6$ O_{2+y}, where Y is Si, Ge, P, and M represents rare earth metal/alkaline earth metal. Figure 8.4 shows the apatite structure where YO_4 tetrahedra are isolated and oxygen ions/M cations form distinct channels.

The larger channels accommodate lanthanum and oxygen ions, whereas small channels contain lanthanum ions and vacancies. One interesting feature of apatites is their tendency to exhibit an isotropic conductivity and the highest conductivity is expected along the *c*-axis. In addition to this, apatites show pure ionic conductivity over a range of oxygen partial pressures, which makes them desirable candidates. Oxygen deficient or excess apatites-like $La_9Sr(SiO_4)_6O_{2.5}$ and $La_{9.33+2x/3}(SiO_4)6O_{2+x}$ ($x > 0$) show high conductivity values. Another noteworthy point of apatites is the conduction process in apatites: i.e., the interstitial oxide ions are involved in a conduction mechanism unlike the oxygen vacancies for perovskite and fluorite-based compounds. It has been found that the oxygen ion conduction occurs through complex sinusoidal migration pathway at the conduction channel periphery and energetically

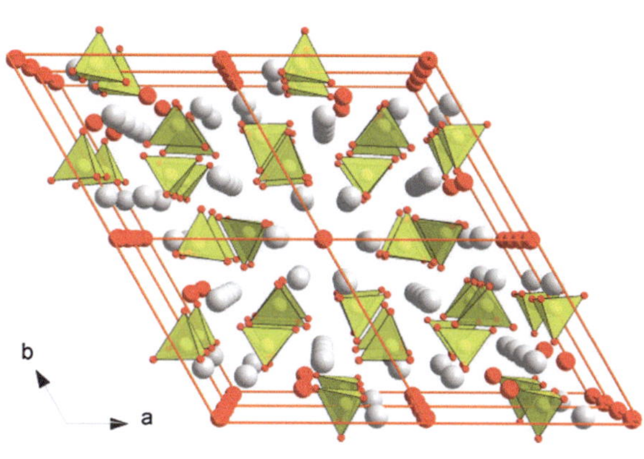

Fig. 8.4 Structure of apatite viewed down the c-axis; the XO_4 tetrahedra are *yellow*, the lanthanum and oxygen atoms are *grey* and *red* spheres (Jacobson 2010)

favourable interstitial site at the channel periphery. Anyhow, better synthesis techniques are required to synthesise the apatites because they require very high temperature for their synthesis. Germanate apatites do not find much applications due to their volatile nature. For the apatites, the M is La, Pr, Nd, Dy, Gd, and Sm usually and the oxygen ionic transport is enhanced upon increased radius of rare earth metal, hence La^{3+}-doped apatites show maximum conductivity. The interstitial migration mechanism is enhanced when the apatite contains a higher concentration of oxygen and highest conductivity for apatites is obtained when it contains 26 oxygens/formula unit. A decrease in ionic conductivity can be seen when the vacancy mechanism dominates in oxygen deficiency. When the *M*-site deficiency occurs, then the doping by suitable metals ion compensates for the oxygen vacancy. $La_{9.33+x/3}Si_{6-x}Al_xO_{26}$ exhibits enhanced ionic conductivity, where aluminium components for the *M*-site vacancy without altering oxygen content.

Another group of electrolytes based on pyrochlore geometry ($A_2B_2O_7$) have been investigated. They can be considered as an (A, B)O_2 flourite structure with unit oxygen vacancy/formula and ordered cation/anion vacancy. The pyrochlores usually disorder into polymorphs of fluorite structure at the high-temperature range of 1650–2500 K. $Gd_{1.8}Ca_{0.2}Ti_2O_{7-\delta}$ has been reported to show the highest conductivity. Calcium incorporation into the $Gd_2Ti_2O_{7-\delta}$ lattice causes increases in *p*-type electronic conduction and decreased *n*-type conduction. These pyrochlores are well suited for their applications as protective layers on to the $LaGaO_3/CeO_2$ electrolytes and thick electrolytes.

8.2 Enhancing the Conductivity of Electrolytes

The conductivity (σ) can be expressed as

$$\sigma = nq\,\mu, \tag{8.1}$$

where

q charge (C)
n carrier concentration (cm^{-3})
μ mobility of charge carrier ($cm^2\ s^{-1}V^{-1}$)

The electrical conductivity of oxides is deeply influenced by the microstructure, grain size, concentration and type of dopants, oxygen partial pressure, and temperature, etc. The following equation is generally used to express the conductivity.

$$\sigma T = A\exp\left[-\frac{E_a}{kT}\right], \tag{8.2}$$

where T is the absolute temperature, k is the Boltzmann constant, E_a is the activation energy, and a is the pre-exponential constant. Activation energy is composed of the oxygen vacancy formation and migration. In this section, we show the effect of different factors on the conductivity.

(a) **Effect of Processing Conditions**
Processing conditions greatly define the microstructure, density, and grain size, hence influencing the conductivity to a larger extent. Processing conditions affect the conductivity through the grain boundaries; e.g., the grain boundary activation energy is observed to be higher for the compression-moulded samples than the dry-pressed samples of SDC (both the samples are made from the same powder of SDC). In addition to this, the processing of samples should be done in such a way, that it should involve fewer fabrication steps in order to minimise the possible introduction of impurities. Fast-quenched CaSZ (calcium stabilised zirconia) exhibited higher conductivity than the slowly quenched zirconia. Initial starting constituent particle morphology also influences the electrical properties of electrolytes. Mori and co-workers obtained higher conductivity for the samples synthesised from spherical particles than the elongated particles. Improvement in the fracture strength and electrical conductivity could be observed for the 3–7 mol% ScSZ (sintered at 1300°C) electrolyte, which is not pressed.

(b) **Effect of the Composition**
By choosing a suitable aliovalent dopant, the ionic conductivity of electrolytes can be enhanced because oxygen vacancies are created which provide the migration pathway. Multiple doping increases conductivity by many fields; i.e., $Ce_{1-x-y}GdA_yO_2$(A = Ca, Fe, Pr, for $y = 0.01$, 0.02 and $x = 0.20$, 0.19, 0.18, 0.10, 0.09) have shown significant increase in conductivity. The oxygen partial pressure limit also extended by two orders of magnitude for $Ce_{0.8}Gd_{0.19}M_{0.1}O_2$ (A = Sm, Pr). In contrast to this, alumina addition (0–10%) to 20% GDC led to overall decrease in ionic conductivity. This is attributed to the fact that the increasing alumina content led to decrease in grain boundary conductivity and increase in grain interior conductivity (due to solid-state reaction between gadolinia and alumina leading to decrease in gadolinia dissolution in ceria). For YSZ system, the doping with calcia has resulted in increased conductivity due to local microstructure. For zirconia co-stabilised with 8 mol% scandia and yttria, the conductivity decreased below 370°C upon the increase in scandia doping levels. When rare earth metals are added to 12% YSZ, then the conductivity was almost comparable to that of high-purity zirconia. The scandia and ceria co-doped ZrO_2 system exhibited higher conductivity than the YSZ system in the temperature range of 300–1100°C. Heterogeneous doping has been implied to increase the conductivity, i.e., the dopants which are soluble/insoluble with low limit in ceria/zirconia bulk structure. Alumina is the best dopant for limited solubility purposes, though its concentration determines its effect on the conductivity. Guo and co-workers obtained an increase in grain boundary resistivity by 600%, whereas the grain

resistivity increased by 3% only upon addition of 0.4 mol% alumina to the electrolyte. Steele and co-workers obtained an increase in the overall conductivity when alumina additions were made to the high-purity YSZ. Feighery and co-workers found that 1 wt% alumina gets dissolved into YSZ structure upon sintering at 1500°C for 24 h. Ceria/zirconia composites have also been developed by either embedding ceria particles in YSZ matrix or YSZ-doped ceria multi-layers. The ionic conductivity of GDC-YSZ composite electrolytes decreased upon 10–50 wt% addition of GDC to YSZ attributed to the formation of a solid solution responsible for the reduced conductivity. Moreover, pore formation at the YSZ/GDC interface leads to decreased conductivity for YSZ/GDC electrolytes. For YSZ–SDC composite the decrease in SDC particle size influenced the temperature for YSZ–SDC reaction.

(c) **Effect of Grain Size, Grain Boundaries, and Microstructure**
Many researchers have observed the grain size effects below 100 nm (grain size) for the polycrystalline solid electrolytes. For the ceria-based electrolytes, the enhancement of electronic conductivity is attributed to the decrease of defect formation energy at the grain boundaries. The researchers also obtained that the microcrystalline cerium oxides exhibit electronic conductivity under the same conditions of temperature, dopant concentrations, and oxygen partial pressure. Upon decreasing size in ceria-based electrolytes, the activated energy decreases, and the electrical conductivity and lower enthalpy in nanocrystalline film of 16 mol% YSZ with average grain size of 20 nm are observed as compared to microcrystalline samples; nanocrystalline specimens exhibit 2–3 order higher electrical conductivity. For $Zr_{0.84}Ca_{0.16}O_{1.84}$, electrical conductivity is observed to increase with decreased grain size below 800°. For large grain size, the activation energy is in the range of 1.29–1.30 eV, whereas for small grains, its value is ≈0.96 eV. Most of the researchers also reported decreased conductivity upon decreasing grain size in the range of 0.2–20 μm. For instance, the conductivity of $GdxZr_{(1-x)}O_{(2-0.5x)}$ [grain size 0.4–20 μm] and 5.7 mol% YSZ [grain size 0.2–18 μm] decreases for lower grain size.

The conductivities of polycrystalline material are deeply influenced by the grain, grain boundary, as well as microstructure. At grain boundaries, the properties are determined by impurities, microcracks, segregation of secondary phases, and mismatch of lattices. Usually, the grain boundaries provide crystallographic mismatch zones which provide rapid mass transport as well. Hence, the quantum of charge carriers and band structure modification occurs at grain boundaries, which largely affects the electrical conductivity. The impedance plot describing the different contributions to conductivity is shown in Fig. 8.5.

The very famous brick layer model is used to correlate the grain/grain boundary conductivity to the microstructure of the specimen. The brick layer model assumes ceramic samples to consist of grains with high conductivity, which are separated by uniform grain boundaries. Positive space change potentials are obtained when

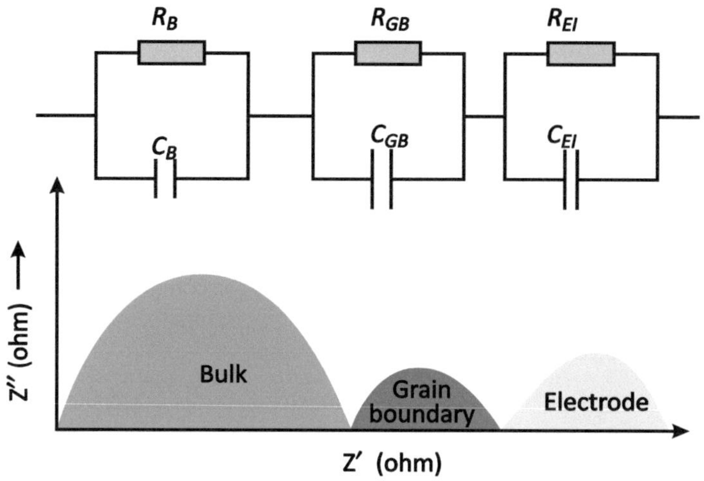

Fig. 8.5 Equivalent circuit and impedance plot for the polycrystalline material

negative charge carriers are enhanced and positive charge carriers are depleted at the grain boundaries. As silica acts as an impurity in zirconia and ceria-based electrolyte it depletes the grain boundary conductivity. By tapping the impurities using second-phase additives, the ionic conductivity could be increased. Other ways to avoid silica segregation at the grain boundaries include preserve variation and optimisation of sintering temperatures. The temperature plays a vital role in determining the grain boundary resistance; i.e., the grain boundary resistance becomes negligible as compared to the grain resistance at higher temperatures. The dopant concentration affects the ground boundary resistance in addition to the ionic conduction. The grain boundary resistance is found to decrease when the dopant concentration increases for the ceria-doped electrolytes [dopants are 9 mol% Gd_2O_3, Sm_2O_3, and La_2O_3]. For ScSZ, the high conductivity is ascribed to the formation of metastable tetragonal phase known as the t'-phase and this t'-phase exhibits a particular bonding microstructure. The bonding occurs due to the formation of twin bands due to strain produced when the cubic phase changes to the tetragonal phase. Upon annealing of the samples, the banding disappears and grain structure becomes mottled which decreases the ionic conductivity. It is also suggested that the ionic conductivity can be increased by creating additional dislocations using irradiation techniques. Composite electrolytes and introduction of a liquid phase in the grain boundaries of ceria and zirconia-based electrolyte lead to enhanced conductivity.

8.3 Interaction of Electrolyte and Glasses

As discussed in Chaps. 6 and 7, the seal must possess gas-tightness, hermeticity, and limited chemical reactivity with the adjacent components of SOFC especially interconnect and electrolytes. With the interconnect, the seal usually tends to make detrimental chromate phases, whereas with electrolytes, the tendency of glass seals is to form a zirconate phase (which is usually not detrimental except in low electric conduction). During SOFC operation, the parent glasses transform to daughter glass–ceramics whose CTE is different from the parent composition. In this section, the focus is to investigate the compatibility of glass/glass–ceramics with high and low temperature electrolytes for SOFC applications.

8.3.1 Alkali Metal-Based Glasses/Electrolytes Interface

Alkali metal-based glasses have been extensively investigated by Smeacetto and co-workers. They studied sodium-based glass/glass–ceramics with Crofer 22 APU, AISI 430 and YSZ electrolytes. SACN1 glass [45% SiO_2, 9% Al_2O, 18% CaO, 18% Na_2O, 10% B_2O_3, mol%], and SACN2 glass [40% SiO_2, 9% Al_2O, 18% CaO, 23% Na_2O, and 10% B_2O_3, mol%] systems are studied extensively for the interaction with YSZ along with their compatibility with the interconnects. SACN1 has relatively lower value of CTE than the steels as well as YSZ electrolytes, and therefore SACN2 glass was investigated for the interaction mechanism with the YSZ electrolyte. Figure 8.6 shows the cross-section of SACN2/YSZ diffusion couple formed at 1000°C [for 30 min] in argon atmosphere. The interface reveals no pores, cracks, or delamination. Furthermore, the composite SACN2 seal (with 10–20 mol% alumina particles, at 1020°C for 30 min in argon atmosphere) revealed a flawless interface with the YSZ after exposure to $H_{2-3\%}$ H_2O atmosphere or 200 h at 800°C. Another composition SACN [53–58% SiO_2, 6–8% Al_2O_3, 24–26% CaO,

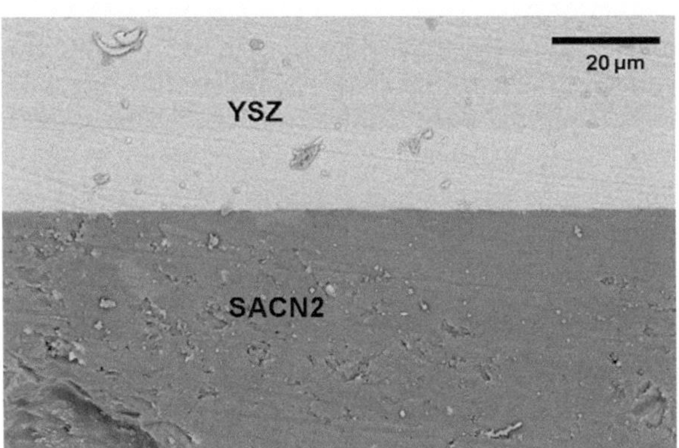

Fig. 8.6 Interface between the SACN2 and YSZ (1000°C, 30 min) (Smeacetto and coworkers 2008a)

and 10–12% Na_2O, mol%] was studied by Smeacetto and co-workers to seal Crofer 22APU to the anode supported electrolyte. The composition SACN exhibit matching CTE of 10.7 × 10^{-6}/K with the interconnect and YSZ electrolyte. Figure 8.7 shows the interface region between SACN glass–ceramic and the YSZ, which clearly depicts the pore/crack free interface. The electrolyte is anode supported, hence the topmost region is anode, whereas the middle region is the YSZ electrolyte (~5 μm in thickness). Even after the mid-term exposure at 800°C for 200 h in reducing atmosphere, the seal remained intact and no interfacial delamination could be observed with YSZ electrolyte.

Usually CaO from the glasses, initiates the formation of monoclinic ZrO_2 as reported by Lahl and co-workers, but no such phase formation could be seen for the SACN/YSZ couple. SACN glass seal was also studied for joining Crofer 22APU to the YSZ electrolyte. Figure 8.8 reveals the Crofer 22APU/SACN glass/YSZ diffusion couple made at 900°C [heating rate 5°C/min with dwelling time of 30 min] after thermal ageing. No structural modifications could be seen at the interface of SACN with the YSZ. The heating rate influences the residual porosity of the SACN glass; i.e., with the higher heating rate of 5°C/min, low residual porosity is observed. This can be attributed to the more effective removal of entrapped gases in the glass sealants upon decreasing the heating rate. Usually, higher heating rates are attributed to overcome concurrent crystallisation, whereas the opposite trend is observed for the SACN glass. It could be due to the effective removal of gases from the joint region with low glass sealant phase viscosity at lower heating rate. After thermal ageing (Fig. 8.8), the CTE and crystal phase composition did not alter indicating the thermomechanical stability of the SACN glass–ceramic seals after exposure at 800°C in air for 400 h heat treatment duration. Another phase composition with low-sodium covalent named SACNZn, SACNLz, and SACNMg were studied which contained 9.1% ZnO, [5% La_2O_3 + 5.7% ZnO], and 0.3% MgO, respectively.

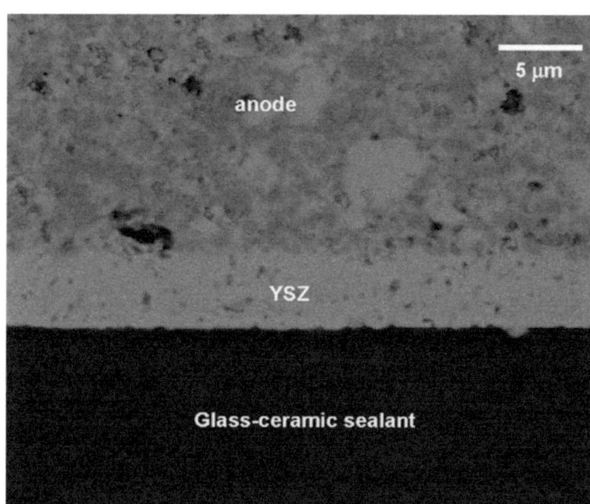

Fig. 8.7 Interface between the SACN and the YSZ side of ASE in Crofer22 APU/SACN/ASE joined sample (Smeacetto and coworkers 2008b)

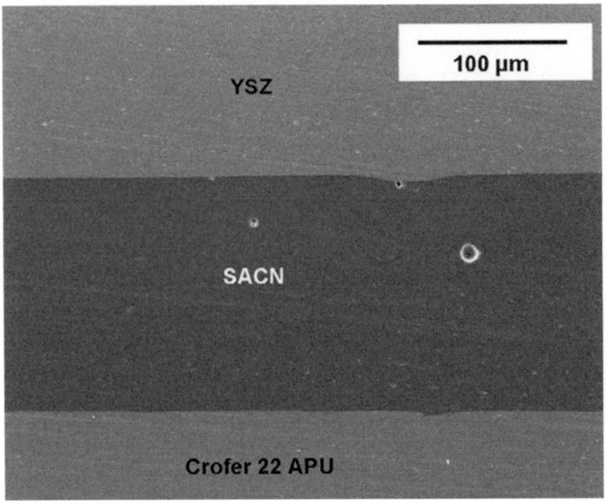

Fig. 8.8 Crofer 22 APU/SACN/YSZ joined structure obtained at 900°C (heating rate 5°C/min), after 400 h of exposure in air at 800°C (Smeacetto and coworkers 2008c)

Upon lowering the sodium content, the glassy materials expected to exhibited higher characteristics temperatures. SACNMg is mostly amorphous, which makes it intrinsically brittle. In contrast to this, the XRD of SACNLZ glass revealed calcium zinc aluminium silicates, calcium silicate, and sodium aluminium silicate phases. SACNLZ revealed very poor adhesion with the YSZ electrolyte along with its poor wettability and sintering capability. At 950°C, the HSM experiments revealed very poor softening and sintering capability of SACNLZ glass samples. SACNZn composition has been prepared from SACNMg composition by adding zinc up to 10 wt%. Addition of ZnO increased CTE and decreased viscosity in the glass transition range. For SACNZn samples, the DTA peaks also shifted towards lower temperature when the particle size decrease from 75 to 38 μm. This indicates the existence of surface crystallisation in the samples. The XRD of SACNZn revealed the calcium zinc aluminium silicate ($Al_{1.25}Ca_{1.94}O_7Si_{1.38}Zn_{0.43}$) and sodium calcium aluminium silicate ($Al_{1.8}Ca_{0.2}Na_{0.8}O_8Si_{2.8}$) phase formation. Crystallisation is a desirable phenomenon which enhances thermomechanical properties of the sealant particularly for high-temperature applications. Figure 8.9 is the SEM image of SACNZn/YSZ interface after heat treatment at 800°C under humidified $H_2{-}3H_2O$ atmospheres. No structural modification could be seen and the interface is free of pores/cracks indicating high thermomechanical compatibility of SACNZn with YSZ. Hence, SACNZn showed excellent sealing capabilities and can be regarded as a potential sealant among SACNZn and SACNMg.

Smeacetto and co-workers also tested the potassium-sodium glass KMBY [54.39% SiO_2, 11.26% Al_2O_3, 9.02% CaO, 13.78% Na_2O, 8.37% MgO, 1.67% K_2O, 0.9% B_2O_3, and 0.61% Y_2O_3, wt%] with spinel-coated interconnects and anode supported electrolyte. Nepheline ($Na_2K(Al_{0.44}Si_{0.36})O_{16}$, diopside ($Ca_{0.8}Mg_{1.11}Si_2O_6$), and akermanite ($Ca_2MgSi_2O_7$) phases precipitated out for KMBY after heat treatment at 850°C for 2 h.

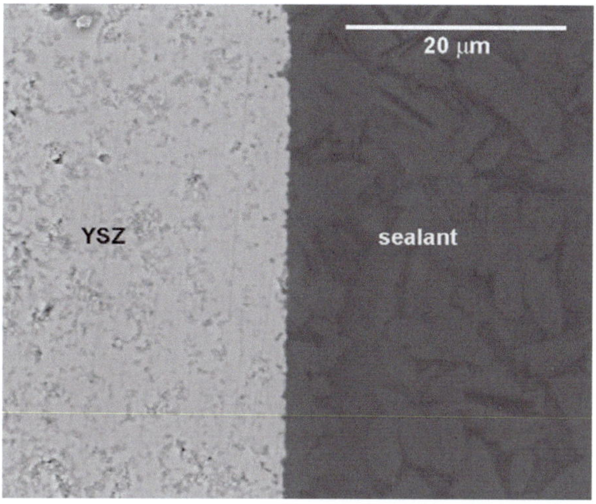

Fig. 8.9 Interface of the YSZ/SACNZn glass–ceramic interface after 300 h of exposure (at 800°C) to H_2–3 vol.% H_2O atmosphere (Smeacetto and coworkers 2010)

Diopside is the main phase, but the nepheline amount grew after longer heat treatment durations. Figure 8.10 shows the SEM micrograph of YSZ/KMBY glass after thermal cycling of 500 h. No cracking could be seen in YSZ or KMBY glass–ceramic along with strong adherence of KMBY to the YSZ. Even after prolonged thermal cycling, no detachment of YSZ from the functional anode layer could be seen. The EDS of different points for Fig. 8.10 is given in Table 8.1.

The elemental analysis from Table 8.1 indicates that the nepheline and diopside phases are present in addition to the amorphous glassy phase, hence revealing stability of the KMBY glass–ceramic after 500 h heat treatment. The sodium, potassium, and lithium (minor constituent) containing glass SCN-1 was studied for the isothermal ageing tests in dual environments of dilute hydrogen inside the test

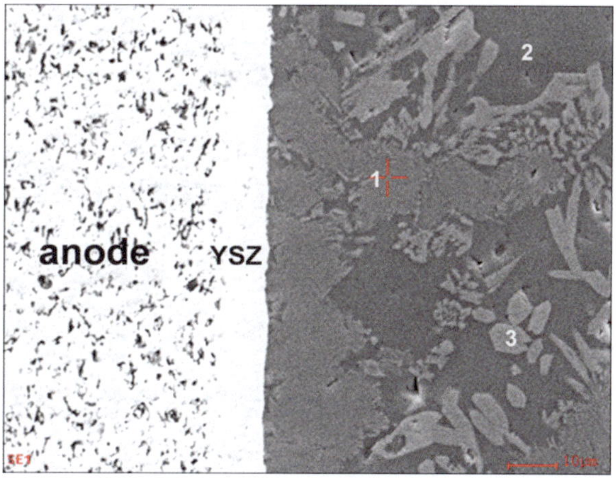

Fig. 8.10 SEM micrograph of the KMBY glass–ceramic/YSZ interface after 500 h of thermal cycling (Smeacetto and coworkers 2014)

Table 8.1 Elemental point analysis for Fig. 8.10

Element Al%	Point 1	Point 2	Point 3
Si	28.7	28.4	25.3
Ca	5.7	0.6	10.5
Al	6.4	6.7	0.5
Mg	6.9	3.3	11.1
Na	5.9	9.9	–
O	51.2	52.0	52.5

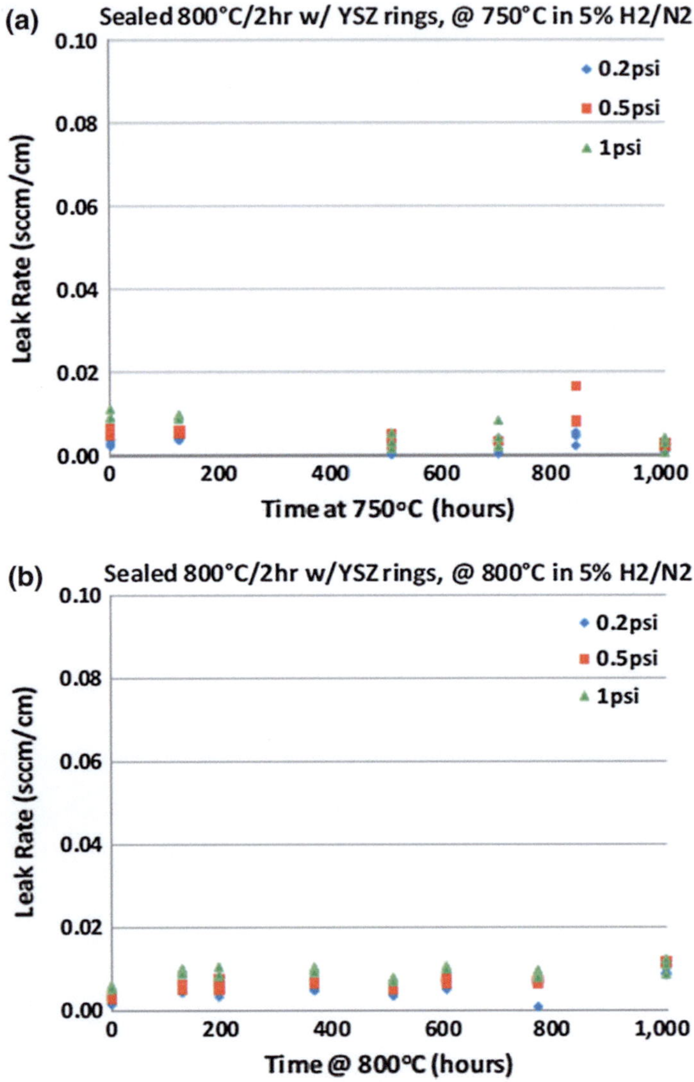

Fig. 8.11 Isothermal-aged SCN-1 glass seals (between ceramic bilayers and aluminised SS441 substrate and contained between concentric YSZ layers) tested at **a** 750°C and **b** 800°C. (Chou and coworkers 2011)

fixture [ambient air surroundings] for 750 and 800°C. Figure 8.11a, b shows the leak rates during isothermal ageing at 750 and 800°C for SCN-1 (with YSZ).

As evident from Fig. 8.11, the leak rates are independent of the helium pressure and the majority of leak rates are less than 0.01 Sccm cm^{-1}. The desired thermal stability could not be demonstrated by the SCN-1 glass as at 800°C, the sample showed increase of leak rates to 0.19, 0.42, 0.58 Sccm cm^{-1} for the He pressure of 1.4, 3.4, and 6.8 kPa, respectively. These leak rates are quite high as compared to the leak rates observed in hybrid micas. A similar increasing trend was also observed for the sample at 750°C, indicating fracture of glass seals/ceramic bilayers. The YSZ electrolyte/SCN-1 glass interface (aged at 800°C for 1000 h in dual environment) reveals good adhesion, though some localised precipitates of grey particles appear in the glass matrix. The grey particles are mainly oxides composed of Si (29 at.%) with minor traces of Al and K. The SCN-1 glass contains 17 wt% of alkalis and is low viscosity glass (10^5 Pas at 800°C). The low viscosity of SCN-1 glass makes it corrosive and the dissolution of 8 YSZ is expected into the glass matrix. Anyhow, no detectable Y or Zr could be observed at the interface which indicates the chemical compatibility of YSZ with SCN-1 glass. The BiBaSi composition doped with Al_2O_3 has also shown promising results with YSZ due to little interdiffusion of its constituents. The glass region contains closed pits and porosity in spite of no delamination at the joint region.

8.3.2 Magnesium-Based Glasses/Electrolyte Interface

ML glass [30% MgO, 40% SiO_2, 20% B_2O_3, 10% La_2O_3, mol%] have been studied for interaction with the high and low temperature electrolytes, i.e., YSZ and titania-doped bismuth vanadate (TBV), respectively. Bismuth vanadate is a favourable oxygen ion conductor with aurivillius structure for SOFC applications. When the vanadium site is doped with different cations, then the electrical conductivity and phase charge behaviour are highly influenced. Usually, Cu, Ni, Ti, Zn, Al, Sb, Nb, Fe, and Pb are common dopants used to increase the conductivity by γ-phase stabilisation. $Bi_2V_{1-x}Ti_xO_{5.5-\delta}$ [for $x = 0.05$, 0.1, 0.15 and 0.2] have been studied as intermediate temperature SOFC electrolyte. For $x = 0.05$, the diffraction pattern consists weak reflection at $2\theta = 24.22°$, whereas doublets appear at $2\theta = 31°$, 39°, 48°, and 54° which clearly indicates the presence of α-phase. In contrast to this for $x = 0.1$ [$Bi_2V_{0.90}Ti_{0.1}O_{5.5-\delta}$], sample shows two supterlattice peaks at 31.8 and 32.4°C with splitting indicating β-phase of this electrolyte.

For $Bi_2V_{0.85}Ti_{0.15}O_{5.5-\delta}$ and $Bi_2V_{0.80}Ti_{0.20}O_{5.5-\delta}$ electrolytes, no superlattice diffraction peaks are observed indicating γ-phase stabilisation. The confirmation of γ-phase stabilisation is also evident from the DTA and TGA thermogram as shown

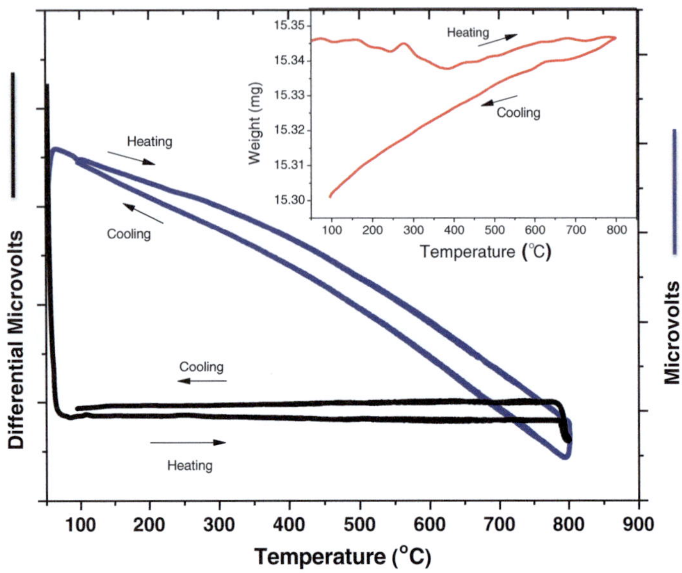

Fig. 8.12 DTA, TGA and differential DTA for the sample $Bi_2V_{0.80}Ti_{0.20}O_{5.5-\delta}$ (Kaur and coworkers 2014a)

in Fig. 8.12. Usually peaks between 400–420 and 520–540°C during the heating cycle signify the α to β phase (α/β) and β to γ (β/γ) phase transition, respectively. During cooling cycle, peaks between 400–420 and 500–520°C represent β/α and γ/β phase transition, respectively.

Differential DTA and DTA show the peak positions in correlation to each other and the advantage of the differential method lies in the fact that the peak detection is easy even when the peaks are overlapped or notdetectable. No evidence of α/β or β/γ phase could be seen during the heating cycle indicating stabilisation of the γ-phase. The increase in crystallographic distortion leads to increased configurational entropy and hence, large oxygen vacancies in V-O layers reduce the ordering of γ to the β/α-phase. The inset of Fig. 8.12 shows the weight change during the heating and cooling cycles indicating reduction during the cooling cycle. The CTE and differential CTE with respect to temperature for $Bi_2V_{0.95}Ti_{0.05}O_{5.5-\delta}$ and $Bi_2V_{0.80}Ti_{0.20}O_{5.5-\delta}$ samples are shown in Fig. 8.13a, b, respectively, and the CTE of all the samples are listed in Table 8.2.

The differential CTE has been calculated from the differential of the expansivity curve, and it gives CTE at any temperature. The technical CTE has been obtained from the slope of the expansivity curve, and it gives the CTE within a defined temperature range. $Bi_2V_{0.80}Ti_{0.20}O_{5.5-\delta}$ oxide exhibits higher CTE values than other samples indicating higher its disorder. It is evident from Fig. 8.13 that two peaks appear at 422 and 552°C for $Bi_2V_{0.95}Ti_{0.05}O_{5.5-\delta}$, which corresponds to α/β and β/γ transitions, whereas no such phase changes could be seen for the $Bi_2V_{0.80}Ti_{0.20}O_{5.5-\delta}$ sample indicating γ-phase stabilisation. In addition to this, the

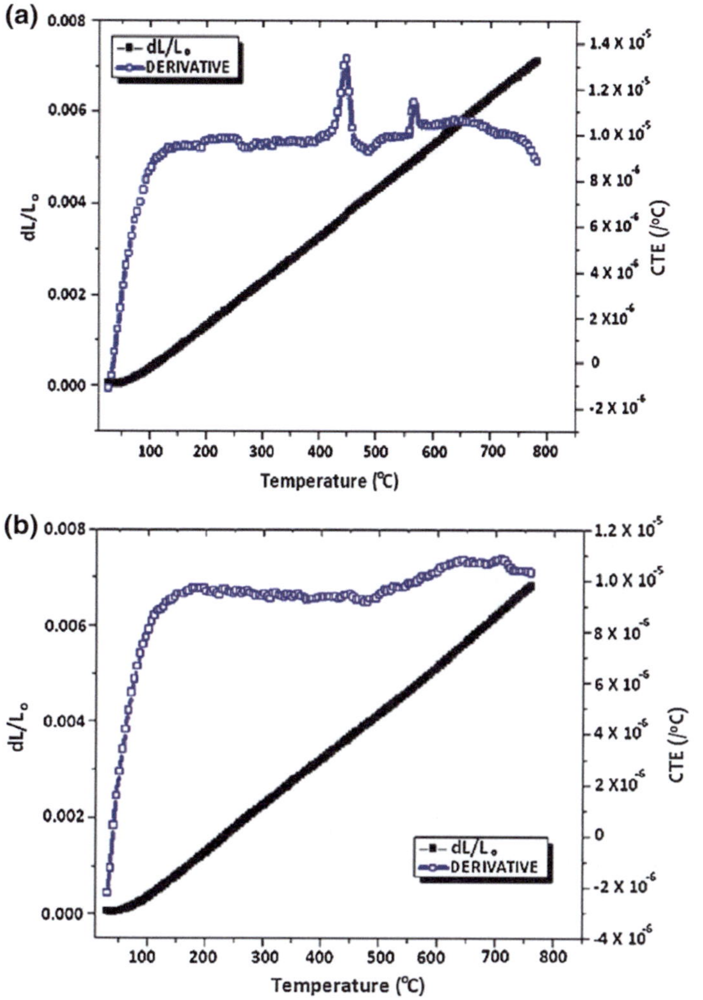

Fig. 8.13 Expansivity and its derivative curves for the sample **a** $Bi_2V_{0.95}Ti_{0.05}O_{5.5-\delta}$ and **b** $Bi_2V_{0.80}Ti_{0.20}O_{5.5-\delta}$ (Kaur and coworkers 2014a)

Table 8.2 CTE during phase transitions for titania-doped bismuth vanadates

Sample name	Technical CTE from RT-800°C (/°C)	Technical CTE from 600–800°C (/°C)	Physical value of CTE at 800°C (/°C)	α/β	β/γ
$Bi_2V_{0.95}Ti_{0.05}O_{5.5-\delta}$	9.56×10^{-6}	9.73×10^{-6}	8.82×10^{-6}	13.8×10^{-6}	11.9×10^{-6}
$Bi_2V_{0.90}Ti_{0.10}O_{5.5-\delta}$	10.12×10^{-6}	9.82×10^{-6}	9.36×10^{-6}	12.2×10^{-6}	11.2×10^{-6}
$Bi_2V_{0.85}Ti_{0.15}O_{5.5-\delta}$	9.93×10^{-6}	9.89×10^{-6}	10.06×10^{-6}	10.4×10^{-6} at 302°C	
$Bi_2V_{0.80}Ti_{0.20}O_{5.5-\delta}$	10.04×10^{-6}	9.91×10^{-6}	10.25×10^{-6}	No transition	

$Bi_2V_{0.80}Ti_{0.20}O_{5.5-\delta}$ sample gives maximum conductivity of 1.17×10^{-3} S/cm at 800°C indicating optimisation of vacancy concentration and O^{2-} pathways for this sample. The sample $Bi_2V_{0.80}Ti_{0.20}O_{5.5-\delta}$ also shows high density along with reduction during the cooling cycle and homogeneous microstructure. Hence, for the interaction purpose, $Bi_2V_{0.80}Ti_{0.20}O_{5.5-\delta}$ has been chosen.

Figure 8.14a, b shows the XRD of coupled glasses with TBV and YSZ electrolytes, respectively indicating no phase formation even after 100 h heat treatment. After 750 h heat treatment at 800°C, coupled TBV showed pyrochlore ($La_2Si_2O_7$) and enstatite ($MgSiO_3$) phase formation which may follow the reactions kinetics as

$$La_2O_3 + 2SiO_2 \longrightarrow La_2Si_2O_7 \tag{8.3}$$

$$MgO + SiO_2 \longrightarrow MgSiO_3 \tag{8.4}$$

Pyrochlore is an ordered phase; no titania rich phase could be observed (Fig. 8.14a). The coupled glass with YSZ also showed enstalite phase formation after prolonged heat treatment of 750 h at 850°C and no zirconate phase could be seen. One thing must be noted: the enstatite phase peak intensity is lower in coupled YSZ glass than the coupled TBV glass indicating lower crystallavity for former coupled glass. No pyrochlore phase could be seen in the coupled YSZ glass, which could be due to suppression of nucleation and retardation of phase separation by ZrO_2 diffusion on to the ML glass.

The central region of the cross-section of ML/TBV and ML/YSZ diffusion couple heat treated for 750 h at 800 and 850°C, respectively, is shown in Figs. 8.15 and 8.16. Figure 8.15a indicates crack formation at the interface along with pits and pores inside the ML glass. No accumulation of elements occurred at the ML/TBV interface region as evident from the electron probe microanalysis (EPMA) (Fig. 8.15b). The line profile is oscillatory is nature and attributed to the low density/concentration of elements, interaction among X-rays of different elements, and composition difference of glassy and devitrified phases. The concentration of every element dropped at the interface indicating discontinuity at the interface region. Vanadium has shown the highest diffusion towards the ML glass side among all the elements, whereas bismuth has shown the least diffusion towards the glass side indicating its sluggish diffusion kinetics (due to larger ionic radii).

Figure 8.16a reveals good adherence of ML glass to YSZ electrolyte even after 750 h heat treatment at 850°C. No delamination of the ML seal could be seen at the interface indicating good bonding characteristics of the ML/YSZ diffusion couple. Anyhow, the pits and pores are seen inside the ML glass and are attributed to the volatility of entrapped gases during the glass formation process. The low CTE of the enstatite phase may be responsible for the tensile stress inside the glass causing pit formation. The dot mapping (Fig. 8.16b) indicates no accumulation of ions at the interface although diffusion has taken place on either side of the interface. No zirconate formation has taken place at the interface, which is in good correlation with the XRD pattern for the couple YSZ glass.

Kumar and co-workers have found that $ZrSiO_4$ phase formation has taken place for coupled MgY glass (30% MgO, 40% SiO_2, 20% B_2O_3, 10% Y_2O_3, mol%) with

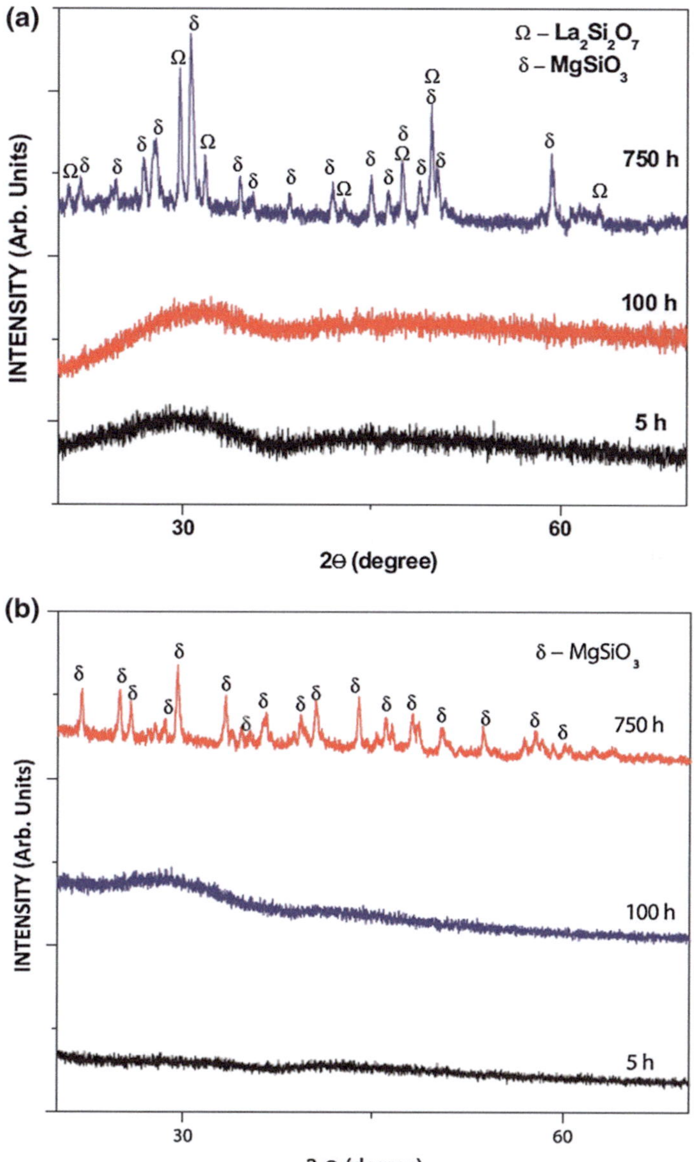

Fig. 8.14 XRD of coupled ML glass with **a** TBV and **b** YSZ electrolyte (Kaur and coworkers 2012b)

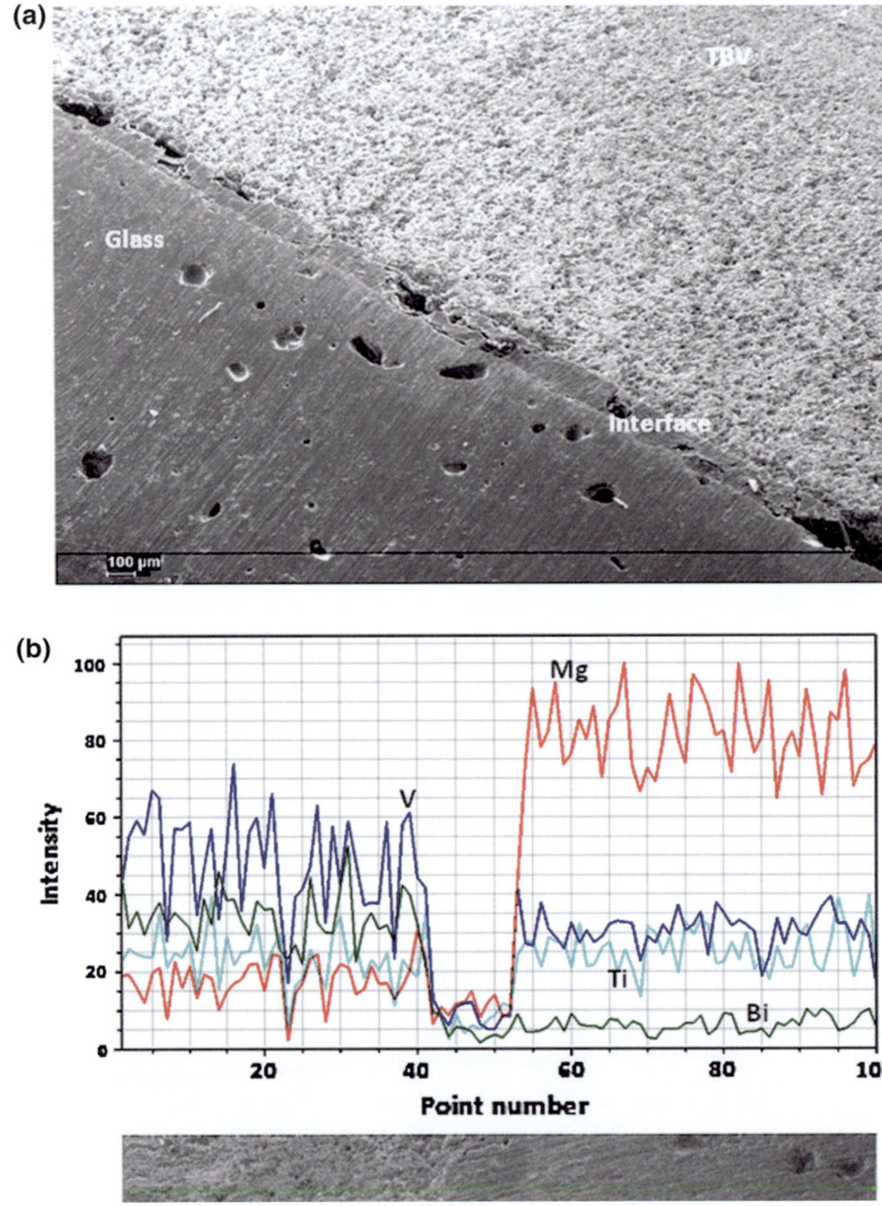

Fig. 8.15 **a** Interface and **b** EPMA of TBV/ML glass diffusion couple after heat treatment at 800°C for 750 h (Kaur and coworkers 2012b)

YSZ electrolyte. Interestingly the, ordered pyrochlore $Y_2Si_2O_7$ phase and low CTE enstatite phase also precipitate for thermally cycled coupled MgY glass (Fig. 8.17) after 500 h heat treatment.

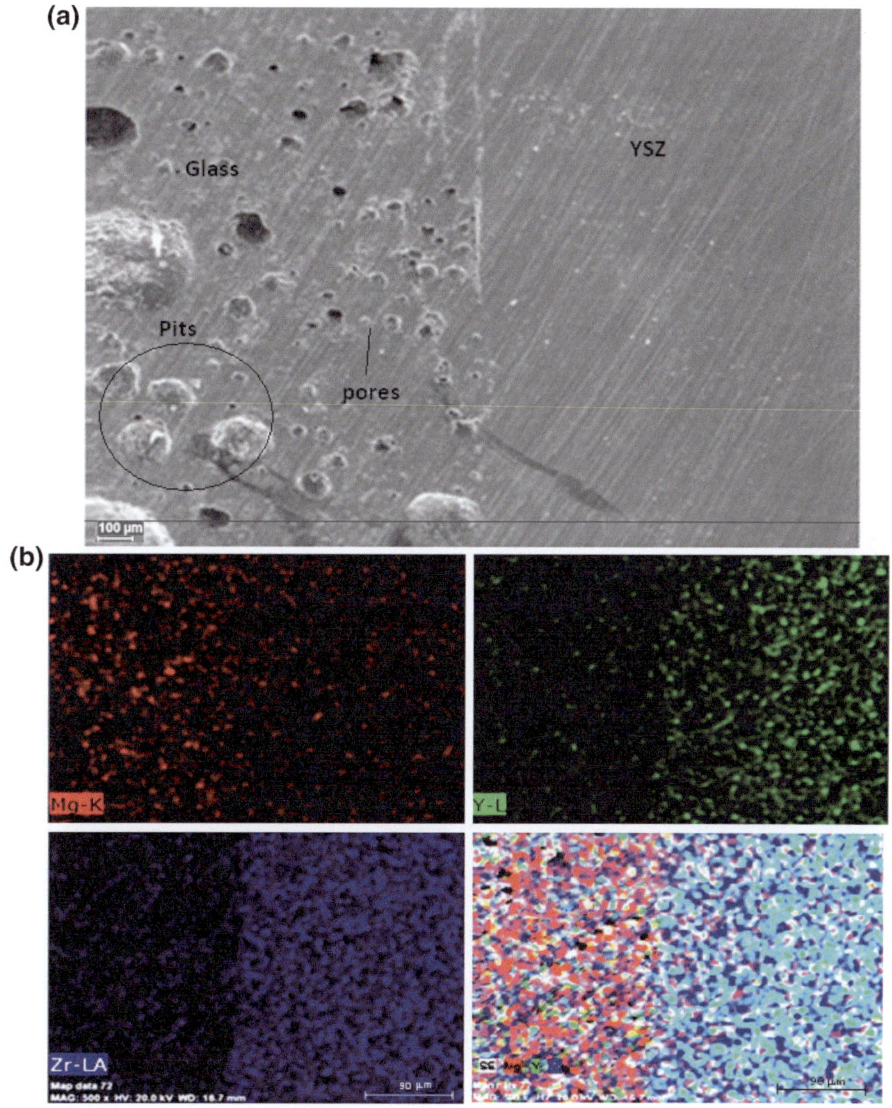

Fig. 8.16 **a** Interface and **b** EPMA of YSZ/ML glass diffusion couple after heat treatment at 850°C for 750 h (Kaur and coworkers 2012b)

$ZrSiO_4$ is a thermally stable phase (with melting point of 2200°C) and may be formed after combination of diffused ZrO_2 and SiO_2 of the glass network. $Y_2Si_2O_7$ phase is a high-temperature polymorph of yttrium disilicate and can withstand temperatures in oxidising atmospheres. Figure 8.18 gives the microstructure of a MgY/YSZ diffusion couple after 1 h, 100 h, 500 h and thermally cycling from RT-850°C (heat treated at 850°C). Intact, pore-free, and continuous interface are

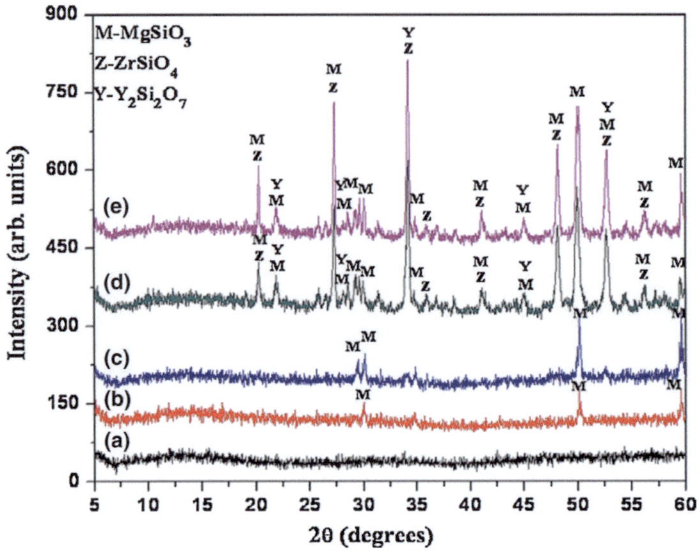

Fig. 8.17 X-ray diffractogram: *a* as prepared MgY glass and MgY/YSZ couple; *b* heat treated for 1 h; *c* thermally cycled; *d* heat treated for 100 h; *e* heat treated for 500 h (Kumar and coworkers 2014)

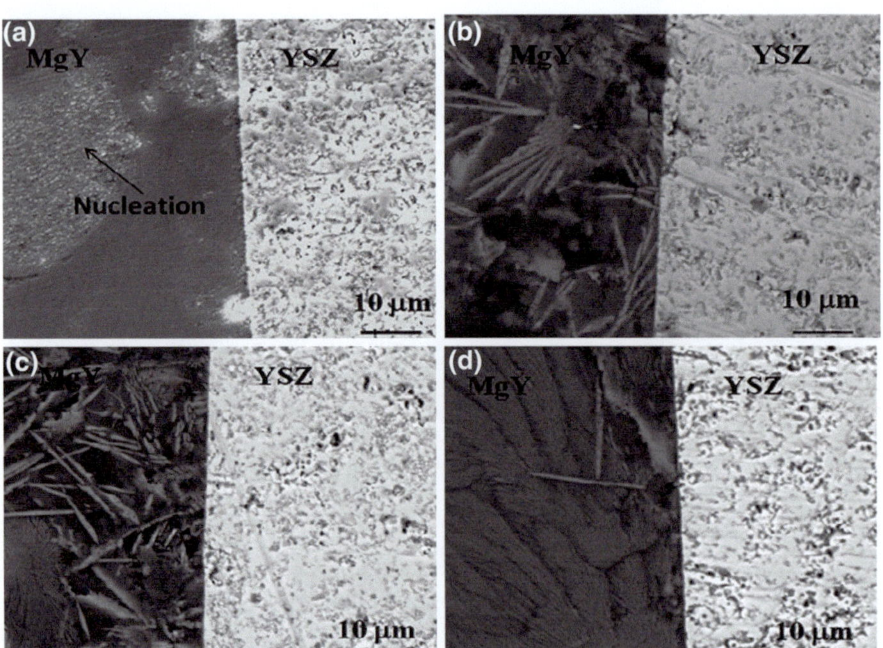

Fig. 8.18 MgY/YSZ couple heat treated at 850°C for **a** 1 h, **b** 100 h, **c** 500 h, and **d** thermally cycled from room temperature −850°C (Kumar and coworkers 2014)

obtained for all the couples, although dendrite formation in glass could be seen for the 100 and 500 h heat-treated couple. When the supercooling is large, then dendrites appear due to the presence of a temperature gradient. In addition to this, the latent heat released during spike formation raises the temperature of the immediate surrounding liquid. Some porosity could be seen in the interior region of the glass (100 and 500 h heat treatment), which could be due to the outgassing of dissolved gases. Interestingly, the porosity decreases in glass for the 500 h heat-treated couple, attributed to the self-heating action of the MgY glass.

For the thermally cycled MgY/YSZ couple, the interface remains smooth and intact indicating good sealing behaviour of MgY glass. The line scans of all the diffusion couples (not shown here) indicate no elemental segregation at the interface though the dendrite regions are found to be rich in yttrium. Although magnesium and silicon are expected to exhibit fast diffusion kinetics on account of this small size, minimum diffusion in either side of the interface is shown by magnesium followed by silicon indicating a strong network of MgY glass. Yttrium from MgY has shown relatively higher diffusion as compared to other elements. Yttrium is a component of YSZ electrolyte as well as MgY glass, though its content is higher in the glass. Two factors, which may be responsible for the yttria diffusion, include concentration gradient and comparable ionic radii to that of zirconium.

8.3.3 Calcium-Based Glasses/Electrolyte Interface

The CAS system does not have most of the sealing requirements, due to their high chemical durability. CAS1 and CAS2 glass compositions yielded the lowest melting points of 1170 and 1265°C, respectively, and subsequently 6 wt% of B_2O_3 was added systematically to these parent glass compositions along with V_2O_5, K_2O, and TiO_2 as minor additives. Figure 6.19a depicts the phase diagram for the CAS ternary system. The CAS1B (22.83% CaO, 12.76% Al_2O_3, 58.36% SiO_2, 6.25% B_2O_3, wt%) and CAS2B (35.8% CaO, 18.24% Al_2O_3, 39.3% SiO_2, 6.03% B_2O_3, wt%) yielded uniform and satisfactory interface quality with YSZ. The CL (30% CaO, 40% SiO_2, 20% B_2O_3, 10% La_2O_3, mol%) composition has also revealed very good interface interaction with the YSZ and TBV after prolonged heat treatment duration. Figure 8.19a, b shows the interface and EPMA of CL/TBV diffusion couple after 100 h heat treatment at 800°C. No interaction products and cracks are observed at the interface. Moreover, the glass region is also uniform without any pits or pores.

The EPMA (Fig. 8.19b) indicates high vanadium concentration at the interface, whereas titanium has diffused to a considerable extent on the glass side of the interface. The calcium follows a *z*-shaped curve and Bismuth follows an 'S' profile. After prolonged heat treatment of 750 h, the CL/TBV diffusion couple shows pit formation in the glass region attributed to the entrapped volatility of gases. Interestingly, the bismuth and calcium diffusion has decreased with the increasing heat treatment duration. The cross-section of CL/YSZ after heat treatment of 750 h

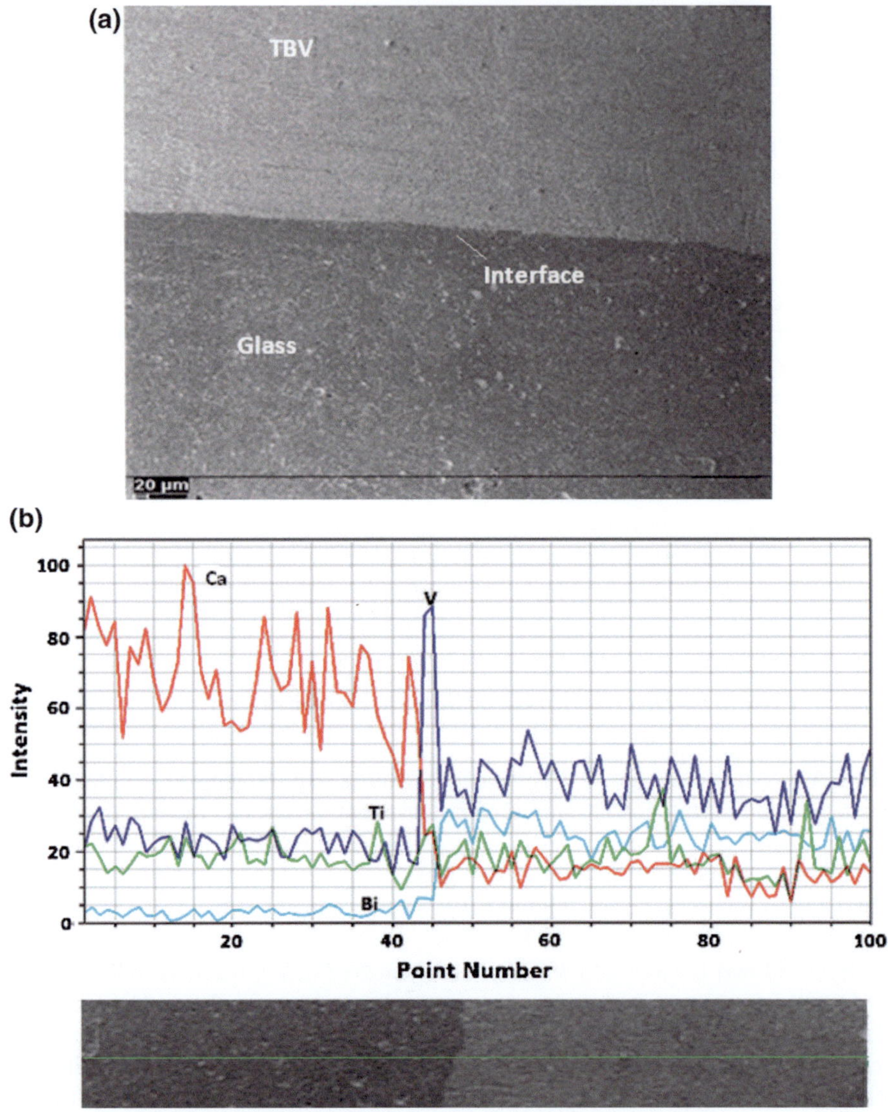

Fig. 8.19 **a** Interface and **b** EPMA of TBV/CL glass diffusion couple after heat treatment at 800°C for 100 h (Kaur and coworkers 2012c)

at 850°C reveals a strong electrolyte/glass bond at the interface without any detrimental reaction product (Fig. 8.20a, b). The interface shows excellent adhesion with preclusion of porosity and undesirable reaction products. As observed in diffusion couples of CL/TBV, the couples of CL/YSZ have also shown slow diffusion of ions with the increasing heat treatment duration. This may be attributed to the hindrance in ion transport process provided by the crystalline phases formed

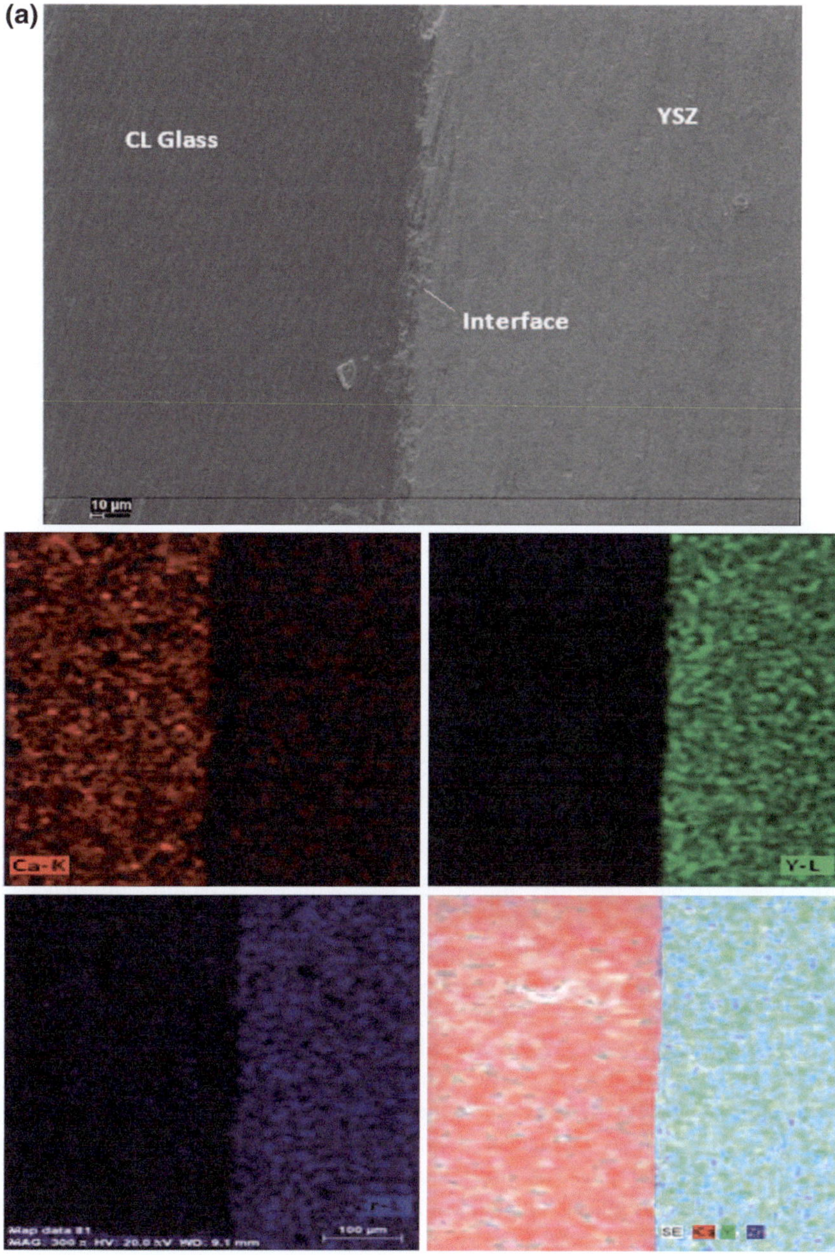

Fig. 8.20 **a** Interface and **b** X-ray dot mapping of YSZ/CL glass interface after heat treatment at 850°C for 750 h (Kaur and coworkers 2012c)

after prolonged heat treatment. The XRD results reveal formation of $CaZrO_3$ and $CaZr_4O_9$ phase after 750 h heat treatment for the coupled glass. The $CaZrO_3$ structure must have formed through the initial diffusion from glass and the following reaction may have occurred,

$$2CaO + 5ZrO_2 \longrightarrow CaZrO_3 + CaZr_4O_9 \tag{8.5}$$

At 921°C, the $CaZr_4O_9$ phase usually dissociates into monoclinic ZrO_2 and $CaZrO_3$.The coupled CL/TBV glass seal revealed formation of a Ca_2SiO_4 phase after heat treatment of 750 h at 800°C. No titanium or vanadium rich phases could be observed, though they diffused inside the glass matrix. The titanium-rich phases are considered harmful for the SOFC functioning. Titanium is highly chemical reactive and gives rise to adherent and dense self-healing oxide. In addition to this, the free energy of titania is low, i.e., −677.8 kJ/mol at 1000°C. The probable reactions of titania with the borosilicate glasses are

$$5Ti + 2B_2O_3 \longrightarrow 2TiB_2 + 3TiO_2 \tag{8.6}$$

$$8Ti + 3SiO_2 \longrightarrow Ti_5Si_3 + 3TiO_2 \tag{8.7}$$

The interfacial silicate phase (Ti_5Si_3) is highly brittle as compared to the boride phase (TiB_2). In contrast to this, the $CaZrO_3$ phase for the CL/YSZ couple is not detrimental as it possesses CTE of 10.4×10^{-6}/K and is close to that of YSZ. This phenomenon leads to good thermomechanical compatibility between YSZ and CL glass. Titanium is usually tetrahedrally coordinated and hence acts as network former, whereas zirconia is generally stable with higher coordination numbers and hence acts as network modifiers.

The Y_2O_3-containing glass CaY [30% CaO, 40% SiO_2, 20% B_2O_3, 10%Y_2O_3, mol%] also yields a pore-free intact interface with the YSZ after thermal cycling from RT-850°C. Ca_2SiO_4 is the first phase to nucleate after 1 h heat treatment (Fig. 8.21). After prolonged heat treatment of 100, 500 h, and thermal cycling, the pyrochlore ($Y_2Si_2O_7$) and $ZrSiO_4$ phase also precipitated out indicating that the yttrium/zirconium primarily interact with the silicate network of the glass. The volume fractions of these phases increased with increase in heat treatment duration. For the CaY/YSZ diffusion couple, yttrium has shown the highest diffusion followed by zirconium, calcium, and silicon without any corrosion products at the interface. The $CaZrO_3$ formation is expected as the lanthanium-based CL glass shows, but no such phase is observed for the CaY glass. The YSZ surface in CaY/YSZ shows irregularly shaped pores and the size of these pores decreases with the increase in heat duration time.

The Al_2O_3 containing CaAl glass (30% CaO, 40% SiO_2, 20% B_2O_3, 10%Al_2O_3, mol%) also revealed very good bonding behaviour with YSZ under oxidising and reducing conditions (Fig. 8.22). Under both reducing and oxidising atmospheres, the interface is intact and homogeneous for the CaAl/YSZ diffusion couple. Only the $ZrSiO_4$ phase precipitates out for coupled CaAl/YSZ diffusion couples under

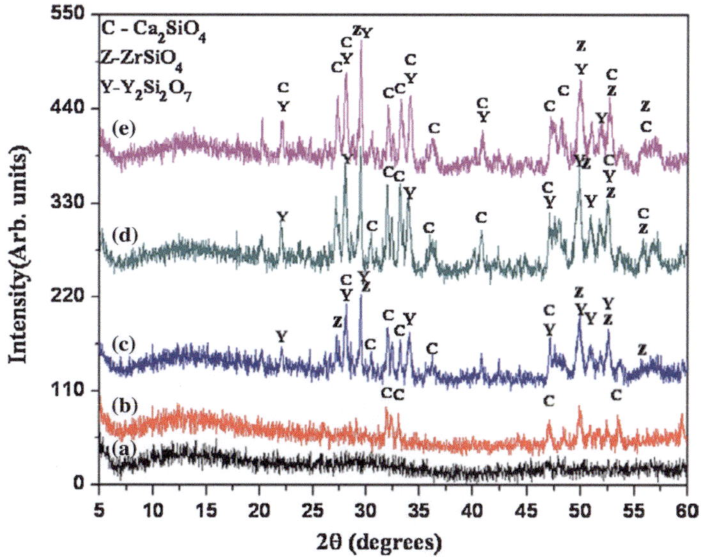

Fig. 8.21 X-ray diffractogram: *a* as prepared CaY glass and CaY/YSZ couple *b* heat treated for 1 h, *c* thermally cycled, *d* heat treated for 100 h, *e* heat treated for 500 h (Kumar and coworkers 2014)

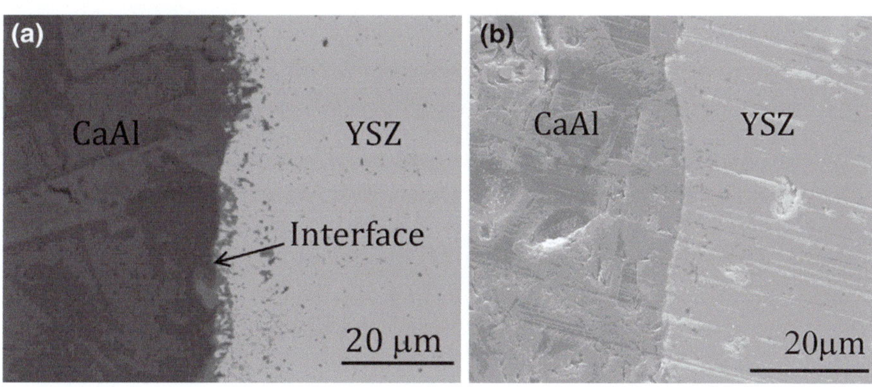

Fig. 8.22 CaAl/YSZ diffusion couples after heat treatment at 850°C for 500 h under **a** oxidising and **b** reducing atmosphere (Kumar and coworkers 2015)

oxidising and reducing conditions. $ZrSiO_4$ is regarded as a thermally stable phase and does not hinder SOFC functioning. Interestingly, the phase fraction of $ZrSiO_4$ is higher under oxidising conditions indicating that the oxidising atmosphere promotes glass crystallisation. Conclusively, calcium-based glasses depict acceptable sealing behaviour with the high and low temperature electrolyte.

8.3.4 Strontium-Based Glasses/Electrolyte Interface

The SZS compositions (51% SrO, 9%ZnO, (40−*x*)% SiO_2, wt%) with various additives such as B_2O_3, Al_2O_3, Cr_2O_3, and V_2O_5 have shown promising results with the YSZ. The SZS-4 composition containing 10% B_2O_3 shows a continuous interface along with elongated microstructure at the joint region (Fig. 8.23a, b). The line scan of the joint region between SZS-4/YSZ indicates the interdiffusion of

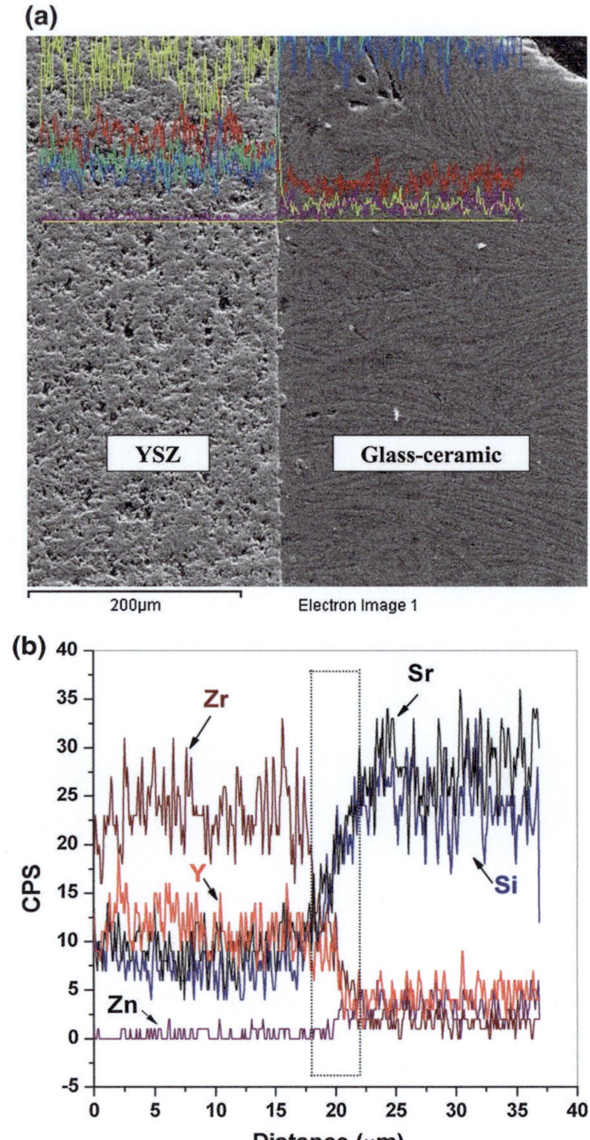

Fig. 8.23 **a** SEM micrograph and **b** EDAX line scans across the SZS-4 glass–ceramic to YSZ interface showing the interdiffusion of Si, Sr, Zr and Y after sealing at 950°C (Tiwari and coworkers 2011)

strontium, yttrium, zirconium, and silicon on either side of the interface. This interdiffusion is also enhanced mechanical locking causing strong bonding. The interfacial joint region thickness is around 3–4 μm. For SZS-4 glass, $Sr_2ZnSi_2O_7$ and $SrSiO_3$ phase precipitate out. For the V_2O_5, Al_2O_3 and Cr_2O_3-doped glasses, only the $Sr_2ZnSi_2O_7$ phase precipitates out, hence the $SrSiO_3$ phase is suppressed.

Anyhow, the B_2O_3-free SiO_2–Al_2O_3–SrO–La_2O_3–ZnO–TiO_2 glass seals yield acceptable CTE and density. The interaction of SALSTi11 (no zinc, 10% TiO_2), SALSZT10 (10% ZnO, 6% TiO_2), and SALSZT12 (6% ZnO, 7% TiO_2) with the GDC ($Gd_{0.2}Ce_{0.8}O_{2-\delta}$) electrolyte revealed good interfaces except the latter one. SALSZT12 have a larger CTE difference with the GDC, which causes extensive cracking along the interfaces. SALSZT10 glass exhibited a small crack parallel to the glass electrolyte interfaces. The main reason of crack initiation is a large pore, which subsequently amplified due to internal stresses. Due to the promising results shown by SALSTi11 and SALSZT10, the leak rate tests were conducted at 700°C for 500 h heat treatment duration after joining at 950°C for 1 h. SALSTi11 glass indicates a low leak rate <0.007 sccm/cm after first thermal cycle, which increased by an order to a value of 0.07 sccm/cm as the temperature decreased to room temperature. This escalation in leak rate is due to the stress induced cracking initiated because of the CTE mismatch. For SALSZT10, the leak rates are maintained at very low values of less than 0.001 Sccm/cm after sealing at 950°C for 1 h, which elevated to 0.07 Sccm/cm. Both glasses provided low leak rates less than 0.01 sccm/cm at 700°C, hence these glasses could be regarded as promising candidates for the SOFC. Zinc-free lanthanum-based SL glass (30% SrO, 40% SiO_2, 20% B_2O_3, 10% La_2O_3, mol%) has revealed $SrZrO_3$ formation at the SL/YSZ interface after 750 h heat treatment at 850°C. Gibbs free energy of $SrZrO_3$ is low and it possesses CTE of 10.9×10^{-6}/K. Except $SrZrO_3$, no other phase could be seen at the interface. Although SL glass has revealed good bonding behaviour with the YSZ, the interface is delaminated and discontinuous for SL/TBV electrolyte after 100 h heat treatment at 800°C. The XRD of coupled SL/TBV and SL/YSZ revealed the formation of $Sr_3Si_3O_9$, $La_2Si_2O_7$, and $SrZrO_3$ phases, respectively, after heating at 800 and 850°C, respectively. The reaction mechanism for coupled SL/TBV glass is

$$3SrO + 3SiO_2 \longrightarrow Sr_3Si_3O_9 \tag{8.8}$$

$$La_2O_3 + 2SiO_2 \longrightarrow La_2Si_2O_7 \tag{8.9}$$

La^{3+} and Sr^{2+} cations have a higher tendency to bond with glass forming cations via non-bridging oxygen atoms. Lanthanum silicate is refractory in nature with a melting point of 1750°C. Again no titanium silicide or boride phases were observed in the coupled SL glass. With YSZ, the interaction of SrO and ZrO_2 led to the $SrZrO_3$ formation. With the increasing heat duration time from 5 to 750 h, the CTE of SL glass increased from (8.96–9.89) $\times 10^{-6}$/K. The general increase in volume with temperature was attributed to the increased atomic vibration amplitude about a mean position. CTE increased more rapidly at lower temperatures as higher atomic separation locuses lattice expansion. The CTE of YSZ and TBV were 10.9×10^{-6}/K and 10.25×10^{-6}/K, respectively. Change in lengths and CTE (α) for TBV and SL are shown in Fig. 8.24, which indicates that the maximum stress occurs at temperature of 731°C. Anyhow, this stress is above T_g of SL glass (722°C) and hence can be tolerated by the glass matrix flow.

The SL glass was also pelletised with aluminium-doped bismuth vanadate ($Bi_4V_{1.8}Al_{0.2}O_{11}$) electrolyte in 3:1 ratio by weight and heated at 800°C for 1, 10, and 100 h to study the chemical interaction and different phases formed. Additionally, Al_2O_3 and Y_2O_3 containing strontium glasses; i.e., 30% SrO–40% SiO_2–20% B_2O_3–10% (Al_2O_3/Y_2O_3) are also pelletised with the aluminium-doped bismuth vanadate. All the samples exhibited the presence of A_2SiO_5, AVO_4, and $SrVSi_2O_7$ phase (where A = Y, La, Al). $SrVSi_2O_7$ has been formed due to interactions of bismuth vanadate with glasses; i.e., V_2O_5 is $Bi_4V_{1.8}Al_{0.2}O_{11}$ and is highly disordered due to which V^{5+} migrates towards the glass network and reacts to form the $SrVSi_2O_7$ phase. Figure 8.25a, b shows the phase fraction of phases formed after different durations of heat treatment for IS1 and IS2 samples. Table 8.3 describes the samples with their composition, sample labels, and phases formed.

IS2 glass shows a high fragility index and lower activation energy of crystallisation than IS1 and IS3. Hence, the La_2SiO_5 and $LaVO_4$ phases might have nucleated during an early stage of heat treatment. The value of the fraction of La_2SiO_5 and $LaVO_4$ decreased with increasing heat treatment duration. In addition to this, $LaVO_4$ and La_2SiO_5 phases exhibited triclinic crystal structure, which is less symmetric. $SrVSi_2O_7$ phase possessed an orthorhombic crystal structure, which is thermodynamically more stable than the La_2SiO_5 and $LaVO_4$ phases. Hence, the volume fraction of $SaVSi_2O_7$ increased after 100 h heat treatment due to more favourable conditions for its growth. Interestingly, for the IS3 sample, the phase fraction of formed phases did not increase considerably after heat treatment.

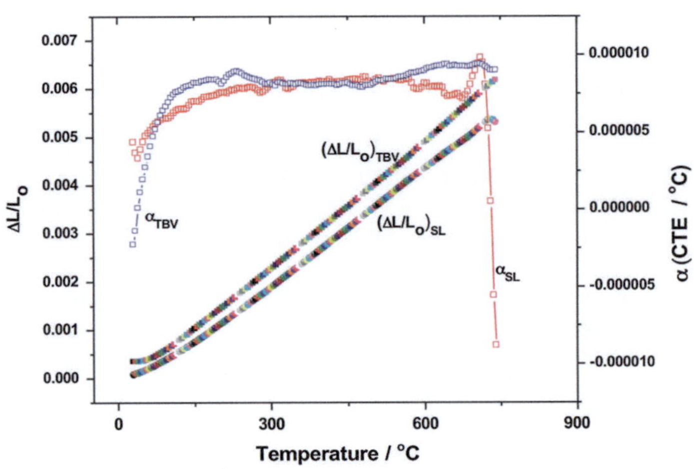

Fig. 8.24 Coefficient of thermal expansion (α) and change in length for TBV and glass with respect to temperature (Kaur and coworkers 2012d)

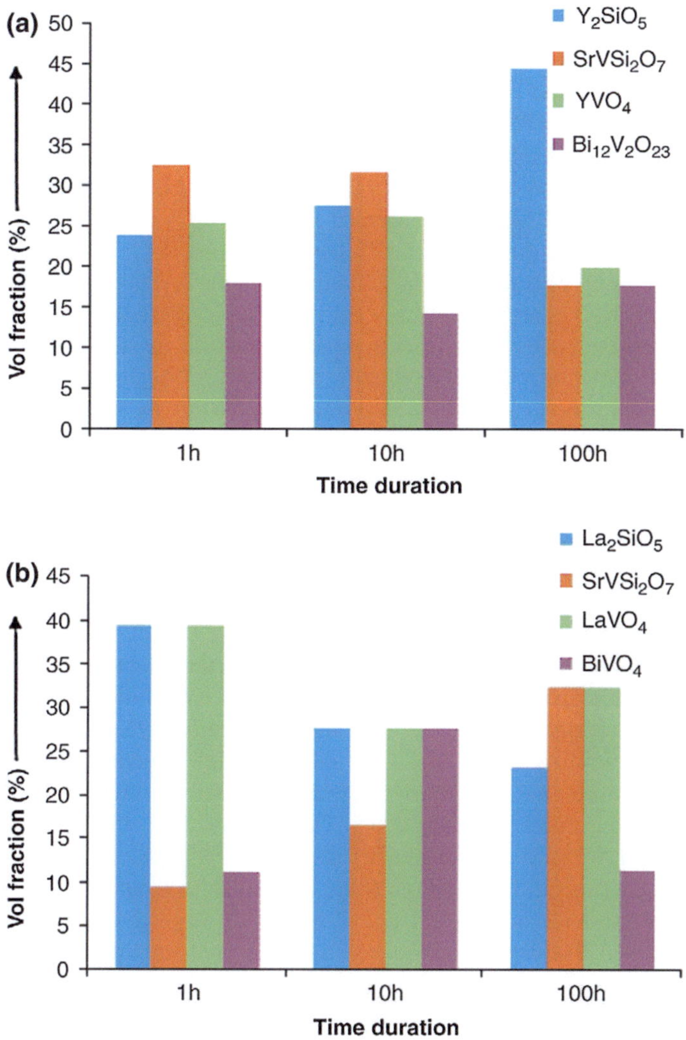

Fig. 8.25 Volume fraction of crystalline phases formed during heat treatment of **a** IS1 and **b** IS2 sample (Kumar and coworkers 2010)

Table 8.3 Sample compositions, labels and phases formed

Sample name	Glass composition/electrolyte	Phases formed
IS1	$30SrO–40SiO_2–20B_2O_3–10Y_2O_3$/ $Bi_4V_{1.8}Al_{0.2}O_{11}$	Y_2SiO_5, YVO_4, $SrVSi_2O_7$, $Bi_{12}V_2O_{23}$
IS2	$30SrO–40SiO_2–20B_2O_3–10La_2O_3$/ $Bi_4V_{1.8}Al_{0.2}O_{11}$	La_2SiO_5, $SrVSi_2O_7$, $LaVO_4$, $BiVO_4$
IS3	$30SrO–40SiO_2–20B_2O_3–10Al_2O_3$/ $Bi_4V_{1.8}Al_{0.2}O_{11}$	Al_2SiO_5, $SrVSi_2O_7$, $AlVO_4$, $Bi_4V_2O_{11}$

8.3.5 Barium-Based Glasses/Electrolyte Interface

As the BAS system had been investigated extensively, Sohn and co-workers have also studied La_2O_3, ZrO_2, and NiO-doped BAS system for its interaction with the YSZ. Lanthanum-containing glasses are designated as L1/L2, whereas nickel containing glass is denoted as N1/N2. The CTE of the glasses increased with the addition of La_2O_3 and NiO but decreased with the ZrO_2 addition. Higher B_2O_3/SiO_2 values influenced the T_g, T_c, and T_s, hence affecting the thermal properties 2% ZrO_2 and 5% La_2O_3 containing glasses indicating stable crystallisation, whereas 2% La_2O_3 and 2% NiO containing glasses exhibited rapid crystallisation. The fluidity of the glass–ceramic deteriorated due to rapid progressive crystallisation because of the decrease of the residual glass matrix. Celsian and hexacelsian phases have precipitated out of the glasses when heat treated at 900°C. In 5% La_2O_3 containing L2 glass, Ba_4La_6 $O(SiO_4)_4$ phase also precipitated out. L2 glass containing 5% La_2O_3, and Z1 containing 2% ZrO_2 exhibited excellent bonding with the YSZ surface, and no cracks or delamination could be observed after heat treatment at 800°C for 1000 h (Fig. 8.26a, b).

B_2O_3 is the main ingredient for these compositions, which usually causes poor stability in a wet reducing atmosphere. SEM observation revealed the formation of a thick surface crystallisation layer, which increased with increasing the heat treatment duration. Anyhow, this layer may improve the chemical stability against wet reducing atmospheres by preventing the boron volatility. Z1 glass showed microstructural changes with the heat treatment duration, whereas L2 showed no such microstructural modifications. This indicated high-structural integrity for the La_2O_3-doped glasses. Sohn and co-workers also focused on other lanthanum-doped compositions such that $B_2O_3/SiO_2 = 0.5$. The compositions with BaO content 20, 25, 30, 35, and 40 mol% are designated as S1, S2, S3, S4, and S5, respectively [10% Al_2O_3, 5% La_2O_3, $B_2O_3/SiO_2 = 0.5\%$, mol%]. The T_g and T_s values for these glasses lie in the range of 668–671 and 735–745°C, respectively. With increasing BaO content the CTE showed an increase S4 glass gives minimum CTE mismatch

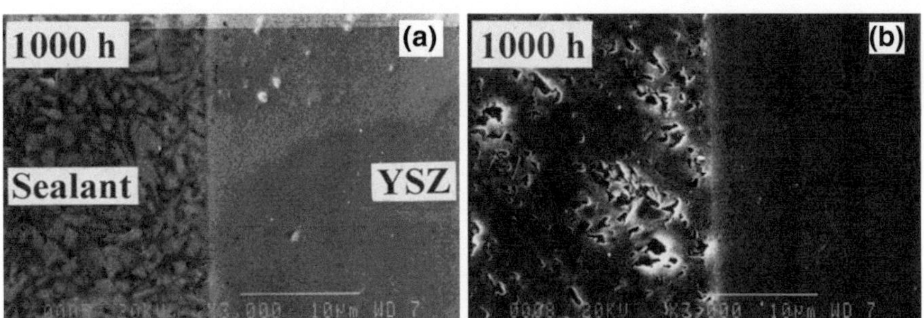

Fig. 8.26 Glass–ceramic/YSZ interface for **a** L2 and **b** Z1 glasses heat treated at 800°C for 1000 h (Sohn and Choi 2004)

of 0.38×10^{-6}/K at temperature below 670°C [$B_2O_3/SiO_2 = 0.5$]. Glass S4 with T_s of 742°C also revealed shrinking behaviour up to 775°C attributed to the liquid-phase sintering. Up to 1130°C, S4 also shows a non-wetting behaviour due to its high surface tension. The S4/YSZ interface at 800 or 850°C after 100 h bonding revealed no cracks or delamination. No interaction between S4 or YSZ had taken place, though the couple heat treated at 850°C revealed higher crystallisation in glass as compared to the couple heat treated at 800°C. During thermal cycling, the electrolyte is susceptible to break down at contact areas due to the thermal stresses. Anyhow, S4 glass exhibited no such breakdowns or delamination even after two thermal cyclings between room temperature 800–850°C indicating its excellent endurence. Kaur and co-workers have reported the barium-based BL composition [30% BaO, 40% SiO_2, 20% B_2O_3, 10% La_2O_3, mol%] to be self-healing in nature. Thermal healing involves crack healing at elevated temperatures. Common mechanisms of crack healing in ceramics involve thermal healing accumulation of reaction products at the tip of the crack and adhesion from intermolecular forces. BL has lower T_g than the working temperature of SOFC which is beneficial because thermal stresses inside the glass develop below the working temperature. The underlying thermodynamic principle for glass formation is that for $\left(\frac{dT}{dt}\right) \rightarrow 0$, the $T_g \rightarrow T_c$ indicating time dependency of properties especially in glass transition region. When the timescale required for structural rearrangement of glasses becomes comparable to the experimental time scale, then the glasses exhibit the phenomena of structural relaxation. During structural relaxation, the glasses try to attain equilibrium. Anyhow, at low temperatures, the timescale is large such that any rapid changes in temperature can 'freeze in' the liquid structure. The BL-coupled glass showed $BaSiO_3$ phase formation after 750 h heat treatment (for both YSZ and BL electrolyte) indicating a low rate of controlled crystallisation. The amorphous glassy phase in glass–ceramic acts as a major factor in deciding flow behaviour of the seal and hence determines reaction kinetics during chemical interaction. Such high amorphous glassy content of BL glass may be responsible for its self-heating behaviour. No zirconate phase was observed for the coupled BL/YSZ glass, though large pores and pits could be seen in the glass. For BL/TBV diffusion couple appreciable vanadium and titanium could be seen on the glass side, through no titania rich zone could be observed at the interface. Similarly, BL exhibited good bonding characteristics with the YSZ electrolyte as well though interior glass regions had big pores of 0.22 μm (area = 37,944 μm^2) after 5 h heat treatment at 800°C as shown in Fig. 8.27. The pits are homogeneously distributed inside the glass–ceramic and population of small pores could also be seen besides the large pits.

The pits exhibit dendrite crystalline phases embedded in glassy matrix. Dendrite usually appears when the temperature gradient is prevented and supercooling is large. Dendrite spikes also remain isolated because the latent heat released during spike formation raises the temperature in the immediate vicinity retarding growth of similar projections. After 100 and 750 h of heat treatment for the BL/YSZ diffusion couple at 850°C, the closed pits decreased to an area of 25,434 and 314 μm^2,

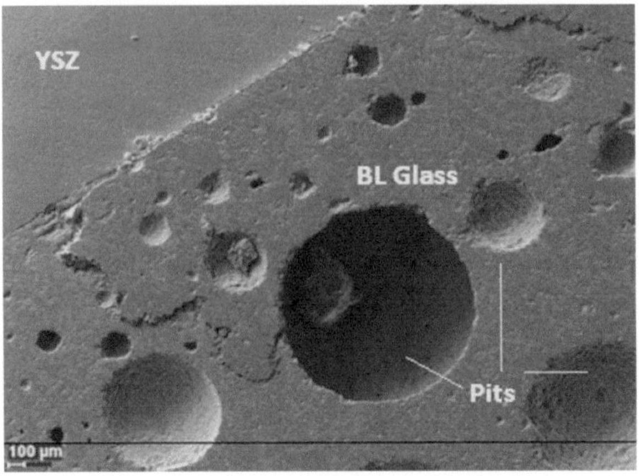

Fig. 8.27 YSZ/BL glass diffusion couple after heat treatment at 850°C for 5 h (Kaur and co-workers 2014b)

respectively. Hence, the considerable decrease in pore size revealed the self-heating behaviour. Pits occur when the dissolved gases evolve out of the glass matrix. Usually gases possess higher solubility in a liquid state than the solid state and their solubility depends upon the applied pressure and temperature; common diatomic gases including H_2, O_2, N_2, etc.

$$C_g = k\sqrt{p} \tag{8.20}$$

where C_g is the dissolved gas (diatomic) solubility, k is constant, and p is gas pressure. Upon the rejection of dissolved gases, they accumulate and may reach a concentration such that it causes nucleation of gas bubbles. The bubble gets traffic inside the solid if the growth rate of the bubble is much slower than the solid growing around it. In the case of BL glasses, such phenomena may have been observed. When the La_2O_3 in BL is replaced by Al_2O_3 then the BaAl glass [30% BaO, 20% B_2O_3, 40% SiO_2, 10% Al_2O_3, mol%] with low CTE of BaAl is obtained. Pits as well as porosity could be observed in the glass side under oxidising conditions for the coupled BaAl/YSZ. Glass exhibited $Ba_2Si_3O_8$ phase formation under reducing as well as oxidising conditions. The barium silicate phases are high CTE phases and hence may be responsible for the large pits inside the glass. Anyhow, no gaps could be observed at the interface under reducing conditions for the BaAl/YSZ diffusion couple. The fast crystallisation of sealing materials can increase the thermal stress. Generally the phase transformation/formation can occur via two processes: i.e., changes that are small in degree but large in spatial extent known as spinoidal decompositions and the changes, which are large in degree but small in spatial extent known as nucleation and growth. For the BaAl and BL glasses, nucleation and growth phenomena are observed (which are usually the cases for most of the glasses). In addition to this, nucleation can be controlled by the nucleating agents and heat treatment temperature/duration. Once nucleation triggers, then higher temperatures are required to maximise the growth of nucleating crystals. Conclusively, the low temperature provides the nucleation site, whereas high temperature is necessary to speed up the growth of the crystals.

Al_2O_3 containing GA8 glass [92 wt%, BiBaSi, 8 wt% Al_2O_3], GA5 [95 wt% BiBaSi, 5 wt% BiBaSi], and GA3 glass [97% BaBiSi, 3 wt% Al_2O_3], yield very well-adhered surfaces with the YSZ without any chemical reaction, though huge porosity could be seen in the glass, which are not interconnects with each other. GA3, GA5, and GA8 show the formation of $Ba_5Si_8O_{21}$, $Ba_5Al_2O_3$, and $Bi_4(SiO_4)_3$ phases indicating that Al_2O_3 forms the network, hence replacing Si^{4+}. Si^{4+} then reacts with Ba^{2+} and Bi^{3+} to form silicate phases. These three glasses exhibit CTE in the range of [10–11] $\times$ 10^{-6}/K, which is acceptable for SOFC working. The pore concentration and pore size are highest for GA3 glass with maximum barium content which indicates that barium enhances pits and pore formation. Huang and co-workers developed YBA glass [60% BaO, 7.21% Al_2O_3, 3.53% SiO_2, 19.26% B_2O_3, 10% Y_2O_3, wt%] with $10^{5.81}$ Pa s viscosity and glass transition range of 893–903 K. The CTE of YBA is 11.64 $\times$ 10^{-6}/K, which is close to YSZ and SDC. YBA showed good adhesion with YSZ and SDC and the glass is relatively dense as compared to BaAl, BL, GA3, GA5, and GA8 glasses. Hence, the addition of Y_2O_3 to the glass may have increased the density and flowability of YBA glass. Zinc and lead-doped barium glasses like AF8 [60% SiO_2, 6% Al_2O_3, 8% $BaCO_3$, 20% B_2O_3, 3% PbO_2, 3% ZnO, mol%] have also yielded a well-adhered surface with YSZ, though the glass is full of closed pores indicating that higher sealing temperature and longer sealing time is required to remove the pores. The AF8 glass exhibit CTE of 10.58 $\times$ 10^{-6}/K and leak rate of 0.81 $\times$ 10^{-7} mbar s^{-1} cm^{-1}, but still the porosity is a factor of concern for sealing applications.

The BAS compositions studied by Wang and co-workers yielded the CTE in the range of (4.92–10.98) $\times$ 10^{-6}/K. The wetting angle ranges from 140 to 145° on a YSZ substrate when the glasses are heated at 700°C. LO9 glass contained the highest BaO of 50 mol%, and it possessed denitrification. In addition to this, an increase in B_2O_3 content of glass improved its wettability on YSZ. The lanthanum-based BAS system studied by Ghosh and co-workers (G1, G2, and G3 glasses–BaO–La_2O_3–Al_2O_3–B_2O_3–SiO_2) yielded CTE in range of (10.8–11) $\times$ 10^{-6}/K and it exhibited a pore and crack free interface with the YSZ.

8.4 Mixed Alkaline Earth Glasses/Electrolyte Interface

When many different alkaline earth modifiers are mixed together, then the synergistic effect of different cations improves the glass properties by tuning the composition. In this section, the focus is on the mixed alkaline earth compositions and their interaction with YSZ electrolyte.

8.4.1 Calcium–Strontium-Based Glasses/Electrolytes Interfaces

The strontium calcium silicate sealing glass system with various amounts of NiO has been investigated by Chou and co-workers. The composite YSO–15(Cr–Ca–B–Y–Si, 15 vol.% NiO) yielded sealing glass and possessed sealing temperatures >950°C, which is higher compared to the sealing temperature of standard sealing glass G18. Glass YSO–15 possessed a smooth interface with the YSZ as shown in Fig. 8.28a, after sealing at 1050°C for 2 h. Interestingly, the YSZ grains showed pronounced grain boundary grooving as compared to the bare YSZ, i.e., without any sealing glass. The grain boundary enhancement is attributed to the fast transport path provided by the molten sealing glass to increase the thermally activated diffusion process. No zirconate formation could be seen at the interface as the zirconate formation is not favoured in that temperature range:

$$SrSiO_3 + ZrO_2 \longrightarrow SrZrO_3 + SiO_2 \tag{8.11}$$

$$\begin{bmatrix} \Delta G_{800\,^\circ C} = 51\ \text{kJ/mol} \\ \Delta G_{1050\,^\circ C} = 49\ \text{kJ/mol} \end{bmatrix}$$

The thermally cycled and fully reduced coupons of YSO-15/YSZ yielded almost similar microstructure (in bulk glass and interface) to the sealed coupons. Darker $SrSiO_3$ phase, lighter silicate phase rich in Sr/Y, and reduced NiO particle (to Ni metal) appeared in the glass microstructure as is clear from Fig. 8.28b. Again, no zirconate rich interface layer could be observed in the YSO-15/YSZ couple. Another glass YSO77 derived from parent glass YSO1 was used in the short stack tests for the commercial NiO-YSZ anode supported YSZ cells. The cells were tested in constant current mode with a total current of 5 A (313 mA/cm^2) for 6000 h. After 6000 h of test, the short stack of 3-cell, including three window frame plates, eight hybrid mica

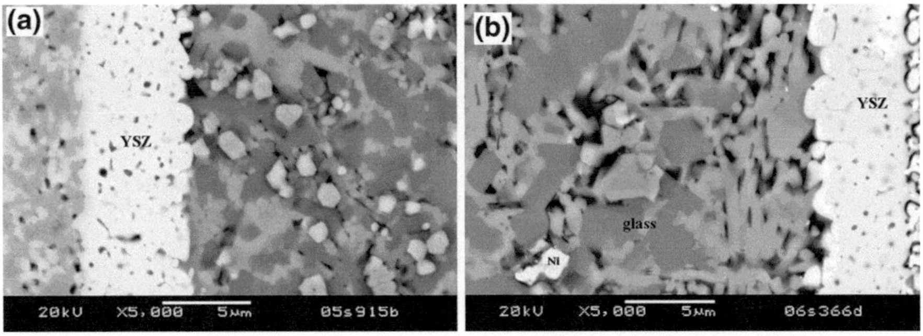

Fig. 8.28 Interfacial microstructure of **a** composite glass (YSO-15)/YSZ electrolyte sealed at 1050°C/2 h and **b** reduced and thermally cycled composite glass (YSO-15)/YSZ electrolyte coupons (Chou and coworkers 2007b)

seals, and four interconnect plates (Fig. 7.7a), was disassembled for the post-mortem analysis. No discoloration was found on the cell or along the sealing edges indicating good bonding of the glass seal to the aluminised window frame plate. The typical grey colour of a reduced state was observed on the cell's anode indicating that YSO77 is a good seal. Fracture could be observed between the LSM contact materials and cathode surface and less bonding could be seen between them, whereas good bonding of LSM was observed with the Ce(Mn, Co) spinel coat formed by ultrasonic spraying and hence possessing rougher surfaces than the screen printed cathode. This structural difference provides mechanical interlocking between the coating and LSM interface. Glass microstructure after 6000 h (800°C) stack testing revealed pores <10 μm with few in the range 20–30 μm.

Two phases are predominant in the glass microstructure: the first phase is the dark phase (Sr, Ba)/Si ratio close to unity, revealed from EDS, which exhibits ~30–50 μm length and ~10 μm thickness. The second phase is the light phase distributed uniformly in between the darker phase, without any distinct morphology [Sr, Y, Si, Ba, and traces of Ca/Al from EDS analysis]. YSO 77 glass is expected to have minimum reactivity with the YSZ electrolyte at an operation temperature of 800°C because of its rapid crystallisation kinetics. The following reactions are the potential reactions likely to occur for ZrO_2, SrO, BaO, and SiO_2 reactions species.

$$\mathrm{BaO} + \mathrm{SiO_2} \longrightarrow \mathrm{BaSiO_3} \quad \Delta G = -161\ \mathrm{kJ/mol} \tag{8.12}$$

$$\mathrm{SrO} + \mathrm{SiO_2} \longrightarrow \mathrm{SrSiO_3} \quad \Delta G = -132\ \mathrm{kJ/mol} \tag{8.13}$$

$$\mathrm{SrO} + \mathrm{ZrO_2} \longrightarrow \mathrm{SrZrO_3} \quad \Delta G = -19\ \mathrm{kJ/mol} \tag{8.14}$$

$$\mathrm{ZrO_2} + \mathrm{SiO_2} \longrightarrow \mathrm{SrSiO_4} \quad \Delta G = -3\ \mathrm{kJ/mol} \tag{8.15}$$

$$\mathrm{SrSiO_3} + \mathrm{ZrO_2} \longrightarrow \mathrm{SrZrO_3} + \mathrm{SiO_2} \quad \Delta G = 51\ \mathrm{kJ/mol} \tag{8.16}$$

$$\mathrm{BaSiO_3} + \mathrm{ZrO_2} \longrightarrow \mathrm{BaZrO_3} + \mathrm{SiO_2} \quad \Delta G = 97\ \mathrm{kJ/mol} \tag{8.17}$$

Equations (8.12)–(8.17) indicate that the silicate is the mostly favoured one during crystallisation and zirconate formation is less likely to occur. All silicates are not detrimental and hence do not possess compatibility with the YSZ electrolyte. Figure 8.29a, b reveals the YSO77/YSZ (anode supported) interface for air and fuel sides, respectively. For both the air and fuel sides, the glass microstructures appear to be the same and consist of dark red-shaped phases with light phases embedded in between. In both cases, fractures in the glass can be observed, which are probably due to the residual stresses, because YSO77 has a matching CTE to that of YSZ. No reaction layer at the interface is observed though glass exhibits a porous microstructure near the interface. Anyhow, at the air side, a large gap can be seen at the interface region. YSO77 sealing glass volatility studies revealed the loss of <0.1 wt% (for life time of 40,000 h). Another sealing glass YSO75 (parent glass

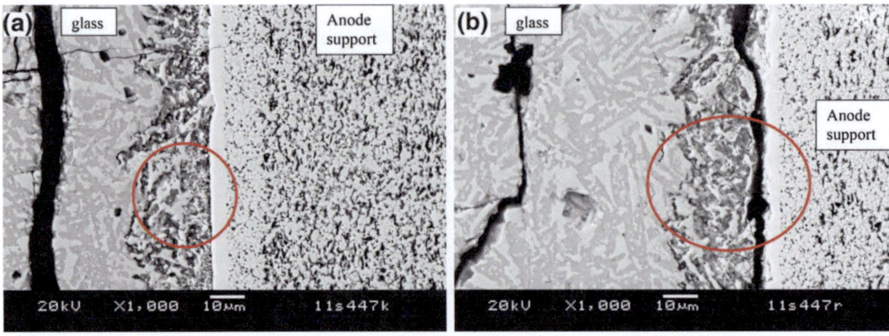

Fig. 8.29 Microstructure of the refractory sealing glass at the YSZ electrolyte interface near the **a** air side and **b** fuel side (Chou and coworkers 2014b)

YSO1) was tested for volatility under flowing wet reducing gas conditions (70% of 2.75% H_2/Ar + 3% H_2O) for almost 1600 h.

Weight loss occurred in two steps and first step (up to 6000 h) yielded weight loss of 3.3×10^{-7} gcm^{-2} h, whereas the second stage (600–1600 h) gave a weight loss of 3.6×10^{-8} gcm^{-2} h, respectively. Usually the B_2O_3 in the outer exposed surfaces is lost as a volatile species, whereas the B_2O_3 inside the glass is surrounded by an amorphous glassy matrix and crystallised phases and does not have much area to come in contact with water. Hence, the major loss due to volatile species such as B_2O_3 occurs at edges. Glass 59 [20% SrO, 20% BaO, 20% B_2O_3, 40% SiO_2] in wet reducing gas [30% H_2O, 10% (H_2/N_2)] yielded a weight loss of 5.5×10^{-6} gcm^{-2} h at 780°C. In contrast to this for dry air conditions, the weight loss for glass 59 is 8.8×10^{-7} gcm^{-2} h (780°C), which indicates higher weight loss under wet conditions.

8.4.2 Barium–Magnesium–Strontium-Based Glasses/Electrolyte Interface

The HGS–1 composite seal [(8.3% B_2O_3, 12.1% BaO, 21.8% BaO, 21.8% MgO, 12.1% ZnO, 45.7% SiO_2, in mol%) + 30% Al_2O_3] revealed gas bonding behaviour with the YSZ and anode (NiO + YSZ) after dwelling at 750°C for 5 h under compressive stress of 20 psi, respectively. The adjacent component exhibits CTE difference, but still the interface remains intact because of the compressive stress loaded on the components. The barium magnesium aluminosilicate glass sealant investigated by Ghosh and co-workers (Table 7.2) yielded CTE in the range of (9–11) $\times 10^{-6}$/K. The interface of alumina-free BM12 glass and alumina containing BMAI glass with YSZ is shown in Fig. 8.30a, b, respectively.

Extensive cracking is observed for the BMA1-YSZ interface and it also propagates within bulk YSZ. Even the glass–ceramic consisted of extensive porosity. In contrast to this, BM12 glass exhibited excellent sealing with the YSZ at SOFC

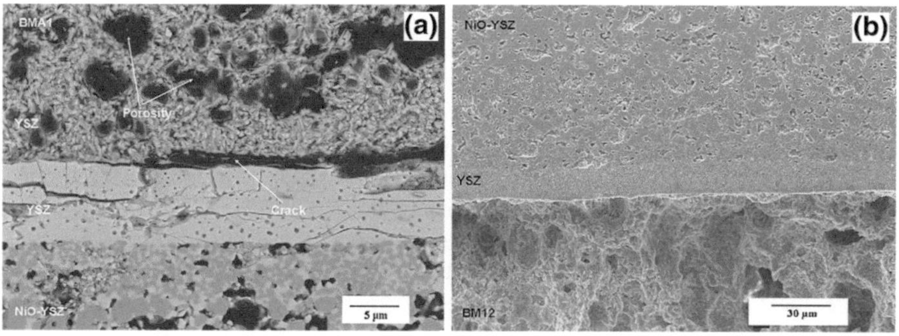

Fig. 8.30 SEM micrograph of cross-section of the half cell for **a** BMA1 and **b** BM12 glass (Ghosh and coworkers 2010)

operating temperature. No chemical reactivity is observed for both the interfaces. The CTE mismatch for BMA1 and BM12 with YSZ is 1×10^{-6}/K and 0.48×10^{-6}/K, respectively. Higher CTE mismatch among BMA1 glass and YSZ may be responsible for extensive cracking inside the glass YSZ microstructure. Brochu and co-workers have used YSZ as unreactive particles in barium–magnesium and magnesium–strontium borosilicate glasses. The composite glasses are expected to possess superior properties such as CTE, viscosity, thermal behaviour, as well as chemical reactivity. The composition of Ba glass and Sr glass is 40% B_2O_3–10% Al_2O_3–20% MgO–20% CaO–10% BaO/10% SrO, respectively. The prepared composites along with their characteristic temperature are listed in Table 8.4.

The YSZ powder decreased the glass transition temperature and increased crystallisation temperatures for the Ba-glass. In contrast to this, addition of YSZ to the Sr-glass did not influence its T_g, but the crystallisation temperature decreased. Barium-based glass exhibited more viscous behaviour indicating its high intrinsic viscosity as compared to the Sr-based glass. Only 10% YSZ is required to maintain the seal geometry for Ba-glass, whereas a minimum of 20% YSZ and Sr-glass composite with 20% YSZ with YSZ after a sealing cycle of 10 min at 950°C, $BaZrO_3$-phase could be seen near the YSZ substrate and the bulk glass exhibited $MgCaB_2O_5$ crystalline phase formation (Fig. 8.31b). In addition to this, a thin reaction layer containing Ca, Al, Zr, and oxygen could be seen. On the YSZ side, the presence of Ca^{2+} and Al^{3+} indicates the elemental diffusion into the electrolyte side. Usually the ZrO_2-rich portion contains ZrO_2 and $BaZrO_3$ phases as suggested by the BaO-ZrO_2 binary diagram. Up to 4 mol%, Al_2O_3 is soluble in the fluorite

Table 8.4 Ba/Sr glass composites and characteristic temperature

Composite	T_g(°C)	T_c(°C)
Ba-glass	584.7	785.2
Sr-glass	588.9	794.1
Ba-glass + 5% YSZ	580.9	793.5
Sr-glass + 5% YSZ	589.2	784.0
Ba-glass + 20% YSZ	581.3	799.1
Sr-glass + 20% YSZ	588.2	766.9

structure as indicated by ZrO_2–CaO–Al_2O_3 phase diagram. Another phase $CaMgB_2O_5$ is also identified (using XRD) in the bulk glass, which indicates the independence of crystallisation behaviour for the glass system. The back-scattered micrograph of the interface between the Sr-glass composite containing 20 vol.% YSZ and YSZ is shown in Fig. 8.31b (after a thermal cycling of 10 min at 950°C). The surface layer is composed of Al, Zr, Ca, and oxygen and the crystalline phase away from the interface but inside the glass contains Ca, Mg, A; Sr and O indicate different interaction mechanism between glass and YSZ particles. Although $SrZrO_3$ possesses lower ΔG_f than the $CaZrO_3$, the presence of $CaZrO_3$ suggests that the reaction products are also driven by kinetic factors (such as CaO and SrO concentration in the glass). As mentioned, $CaZrO_3$ and $SrZrO_3$ have CTE of 10.4×10^{-6}/K and 10.9×10^{-6}/K, respectively, hence they are not detrimental. $BaZrO_3$ has low CTE of 7.9×10^{-6}/K, which is low as compared to the CTE of YSZ. The pure Ba-glass and Sr-glass possess CTE of 9.1×10^{-6}/K and 6.78×10^{-6}/K, respectively. Surprisingly, a greater reduction in CTE is observed for the Ba composite; i.e., Ba–10% YSZ composite yields CTE of 7.57×10^{-6}/K, which could be due to the low CTE of the $BaZrO_3$ phase.

The composites prepared by Kaur and co-workers by the ball milling technique have yielded excellent adhesion with the YSZ substrate (Table 7.2). All the glass composites except GG5 exhibit good binding characteristics with high-temperature solid electrolyte YSZ. Extensive cracking could be seen inside the GG5 composite attributed to its low CTE of 7.05×10^{-6}/K. GG1 composite (with Mg^{2+} and Ca^{2+} ions) exhibited good adhesion with YSZ even after prolonged heat treatment of 1000 h at 850°C and no reaction products have been observed at the interface. GG2 (with Ca^{2+} and Sr^{2+} ions), GG3 (with Sr^{2+} and Ba^{2+} ions), GG4 (with Ba^{2+} and Mg^{2+} ions), and GG6 (with Mg^{2+} and Sr^{2+} ions) have all yielded a very smooth and well-bonded interface with the YSZ after 100 h heat treatment at 850°C. The GG7 composite (with Mg^{2+}, Ca^{2+}, Sr^{2+}, and Ba^{2+}ions) consists of all four modifier alkaline earth metal ions and it has also revealed very good bonding characteristics without any delamination at the joint region as is clear from Fig. 8.32a. Elemental

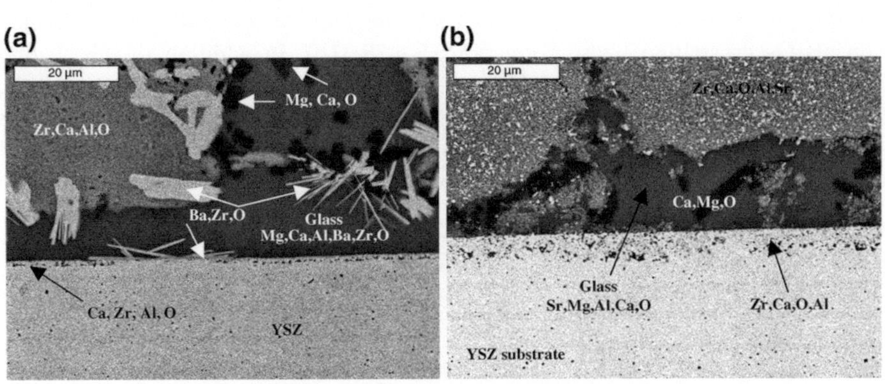

Fig. 8.31 Back-scattered micrograph of interface between **a** Ba-glass 20% YSZ composite and **b** Sr-glass 20% YSZ composite and YSZ after a sealing cycle of 10 min at 950°C (Brochu and coworkers 2006)

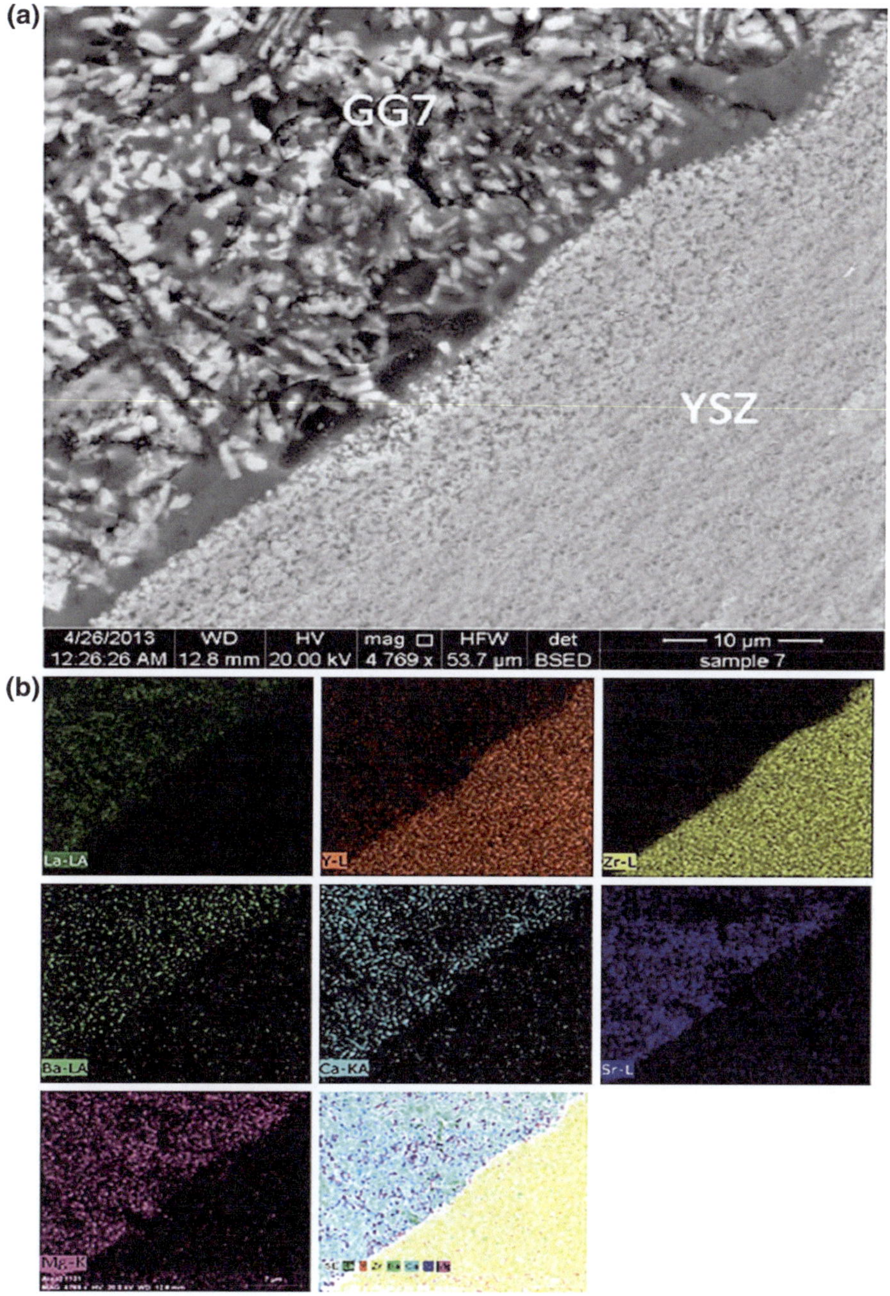

Fig. 8.32 **a** Microstructure and **b** X-ray dot mapping of GG7/YSZ diffusion couple after heat treatment at 850°C for 1000 h (Kaur and coworkers 2015)

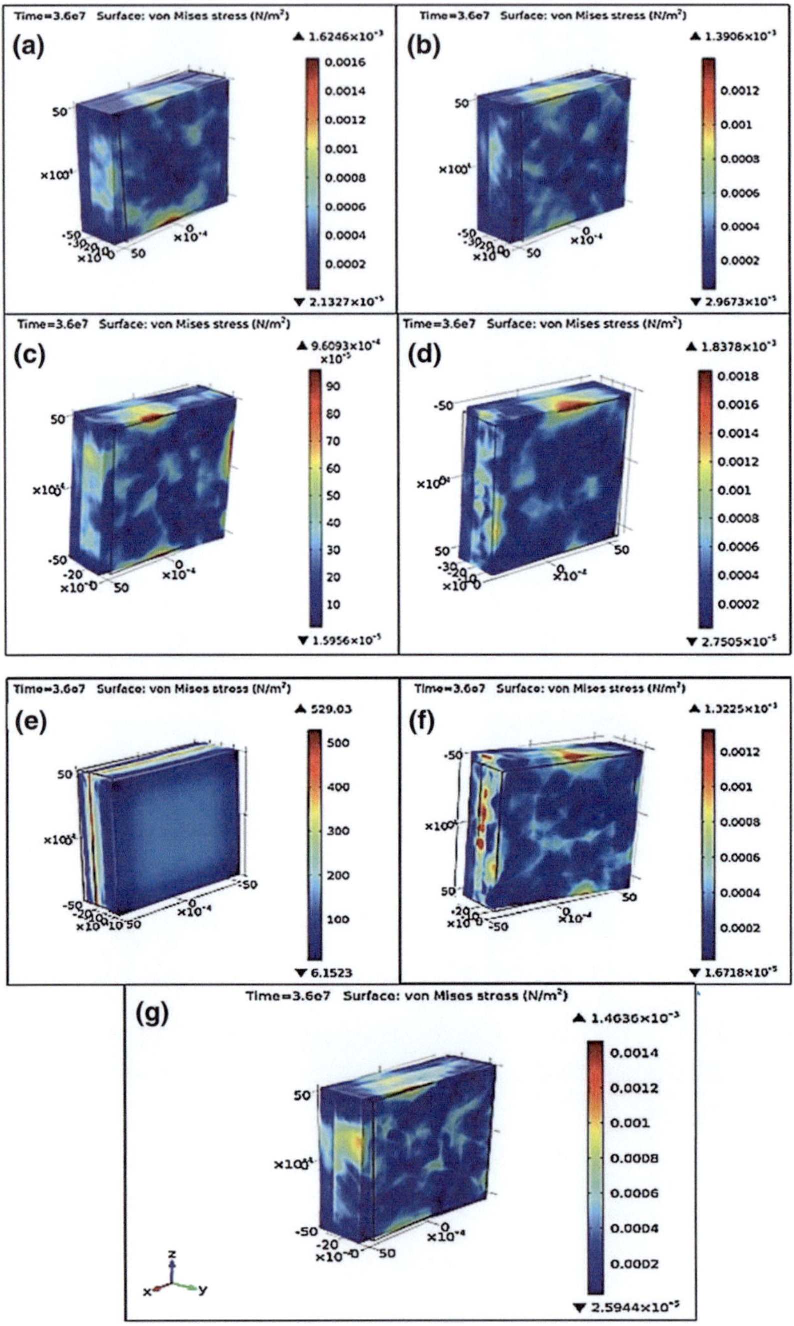

Fig. 8.33 The stress contours and their magnitude for diffusion couples of **a** GG1 **b** GG2 **c** GG3 **d** GG4 **e** GG5 **f** GG6 and **g** GG7 composite with YSZ (simulated up to 10,000 h, step size 1) (Kaur and coworkers 2015)

segregation is completely precluded at the sealing region, which is supported by the dot mapping (Fig. 8.32b). Although diffusion of alkaline earth metal ions has taken place towards the YSZ electrolyte sides, no zirconate is observed at the interface. Figure 8.33a, b, c gives the stress distribution for all the diffusion couples with a YSZ surface. The GG5 composite/YSZ diffusion couple exhibits maximum stress due to the highest mismatch among all other composites. $\Delta\alpha$ for GG5 with YSZ is 3. 75 $\times$ 10^{-6}/K, which is not within the acceptable limit for SOFC functioning. The high-stress regions for the GG5 composite/YSZ diffusion couple are 529 and 6.15 N/m^2, respectively. Sample GG6 have a stress region between 1.67 $\times$ 10^{-5} and 1.32 $\times$ 10^{-3} N/m^2 and sample GG7 have stress region between 1.67 $\times$ 10^{-5} and 1.32 $\times$ 10^{-3} N/m^2 and sample GG7 has the stress regions concentrated on the edges. Even for samples GG1, GG2, and GG3 composites the surface is more or less stress free and high-stress regions are concentrated on the edges. The simulation results indicate that the stress follows the trend GG5 > GG4 > GG1 > GG7 > GG2 > GG6 > GG3 which is in correlation with the thermal expansion studies.

Goel and co-workers obtained a pore-free and continuous interface of 7A glass with the 8YSZ electrolyte. The same group also investigated the 10B–8YSZ pseudocell interface after thermal shock experiments in air (Fig. 8.34a) 10B glass (16.48% MgO, 18.34% CaO, 6.27% BaO, 46.67% SiO_2, 2.08% Al_2O_3, 6.66% La_2O_3, 2% B_2O_3, 0.5 Cr_2O_3, and 1% NiO, wt%) bonded well with the YSZ substrate. Sealing was observed to be gas tight without cracks, hence demonstrating an excellent thermal shock resistance. Both 7A and 10B are Ba–Ca–Mg-based (BCM) compositions and yielded good adhesion results. Another BCM compositions BaCMAS1 (14.54% MgO, 24.03% CaO, 3.46% BaO, 48.78% SiO_2, 9.20% Al_2O_3, wt%) also depicted strong interfacial adhesion with the 8YS after heat treatment at 900°C for 100 h in air (Fig. 8.34b).

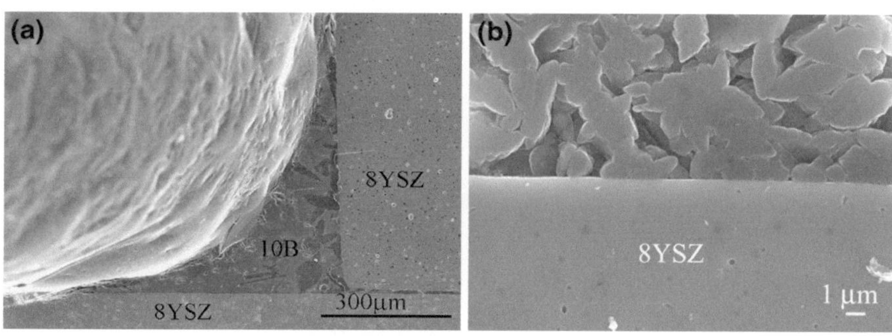

Fig. 8.34 a SEM image of the 8YSZ-GC 10B pseudocell after thermal shock resistance experiments in air after 15 quenching cycles (Goel and coworkers 2010b) and **b** interface of 8YSZ/BaCMAS1, developed after heat treatment at 900°C for 100 h in Air (Goel and coworkers 2007)

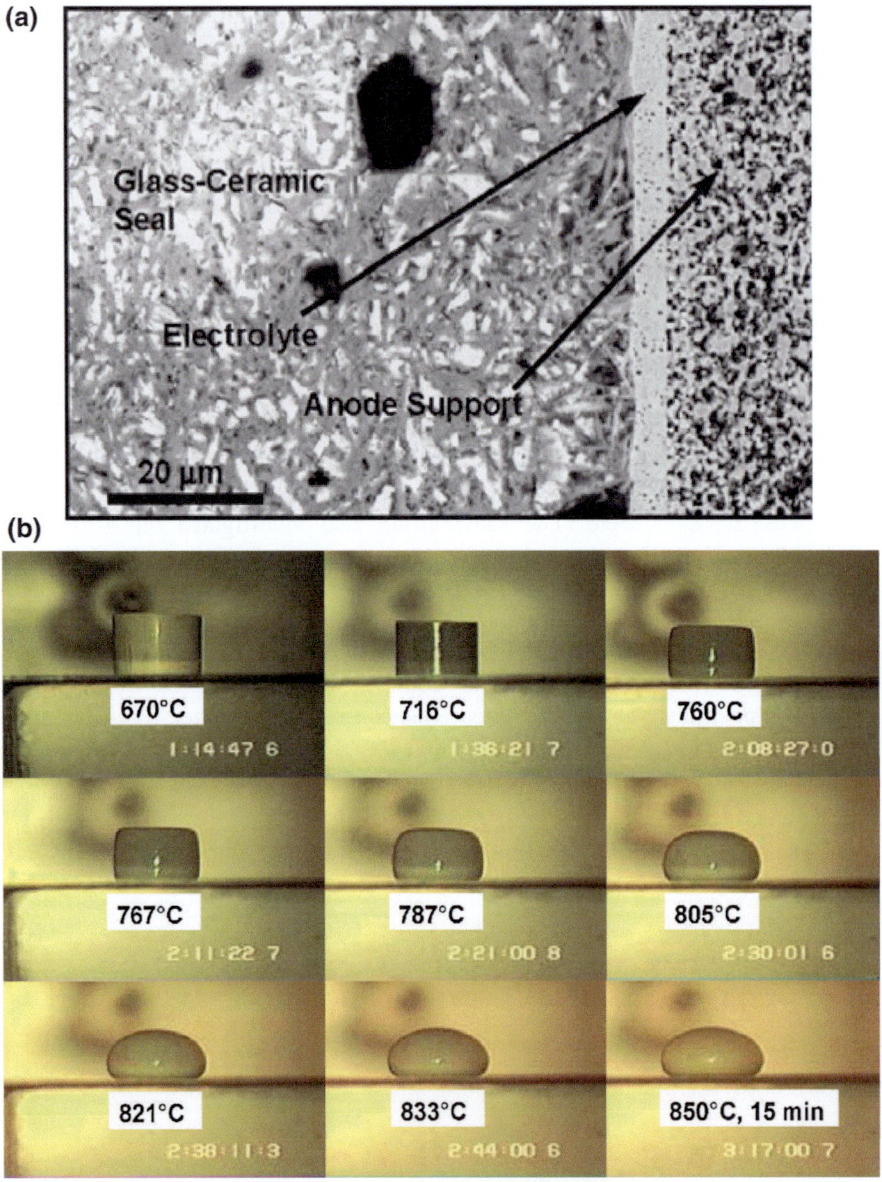

Fig. 8.35 SEM micrographs of G-18 in contact with: **a** 5YSZ after ageing in 750°C air for 1200 h and **b** photographs of a G-18 pellet as it is heated to the temperature typically used in joining (Meinhardt et al. 2008)

Ba–Ca–Sr (BCS) composition G73 (Ba–Ca–Sr–Al–Si–B) also revealed complete wetting behaviour with the YSZ interface. The seals remained hermetic even after rapid quenches ($\sim$25°C s^{-1}) after 100 thermal cycles between 750°C and room temperature. No zirconium or yttrium was detected in the glass or at the interface. The barium calcium composition H [(48.2% BaO, 29.8% SiO_2, 6.1% CaO), balance by (Al_2O_3, B_2O_3, V_2O_5 and ZnO in wt%)], is a typical invert glass with low silica content. The composite of H glass with 10, 20, and 30 wt% of ZrO_2 powder was prepared which yielded CTE in the range of (9–10) × 10^{-6}/K. The 20% wt filler material proved to be the optimised composition and the crystallisation process was changed dramatically. Zirconia was also added to the G-18 glass composition in 10, 15, and 20 vol.%. Addition of ZrO_2 particles has enhanced the mechanical strength of cordierite spodumen and apatite–wollastonite glass–ceramics. The glass–ceramic composite has shown the formation of $BaSiO_4$, $BaCaSiO_4$, hexacelsian $BaAl_2Si_2O_8$, ZrO_2, and $ZaSiO_4$ phases. Upon increasing the heat treatment time, the glass viscosity decreased, whereas ZrO_2 solubility in the glass matrix increased. Increasing the amount of ZrO_2 increased the microhhardness of composites, though increase in heat temperature duration did not affect the microhardness, flexural strength, or fracture toughness.

The composition G-18 glass is a barium calcium glass with SiO_2/B_2O_3 ratio of 3.5 and CTE of 11.8 × 10^{-6}/K. Glass G-18 exhibits less interaction with the anode supported 8YSZ after 1200 h at 750°C in air (Fig. 8.35a). $BaZrO_3$ is formed at the interface, and if the interaction layer is sub-micron thick then the mechanical interlocking at interface may enhance the bonding mechanism. Figure 8.35b indicates the wetting characteristics of G-18 glass, which reveals that the contact angle is greater than 90° at the sealing temperature. In order to felicitate the joining process, pressure is required to achieve a wetting angle <90°C.

8.5 Interaction of Cathode and Electrolytes

Doped lanthanum manganite cathode materials remain stable in oxidising atmospheres, matching CTE and sufficient electrical conductivity. The performance of cathodes can be increased by either adding a second ionically conducting phase to LSM hence increasing the surface area, or by fabricating the composite cathode materials. Doping the A-site by an ion that promotes oxygen vacancies, also increased the conductivity of cathode materials. Gallium and aluminium doping onto the B-site also improved the performance of the cathode because both cations prefer tetrahedral coordination, which provides an alternative to Mn oxidation for stabilising small vacancies. When the cathode material is used with the electrolyte materials, then its use may be limited by the formation of an insulating $SrZrO_3$ layer. In order to reduce the reactivity of the cathode and electrolytes, its becomes very important to focus on the interface reactivity and the dopant nature and concentration. This section deals with the cathode/electrolyte reactivity and interface of the reaction kinetics in order to investigate any detrimental phase formations and measures to reduce them. Perovskite oxides are considered to be chemically active because they react with the electrolyte especially YSZ and chromia containing ferritic interconnects. Hence, the stabilising as well as chemical compatibility of materials is very important, particularly for high-temperature applications. The oxygen exchange takes place at triple-phase boundaries (TPB) of the cathode with the gas phase, cathode, and electrolyte. Hence, any small change at the boundary may lead to considerable effect on oxygen exchange rates and SOFC performance. Grosjean and co-workers studied the reactivity and diffusion between $Lr_{0.8}Sr_{0.2}MnO_3$ and ZrO_2 using TEM analysis. A significant amount of manganese diffusion may cause structural changes in YSZ along with decomposition of LSM and growth of a parasite phase. The manganese diffusion increased the grain boundaries' resistance along with increasing the electronic conduction. Manganese diffusion also caused decomposition of LSM into lanthanum and strontium oxides, which reacted with YSZ to form $La_2Zr_2O_7$ and $SrZrO_3$ phases. Both these phases are known to exhibit low conductivity. Suzuki and co-workers studied the fuel cell behaviour for a composite cathode $(La_{0.8}Sr_{0.2})_{0.9}MnO_3$–Y doped ZrO_2. The symmetrical cell test showed the ASR for overpotential to be 0.14 and 0.3 Ωcm^2 at 800 and 700°C, respectively. At temperature of 850°C, the maximum power and current density of 0.26 Wcm^{-2} and 0.65 Acm^{-2} were obtained.

Ricoult studied the interface chemistry for the LSM-3YSZ composite where the average composite cathode composition is 61ZrO_2, 4.2% Y_2O_3, 11% Mn_2O_3, 2.7% SrO, and 20% La_2O_3 (in wt%). During annealing at 1200°C, the LSM-3YSZ composite cathode did not exhibit structural modifications and no pyrochlore $La_2Zr_2O_7$ could be observed as reaction products. Figure 8.36 gives the high-resolution TEM of the LSM-3YSZ interface. The interface is free of any amorphous film even though the relative orientations of the two lattice planes are not perfect. A transition zone can be observed at the perovskite and tetragonal interface indicating that the low planes of zirconia and perovskite are not perfectly parallel to each other. The near edge fine structure was obtained through ELNES line scans across interfaces indicating high manganese and lanthanum content in the zirconia solid solution at the interface. At 1250°C, manganese and lanthanum concentrations reached up to 20 and 18%, respectively. High dopant concentration yields a strong lattice distortion as well as a very defective fluorite structure. High concentration of Mn^{2+} was due to cathode polarisation, which made the interface

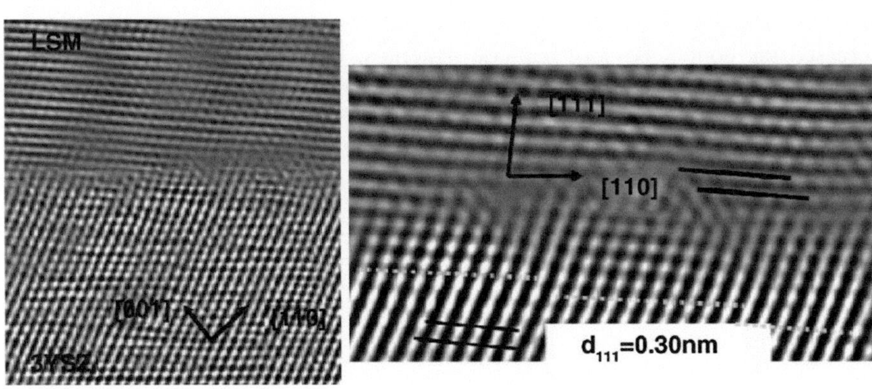

Fig. 8.36 LSM/3YSZ interface in the composite cathode (Backhaus-Ricoult 2006)

richer in manganese than the adjacent zirconia solid solution. Oxygen incorporation at the cathode is promoted due to manganese and zirconia concentrations as illustrated by reaction:

$$4Mn^{2+} + 2V_o + \frac{1}{2}O_{2(g)} \longrightarrow 4Mn^{3+} + 2O^{2-} \tag{8.18}$$

The continuum interfacial defect modeling approach is used to evaluate the interfacial defect concentrations with oxygen chemical potential. It is obtained that with decreasing oxygen activity and negative polarisation, the Mn^{2+} interfacial cluster increases.

Lanthanum deficient manganite $La_{1-x}MnO_3$ ($x = 0$–0.2) has shown less reactivity with YSZ. For the $La_{0.9}MnO_3$ and YSZ interface, $La_2Zr_2O_7$ formation takes place, which increases with the increase in yttrium context of YSZ. When Mn excess occurs in LSM, then the Mn_3O_4 does not equilibrate with $La_2Zr_2O_7$, hence inhibiting the formation of this pyrochlore. The La/Sr ratio on the A-site and temperature of sintering affects the interaction between LSM and YSZ. Tricker and co-workers obtained $La_2Zr_2O_7$ at the interface of $La_{0.85}Sr_{0.15}MnO_3$/YSZ at 1000°C. Hang and co-workers also obtained the pyrochlore $La_2Zr_2O_7$ formation for $La_{0.80}Sr_{0.2}MnO_3$/YSZ (1150°C) but not for $(La_{0.80}Sr_{0.2})_{0.9}MnO_3$/YSZ (1150°C) interfaces. Yokokawa and co-workers suggested that when (La + Sr) content (A site) > Mn^{3+} ions (B-site) and Sr/La < 0.43, then the formation of $La_2Zr_2O_7$ occurs at the electrode–electrolytes interface. $SrZrO_3$ formation is enhanced if high Sr context is present in the LSM. Mitterdorfer and co-workers described that excess La_2O_3 within the perovskite structure is responsible for $La_2Zr_2O_7$ formation because it reacts readily with YSZ. It is also observed that 75% YSZ/25%LSM yields a pyrochlore phase but not for the 25% YSZ/75% YSZ after heat treatment at 1300°C for 2 h. This implies that excess Mn in LSM can either enhance or delay the La_2O_3 formation at the interface. The electrical conductivity of $La_2Zr_2O_7$ layer is 2×10^{-4} S/cm (1000°C), whereas strontium zirconates ($Sr_3Zr_2O_7$, Sr_2ZrO4, $SrZrO_3$) yield lower conductivity of 10^{-4}–10^{-5} S/cm (1000°C). Hence, the conductivities of these phases are lower than that of conductivity of YSZ (0.185 S/cm at 1000°C). The yttria content in ZrO_2 also influences the reactivity of LSM with YSZ. Jiang and co-workers obtained two distinct layer formations between the LSM coating and 3YSZ electrolyte: one layer is due to the $La_2Zr_2O_7$ and the second layer consists of a cubic zirconia solid solution precipitated at the LSM/3YSZ interface. Yamamoto and co-workers obtained that the lower yttria context enhances the reactivity of YSZ with $La_{0.9}MnO_3$; i.e., 12YSZ has lower reactivity and 8 YSZ has higher reactivity with $La_{0.9}MnO_3$. Similar results were obtained by Van Roosmalen and co-workers indicating 3YSZ to be more reactive than 8YSZ with LSM. Hildenbrand and co-workers worked on improving the adherence of a porous electrode to the electrolyte by applying a dense $La_{0.6}Sr_{0.4}Co_{0.2}Fe_{0.8}O_{3-\delta}$ (LSCF) paste before the screen printing process. 3YSZ was used as an electrolyte and yttria-doped ceria (YDC) as an interlayer in order to form a YDC/3YSZ/YDC substrate and LSCF was screen printed on both sides (Fig. 8.37) and the electrolyte layer was 92 μm thick.

Fig. 8.37 SEM micrograph of a cross-section of the symmetrical cell (Hildenbrand and coworkers 2011)

The YDC interlayer is ~3 μm thick and the porous LSCF electrode has mean grain size of 0.50 μm. A dense LSCF layer decreases the polarisation resistance of cathode by a factor of THREE. $La_{0.6}Sr_{0.4}CoO_{3-\delta}$ (LSC) and $La_{0.8}Sr_{0.2}FeO_{3-\delta}$ (LSF) are used as alternatives to the LSM cathode along with a doped ceria interlayer to avoid detrimental reactions.

$(CeO_2)_{0.01}$–$(Sc_2O_3)_{0.1}$–$(ZrO_2)_{0.89}$ (1Ce10ScZr) electrolyte powders were implanted on the YSZ surface to optimise the electrolyte/cathode interface. The interface between 1Ce10ScZr modified cell is shown in Fig. 8.38. No pinholes/cracks are observed at the interfaces and the electrolyte is dense. Two electrolytes sintered very well because of matching CTE and good chemical compatibility. The YSZ electrolyte structure is not destroyed due to incorporation of 1Ce10ScZr in its matrix and hence the 1Ce10ScZr grain discrimination is difficult. Figure 8.38b gives the impedance behaviour of cells at 650°C under open circuit conditions (for big modified and unmodified cells). For both the cells, high

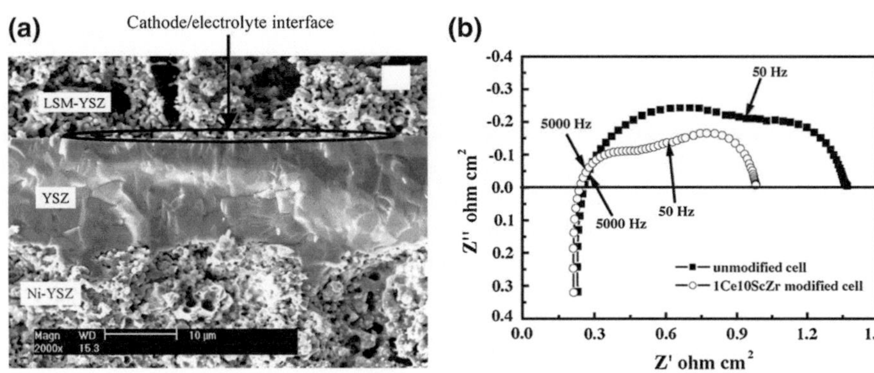

Fig. 8.38 **a** SEM graph of the cross-section of the 1Ce10ScZr modified cell and **b** impedance spectra of the single cells with and without the 1Ce10ScZr modification (at 650°C under open circuit conditions) (Wang and coworkers 2006)

frequency arc is suppressed from the frequency range of 50 kHz–50 Hz. The charge transfer process is promoted due to 1Ce10ScZr modification on YSZ films and Ce^{4+}/Ce^{3+} modification on YSZ films and Ce^{4+}/Ce^{3+} redox reactions are also favoured at the reaction sites. No deleterious interfacial reaction between 1Ce10ScZr grains and YSZ electrolyte occurs during the co-firing process at 1400°C. The modifications of the YSZ electrolyte by 1Ce10ScZr grain incorporation provide a simple method to improve cell performance.

Anderson and co-workers worked on the possibility of reducing lanthanide ferrite cathodes—YSZ interaction with A-site dopant concentration (La with Nd, Sm, Pr, and Sr with Ba, Ca). Table 8.5 yields the reaction products formed when the cathode interacted with YSZ.

Ralph and co-workers obtained pyrochlores and zirconates when the cathode reacted with the YSZ. No reaction products were obtained when the cathode interacted with the GDC. Table 8.6 gives the different compositions, their reaction products, and activation energy with YSZ and GDC interfacial.

Table 8.5 XRD phase analysis for cathode YSZ reactions (1200°C)

Cathode	Reaction product	Cathode	Reaction product
$La_{0.8}Sr_{0.2}FeO_3$	<2 wt% $SrFe_{12}O_{19}$	$Sm_{0.8}Sr_{0.2}FeO_3$	>10 wt% $Sm_{0.8}Zr_{0.2}O_{16}$
$La_{0.8}Ca_{0.2}FeO_3$	<2 wt% Fe_2O_3 and Y_2O_3	$Pr_{0.8}Sr_{0.2}FeO_3$	NA
$La_{0.8}Ba_{0.2}FeO_3$	NA	$Nd_{0.8}Sr_{0.2}FeO_3$	NA
$Ln_{0.8}Sr_{0.2}FeO_3$	<2 wt% $La_2Zr_2O_7$		
	>5 wt% Fe_3O_4		
	<2 wt% $La_2SrFe_2O_7$		
	>2 wt% Y_2O_3		
	<2 wt% $La_{0.5}Sr_{0.5}Fe_2O_3$		
	2–5 wt% $Sr_2Fe_2O_3$		

Table 8.6 Cathode compositions, their reactivity, and interfacial activation energy with YSZ and GDC

Composition	YSZ		GDC	
	Reaction product	Activation energy (eV)	Reaction product	Activation energy (eV)
$La_{0.8}Sr_{0.2}CoO_3$	$La_2Zr_2O_7$, $SrZrO_3$	1.7	None	1.7
$LaNiO_3$	$La_2Zr_2O_7$, $SrZrO_3$	1.3	None	1.2
$La_{0.8}Sr_{0.2}FeO_3$	None	1.7	None	1.9
$Sm_{0.8}Sr_{0.2}CoO_3$	$Sm_2Zr_2O_7$, $SrZrO_3$	1.7	None	2.0
$La_{0.8}Sr_{0.2}Co_{0.8}Ni_{0.15}Cu_{0.05}O_3$	$La_2Zr_2O_7$, $SrZrO_3$	1.8	None	1.7
$La_{0.8}Sr_{0.2}Co_{0.8}Fe_{0.2}O_3$	$La_2Zr_2O_7$, $SrZrO_3$	1.7	None	2.1
$La_{0.8}Sr_{0.2}Co_{0.8}Ni_{0.2}O_3$	$La_2Zr_2O_7$, $SrZrO_3$, NiO	1.9	None	2.4
$La_{0.8}Sr_{0.2}Fe_{0.8}Ni_{0.2}O_3$	$La_2Zr_2O_7$, $SrZrO_3$	1.9	None	1.7
$La_{0.8}Sr_{0.2}Fe_{0.8}Ni_{0.15}Cu_{0.05}O_3$	$La_2Zr_2O_7$, $Sr_2Fe_2O_5$	2.1	La_2NiO_4, NiO	1.9

Letilly and co-workers prepared a $La_{0.58}Se_{0.4}Co_{0.2}Fe_{0.8}O_{3-\delta}$ (LSCF)/ $BaIn_{0.3}TiO_{0.7}O_{2.85}$(BIT07) cathode/electrode assembly. An interlayer is formed between the two compounds, which seem to be composed of $Ba_xLa_{0.58(1-x)}$ $Sr_{0.4(1-x)}$ $In_{0.3x}Co_{0.2(1-x)}Fe_{0.8(1-x)}O_{3-\delta}$ (BLSITCFx) phase, LSCF/BIT07 composite, BLSITCFx/BIT07 composite, and LSCF/BLSITCFx composites. Most of these cathodes exhibit good performance for a range of thermal treatments and cathode grain sizes. $La_{0.8}Sr_{0.2}FeO_{3-\delta}$ (LSF–20) has been a promising cathode and reaction mixture with the Y_2O_3 and ZrO_2, respectively, at 1200°C, subjected to XRD as shown in Fig. 8.39a. The ZrO_2 reaction mixture exhibited an identical shift to that observed for the LSF–YSZ reaction mixture, whereas Y_2O_3 had little effect on the XRD peak position. Hence, the Zr^{4+} cation is responsible for the unit cell expansion due to its higher concentration and small ionic radii as compared to Y^{3+}. For further investigation, the SEM and EDS was done for the LSF cathode/8-YSZ electrolyte (fixed at 1150°C/2 h) as shown in Fig. 8.39b. Table 8.7 gives the spot analysis for the microstructure. The bright phase at the YSZ-LSF interface is platinum, which usually appears from the Pt cathode current collection and is seen after long-term testing duration. Zr diffusion occurred at a depth of 20 μm into the cathode.

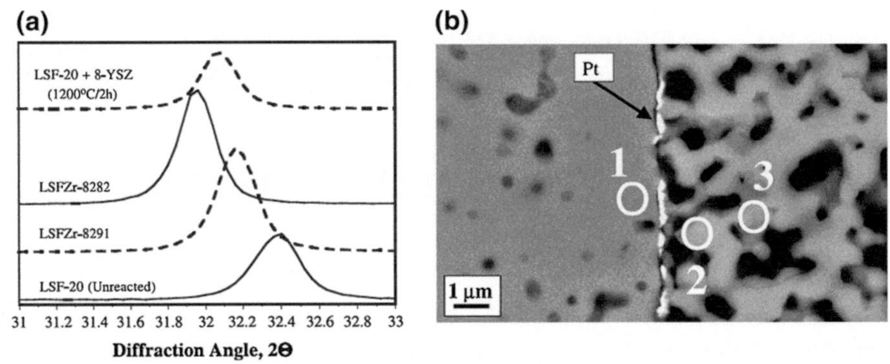

Fig. 8.39 **a** XRD patterns for $La_{0.8}Sr_{0.2}Fe_{1-y}Zr_yO_{3-d}$, y = 0.1 (LSFZr-8291) and y = 0.2 (LSFZr-8282) calcined at 1200°C and **b** SEM and EDX regions for the $La_{0.8}Sr_{0.2}FeO_3$/8YSZ interface (Simner and coworkers 2003b)

Table 8.7 Elemental analysis of spots in Fig. 8.39b

Element	Portion 1	Portion 2	Portion 3
O	56.1	50.6	46.9
Sr	0	6.2	5.3
Fe	0	20	24.5
La	0	17.1	21.8
Zr	38	6.1	1.5
Y	5	0	0
Ni	0.9	0	0

The XRD of LSF-YSZ indicated no reaction mixture of poorly conducting La/Sr zirconate. Zr^{4+} have shown dissolution into the perovskite structure, though Fe ions are not displaced by the Zr resulting in an A-site deficient structure. Huang and co-workers have studied the interaction of $La_{0.9}Sr_{0.1}Ga_{0.8}Mg_{0.2}O_3$ with $La_{0.5}Sr_{0.5}CoO_3$ (LSC) and $La_{0.84}Sr_{0.16}MnO_3$ cathodes. Ga diffused into LSC, whereas Co migrated to the LSGM electrolyte. The small content of Ga into LSM and Mn into LSGM could be seen even at a firing temperature of 1470°C $Nd_{1-x}Sr_xMnO_3$ and LSM have shown good compatibility with the GDC electrolyte. Even the $Gd_{1-x}Sr_xMnO_3$ (GSM)/GDC interface was found reaction free.

8.6 Interaction of Cathodes and Interconnect

During the development of SOFC, the interaction between cathode and interconnect is also a challenging issue and has been addressed by many research groups. Jiang and co-workers studied the interaction between the LSM/YSZ composite cathode and Fe–Cr alloy interconnect under SOFC operating conditions. No Cr accumulation was observed on the surface or interface of the composite cathode/interconnect alloy. Quadakkers and co-workers studied the interaction of LSM, $LaCoO_3$, and (La, Sr) CoO_3 films coated on alumina and chromia forming alloys. $CoCr_2O_4$ and $MnCr_2O_4$ spinel phases were deposited for the (La, Sr) CoO_3 and LSM coating, respectively, on the chromia forming alloy. Taniguchi and co-workers obtained an increase in cathodic polarisation attributed to an increase in the Cr amount at the LSM/Inconel600 interface. Usually, the chromium deposition leads to the spinel phase formation at the triple-phase boundaries, hence blocking the active sites. The process can be explained as

$$CrO_{3(g)} + Cr(OH)_2O_{2(g)} + 6e^- \longrightarrow Cr_2O_{3(g)} + 3O^{2-}_{YSZ} + H_2O_{(g)} \tag{8.19}$$

$$3Cr_2O_{3(s)} + 6(La, Sr)MnO_3 \longrightarrow 2(Cr_xMn_{1-x})_3O_4 + 6(La, Sr)(Mn_xCr_{1-x})O_3 + \frac{1}{2}O_2 \tag{8.20}$$

Zhen and co-workers investigated the LSM coating/Mn_2O_3/SrO/La_2O_3 and Fe–Cr alloy interconnect. When the Fe–Cr alloy reacts with SrO oxide, then Cr_2O_3 is the main product. In contrast to this, $LaCrO_3$ and $(Cr, Mn)_3O_4$ are the primary reaction products when the Fe–Cr alloy reacts with La_2O_3 and Mn_2O_3.

8.7 Interaction of Anode and Electrolyte

The anode materials must be compatible with the electrolyte and transport oxygen ions from the electrolyte, and electrons from the interconnect and gas from the fuel. Nickel and YSZ cement are the most widely used anode materials due to their catalytic properties, and thermomechanical compatibility with the fuel cell components. The carbon deposition on the electrode, although using carbon-rich fuels remains the potential problem for the cermet as the higher carbon content may block the anode reactions. Anodes are typically mixed conductors and hence contain transition metals. The perovskite structure is preferred for the promising anodes such as titanates and chromites. Apart from them, the cubic fluorite structures such as zirconia and ceria-based structures are also desirable for the anode materials. $SrTiO_3$ has been investigated as a very promising anode material and its conductivity increases with yttrium doping; i.e., conductivity of 64 S cm^{-1} is obtained for $Sr_{0.88}Y_{0.08}TiO_3$. (La, Sr) TiO_3 anode material is sulphur resistant, and the conductivity shows an increasing trend with increasing lanthanum content (is stabilised for La = 0.4). When cerium is used as a dopant for the (La, Sr)TiO_3, then polarisation resistance is decreased. Strontium-doped lanthanum chromite provides direct oxidation of the methane and does not catalyse carbon deposition, hence acting as a potential anode material. The B-site of strontium-doped lanthanum chromite has been doped with Fe, Co, Ni, Cu, and Mn ions for creating oxygen vacancies, and Ni dopant has shown more promising results. Vanadium additions in (La, Sr)CrO_3 have yielded desirable outputs; i.e., (La, Sr)VO_3 possesses high conductivity and nickel doping enhances it further (Fig. 8.40).

Strontium cobalt ferrites are good mixed conductors but are more suitable to be used as cathodes due to their p-type conduction. Pervoskite gallates/niobates/barium cerate and chromium-doped lanthanum–strontium ferrite have also been studied as the anode material. Although the gallates and niobates exhibit low conductivity, they exhibit good compatibility with the LSGM electrolytes. The conductivity of yttrium-doped barium cerates increases by increasing yttrium content. For barium cerates, the increase in electronic conduction due to the reduction of cerium at low

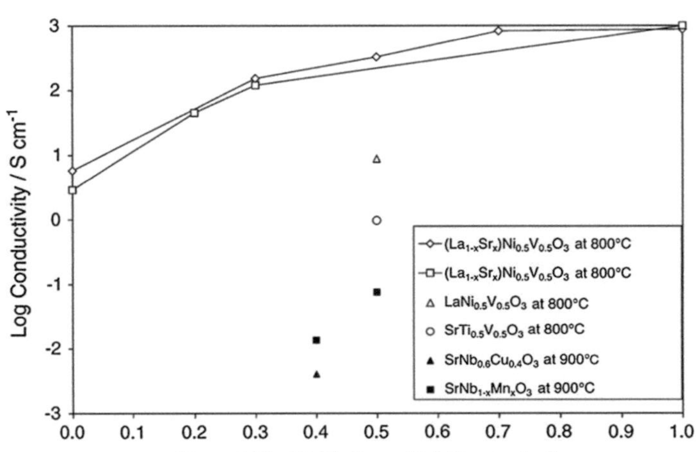

Fig. 8.40 Conductivity of (La,Sr)(Ti,Ni,V)O_3 and Sr(Nb,Mn,Cu)O_3 in fuel atmospheres at 800 and 900°C, respectively (Fergus 2006a)

oxygen partial pressures occurs. For zirconia-based anodes, doping with titania has increased the electronic conductivity that the total conductivity. The Ti^{4+}/Ti^{3+} polaron hopping mechanism may result in the increase of electronic conductivity. For large dopant levels, the titanium defects with oxygen vacancies lead to large polaron mechanisms. For niobium-doped zirconia anode materials, niobium does not get reduced easily and hence the electronic conductivity does not get increased much. Ceria is also used in cermets with nickel/copper as an addition to Ni-YSZ cermets; pyrochlore anodes such as $Gd_2Ti_2O_7$ show and increase in conductivity upon replacement of Ti with Mo/Mn. The compounds with tungsten bronze structure such as $Sr_{0.2}Ba_{0.4}Ti_{0.2}Nb_{0.8}$ exhibit high conductivity of 10 S/cm but have low thermal expansion in (2.3–6.5) × 10^{-6}/K. Bi_2O_3–Ta_2O_5 mixtures have shown good performance for direct reaction of butane, but bismuth possesses high vapour pressure. For the anode/electrolyte reactions, H_2O plays an important role and the different reaction paths for the electrolyte/anode are shown in Fig. 8.41.

Jiang and co-workers gave the following reaction mechanism for the surface reactions on the YSZ surface.

$$O^{2-}_{(\text{oxide})} + 2H_{\text{ad, oxide}} \longrightarrow H_2O + V^{\bullet\bullet}_{\text{o(oxide)}} + 2e^-_{(\text{near TPB})} \tag{8.21}$$

where H_{ad} is the adsorbed H on the active metal surface site. The charge transfer reaction is deeply influenced by the adsorption of hydrogen on the metal surface. The dissociation adsorption of hydrogen occurs preferably on the metal, which contains a high density of d-electronic states. Ceria-based materials added to the Ni-YSZ cermets also promote the reaction mechanism in a H_2–H_2O atmosphere due to the surface catalytic activity of ceria. Ceria-based oxides possess affinity for the hydrogen which promotes anodic reaction. Liu and co-workers have investigated the microstructure degradation of the anode/electrolyte interface using TEM and focused ion beam (FIB) techniques. Figure 8.42 shows FIB image of the bulk

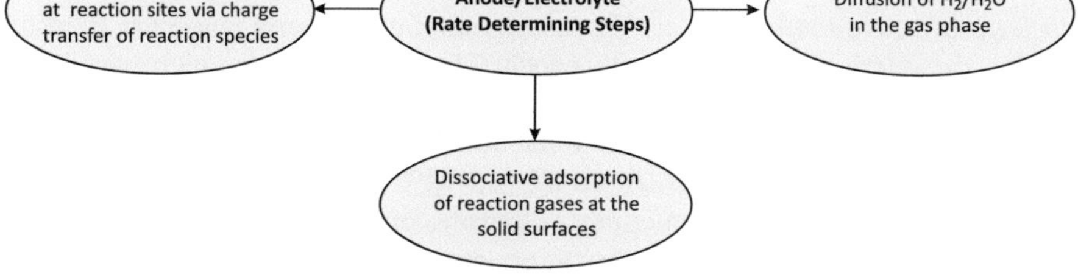

Fig. 8.41 Different reaction mechanisms at the anode/electrolyte interface

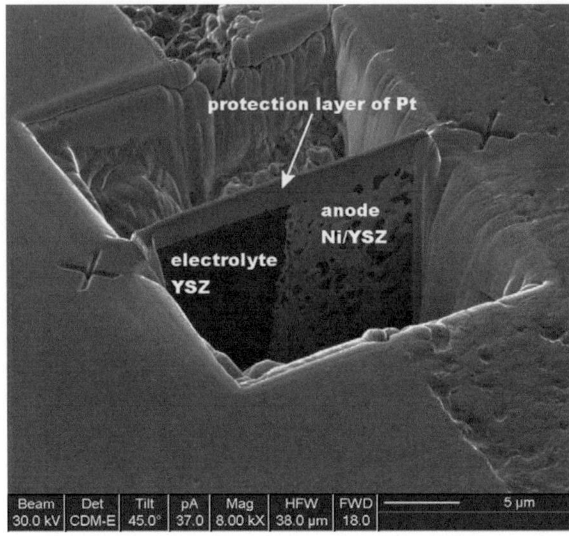

Fig. 8.42 An FIB image under ion-induced secondary electron mode (sample thickness around 250 nm) (Liu and Jiao 2005)

sample (before polishing). The black region reveals the non-conducting YSZ electrolyte and the grey porous region reveals the Ni/YSZ anode cermet.

Figure 8.43a, b gives the bright-field and dark-field micrograph for the anode/electrolyte interface. The interfacial region is ~7 μm in length for the anode/YSZ. The Ni grains are seen with size range of ~200–700 nm and are round in shape. Interestingly, a Si containing phase is formed which covers a part of the electrolyte surface (~400–600 μm thick layer). The diffraction patterns confirmed that it is a glass phase due to its amorphous nature (area A). The EDS spectrum of the Si-phase (not given here) proved that the phase is SiO_2.

The area C has an irregular shape and these clusters contains 2–3 at.% Si, 20 at.% of Ni and YSZ each. Area B and C contain Ni and YSZ-rich phases. As SiO_2, Na_2O, ZrO_2, NiO, Fe_2O_3, Al_2O_3, and Y_2O_3 etc., are used as raw materials for fabricating cermet and electrolyte, hence for a particular region the glass formation may have taken place. For the SiO_2–Na_2O–ZrO_2 system, the glass formation region exists for $ZrO_2 < 20$ mol%. For the present system, the ZrO_2 concentration in the glass is ~5 mol%. A small amount of Ni and Cr is found in the YSZ grains attributed to the Ni solubility in the matrix. As the silicate glass phase is distributed along the anode/electrolyte interface, hence the decrease in reaction rate is observed due to blockage of active reaction zones. The composition of the glass phase has been the same at the interface and inside bulk anode. The glass possesses high wettability, which is evident from the formation of thin films around the Ni/YSZ grain boundaries.

Knibbe and co-workers used alumina additions to 10YSZ for cleaning the interface boundaries from silica. At CFCL, 15 wt% Al_2O_3 addition to the 10YSZ resulted in increased strength and grain boundary conductivity. Anyhow during the anode sintering, $NiAl_2O_4$ phase formation may take place, if NiO reacts with Al_2O_3. $NiAl_2O_4$ possesses very low oxygen diffusivity and hence hinders the oxide

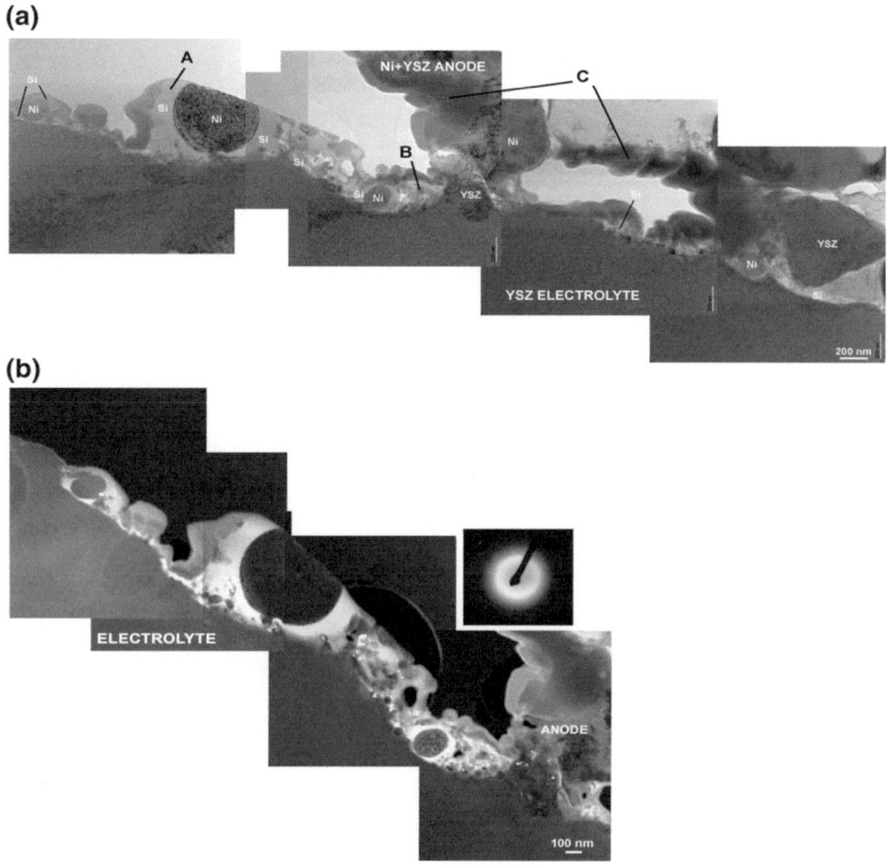

Fig. 8.43 TEM **a** bright-field micrograph showing the whole length of the anode/electrolyte interface and **b** dark-field micrograph of a part of the anode/electrolyte interface (Liu and Jiao 2005)

ion conductivity of the electrolyte. The reduction of $NiAl_2O_4$ to Ni and Al_2O_3 may occur accompanied by an 18% volume decrease, ultimately leading to formation of microcracks which affect structural integrity. Figure 8.44a–c gives the microstructure of an unreduced fuel cell interface with corresponding Al and Ni elemental maps. Al_2O_3 grain appears the darkest and YSZ grain the lightest. At the interface, the light grey phase ~4 μm width is seen in the electrolyte region. The interface is rich in both Al and Ni. These results are in support of XRD, according to which $NiAl_2O_4$ phase formation is expected. Figure 8.45a–c shows the SEM micrographs of an anode/electrolyte interface after 500 h of fuel cell operation and the corresponding elemental maps of Al and Ni. $NiAl_2O_4$ is still found within several microns of the electrolyte. The $NiAl_2O_4$ exhibits an interconnected nature in the electrolytes, still the effect of $NiAl_2O_4$ should be addressed in regard to the fuel cell performance.

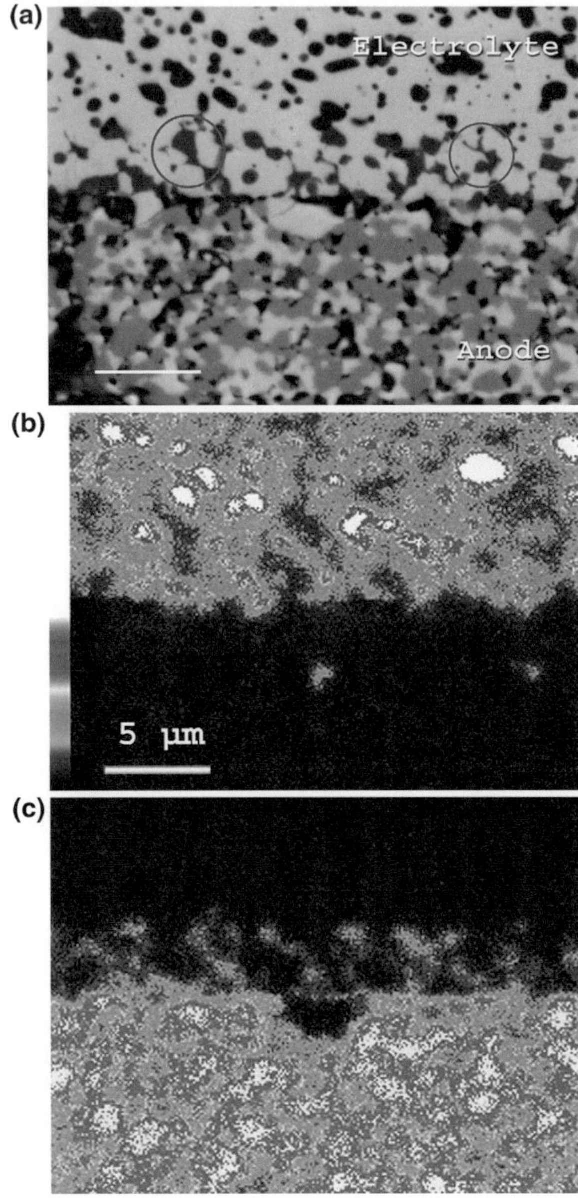

Fig. 8.44 **a** SEM backscattered micrograph of typical anode electrolyte interface after anode sintering, with corresponding Al **b** and Ni **c** elemental maps (Knibbe and coworkers 2008)

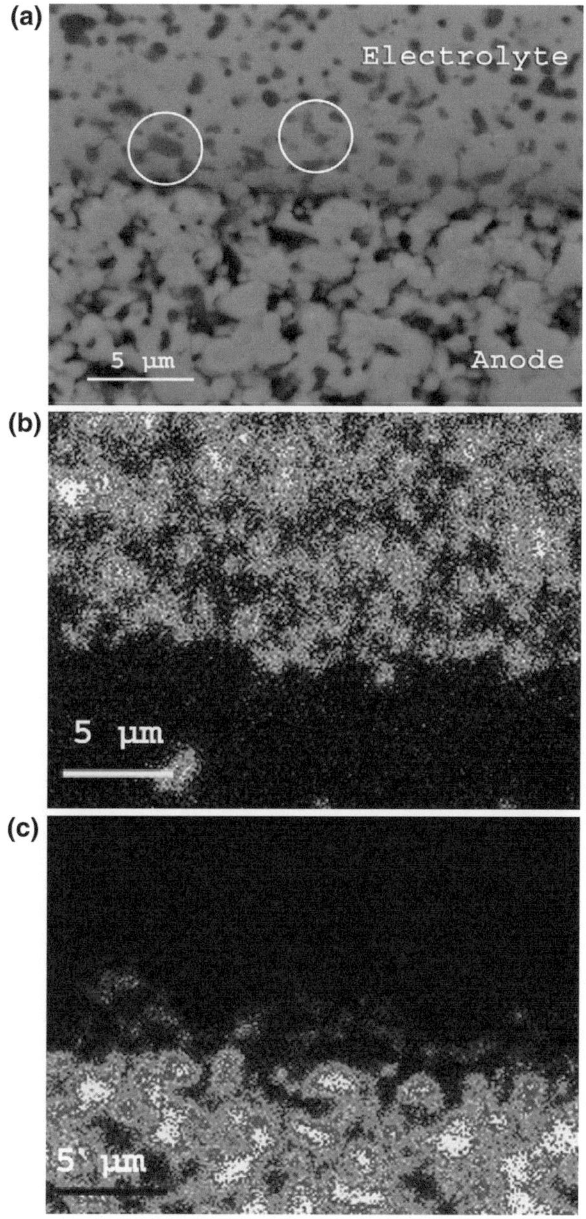

Fig. 8.45 **a** SEM micrographs of anode electrolyte interface after 500 h of fuel cell operation, with corresponding Al **b** and nickel **c** maps (Knibbe and coworkers 2008)

Bibliography

Backhaus-Ricoult M (2006) Solid State Ionics 177:2195
Brochu M, Gauntt BD, Shah R, Miyake G, Loehman RE (2006) J Eur Ceram Soc 26:3307
Chou YS, Stevenson JW, Gow RN (2007b) J Power Sources 170:395
Chou YS, Thomsen EC, Williams RT, Choi JP, Canfield NL, Bonnett JF, Stevenson JW, Shyam A, Curzio EL (2011) J Power Sources 196:2709
Chou YS, Stevenson JW, Choi JP (2014b) J Power Sources 250:166
Fergus JW (2006a) J Power Sources 162:30
Ghosh S, Sharma AD, Mukhopadhyay AK, Kundu P, Basu RN (2010) Int J Hydrogen Energy 35:272
Goel A, Tulyaganov DU, Kharton VV, Yaremchenko AA, Agathopoulos S, Ferreira JMF (2007) J Am Cera Soc 90:2236
Goel A, Tulyaganov DU, Kharton VV, Yaremchenko AA, Ferreira JMF (2010b) J Power Sources 195:522
Hildenbrand N, Boukamp BA, Nammensma P, Blank DHA (2011) Solid State Ionics 192:12
Jacobson AJ (2010) Chem Mater 22:660
Kaur G, Pandey OP, Singh K (2012b) Int J Hydrogen Energy 37:17235
Kaur G, Pandey OP, Singh K (2012c) J Electrochem Soc 159:F717
Kaur G, Pandey OP, Singh K (2012d) Fuel Cells 12:739
Kaur G, Pandey OP, Singh K (2014a) Indian J Pure Appl Phys 52:262
Kaur G, Pickrell G, Cheng Y (2015) Int J Energy Res 39:681
Knibbe R, Drennan J, Dicks A, Love J (2008) J Power Sources 179:511
Kumar V, Sharma S, Pandey OP, Singh K (2010) Solid State Ionics 181:79
Kumar V, Kaur G, Lu K, Pandey OP, Singh K (2014) Int J Appl Glass Sci 5(4):329
Kumar V, Kaur G, Lu K, Pickrell G (2015) Int J Hydrogen Energy 40:1195
Liu YL, Jiao C (2005) Solid State Ionics 176:435
Meinhardt KD, Kim DS, Chou Y-S, Weil KS (2008) J Power Sources 182:188
Simner SP, Shelton JP, Anderson MD, Stevenson JW (2003b) Solid State Ionics 161:11
Smeacetto F, Salvo M, Ferraris M, Casalegno V, Asinari P (2008a) J Euro Cera Soc 28:611
Smeacetto F, Salvo M, Ferraris M, Casalegno V, Asinari P, Chrysanthou A (2008b) J Euro Cera Soc 28:2521
Smeacetto F, Salvo M, Ferraris M, Cho J, Boccaccini AR (2008c) J Euro Cera Soc 28:61
Smeacetto F, Salvo M, D'Hérin Bytner FD, Leone P, Ferraris M (2010) J Euro Cera Soc 30:933
Smeacetto F, Miranda AD, Chrysanthou A, Bernardo E, Secco M, Bindi M, Salvo M, Sabato AG, Ferraris M (2014) J Am Ceram Soc 97(12):3835
Sohn SB, Choi SY (2004) J Am Cera Soc 87:254
Tiwari B, Dixit A, Kothiyal GP (2011) Int J Hydrogen Energy 36:15002
Wang Z, Chenga M, Dong Y, Zhang M, Zhang H (2006) J Power Sources 156:306

Suggested Bibliography

Anderson MD, Stevenson JW, Simner SP (2004) J Power Sources 129:188
Arashi H, Naito H (1992) Solid State Ionics 53–54:436
Badwal SPS, Deller R, Foger K, Ramprakash Y, Zhang JP (1997) Solid State Ionics 99:297.
Berenov A, Wood H, Atkinson A (2007) ECS Trans 7:1173.
Chou YS, Stevenson JW, Choi JP (2014a) J Power Sources 255:1
Chou YS, Stevenson JW, Gow RN (2007a) J Power Sources 168:426
Clausen C, Bagger C, Bildesorensen JB, Horsewell A (1994) Solid State Ionics 70:59
Colomer MT, Durán P, Caballero A, Jurado JR (1997) Mater Sci Eng A A229:114
Donald JW, Mallinson PM, Metcalfe BL, Gerrard LA, Fernie JA (2011) J Mater Sci 46(7):1975

Durán P, Capel F, Moure C, González-Elipe AR, Caballero A, Bañares MA (1999) J Electrochem Soc 146(7):2425
Feighery AJ, Irvine JTS (1999) Solid State Ionics 121:209
Fergus JW (2006b) Solid State Ionics 177:1529
Fergus JW (2005) J Power Sources 147:46
Ghosh S, Kundu P, Sharma AD, Basu RN, Maiti HS (2008) J Eur Ceram Soc 28:69
Goel A, Tulyaganov DU, Kharton VV, Yaremchenko AA, Ferreira JMF (2008) Acta Mater 56:3065
Goel A, Tulyaganov DU, Pascual MJ, Shaaban ER, Muñoz F, Lu Z, Ferreira JMF (2010a) J Non-Cryst Solids 356:1070
Goodenough JB, Manthiram A, Paranthaman M, Zhen YS (1992) Mater Sci Eng B12:357
Grosjean A, Sanséau O, Radmilovic V, Thorel A (2006) Solid State Ionics 177:1977
Gross SM, Koppitz T, Remmel J, Bouche JB, Reisgen U (2006) Fuel Cells Bull 2006:12
Guo X, Sigle W, Fleig J, Maier J (2002) Solid State Ionics 154/155:555
Hashimoto S, Iwahara H (2000) J Electroceram 4:225
Heydari F, Maghsoudipour A, Hamnabard Z, Farhangdoust S (2012) Mater Sci Eng A 552:119
Hibino Y, Hashimoto A, Suzuki M, Sano M (2002) J Electrochem Soc 149(11):A1503
Horita T, Kishimoto H, Yamaji K, Xiong Y, Sakai N, Brito ME, Yokokawa H (2006) Solid State Ionics 177:1941
Hsu JH, Kim CW, Brow RK (2014) J Power Sources 250:236
Huang S, Lu Q, Wang C (2011) J Alloy Compd 509:4348
Huang J, Xie F, Wang C, Mao Z (2012) Int J Hydrogen Energy 37(1):877
Hui S, Roller J, Yick S, Zhang X, Deces-Petit C, Xie Y, Maric R, Ghosh D (2007) J Power Sources 172:493
Ishihara T, Tabuchi J, Ishikawa S, Yan J, Enoki M, Matsumoto H (2006) Solid State Ionics 177:1949
Jiang SP (2008) J Mater Sci 43:6799
Jiang SP, Chan SH (2004) J Mater Sci 39:4405
Jiang SP, Zhang JP, Ramprakash Y, Milosevic D, Wilshier K (2000) J Mater Sci 35:2735
Jiang SP, Zhang JP, Foger K (2003) J Eur Ceram Soc 23:1865
Kaiser A, Feighery AJ, Irvine JTS (1999) Electrochem Soc Proc 99–19(SOFC VI):541
Kaur G, Kumar V, Pandey OP, Singh K (2012a) J Electrochem Soc 159:B277
Kaur G, Singh K, Pandey OP, Homa D, Scott B, Pickrell G (2013a) J Power Sources 240:458
Kaur G, Homa D, Singh K, Pandey OP, Scott B, Pickrell G (2013b) J Power Sources 242:305
Kaur G, Pandey OP, Singh K (2014b) Int J Appl Ceram Technol 11(1):136
Kawada T, Sakai N, Yokokawa H, Dokiya M, Anzai I (1992) Solid State Ionics 50:189
Kharton VV, Marques FMB, Atkinson A (2004) Solid State Ionics 174:135
Khedim H, Nonnet H, Méar FO (2012) J Power Sources 216:227
Kostogloudis GC, Ftikos C (1999) J Eur Ceram Soc 19:497
Kostogloudis GC, Vasilakos N, Ftikos C (1997) J Eur Ceram Soc 17:1513
Kumar V, Kaur G, Pandey OP, Singh K (2011) Phys Chem Glasses 52:212
Labrincha JA, Frade JR, Marques FMB (1993) J Mater Sci 28:3809
Lahl N, Bahadur D, Singh K, Singheiser L, Hilpert K (2004) J Electrochem Soc 149:A607
Lara C, Pascual MJ, Duran A (2004) J Non-Cryst Solids 348:149
Letilly M, Joubert O, La Salle ALG (2012) J Power Sources 212:161
Matsuzaki Y, Yasuda I (2001) J Electrochem Soc 148:A126
Mitterdorfer A, Gauckler LJ (1998) Solid State Ionics 111:185
Mizusaki J, Yonemura Y, Kamata H, Ohyama K, Mori N, Takai H et al (2000) Solid State Ionics 132:167
Mori M, Hiei Y, Sammes NM, Tompsett GA (1999a) In: Singhal S, Dokiya M (eds) SOFC-VI. The electrochemical society, Inc., Pennington, p 347
Mori M, Abe T, Itoh H, Yamamoto O, Shen GQ, Takeda Y et al (1999b) Solid State Ionics 123:113
Naoumidis A, Ahmad-Khanlou A, Samardzija Z, Kolar D, Fresenius J (1999) Anal Chem 365:277

Nasrallah MM, Carter JD, Anderson HU, Koc R (1991) In: Grosz F, Zegers P, Singhal SC, Yamamoto O (eds) SOFC-II. Commission of the European Communities, Luxemburg, p 637
Park JY, Choi GM (2005) Solid State Ionics 176:2807
Pascual MJ, Kharton VV, Tsipis E, Yaremchenko AA, Lara C, Duran A, Frade JR (2006) J Euro Cera Soc 26(15):3315
Paulson SC, Birss VI (2004) J Electrochem Soc 151:A1961
Pernot E, Anne M, Bacmann M, Stroble P, Fouletier J, Vannier RN, Mairesse G, Abraham F, Nowogrocki G (1994) Solid State Ionics 70–71:259
Phillipps MB, Sammes NM, Yamamoto O (1999) Solid State Ionics 123:131
Quadakkers WJ, Greiner H, Hansel M, Pattanaik A, Khanna AS, Mallener W (1996) Solid State Ionics 91:55
Ralph JM, Schoeler AC, Krumpelt M (2001) J Mater Sci 36:1161
Sakaki Y, Takeda Y, Kato A, Imanishi N, Yamamoto O, Hattori M et al (1999) Solid State Ionics 118:187
Shelby JE (2005) Introduction to glass science and technology, second ed. The Royal Society of Chemistry, Cambridge
Simner SP, Bonnett JF, Canfield NL, Meinhardt KD, Shelton JP, Sprenkle VL, Stevenson JW (2003a) J Power Sources 113:1
Smeacetto F, Salvo M, Leone P, Santarelli M, Ferraris M (2011) Mater Lett 65:1048
Sohn SB, Choi SY, Kim GH, Song HS, Kim GD (2002) J Non-Cryst Solids 297:103
Steele BCH, Butler EP (1985) Proc Br Ceram Soc 36:45
Sun C, Hui R, Roller J (2010) J Solid State Electrochem 14:1125
Takeda Y, Tu HY, Sakaki H, Watanabe S, Imanishi N, Yamamoto O et al (1997) J Electrochem Soc 144:2810
Taniguchi S, Kadowaki M, Kawamura H, Yasuo T, Akiyama Y, Miyake Y et al (1995) J Power Sources 55:73
Tong JJ, Han MF, Singhal SC, Gong Y (2012) J Non-Cryst Solids 358:1038
Tricker DM, Stobbs WM (1993) In: Poulsen FW, Bentzen JJ, Jacobsen T, Skou E, Ostergard MJL (eds) 14th Riso international symposium on materials science: high temperature electrochemical behaviour of fast ion and mixed conductors. Riso National Laboratory, Roskilde, p 453
Tsai T, Barnett SA (1997) Solid State Ionics 93:207
Ullmann H, Trofimenko N, Naoumidis A, Stover D (1999) J Eur Ceram Soc 19:791
van Roosmalen JAM, Cordfunke EHP (1992) Solid State Ionics 52:303
Wang R, Lu Z, Liu C, Zhu R, Huang X, Wei B, Ai N, Su W (2007) J Alloy Compd 432:189
Wang SF, Wang YR, Hsu YF, Chuang CC (2009) Int J Hydrogen Energy 34:8235
Wang SF, Hsu YF, Lu HC, Lo SC, Cheng CS (2012) Int J Hydrogen Energy 37:5901
Yamamoto O, Shen GQ, Takeda Y, Imanishi N, Sakaki Y (1991) In: SOFC-II. The electrochemical society, Inc., Pennington, p 158
Ye Y, Yan D, Wang X, Pu J, Chi B, Jian L (2012) Int J Hydrogen Energy 37(2):1710
Yokokawa H, Sakai N, Kawada T, Dokiya M (1991) J Electrochem Soc 138:2719
Yoon HS, Choi SW, Lee D, Kim BH (2001) J Power Sources 93:1
Zhen YD, Jiang SP, Zhang S, Tan V (2006) J Eur Ceram Soc 26:3253
Zhu WZ, Deevi SC (2003) Mater Sci Eng A362:228

Chapter 9
Fuel Cell Status

Vishal Kumar

9.1 Fuel Cell Design

Fuel cell developers and manufacturers are focusing on limiting the cost of materials and manufacturing of their systems; their biggest goal is addressing issues of system design and construction. Although fuel cells promise green technology compared with other technologies, most fuel cell-based systems face tough commercial competition. These competitors are in the most advanced stage on the market. In the automobile sector, the internal combustion engine is a piece of highly engineered and evolved technology, which is of relatively low cost compared to current fuel cell systems. In the stationary power sector, heavy duty vehicles running on gas or steam turbine technology can deliver electricity at lower cost than most fuel cell systems. The commercial viability of the fuel cell depends on its ability to supply power at a cost similar to its competing technologies. In recent years several economic studies of fuel cell systems have been conducted, but these are not enough. Much of the information on fuel cell component costs is proprietary; a detailed study of the economics of fuel cell systems is required.

9.2 Global Scenario of Fuel Cells

Fuel cells are basically energy systems which have inputs and outputs. These energy systems are a part of a huge energy system, the world's energy system, with its inputs–all renewable and non-renewable energy sources, and its output–the consumer needs and energy services that mankind requires, such as heating, lighting, transport, etc. (Fig. 9.1).

G. Kaur, *Solid Oxide Fuel Cell Components*,
DOI 10.1007/978-3-319-25598-9_9

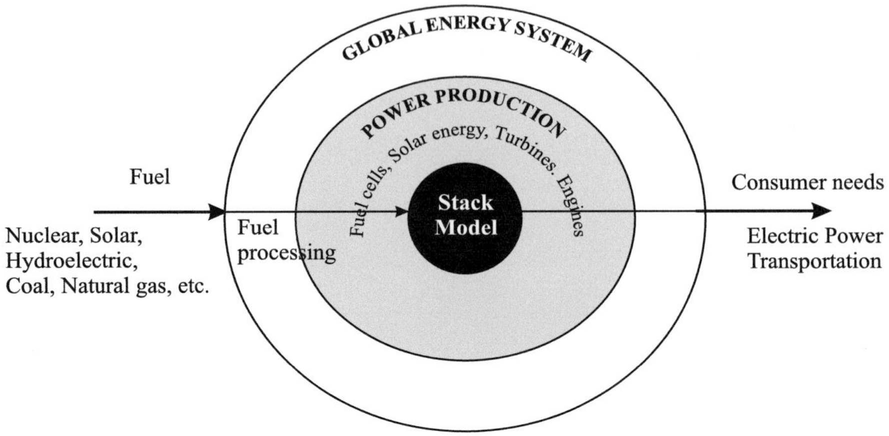

Fig. 9.1 Global energy scheme

Different technologies are being utilised to convert raw fuel into energy services; some are more efficient than others and some play an important role in our day-to-day life. The major concern at this time is the impact of energy sources on the environment; fuel cells as a part of much larger energy systems have little effect on the atmosphere. It becomes necessary to have knowledge about new technologies and manufacturing processes with respect to their societal and environmental effects. The automobile industry needs to consider the entire life cycle cost of its products, including cost of disposing of components such as plastics, glass, rubber, and steel used in constituents. So while using a fuel cell system as an energy source, it becomes important to analyse systematically the inputs and outputs that may be involved. These are shown in Fig. 9.2.

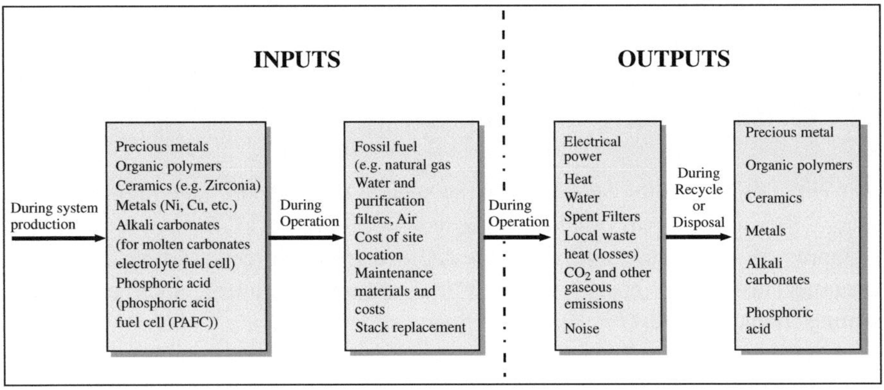

Fig. 9.2 Impact on environment during fuel cell operation

9.3 Source to Speed Analysis

The primary fuel used during working of fuel cells is hydrogen. Currently, hydrogen is generated by electrolysis with the help of electricity. However, in the near future with improvement and development of more renewable sources such as wind and solar power, hydrogen can be produced by electrolysis using these renewable energy sources. Generating hydrogen from renewable energy sources also releases oxygen during electrolysis of water. In the case of hydrocarbon from fossil fuel, carbon dioxide is also produced during conversion of fossil fuel into hydrogen. This is the biggest challenge which cannot be avoided, irrespective of the route for fuel processing followed except pyrolysis in which the major by-product is carbon. Critical planning is required for utilisation of carbon containing fuels for production of hydrogen; the situation is grimmer because of increasing pressure to reduce global emissions of carbon dioxide into the atmosphere. Although cars and buses running on fuel cells have no harmful gases coming out of their exhaust and are termed zero emission vehicles, what about the hydrogen stored aboard? There is a strong possibility that during generation of hydrogen for these automobiles, carbon dioxide may have been released into the atmosphere, where reforming of hydrogen from natural gas was carried out. It simply implies carbon dioxide has been released into the atmosphere on the link between oil or gas source to the vehicle electric drivetrain.

The emission criteria for a fuel cell vehicle (FCV) cannot be accomplished in isolation; vehicle manufacturers need to look at the whole system which begins with extraction of natural gas or fuel from the source and ends at the vehicle tank which provides power and speed to the vehicle. Source to speed analysis is therefore a key aspect of fuel cell systems and more fuel cycle studies are being published every year. The Office of Transportation Technologies under the US Department of Energy's Office (2001) is supporting such programmes carried out by companies including General Motors (GM), Argonne National Laboratory, BP, Exxon Mobil, and Shell. These programmes aim at evaluating innovative propulsion technologies, that is, hybrid electric vehicle (HEVs), FCVs, and advanced internal combustion engines.

9.3.1 Source to Supply Analysis

Although there are several different routes for production of hydrogen, the major challenge for hydrogen fuelled vehicles is the infrastructure needed to deliver hydrogen. There can be some other alternatives for hydrogen supply, including:

(i) Generation of hydrogen by electrolysis at refuelling stations
(ii) Generation of hydrogen as a by-product of industrial ammonia plant, oil refineries, etc.

(iii) Natural gas fuelled steam reformers for production of hydrogen at filling stations
(iv) Steam reforming of natural gas in a large facility to generate hydrogen, and then transporting it in liquid state to filling stations

Economic analyses and environmental assessments of different routes for hydrogen synthesis have been studied by several groups. The collective view of these studies conducted in the United States is that currently, hydrogen generation by electrolysis might be preferred; in the future it might be economical to generate hydrogen in a centralised facility with the added advantage of isolating carbon dioxide generated from the reformer.

9.3.2 Source to Speed Analysis in North America

GM carried out this analysis in North America; the results can vary depending on the geographic location and nature of fuel feedstock available. Depending on the local condition, the nature of the vehicles or transportation system varies from continent to continent; however, studies conducted in North America are worth monitoring because of the impact they will have on the development of transportation systems.

Fuel Economy: Hydrogen-based FCVs and hybrid systems offer better fuel mileage compared to conventional vehicles.

Tank to Speed Efficiency: The power train working in conventional vehicles requires more energy compared to fuel cell systems. The latter have intrinsic high efficiency of the stacks. The study also exhibited higher fuel economy for hydrogen-based FCVs compared to those processing fuel on board.

Harmful Gas Emission: It was observed that a large amount of greenhouse gas emissions are associated with compressed gaseous hydrogen produced by electrolysis. Liquid hydrogen generated in a centralised facility or at filling stations also contributes large amounts of greenhouse gas (GHG) emissions and is also rather energy inefficient.

Energy Efficiency: Considering the efficiency on the basis of energy loss during delivery from source to vehicle tank methanol, hydrogen from natural gas and corn-based ethanol show moderate energy losses. Liquid hydrogen from natural gas or from electrolysis is subject to large energy losses. Petroleum-based fuels and compressed natural gas exhibit the lowest energy loss from source to vehicle tank.

9.4 Power Circuit Analysis

Comparing an FCV with the internal combustion engine, there is a big difference in the way power transmission takes place. In an internal combustion engine, energy from gasoline or diesel is transmitted from the engine to the wheels through a mechanical power circuit (Fig. 9.3).

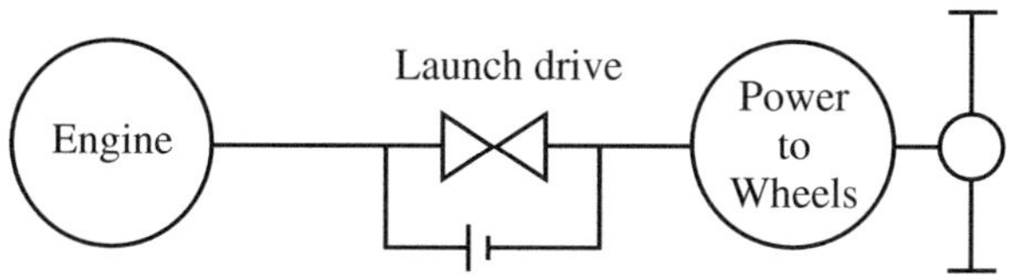

Fig. 9.3 Power transmission in internal combustion engine

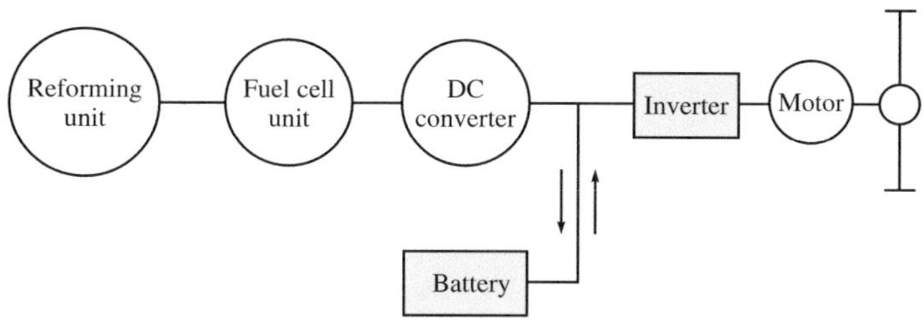

Fig. 9.4 Fuel cell parallel hybrid power circuit

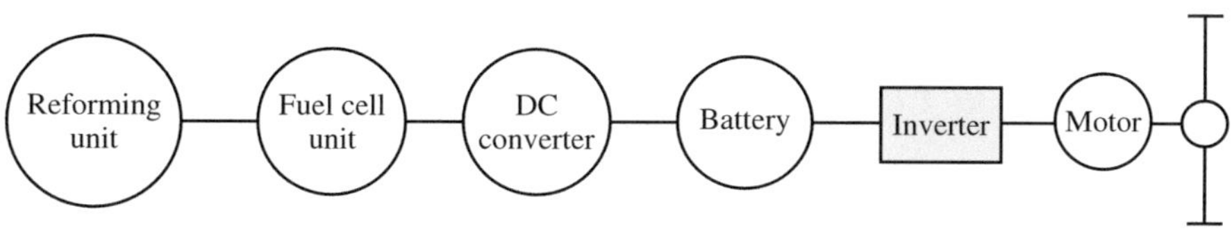

Fig. 9.5 Fuel cell series hybrid power circuit

FCVs are special as they have electrical power transmission, whereas a hybrid vehicle may be a combination of electrical and mechanical power transmission. Hybrid vehicles are further categorised as FCV with parallel hybrid power circuit and FCV with series hybrid power circuit as shown in Figs. 9.4 and 9.5, respectively.

As shown in Fig. 9.4, in parallel hybrid power circuits the fuel cell delivers power to the motor and charges the battery when required. It also provides extra power to the motor for acceleration once the battery is fully charged. In the case of a series hybrid power circuit, all the power is delivered by battery so it needs to be large. The function of the fuel cell is only to charge the battery; depending on driving conditions it can be of relatively low power. Such vehicles generally use alkaline fuel cells.

9.5 Reforming

In the case where the fuel type is hydrocarbon for a solid oxide fuel cell (SOFC) unit, it can be catalytically converted into hydrogen and carbon monoxide within the cell stack, known as internal reformation. Carbon monoxide and hydrogen thus produced are then electronically oxidised to carbon dioxide and water at the anode with the production of heat and electric current. Internal reforming can be of two types, as shown in Fig. 9.6:

(i) Indirect internal reforming which utilises a separate fuel reforming catalyst within the SOFC stack
(ii) Direct reforming which takes place on the anode

The elevated operating temperatures of SOFC provide it a key advantage of fuel flexibility over other types of fuel cells. SOFCs show greater tolerance to carbon monoxide, and other impurities and variations in fuel composition. This allows for the possible use of a variety of renewable fuel sources such as biogas, vegetable matter, etc., which are not viable in other types of fuel cells.

9.5.1 Types of Fuels

As already mentioned, fuel cells have a broad spectrum of fuel input sources (as shown in Fig. 9.7); the type of fuel source can be selected depending on the local conditions. With the supply infrastructure already existent, natural gas can be the most common fuel, especially for stationary applications due to its easy availability, abundance, and low cost. Another benefit of using natural gas as a fuel is it can be internally reformed within the SOFC stack at temperatures as low as 600°C by steam or oxygen. However, natural gas can have sulphur-containing compounds and higher hydrocarbons which can cause sulphur poisoning and carbon deposition, respectively.

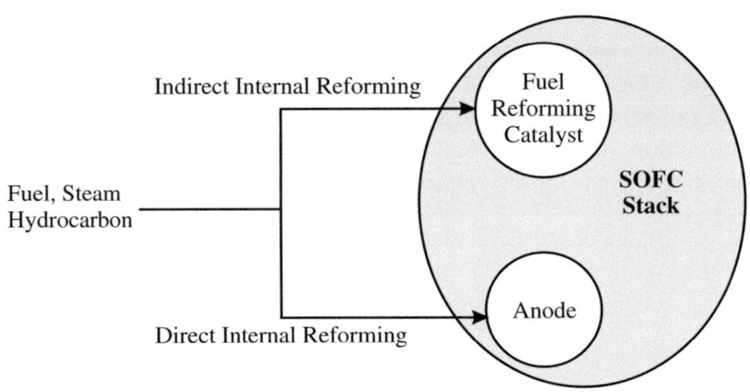

Fig. 9.6 Flowchart of direct and indirect internal reforming

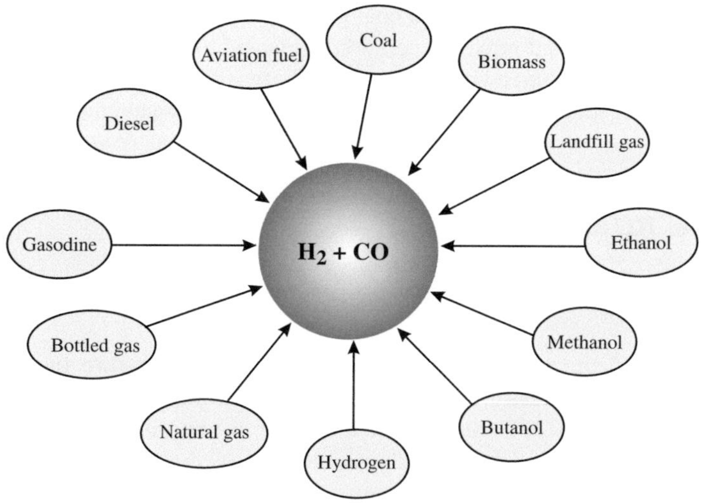

Fig. 9.7 Different fuel sources for a fuel cell

An alternative for areas having a poor supply of natural gas or where diesel is generally used, bottled propane/butane provides clear energy. A small tubular derive powered by butane has already been demonstrated by D. Hart and coworkers. This bottled gas can be internally reformed within the SOFC stack similar to natural gas. However, SOFCs operating on bottled gas (Fig. 9.8) are prone to carbon deposition leading to deactivation. Thus, an SOFC system using bottled gas can provide an alternative route to power generation in emergencies or in the event of interruption to the natural gas supply.

SOFC systems have demonstrated usage of waste biogas, landfill gas, and output from coal gasification systems for producing power. The biggest challenge for these gases is low level of methane and non-uniform composition, which results in wastage of a clean renewable energy resource. For the output from coal gasification systems, the sulphur content is high which can result in poisoning and poor cell durability. Sulphur needs to be removed from gas before it is fed to the SOFC stack. Although SOFCs operating on natural gas have better efficiency compared to SOFC

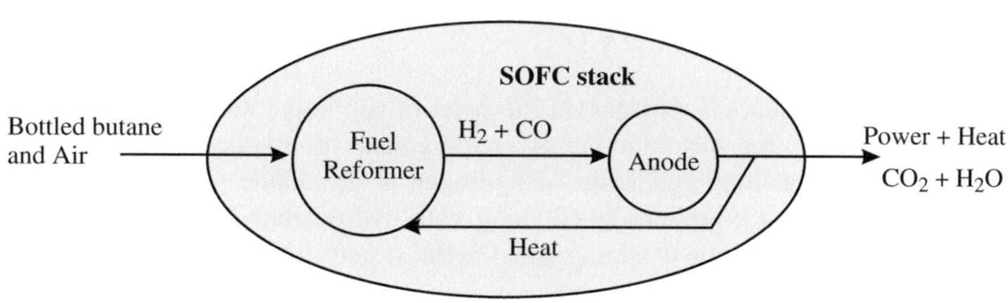

Fig. 9.8 SOFC fuelled by butane for small-scale application

operating on biogas or output from a coal gasification system, the important thing is cleaner and more efficient power generation compared to diesel operated generators. An important energy source (biogas) which otherwise is wasted is now capable of providing power, though at low efficiency, which is an achievement.

9.6 Fuel Reforming

9.6.1 Internal Reforming

The major significant advantage of SOFC over low temperature fuel cells is its ability to electrochemically oxidise CO, internal reformation of a variety of hydrocarbons within SOFC, and higher tolerance to impurities such as sulphur in the fuel used. Internal reforming reduces the cost of the system significantly by eliminating the use of an external reformer and associated equipment. SOFC systems' internally reforming hydrocarbons also show increased efficiency by channeling waste heat from the stack into the fuel supply. SOFC technology offers significant cost benefits over other fuel cells as it is more flexible, simple, and more efficient. However, internal reforming does have some limitations which can reduce the performance of a cell or cause its deactivation, resulting in poor cell durability. The biggest challenge for internal reforming is carbon deposition on active SOFC components, especially the nickel cermet anode during hydrocarbon pyrolysis. This is shown in Eq. (9.1):

$$C_xH_{2x+2} \rightarrow xC + (x+1)H_2 \tag{9.1}$$

This carbon deposition can be inhibited by adding steam to the fuel. This reaction (Eq. 9.2) is strongly endothermic in nature, whereas fuel cell reactions in Eqs. (9.3) and (9.4) are slow exothermic reactions. This can create instability in the coupling between fast endothermic and slow exothermic reactions.

$$C_xH_{2x+2} + xH_2O \rightarrow xC + (2x+1)H_2 \tag{9.2}$$

$$H_2 + O^{2-} \rightarrow H_2O + 2e^- \tag{9.3}$$

$$CO + O^{2-} \rightarrow CO_2 + 2e^- \tag{9.4}$$

A self-sustaining internal reformer is the need of the hour, where heat from the stack is sufficient for the endothermic requirement of hydrocarbon reforming. During fuel cell reaction oxidation of hydrogen at the anode (Eq. 9.3) water is produced, which can be reintroduced along with hydrocarbon fuel in the form of steam instead of continuously adding water to the system. This is shown in Fig. 9.9.

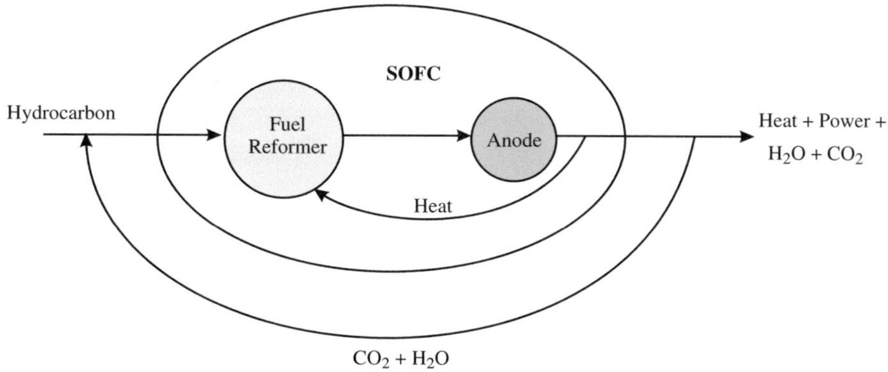

Fig. 9.9 Reintroduction of steam along with fuel in an internally reforming SOFC

9.6.1.1 Direct Internal Reforming

The simplest and most cost-effective design for internal reforming is direct reforming of the fuel on the anode; it provides the greatest efficiency with minimal losses. Developing a suitable material for the anode is a major challenge as it has to fulfil three stringent conditions: first, act as catalyst for conversion of hydrocarbons to hydrogen and carbon monoxide; second, to carry out oxidation of H_2 and CO to water and CO_2 as an electrocatalyst; and, finally, to work as an electrically conducting anode. The exothermic fuel cell reactions occurring inside the SOFC stack provide the heat required for reforming the hydrocarbon fuel; this in turn increases overall efficiency of the cell (Fig. 9.10). In addition, during the fuel cell reaction hydrogen is converted into steam which is again used for reforming hydrocarbon fuel, thus electrochemical reaction propels the reforming reaction to completion. In spite of these advantages, direct reforming has the following limitations.

(a) The strong endothermic hydrocarbon reforming occurring at the cell inlet causes a temperature drop which results in a sharp temperature gradient along the length of the anode, which becomes tedious to control and can cause cracking of the electrolyte and anode.

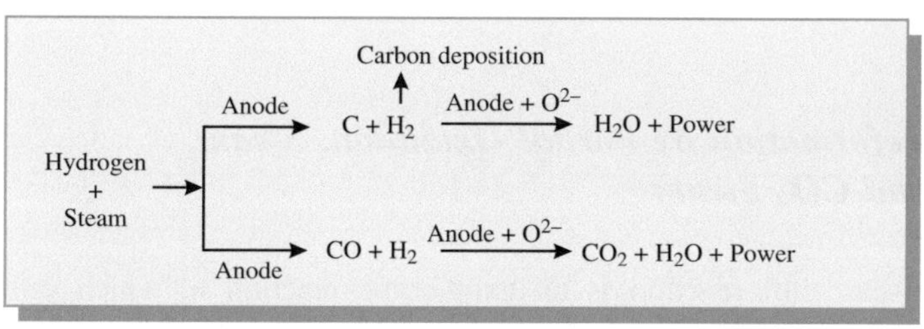

Fig. 9.10 Reaction mechanism occurring at anode during direct reforming

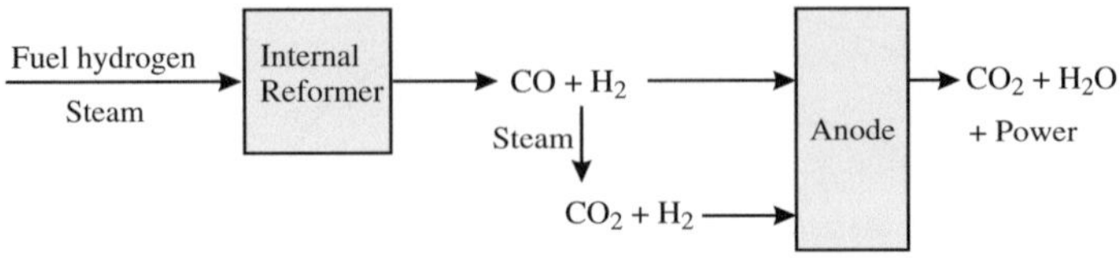

Fig. 9.11 Indirect internal reforming in SOFC

(b) Nickel anode is susceptible to catalyse the pyrolysis of hydrocarbon (Eq. 9.1) which results in carbon deposition, causing quick deactivation of the cell. The probable reaction mechanism occurring at the anode during direct reformation may be given, as some literature reports have suggested, doping of nickel anode with small amounts of copper, gold, and molybdenum that can sufficiently activate hydrocarbon reforming and resist carbon deposition for low steam/carbon ratios in the fuel.

(c) Direct hydrocarbon reforming on the anode exposes it to high steam partial pressures which can sinter nickel particles of the anode, which is detrimental for cell performance and also reduces the catalysing ability of the anode.

9.6.1.2 Indirect Internal Reforming

During indirect reforming the SOFC unit has a special provision for a separate catalyst which reforms hydrocarbon for producing gas which is oxidised at the anode. The scheme of an indirect reforming path is shown in Fig. 9.11.

Compared with direct reforming, the efficiency is less and also indirect reforming is a little more complex but is still economical, efficient, and easy rather than using an external reformer.

On the basis of thermodynamic parameters, indirect reforming is easier to control and also provides the freedom to use highly dispersed catalyst as it does not need to be electrically conducting. Although expensive compared to nickel, platinum and rhodium catalysts in highly dispersed form show greater resistance to sulphur poisoning and carbon depositions. Currently, the major emphasis is on the development of designs which have provision of a separate catalyst within the SOFC unit prior to the anode for indirect reforming of hydrocarbon fuel.

9.6.2 *Reformation by Partial Oxidation, Steam, and CO_2 Steam*

The water–gas shift reaction is an exothermic reaction in which one mole of hydrogen is produced for each mole of CO consumed; some of the CO is converted into CO_2.

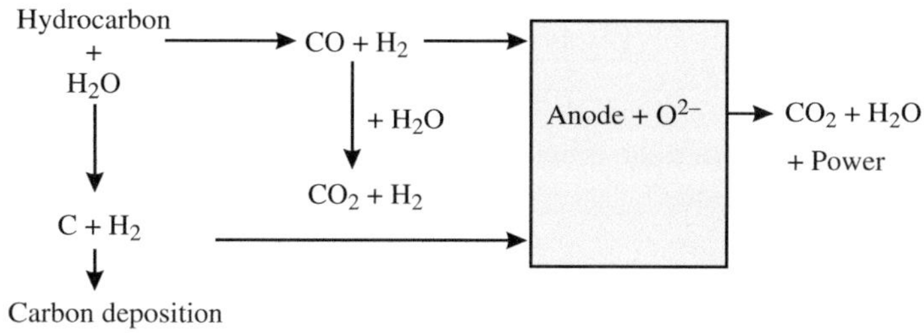

Fig. 9.12 Reaction mechanism for internally reforming SOFC operating on natural gas and steam

$$H_2O + CO \rightarrow CO_2 + H_2 \tag{9.5}$$

The reaction products H_2 and CO are then oxidised to H_2O and CO_2 at the anode. Extensive research has been done on steam reforming catalysis. During steam reforming catalysis, steam to carbon ratios of approximately 2.5–3 are used which is more than the required stoichiometry so that the equilibrium of the water–gas shift reaction points to the right favouring H_2 production and minimum carbon deposition. Figure 9.12 shows the reaction mechanism for SOFC utilising natural gas and steam.

Using large quantities of steam may sometimes lead to sintering of nickel anode particles of the SOFC unit; it also increases the cost and the system becomes more complex. This has created the need for reforming a system requiring a lower steam to carbon ratio. Consequently, there is greater need for developing catalyst and anode formulation resistant to carbon deposition at lower steam contents.

9.6.2.1 Partial Oxidation

Some researchers have successfully demonstrated partial oxidation in SOFC operating on natural gas and butane. Running an SOFC on air and oxygen can be economical and easy to assemble, but overall efficiency of the cell is reduced as some of the energy is utilised in oxidising hydrocarbon. If the anode selectively carries out partial oxidation of hydrocarbon to produce CO and H_2 which is an exothermic reaction, the output from SOFC will increase. Equation (9.6) shows partial oxidation of hydrocarbon.

$$C_xH_{2x+2} + x/2O_2 \rightarrow xCO + (x+1)H_2 \tag{9.6}$$

Extensive studies of partial oxidation point out that it is tough to develop a catalyst and operating region where partial oxidation products show high activity with minimum carbon deposition. Even if the supply of oxygen is increased to reduce carbon deposition, it has the tendency to oxidise hydrocarbon producing CO_2 and H_2O completely; see Eq. (9.7):

$$C_xH_{2x+2} + \frac{(3x+1)}{2}O_2 \rightarrow xCO_2 + (x+1)H_2O \tag{9.7}$$

This can further deteriorate the efficiency of the SOFC as CO_2 and H_2O cannot be electrochemically oxidised.

9.6.2.2 CO_2

Carbon dioxide is produced by a water–gas shift reaction and electrochemical oxidation reaction taking place at anode as shown in Eqs. (9.5) and (9.4), respectively. CO_2 thus produced is fed back into the fuel supply at the cell inlet; using a suitable catalyst CO_2 can be made to oxidise hydrocarbons known as dry reforming.

$$C_xH_{2x+2} + xCO_2 \rightarrow 2xCO + (x+1)H_2 \tag{9.8}$$

Similar to other techniques of reforming, carbon deposition is the biggest challenge for dry reforming in the presence of nickel-based catalysts. As stated above, platinum and rhodium which are more expensive have shown better resistance to carbon deposition.

9.6.2.3 Carbon Deposition

The anode and the reforming catalyst are prone to carbon deposition due to reduction of CO by H_2 by the reaction:

$$CO + H_2 \rightarrow C + H_2O \tag{9.9}$$

The carbon once deposited can reduce cell performance and durability by blocking active sites on the reforming catalyst or anywhere in the fuel supply inlet. Higher hydrocarbons have a greater tendency to cause carbon deposition compared to methane. For SOFCs operating on natural gas, it is the presence of higher hydrocarbons in gas, rather than methane, which is responsible for carbon build-up. Several research groups have studied deeply carbon deposition on reforming catalysts and SOFC anodes so as to chalk out a strategy in order to avoid carbon build-up on the anode or on reforming catalyst at operating temperature. Some researchers have demonstrated that doping of nickel anodes with gold, copper, molybdenum, and ruthenium can improve resistance to carbon deposition while working at low oxidant/carbon ratios.

In order to determine carbon deposition on the anode and catalyst a new technique has been introduced, temperature-programmed oxidation (TPO). This technique is beneficial in determining a safe operating temperature for avoiding carbon deposition at low oxidant/carbon ratios. The technique allows determining the quantity of carbon deposited on the anode or catalyst with respect to operating time.

In addition, it helps to determine the temperature at which carbon is removed from the anode thus giving an idea of the interaction strength of carbon with the anode or catalyst material.

9.6.2.4 Sulphur Compounds and Their Removal

In addition to hydrogen sulphide (H_2S) and carbonyl sulphide (COS), which are normally present in natural gas, there are other compounds such as tetrahydrothiophine (C_4H_8S), diethyl sulphide ($(C_2H_5)_2S$), ethyl-mercaptan ($(C_2H_5)HS$), and butyl mercaptan ($(CH_3)_3CHS$ which are added as odorants at the level of approximately 5 ppm of sulphur. During a majority of reactions sulphur is removed from natural gas before it poisons the anode and the reformer. High operating temperature of SOFC also works against sulphur poisoning as reformer and nickel anode are able to show some tolerance to sulphur. Tolerance of reformer and anode increases with temperature; at small concentrations of sulphur containing compounds, sulphur adsorbs on nickel reversibly. However, the chances of irreversible sulphidation of the catalyst or anode increase with the increase in sulphur concentrations.

High temperature desulphurisation is employed to remove sulphur from organic compounds. A small quantity of hydrogen is added to the fuel and this is passed over cobalt catalyst or molybdenum. Hydrogen reduces sulphur containing organic compounds to H_2S and the corresponding organic compound as shown in the equation below.

$$(C_2H_5)_2S + 2H_2 \xrightarrow{CO} 2C_2H_6 + H_2S \tag{9.10}$$

$$2H_2S + ZnO \rightarrow ZnS + H_2O \tag{9.11}$$

Carbon activated by thermal treatment can also adsorb H_2S. Sulphur can be removed by using activated carbon for smaller SOFC systems. However, in the case of large-scale SOFC units large quantities of adsorbent will be required and it also becomes difficult to reactivate the adsorbent and dispose of the desorbed sulphur.

9.7 Designs of SOFC System

A typical SOFC system takes fuel and air as the input and electricity; hot water or steam is the main output. Major commercial organisations have shown interest and made large investments in this technology. They have come up with their own designs for different applications. Different SOFC systems include residential and auxiliary power systems, oxygen separating systems, pressurised SOFC or turbine hybrid systems, and atmospheric SOFC combined heat and power (CHP) systems.

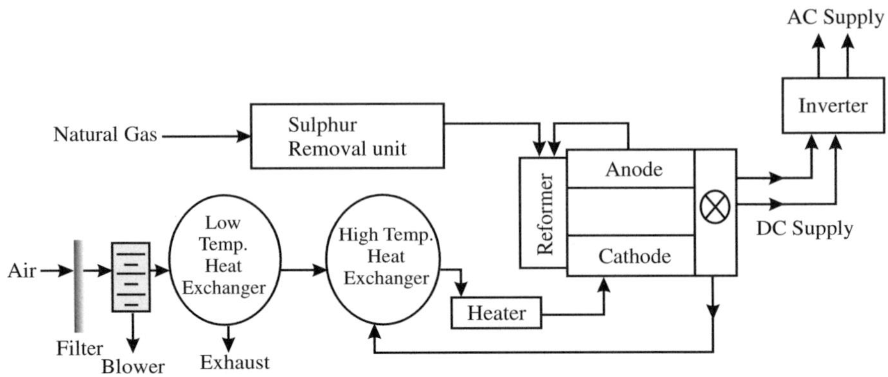

Fig. 9.13 Siemens Westinghouse atmospheric SOFC systems

9.7.1 *Siemens Westinghouse Distributed Power Generations SOFC System*

Siemens Westinghouse Power Corporation has built a tubular SOFC which has its stack integrated with a reformer for reforming higher hydrocarbons present in natural gas.

Figure 9.13 shows a simple layout of an atmospheric SOFC. In order to keep the sulphur level below 0.1 ppm in natural gas, a desulphuriser unit is attached to the SOFC unit which desulphurises fuel before its enters the SOFC. The exhaust gas is moved through the heat exchanger to heat the air before it enters the SOFC. The SOFC stack could be damaged by thermal shock if the temperature of air is less than 500°C. The DC current produced by the SOFC stack is converted into grid quality alternating current (AC) by an inverter. These atmospheric CHP systems work at an electrical efficiency of 45–50% and considering thermal and electrical efficiency together the efficiency is 85–90%. Such natural gas fuelled systems work on desulphurised gas, hence emissions from them are negligible except for a small amount of CO_2. Also, higher hydrocarbons are converted into carbon monoxide and hydrogen at the cell input and the spent fuel is catalytically burnt at the cell exit removing all carbon monoxide.

9.7.2 *Auxiliary Power SOFC System*

Auxiliary Power Units (APUs) are another type of atmospheric system with power capacity of 1–10 kW. A typical small system APU fuelled by gasoline is shown in Fig. 9.14.

Gasoline fuelled APUs can provide electrical efficiency at a maximum 30%, whereas residential systems working on natural gas can have efficiency up to 40%.

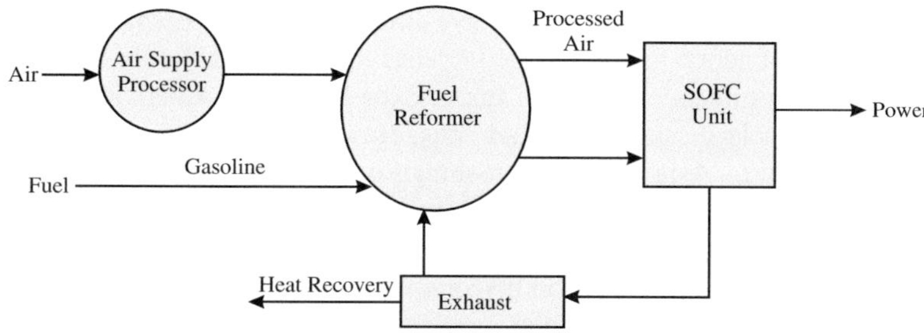

Fig. 9.14 Small auxiliary/residential unit fuelled by gasoline

These SOFCs can also act as commercial sources of syngas (mixture of hydrogen and carbon monoxide); however, separation of oxygen from air is a major cost determining factor. In this way the SOFC can be an efficient technology if it can generate power and syngas or combined production of heat. Another important advantage of this small-scale SOFC could be carbon dioxide sequestering and utilising it for other applications. The exhaust gas from the anode consists of steam and CO_2, thus cooling the exhaust separates water vapour from carbon dioxide and there is hardly any loss of efficiency.

9.8 Current Status of Fuel Cell Industry

Fuel cells are green sources of energy. They find a strong foothold in the market because of their better technology than conventional internal combustion engine generators or batteries. Fuel cell industries are growing and increasing their output; shipments of fuel cell systems almost doubled in 2012. The industry is moving at a rapid pace showing growth in 2013 up to 46% and shipping of 67,000 units has been achieved. The reason for the success of fuel cells is their ability to cater to three major areas of power consumption:

(a) **Portable Fuel Cells**: These are APUs which are built into, or charge up products that are designed to be moved.
(b) **Stationary Fuel Cells**: These units are not designed to be moved; they provide electricity and sometimes heat.
(c) **Transport**: Fuel cells in this application provide propulsive power or range extension to a vehicle.

Till date, stationary fuel cells have been the best performers for fuel cell technology. These are finding applications across all scales: from small-scale grid connected micro-CHP units for residential use, to main power for buildings and even to megawatt-scale installations designed as grid connected power stations to off-grid backup power systems providing an uninterrupted power supply for important infrastructure. Fuel cell technology has found its strongest foothold in Asia. Fuel cells manufactured in the United States have almost 61% of units shipped to Asia in 2012. Asia's dominance is going to increase in the near future. In terms of megawatts, this dominance has been taken over by North America which has a sizeable market for large stationary fuel cells. However, still 50% of megawatts in 2012 were shipped to Asia and the percentage is increasing every year.

In terms of popularity, PEM fuel cell technology is leading the market with regard to unit shipments. Its suitability for use at small and large scales and widespread applications is the reason it scores over other fuel cells. Talking in terms of megawatts, the scene is distributed more evenly between PEMFC, MCFC, and SOFC. The contribution of the latter two is due to their use in the stationary sector for large-scale prime power developments. The level of interest shown by the automotive sector in fuel cell technology is phenomenal. Several high-level partnerships are being forged and significant technology acquaintance is taking place in the sector. It is an extremely exciting time for fuel cell and hydrogen technologies. This new endeavour is guided by three forces: the recognition of hydrogen as an important and attractive energy storage platform by energy utilities; the interest of major global telecoms in fuel cell backup powers; and an initiative by major automakers to commercialise fuel cell electric vehicles (FCEV). Global companies coming forward and investing in fuel cell technology have severely changed the scenario, with improvement in supply chain and technology recognition which will come along with mass manufacture of passenger vehicles; the fuel cell industry is all set for a global success. The following section discusses major applications (portable, stationary, and transport) and an overview of development in fuel and infrastructure.

9.8.1 Portable Applications

The fuel cell units in this category address the demand for consumer electronics. Horizon Fuel Cells entered the market but the initial response to these fuel cells was lower than expected, yet Horizon and MCFC are continuing their efforts to commercialise the systems. Drawing a comparison in terms of applications, the figure for portable fuel cells is small in terms of megawatts. This is due to the small size of the fuel cell unit. However, these portable fuel cells have good sales for educational purposes. Apart from training future engineers on the concept of fuel cells, these educational kits and toys are an important source of cash for the companies. Horizon Fuel Cell Technologies is one of the largest suppliers of these systems.

9.8.1.1 Auxiliary Power Units

In March 2013, Advanced Manufacturing Research Unit (AIST) in Japan announced the release of a portable SOFC that can operate on a range of liquid fuels. It is a microtubular SOFC technology; the system can directly use

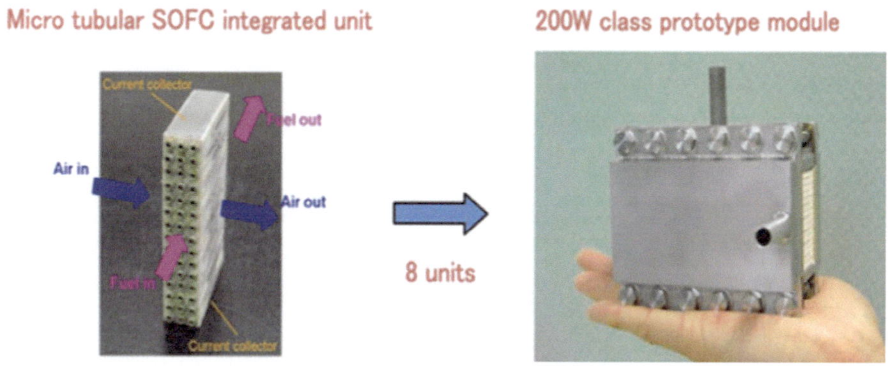

Fig. 9.15 Micro-tubular SOFC developed by AIST (Photo courtesy of AIST Japan)

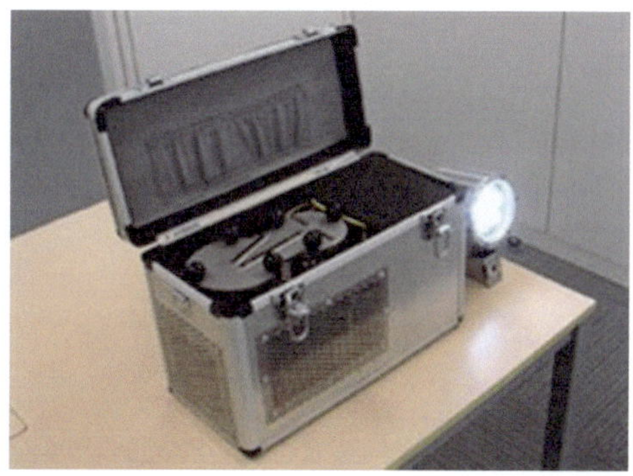

Fig. 9.16 A 33 W portable fuel cell by bio coke (Photo courtesy of fuel cell today)

general-purpose hydrocarbon fuels, including liquefied petroleum gas (LPG) (Fig. 9.15). This system has been developed by AIST for disaster and emergency situations, or for outdoor use.

Figure 9.16 shows a 33 W portable fuel cell system developed by an environmental technology company, Bio Coke Lab Co. in Japan. The company plans to commercialise the system at a price of around \$4700. This portable fuel cell system will be marketed as an emergency power supply; initially it will be supplied to office buildings, families, and the municipal government. The company has collaborated with Horizon Fuel Cell Technologies for stacks and hydrogen will be stored in the company's own magnesium hydride storage alloy. The prototype system weighs 8.5 kg and each 40 Wh fuel cartridge can supply 72 min of power.

9.8.1.2 Military Applications

Fuel cells have attracted the attention of military organisations around the world by providing an effective alternative which reduces the weight carried by soldiers in

Fig. 9.17 XX55 microbial fuel cell by ultra cell (Photo courtesy of Google)

Fig. 9.18 JENNY fuel cell by SFC energy (Courtesy of Google)

the field. Figure 9.17 shows XX55, a microbial fuel cell developed by Ultra Cell, a California-based company. It was evaluated by the New Zealand Defense Forces (NZDF).

During the testing phase XX55 powered critical communications equipment including laptops, radios, and other devices compatible with 5 V USB, 12 V DC cigarette lighters. Ultra Cell received grants of $1 million to test solid acid fuel cell stacks from SAF cell. Protonex, a privately owned military fuel cell developer, is showing successful results. It received a grant of $2 million to expand its sales facilities. SFC Energy (SFC), another fuel cell company which has shown promising results, has been awarded $1 million to qualify its 50 W JENNY system for the US Air Force and is shown in Fig. 9.18.

9.8.1.3 Stationary Applications

In the past few years, stationary fuel cells have shown a continuous increase in unit shipments, with support from large and small-scale applications. The annual shipment in terms of megawatts is largest in the case of the stationary sector. In 2012, 124.9 MW were shipped and the figures are increasing every year; more support is coming for installation of large megawatt fuel cells as new regional markets are opening for both telecommunications backup power and residential micro CHP.

9.8.1.4 Prime Power

There are two companies which dominate the scene for prime power applications.

9.8.1.5 Fuel Cell Energy

Fuel Cell Energy is an integrated fuel cell company that designs, manufactures, installs, operates, and services stationary fuel cell power plants. They generate energy for electric utilities, commercial and industrial companies, universities, government entities, and other customers throughout the world. The direct fuel cell shown in Fig. 9.19 manufactured by Fuel Cell Energy can utilise a variety of fuels including renewable biogas from wastewater treatment, directed biogas, and propane. These power plants produce power electrochemically without burning fuels, making them quiet, clean, and environment friendly. These power plants have already generated 3 billion kWh of electricity, equivalent to power usage of one year in 245,000 average size US homes.

Fig. 9.19 Direct fuel cell manufactured by fuel cell energy (Photo courtesy of Google)

9.8.1.6 Bloom Energy

Bloom Energy is a West Coast based manufacturing facility in California; the energy server supplied by Bloom has a patented SOFC technology. A typical Bloom fuel cell is flat and square in shape made from ceramic; thousands of such fuel cells constitute one energy server. These fuel cells are sandwiched along with metal interconnect plates into a fuel cell stack for higher power output. A few stacks together about the size of a loaf of bread are enough to power a home. A typical energy server manufactured by Bloom is shown in Fig. 9.20. This energy server has multiple stacks arranged together into a "power module" and then multiple power modules along with a common fuel input and electrical output are assembled as a complete system.

Bloom energy has supplied two energy sources to Target Corporation, the second largest retailer in the United States to power its stores in San Francisco and Pasadena. It will also supply 1 MW energy server to power the offices of American Honda Motor Co.

9.8.1.7 Micro-Combined Heat and Power

The year 2012 marked the arrival of a fuel cell micro-CHP which outsold conventional engine-based micro-CHP (Fig. 9.21), accounting for 64% of worldwide sales. The emergence of this dominant technology for micro-CHP is an initiative of Japanese Ene-Farm Scheme. This scheme is growing every year resulting in deployment of the largest number of fuel cells for a single purpose in the world. The latest generation of the Ene Farm fuel cell can operate in a range of electrical output modes, providing between 250 and 750 W of electricity.

Two fuel cell technologies are being promoted through this endeavour, the proton exchange membrane fuel cell (PFMFC) and SOFC; their contribution is currently 80% PEMFC and 20% SOFC. The SOFC units have higher maximum efficiency; the only challenge is they require continuous operation, so at night the efficiency is slightly reduced. In the German market also a series of successful installations under the European SOFT-PACT Project have been achieved by Ceramic Fuel Cells Limited (CFCL). This project involved collaboration of CFCL with utility E.ON, heating system manufacturers

Fig. 9.20 Energy server by BLOOM (Photo courtesy of Bloom Energy)

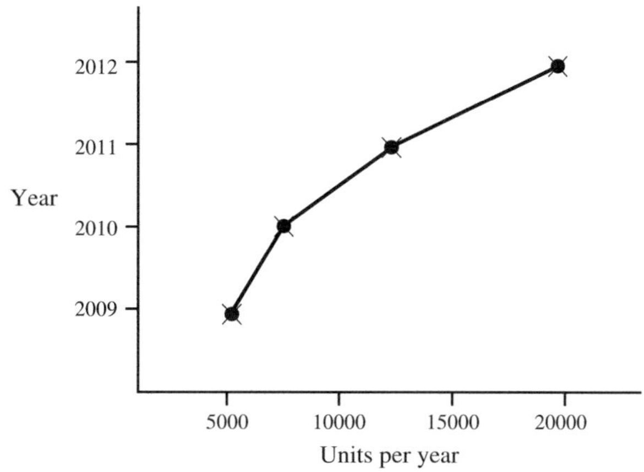

Fig. 9.21 Units sold per year (Photo courtesy of fuel cell today)

and software control company HOMA to install up to 100 fuel cell micro-CHP systems in various configurations in Germany and the United Kingdom. The company has a manufacturing facility (Blue Gen) in Heinsberg, Germany with a production capacity of 1000 systems per year. CFCL has also launched a scheme in the United Kingdom offering fully financed BlueGen units for small business, housing, and schools.

9.8.1.8 Grid-Support and Off-Grid Powers

The US government is promoting the use of fuel cells as power backup systems in the telecom market. Customers around the world are experiencing the benefits and cost savings that can be achieved through fuel cell backup powers. Fuel cells can be reliable sources of energy. In October 2012, Hurricane Sandy passed over the east coast of the United States with devastating effect, but during this time fuel cell powered cell phone towers remained in operation for extended periods. China represents a huge market for adoption of fuel cell technology in mobile phones. Major telecommunications providers are running trials with a number of companies, including ReliON, Future E, and Ballard. Trials in the United States have been successful and the same is expected in China where two giant mobile network providers China Mobile and China Telecom are trialing fuel cells.

9.8.2 Transport Industry

The biggest issue throughout the world in the transport industry is fulfilment of stringent GHG emission targets. Road transportation is the major contributor to greenhouse gas emission. The European Union has made commitments to reduce GHG emissions by 80% by 2050 and this would require a 95% reduction in emission from road transport. This tough criterion can only be realised by increasing the use of public transport and promoting the use of low to zero emission vehicles. In North America fuel cell systems in transportation are getting popular among customers who buy them without government subsidies. Fuel cell buses operating in cities around the world are increasing in number and for the first time fuel cell cars are rolling off a production line. Automakers are adopting aggressive strategies for cost reduction as they are trying to bring the total cost of ownership (TCO) in the range of internal combustion engine vehicles. This will particularly benefit manufacturers and suppliers of PEMFC, which operates with the help of expensive catalyst materials such as platinum.

9.8.2.1 Automobile Industry

The years 2011 and 2012 witnessed huge efforts made by automakers, industries, and government; these activities were further carried out in 2013, as different collaborations prepared themselves for the market introduction of FCEV. Automakers are coming together through collaboration to pool their resources, so as to ease mass production of fuel cell systems which will subsequently reduce manufacturing cost. These major partnerships started in 2013: BMW with Toyota, Renault–Nissan with Automotive Fuel Cell Corporation (AFCC-Daimler and Ford), and GM with Honda. In a joint announcement made by BMW and Toyota in January 2013, technology sharing and codevelopment of fundamental FCV platform by 2020 has been proposed. This joint venture will not only develop a fuel cell system, but also a hydrogen tank supporting the battery system and electric motor. Toyota has years of experience and expertise in the development of fuel cell and battery powered drivetrains. It has also signed an agreement in 2011 amongst Japanese automakers, oil, and energy companies to popularise FCEV and begin joint efforts to set up a network of approximately 100 hydrogen stations across four major metropolitan areas by 2015. Toyota is working on the production of FCEV, for a market launch in the United States, Japan, and Europe starting in 2015 (Fig. 9.22).

The Renault–Nissan Alliance collaboration with AFCC as a technology partner is mainly focused on jointly developing a common fuel cell system for use in separate mass market cars from 2017. Honda and GM signed a collaboration agreement in

Fig. 9.22 Toyota FCEV (Photo courtesy of fuel cell today)

Fig. 9.23 Honda fuel cell stack (Photo courtesy of Honda)

2013 agreeing to develop next-generation fuel cell systems and hydrogen storage technologies. Honda plans to launch a successor to the FCX clarity, which will have significant technological advancements to the fuel cell stack (Fig. 9.23) with a 60% increase in power density and size reduction of stack by 33%. American Honda joined the public–private partnership H2USA, which collaborates with government agencies, gas suppliers, automakers, and the hydrogen and fuel cell industries to coordinate research and provide economical solutions to deploy infrastructure that can deliver affordable, clean hydrogen fuel in the United States.

Hyundai is also showing its commitment to planet saving technology with a hydrogen-electric version of its car, also known as IX35. This vehicle has a high pressure tank which can hold 12.3 pounds of hydrogen, which passes through a fuel cell stack to generate electricity. Hyundai has predicted a top speed of 99 mph and maximum range of just above 300 miles per tank of hydrogen. The 2015 Hyundai Tucson Fuel cell will be leased to customers at $499 a month for 36 months, which includes fuel and maintenance. Hyundai plans to build 1000 hydrogen Tucsons annually at its primary production facility in South Korea.

9.8.2.2 Buses

A significant air quality improvement can be provided by fuel cell buses and this could be the first interaction of the public with fuel cell transportation. These buses will be operating on conventional routes in major cities; passengers using these buses during their daily routine will experience the benefits of this technology. Once familiar with this pollution-free technology, they will be encouraged to consider fuel cells as they emerge in consumer markets. The current emphasis is on increasing the number of buses running on regular service in prominent cities. The fuel cell buses are emerging from a demonstration phase to a commercial aspect; this a positive development though slow moving. European projects are financing this development across the Atlantic; CT transit has procured the world's first unsubsidised fuel cell bus. European Union-backed demonstration projects as well as ongoing projects across the United States have increased bus shipments in the past few years. The majority of the shipments were in European locations, which were funded by the fuel cell and hydrogen joint undertaking (FCH-JU). FCH-JU published a report 'Urban Buses' alternative powertrains for Europe. It summarises findings from 40 companies and government organisations on eight powertrain technologies available to urban buses from 2012 to 2030. It suggests that it will be very difficult for diesel engine manufacturers to meet tightening heavy duty vehicle emissions legislation beyond 2015 and the hydrogen fuel cell bus is the only articulated bus expected to decrease TCO until 2030.

In 2010 based on European fuel cell demonstration programmes, Clean Hydrogen in European cities (CHIC) began a goal of integrating 26 fuel cell buses into daily transport operations across five European locations until 2016. The entire fuel cell bus value proposition, including maintenance and hydrogen infrastructure build-up will be monitored by the High VLO-City Project. Centre for Transportation (CT) and the Environment, a non-profit organisation in North America, which finalised a contract with CT transit in 2013 for the delivery of 40-foot heavy duty fuel cell powered transit buses in collaboration with bus manufacturer El Dorado National, Systems Integrator BAE Systems, and Ballard Power Systems. This is an important step towards full commercialisation of fuel cell buses in the United States. The Federal Transit Administration's (FTA's) National Fuel Cell Bus Program (NFCBP) is responsible for funding the majority of existing buses operating in the United States.

Azura Hydrogen from China is collaborating with Ballard for manufacture of fuel cell buses. Private and government agencies are providing funding for this initiative to develop fuel cell bus capabilities in China with technical support provided by Ballard. India's largest automaker and world's fourth largest bus manufacturer, Tata Motors, and the Indian Space Research Organization (ISRO) have been working together on fuel cell bus development MoU since 2006 and in July 2013 the first fuel cell bus from this effort was undergoing testing at an ISRO

Fig. 9.24 Star bus by TATA and Ballard (Photo courtesy of TATA Motors)

facility (Fig. 9.24). Tata Motors' clear interest in fuel cells is backed up by the experience of ISRO in development of cryogenic technologies and production, handling, and storage of gaseous and liquid hydrogen. The company is accepting delivery of 12 Ballard fuel cell bus modules for integration into demonstration buses to be used across Indian cities.

Ballard has developed an FC velocity-HD7 product, a next-generation fuel cell power module offering improved durability and reliability along with significant cost reduction. The first two buses powered by Ballard's prototype FC velocity HD7 fuel cell power module were presented to Hamburgers Hochbahn AG-transit operator for the city of Hamburg, Germany. Currently these two buses are part of the trial and evaluation of alternative drive technologies being undertaken by the city of Hamburg. These next-generation modules also begin to bring lifetimes closer to those of diesel engine buses, thus enhancing their value and reducing the need for financial support.

Suggested Bibliography

Achenbosch E, Riensche E (1994) J Power Sources 52:283
Appleby AJ, Foulbar FR (1989) Fuel cell handbook. Van Abstrand Reinhold
Ashroft AT, Cheetham AK, Green MLH (1991) Nature 352:225
Bartholomew CH (1982) Catal Rev Sci Eng 24:67
Cheu KJ, Worg YS, Chan CC (1999) J Energy Convers Manag 40:1021
Choudhary VR, Rajput AM, Ind Eng Chem Res 35:3934
Clark SH, Dicks AL, Pointon K, Smith TA, Swann A (1997) Catal Today 38:411
Coe NJ, Cunningham RH, Ormerod RM (1997) Catal Lett 49:189
Cunningham RH, Finnerty CM, Ormored RM (2002) Solid Oxide Fuel Cells V. eds. Stimming U, Singhal SC
de Beer VHJ, Van Sint Fiet THM, Van der Stress GHAM, Zwaga AC, Schuit GCA (1974) J Catal 35:297
Deeprez D, Demichelli MC, Marrcot P, Barbier J, Ferrralti OA, Ponzi FN (1990) J Catal 124:324
Dicks AL (1996) J Power Sources 61:113
Dyer PN, Richards RE, Rusek SL (1999) Solid oxide fuel cells VI. In: Singhal SC, Dokiya M (eds) The Electrochemical Society, Pennington, NJ, PV 99-19, pp 1173
EDB/ELSAM 100 KW, SOFC field test, Executive summary of final report January 2001, EDB project SOFC 800 ELSAM Project 400, 3.1 Novem project 219.301-00.50
Finnerty CM, Coe NJ, Cunningham RH, Ormored RM (1998) Catal Today 46:137
Finnerty CM, Cunningham RH, Ormerold RM (1999) In: Singhal SC, Dokings M (eds) Proceedings of 3rd European SOFC forum. The Electrochemical Society, Pennington, NJ, PV99-19, pp 583
Finnerty CM, Ormerod RM (2000) J Power Sources 86:390
Foger K, Godfrey B (2000) In: McEvoy AJ (ed) Proceedings of 4th European SOFC forum, Switzerland, p 167
Finerty CM, Staniforth J, Kendall K (2000) In: McEnvoy AJ (ed) Proceeding of the 4th European SOFC forum, Switzerland, p 151
Fuel Cell Today (2013) Fuel Cell Industry Review
Gadalla AM, Summer ME (1989) Chem Eng Sci 44:2825
Hart D (1998) Chem Ind 344

Heodig WK, Haines MR, Li KJ (2000) Economical carbon dioxide recovery for sequestration from a solid oxide fuel cell power plant. In: 5th International conference on greenhouse gas technologies, Australia
Hirschenhofer JH, Stauffer DB, Engleman RR, Klett MG (1995) Fuel cell handbook, 4th edn. US Department of Energy
Iruine JTS, Emgg DP, Labrincha J, Marques EMB (1997) Catal Today 38:467
Janssen GJM, Jong JP Jr, Harihsmans JPP (1996) In: Thorntensen B (ed) Proceedings of 2nd European forum, Switzerland, p 163
Kinashita K, McLarnon FR, Cairns EJ (1998) Fuel cells. A handbook, DoE/METC-88/6096
Lloyd AC (1999) The power plant in your basement. Sci Am 281:64
Mukerjee S et al. (2000) Fuel cell seminar abstracts, courtesy associates, Washington, DC
Murray E, Tsai T (1999) Nature 400:649
Ormored RM (2003) Chem. Soc. Rev 32:17
Ormored RM, Steed (1999) Surf Sci Catal 122:35 (Refs Thesis)
Otsuka K, Vokogama S, Mosibawa A (1984) Bull Chem Soc Jpn 57:3286
Park S, Grasium R, Radu V, Gorte RJ (1999) J Electrochem Soc 146:3603
Proctor IA, Hopkin AL, Ormerold RM (2003) 9(3–4):242
Putna ES, Stubenrauh J, Voks JM, Gorte RJ (1995) Langmuir 11:4832
Rastrap-Nielson JR (1993) Production of synthesis gas. Catal Today 18:305
Rastrup-Nielson JR, Christiansen LJ (1995) Appl Catal A: Gen 126:381
Ridler DE, Twigg MV (1996) Chapter 5. In: Twigg MV (ed) Catalyst handbook. Manson, London
Rostrup-Nielson JR (1984) Catalytic steam reforming. In: Andersson JR, Boudart M (eds) Catalysis, science and engineering, vol 5, Springer, Berlin
Sandler YL (1971) J Electrochem Soc 118:1378
Satterfield CN (1991) Heterogeneous catalysis in industrial practice. McGraw Hill, New York
Shockling LA, Reichner P (1989) Fuel cells. In: White RE, Appleby AJ (eds) The Electrochemical Society Proceedings, Pennington, NJ, PV89-14, p 106
Siemens Westinghouse Power Company (2000) Pressurized solid oxide fuel cell/gas turbine power system. Final report DE-AC26-98FT40355 by Siemens Westinghouse Power Company for the US Department of Energy, 2000
Siemens Westinghouse Power Company (2001) A high efficiency PSFOC/ATS—Gas turbine power system, Final report DF-AC26-98FT40455, by Siemens Westinghouse Power Company for the US Department of Energy
Simwonis D, Tietz F, Strover D (2000) Solid State Ionics 132:241
Staniforth J, Kendall K (2000) J Power Sources 86:401
Staniforth J, Ormored RM (2001) Green Chem 3:661
Staniforth J, Ormored RM (2002) Catal Lett 81:19
Steele BCH, Heinzel A (2001) Nature 414:345
Steele BCH, Kally I, Middleton H, Rudking E (1988) Solid State Ionics 28–30:649
Tagawa H, Lehart W (1997) The Electrochemical Society Proceedings, Pennington, NJ, PV97-40, p. 973
Topspe H, Clarsen BS, Massoth FE (1996) In: Hydroterating catalysis and technology. Springer Verlag, Berlin
van Berbel FPF, Schipper GS, Christie GM (1999) Solid oxide fuel cells VI. In: Singal SC, Dobiya M (ed) The Electrochemical Society Proceedings, Pennington, NJ, PV99-1999, pp 1177
Vendubosch HW (1996) Survey desulpurisation of natural gas for fuel cell systems, KEMA report 34051-FPP95-46/2
Westervoort SOFC (1998) Modern Power Systems, pp 29
http://www.aist.go.jp/aist_e/latest_research/2009/20091204/20091204.html
http://www.bloomenergy.com/fuel-cell/energy-server-architecture/
http://www.buses.tatamotors.com

http://www.fuelcelltoday.com
http://www.hondanews.ca/
https://www.google.co.in/search
www.fuelcelltoday.com/analysis
www.toyota-global.com

Index

G. Kaur, *Solid Oxide Fuel Cell Components*,
DOI 10.1007/978-3-319-25598-9

MIX
Papier aus verantwortungsvollen Quellen
Paper from responsible sources
FSC® C105338
FSC
www.fsc.org

If you have any concerns about our products,
you can contact us on
ProductSafety@springernature.com

In case Publisher is established outside the EU,
the EU authorized representative is:
Springer Nature Customer Service Center GmbH
Europaplatz 3, 69115 Heidelberg, Germany

Printed by Libri Plureos GmbH
in Hamburg, Germany